# High Value Fermentation Products
# Volume 1

**Scrivener Publishing**
100 Cummings Center, Suite 541J
Beverly, MA 01915-6106

*Publishers at Scrivener*
Martin Scrivener (martin@scrivenerpublishing.com)
Phillip Carmical (pcarmical@scrivenerpublishing.com)

# High Value Fermentation Products Volume 1

## Human Health

Edited by
**Saurabh Saran, Vikash Babu and Asha Chuabey**

Scrivener Publishing

WILEY

This edition first published 2019 by John Wiley & Sons, Inc., 111 River Street, Hoboken, NJ 07030, USA and Scrivener Publishing LLC, 100 Cummings Center, Suite 541J, Beverly, MA 01915, USA

For more information about Scrivener publications please visit www.scrivenerpublishing.com.



Wiley Global Headquarters
111 River Street, Hoboken, NJ 07030, USA

For details of our global editorial offices, customer services, and more information about Wiley products visit us at www.wiley.com.



***Library of Congress Cataloging-in-Publication Data***

ISBN 978-1-119-46001-5

Cover image: Lab - James Steidl | Dreamstime.com, Petri Dish Bacteria - Guntars Grebezs | Dreamstime.com
Petri Dish Virus - Andrey Burmakin | Dreamstime.com, Lab Tech - Motortion | Dreamstime.com
Cover design by Kris Hackerott

Set in size of 11pt and Minion Pro by Exeter Premedia Services Private Ltd., Chennai, India

10 9 8 7 6 5 4 3 2 1

# Contents

# Foreword

From last two decades we have witnesses unprecedented growth and development in biotechnology positioning the bioeconomy as a major indicator of advancement. Today, the global fermentation-based industry is already worth over 127 billion dollars. Based on the experience and expertise in this filed, we are trying to collect the different technologies advancement and products developed in biotechnology. This book 'High Value Fermentation Products-Volume 1 (Human Health) is divided into various important sections related to Human Health like antibiotics, sugar & sugar alcohols, enzymes, nutraceuticals, metabolic engineered derived products, this will help the readers to understand the importance of fermentation derived product for the betterment of human health. This book will also help to overcome of various bottle necks of the Industry/ scientific community and shall be useful for the betterment of the society and environment. This book will also shares an insight into the recent research, cutting edge technologies, high value products, industrial demand which bring immense interest among young and brilliant researchers, cultivated scientists, industry personnals and talented student communities. The contents of the book have been designed in such a way that it is providing extensive coverage of new developments, state of the art technologies, current and future trends in biotechnology and fermentation. The reader will be introduced with basic and advanced methodologies on industrial microbiology and fermentation technology. The main goal of this book is to share and enhance the knowledge of each and every individual in the fermentation world.

Ram A Vishwakarma
Director, CSIR-IIIM

# About the Editors

**Dr. Saurabh Saran**, PhD, is a Fermentation Scientist having experience in Industrial microbiology, Biotechnology and Fermentation Technology for more than 15 years. Dr. Saran has completed his PhD from Delhi University, India. Dr. Saran has got hands-on experience in working both industries and academic. He has worked in the industries like Reliance Industries Ltd., India. Later he was appointed as a Research Professor at Republic of Korea, South Korea. He has also worked as a Coordinator at the Technology Based Incubator, Delhi University South Campus, Delhi, Inida. Presently, he is working as a Senior Scientist, Fermentation technology division, CSIR-IIIM, Jammu, India. He has an expertise on the screening, isolation, production and scale up of Industrial Enzymes, Biochemicals & Biofuels. Expert in process development/engineering, scale up to 5L, 10L, 30L, 100, 300 L & 500L fermentation size, (batch, fed batch and continuation fermentation) strain engineering, downstream processing and applications of industrially important biomolecules. To my credentials, I have 3 patents and more than 25 international publications in peer reviewed international journals on fermentation technology.

**Dr. Vikash Babu,** PhD was born in Bulandshahr district of Uttar Pradesh, India on 1st September 1981. He did his Bachelor's degree from I.P (PG) College Bulandshahr, India. After qualifying all India combined entrance exam for biotechnology conducted by JNU, New Delhi, India, he did his degree in Biotechnology from Kumaun University, Nainital. After completing his M.Sc degree, he qualified many national level competitive exams such as DBT-JRF- 2005, CSIR-UGC NET for lecturership- Dec. 2004 & June 2005 and GATE-2005. In Nov. 2005, he joined as a DBT-JRF in the Department of Biotechnology, Indian Institute of Technology, Roorkee under the superivision of Dr. Bijan Choudhury and registered for the Ph.D in the same department and Institute and completed his Ph.D degree in the year 2011. After finishing his Ph.D research work he joined Mangalayatan University, Beswan, Aligarh (India) as a lecturer. He left Manglayatan University in the year 2012 and joined Graphic Era University, Dehradun (India) as an assistant professor where he worked till June 2013. Currently, he is working as a scientist in CSIR-IIIM.

**Dr. Asha Chaubey**, Ph.D is Principal Scientist and Head of Fermentation Technology Division, CSIR-Indian Institute of Integrative Medicine, Jammu, India. She has about 15 years of research experience in the area of enzymology and fermentation technology. She is actively engaged in development of indigenous process development. Her research interests include exploration and exploitation of microorganisms for production of enzymes and bioactives in special reference to industrial applications. She has published research articles in the area of bioactives production, enzyme immobilization, biotransformation, kinetic resolution of racemic drug intermediates. She has also published several review articles and has been actively involved in the development of biosensors for health care and environmental monitoring and has several patents on lactate and cholesterol biosensors.

# List of Contributors

**Akankszha Dwivedi,** *Biochemical Engineering Research and Process Development Centre CSIR- Institute of Microbial Technology, Chandigarh, (India)*
**Alok Malaviya,** *Department of Life Sciences, CHRIST (Deemed To Be University), Hosur Road, Bengaluru, (India)*
**Anil Agarwal,** *Department of Chemistry, CHRIST (Deemed To Be University), Hosur Road, Bengaluru, (India)*
**Anshela Koul,** *Fermentation Technology Division, CSIR-Indian Institute of Integrative Medicine, Canal Road, Jammu*
**Apurwa Kestwal,** *National Institute of Biologicals (Ministry of Health & Family Welfare) Government of India Plot No.A-32, Sector-62 Institutional Area, NOIDA, (U.P.), (India)*
**Aravind Madhavan,** *Rajiv Gandhi Center for Biotechnology, Thiruvananthapuram, Kerala, (India)*
**Asha Chaubey,** *Fermentation Technology Division, CSIR-Indian Institute of Integrative Medicine, Canal Road, Jammu Tawi, (India)*
**Ashok Pandey,** *CSIR- Indian Institute for Toxicology Research (CSIR-IITR), 31 MG Marg, Lucknow, (India)*

**Babbal, Adivitiya,** *Department of Microbiology, University of Delhi South Campus, New Delhi, (India)*
**Bhabatosh Chaudhuri,** *CYP Design Ltd, The Innovation Centre, 49 Oxford Street, Leicester, LE1 5XY, (UK)*
**Bhumica Agarwal,** *Department of Biotechnology, Meerut Institute of Engineering and Technology, Meerut, (India)*

**Cristóbal N. Aguilar** *Group of Bioprocesses. Food Research Department, School of Chemistry, Universidad Autónoma de Coahuila, Saltillo, Coahuila, (México)*

**Debendra K. Sahoo,** *Biochemical Engineering Research and Process Development Centre CSIR- Institute of Microbial Technology, Chandigarh, (India)*
**Dignya Desai,** *Amity Institute of Biotechnology, Amity University Rajasthan, Jaipur, (India)*
**Divya Gupta,** *Department of Life sciences, Uttarakhand Technical University, Dehradun, (India)*

**Edgard Gnansounou** *Ecole Polytechnique Federale de Lausanne, ENAC GR-GN, GC A3, Station 18, CH-1015, Lausanne, (Switzerland)*

**Girijesh K. Patel,** *Department of Oncologic Sciences, University of South Alabama, Mobile, (USA)*

**Gurdeep Singh,** *Biochemical Engineering Research and Process Development Centre CSIR-Institute of Microbial Technology, Chandigarh, (India)*

**Harikrishna Reddy** , *High Value Chemicals, Breakthrough R & D, Reliance Industries Limited. Reliance Corporate Park. Thane Belapur Road, GhansoliNavi Mumbai, (India)*
**Himanshu Verma,** *Biochemical Engineering Research and Process Development Centre CSIR-Institute of Microbial Technology, Chandigarh, (India)*
**Hitesh Sharma,** *Fermentation Technology Division, CSIR-Indian Institute of Integrative Medicine, Canal Road, Jammu(India)*

**Ibidapo Stephen Williams,** *CYP Design Ltd, The Innovation Centre, 49 Oxford Street, Leicester, LE1 5XY, (UK)*
**Ivanoe García,** *Group of Bioprocesses. Food Research Department, School of Chemistry, Universidad Autónoma de Coahuila, Saltillo, Coahuila, (México)*

**Jayesh Varavadekar,** *High Value Chemicals, Breakthrough R & D, Reliance Industries Limited. Reliance Corporate Park. Thane Belapur Road, GhansoliNavi Mumbai, (India)*
**Jitender Nandal,** *Biochemical Engineering Research and Process Development Centre CSIR-Institute of Microbial Technology, Chandigarh, (India)*

**Kakoli Dutt**, *Department of Bioscience and Biotechnology, Banasthali Vidyapith, Rajasthan (India)*
**Kanti N. Mihooliya,** *Biochemical Engineering Research and Process Development Centre CSIR-Institute of Microbial Technology, Chandigarh, (India)*
**Karan Malhotra,** *National Centre for Biological Sciences, Tata Institute of Fundamental Research, GKVK Campus, Bellary Road, Bengaluru, (India)*
**Katherine Saikia,** *Department of Life Sciences, CHRIST (Deemed To Be University), Hosur Road, Bengaluru, (India)*
**KB Arun,** *Rajiv Gandhi Center for Biotechnology, Thiruvananthapuram, Kerala, (India)*

**Lalit Kumar Singh**, *Department of Biochemical Engineering, School of Chemical Technology, Harcourt Butler Technical University Kanpur, (India)*
**Leonardo Sepulveda,** *Group of Bioprocesses. Food Research Department, School of Chemistry, Universidad Autónoma de Coahuila, Saltillo, Coahuila, (México)*
**Liliana Londoño-Hernández,** *Group of Bioprocesses. Food Research Department, School of Chemistry, Universidad Autónoma de Coahuila, Saltillo, Coahuila, (México)*
**Lipsy Chopra,** *Biochemical Engineering Research and Process Development Centre CSIR-Institute of Microbial Technology, Chandigarh, (India)*

**M. Sudhakara Reddy,** *Department of Biotechnology, Thapar University, Patiala 147004 (India)*
**Mahendra Pal Singh,** *National Institute of Biologicals (Ministry of Health & Family Welfare) Government of India Plot No.A-32, Sector-62 Institutional Area, NOIDA, (U.P.), (India)*
**Manali Datta,** *Amity Institute of Biotechnology, Amity University Rajasthan, Jaipur, (India)*
**Manmeet Ahuja,** *High Value Chemicals, Breakthrough R & D, Reliance Industries Limited. Reliance Corporate Park. Thane Belapur Road, GhansoliNavi Mumbai, (India)*

**Mansi Vora,** *High Value Chemicals, Breakthrough R & D, Reliance Industries Limited. Reliance Corporate Park. Thane Belapur Road, GhansoliNavi Mumbai, (India)*

**Maryam Faiyaz1,** *Department of Bioengineering, Integral University, Lucknow, Dasauli, (India)*

**Parameswaran Binod ,** *Microbial Processes and Technology Division, CSIR-National Institute for Interdisciplinary Science and Technology, Thiruvananthapuram, Kerala (India)*

**Piyush Sethia,** *High Value Chemicals, Breakthrough R & D, Reliance Industries Limited. Reliance Corporate Park. Thane Belapur Road, GhansoliNavi Mumbai, (India)*

**Pradeep Kumar,** *Biochemical Engineering Research and Process Development Centre CSIR-Institute of Microbial Technology, Chandigarh, (India)*

**Puja Tandon,** *School of Environmental Science and Engineering, Shanghai Jiao Tong University, Shanghai,PR (China)*

**Rahul Vikram Singh,** *Fermentation Technology Division, CSIR-Indian Institute of Integrative Medicine, Canal Road, Jammu(India)*

**Ramita Taggar,** *Biochemical Engineering Research and Process Development Centre CSIR-Institute of Microbial Technology, Chandigarh, (India)*

**Ramón Larios-Cruz,** *Group of Bioprocesses. Food Research Department, School of Chemistry, Universidad Autónoma de Coahuila, Saltillo, Coahuila, (México)*

**Raúl Rodríguez-Herrera** *Group of Bioprocesses. Food Research Department, School of Chemistry, Universidad Autónoma de Coahuila, Saltillo, Coahuila, (México)*

**Raveendran Sindhu,** *Rajiv Gandhi Center for Biotechnology, Thiruvananthapuram, Kerala, (India)*

**Ravi S. Manhas ,** *Fermentation Technology Division, CSIR-Indian Institute of Integrative Medicine, Canal Road, Jammu, (India)*

**Ricardo Gómez-García,** *Group of Bioprocesses. Food Research Department, School of Chemistry, Universidad Autónoma de Coahuila, Saltillo, Coahuila, (México)*

**Richa Sharma,** *Fermentation Technology Division, CSIR-Indian Institute of Integrative Medicine, Canal Road, Jammu, (India)*

**Richi V Mahajan ,** *National Institute of Biologicals (Ministry of Health & Family Welfare) Government of India Plot No.A-32, Sector-62 Institutional Area, NOIDA, (U.P.), (India)*

**Ruchika Goyal,** *Department of Biotechnology, Graphic Era University, Dehradun, (India)*

**Sanjog Garyali,** *Key Laboratory of Synthetic Biology, Shanghai Institutes for Biological Sciences, Chinese Academy of Sciences, Shanghai, (China).*

**Saurabh Saran,** *Fermentation Technology Division, CSIR-Indian Institute of Integrative Medicine, Canal Road, Jammu Tawi, (India)*

**Shikha Gupta,** *Gujarat State Biotechnology Mission, Gandhinagar, (India)*

**Shilpa Mohanty,** *Department of Microbiology, University of Delhi South Campus, New Delhi, (India)*

**Subhash Chand,** *National Institute of Biologicals (Ministry of Health & Family Welfare) Government of India Plot No.A-32, Sector-62 Institutional Area, NOIDA, (U.P.), (India)*

**Surinder Singh,** *National Institute of Biologicals (Ministry of Health & Family Welfare) Government of India Plot No.A-32, Sector-62 Institutional Area, NOIDA, (U.P.), (India)*

**Syed M. Waheed,** *Department of Biotechnology, Graphic Era University, Dehradun, (India)*

**Vidhya Rangaswamy,** *High Value Chemicals, Breakthrough R & D, Reliance Industries Limited. Reliance Corporate Park. Thane Belapur Road, GhansoliNavi Mumbai, (India)*

**Vikash Babu**, *Fermentation Technology Division, CSIR-Indian Institute of Integrative Medicine, Canal Road, Jammu*

**Yogender Pal Khasa,** *Department of Microbiology, University of Delhi South Campus, New Delhi, (India)*

**Yong Wang,** *Key Laboratory of Synthetic Biology, Shanghai Institutes for Biological Sciences, Chinese Academy of Sciences, Shanghai, (China).*

# Preface

A century after the pioneering work of Louis Pasteur, the science of microbiology has reached to its zenith. In a short span of time, modern biotechnology has grown up drastically from a laboratory scale to a commercial level. Advances in Fermentation Technology have created a favorable niche for the development of fermentation based products to facilitate their applications and to provide a sustainable environment for mankind and to improve the quality of human life. The modern day biotechnology offers many opportunities and effective techniques to address the human concerns in the areas of pharmaceuticals, diagnostics, polymers, textiles, aquaculture, forestry, chemicals, household products, environmental cleanup, food processing, feed and forensics etc.

The book entitled "High Value Fermentation Products" has been divided in different volumes namely, Human Health and Human Welfare. The Volume 1 of the book has 18 chapters focussed on basics to fermentation technology, antibiotics & immunosuppressants, antibodies, peptides & proteins, sugars & sugar alcohols and metabolic engineering derived products. The fist chapter entitled 'Introduction, scope and significance of fermentation technology' aims to provide the insights on the basics of fermentation technology to the readers. The second chapter on 'Extraction of bioactive molecules through fermentation and enzymatic assisted technologies' elaborates the techniques involved in isolation of bioactive molecules and other chemo-enzymatic approaches for bioactives production. Third and fourth chapters compile the important antibiotics against Gram positive and Gram negative bacteria discovered so far. Fifth chapter emphasizes on the role of antifungal drugs in combating invasive fungal diseases. Sixth chapter provides the update on rapamycin production and its potential clinical implications. The seventh chapter on 'Advances in production of therapeutics monoclonal antibodies' highlights the methodologies involved in the production of monoclonal antibodies for therapeutics. The eighth and ninth chapters focus on the antimicrobial peptides of microbial origin and their mechanism of production. The chapter tenth provides the insight on the therapeutic enzymes for human disease management. Chapters eleven to thirteen focus on the strategies involved in the production of natural sweeteners erythritol, xylitol and trehalose. Chapter fourteen describes how production of yeast derived microsomal human CYP450 enzymes (Sacchrosomes) with high yields and activities are superior to other commercially available microsomal enzymes. Chapters fifteen and sixteen compile the methodologies for production of artemisinin and flavonoids respectively. The chapter seventeen provides the information on how advances in metabolic engineering of the carotenoids can be useful for production of astaxanthin. The last chapter of the volume 1 describes how fungal endophytes can be exploited as biofactories for production of functional metabolites through metabolic engineering, with special reference to taxol production.

This book provides deep insights on the strategies decisive factors involved in the fermentation based high value products like antibiotics, enzymes and other therapeutic secondary metabolites in the area of human health. We are confident that this book will be useful to students, researchers, academicians, and industry professionals interested in studying fermentation technology.

Editors

# Acknowledgement

The Editors take this opportunity to gratefully acknowledge the assistance and contribution of the people who have faith in us in this undertaking for compiling of the Book "“High Value Fermentation Products”-Volume 1 (Human Health).

We are in debt of Dr. Ram A. Vishwakarma, Director, CSIR-Indian Institute of Integrative Medicine, Jammu for his valuable and esteemed guidance to carry out this task. His scholarship and authorative knowledge has been a great source of motivation and inspiration.

First and foremost, it is not enough to express our gratitude in words to all the contributors for devotion and providing excellent matter of chapters on time.

The help and support provided by Mr. Chand Ji Raina, Mr. R.K. Khajuria and Mrs. Urmila Jamwal, was important and we acknowledge all of them with sincere thanks.

We are also thankful to the students of Fermentation Technology Division, CSIR-IIIM for their sincere efforts, dedication and determination to achieve objectives for the completion of this task in a given time.

Where emotions are involved, words cease to mean for our family members for the consistent motivation during the planning and edition of this book.

We avail the opportunity to express our heartiest thanks to 'Almighty' for pouring His care and blessings throughout and making this work a success.

Saurabh Saran
Vikash Babu
Asha Chaubey

# 1

# Introduction, Scope and Significance of Fermentation Technology

**Saurabh Saran[1]*, Alok Malaviya[2] and Asha Chaubey[1]**

*[1]Fermentation Technology Division, CSIR-Indian Institute of Integrative Medicine, Canal Road, Jammu Tawi, (India)*
*[2]Department of Life Sciences, CHRIST (Deemed To Be University), Hosur Road, Bengaluru, (India)*

***Abstract***
Fermentation technology is a field which involves the use of microorganisms and enzymes for production of compounds that have applications in the energy, material, pharmaceutical, chemical and food industries. Though fermentation processes have been used for generations as a requirement for sustainable production of materials and energy, today it has become more demanding for continuous creations and advancement of novel fermentation processes. Efforts are directed both towards the advancement of cell factories and enzymes, as well as the designing of new processes, concepts, and technologies. The global market of microbial fermentation technology was valued at approximately USD 1,573.15 million in 2017 and which is expected to generate revenue of around USD 2,244.20 million by end of 2023. However, regular supply of materials, such as nutrients, microorganisms, the complex nature of production process, and high manufacturing cost hinder the market growth.

***Keywords:*** Fermentation, world market, fermenter design, submerged fermentation, solid state fermentation

## 1.1 Introduction

In the present century, we are witnessing a revolution in biotechnology that has far-reaching implications in different industries like pharmaceuticals, food and feed, polymer, oleo-chemicals, textiles, leather, cosmetics and agriculture, consequently resulting in the betterment of the society beyond anything previously imaginable in the history of science. The present scenario defines biotechnology as "any technique that uses living organisms or substances to make or modify a product, to improve plants and animals, or to develop microorganisms for specific and beneficial uses" [1]. Thus, biotechnology encompasses tools and techniques, including those of recombinant DNA technology for improving the living organisms, which may be plant, animal

**Corresponding author*: ssaran@iiim.ac.in

Saurabh Saran, Vikash Babu, and Asha Chaubey (eds.) High Value Fermentation Products, Volume 1, (1–26) © 2019 Scrivener Publishing LLC

or microorganism. The product formed can be new or rare, that is, not having existed before naturally, or being less abundant than for certain needs or purposes. Thus, biotechnology is a multidisciplinary pursuit involving a variety of natural sciences such as cell and molecular biology, microbiology, physiology, biochemistry and genetics.

In terms of microbiology, biotechnology refers to the use of microorganisms such as bacteria, yeast and fungi or other biological substances produced from them, to perform important industrial processes. Although, era of advanced technology is not very new, the roots of fermentation technology are known for over 6,000 years, when beer was first fermented. Today, biotechnology has intersected and redefined our lives by producing a large variety of value-added products and biomolecules such as antibiotics, enzymes, hormones, organic acids and other metabolites [2].

Broadly, fermentation is a process used to produce a specific product by living organisms. Examples of fermentation processes include the production of simpler products such as baker's yeast and alcohols, as well as complex products such as therapeutic proteins, antibiotics, enzymes, and genetically engineered materials. Fermentation processes should be carefully and critically monitored with regards to the culture conditions and time of harvest depending on the desired product. Typically, fermentation is a natural process. People applied fermentation to make products such as cheese, wine, meat, and beer long before the biochemical process was understood.

## 1.2 Background of Fermentation Technology

In the 1850s and 1860s, Louis Pasteur became the first scientist to study fermentation, when he demonstrated that fermentation was caused by living cells, i.e., yeast. His work was influenced by the earlier work of Theodor Schwann, the German scientist who helped to develop the cell theory. Around 1840, Schwann had concluded that fermentation is the result of processes that occur in living things. In 1857, Pasteur showed that lactic acid fermentation is caused by living organisms. In 1860, he also demonstrated that souring in milk is caused by bacteria. This process was earlier thought to be a chemical change. Pasteur's work to identify the role of microorganisms in food spoilage, thus led to the process of pasteurization. While working to improve the French brewing industry, Pasteur published his famous paper on fermentation, *Etudes sur la Biere* in 1877, which was translated into English in 1879 as *Studies on Fermentation.* He defined fermentation (incorrectly) as "Life without air," but correctly showed specific types of microorganisms cause specific types of fermentations and specific end products.

Eduard Buchner, a German chemist received the Nobel prize in 1907 for showing that enzymes in yeast cells cause fermentation. About two decades later, two other scientists namely Arthur Harden and Hans Euler-Chelpin won the Nobel prize in 1929 for their work who showed how enzymes cause fermentation. Further by 1940s, fermentation based antibiotics production technology was established.

A British scientist, Chain Weizmann (1914-1918) developed a fermentor for the first time for the production of acetone during First World War. But, the first large scale fermentor (above 20 litre capacity) was used for the production of yeast in 1944 (3). Later, importance of aseptic conditions for fermentation process was recognised, which led to the designing and construction of piping, joints and valves in which sterile conditions could be achieved. The large scale aerobic fermentors consisting of a large cylindrical tank with air introduced at the base via network of perforated pipes were used for the first time, in central Europe in 1930's for the production of compressed yeast. Later, modifications were made to design of mechanical impellers to increase the rate of mixing and to break up and disperse the air bubbles. Baffles on the walls of the vessels were useful to prevent formation of vortex in the liquid. A system in which the aeration tubes were introduced with water and steam for cleaning and sterilization was patented by Strauch and Schmidt in 1934.

After the decision of British Govt. in 1934 on inadequate surface fermentation processes, use of submerged culture technique was realized for penicillin production. Essential aseptic conditions, with good aeration and agitation were probably the most important factors, which led to the development of carefully designed and purpose-built fermentation vessels. Hindustan Antibiotic Ltd., Pimpri, Pune established the first pilot fermentor in India at in the year 1950.

## 1.3 Market of Fermentation Products

### Market Definition

Fermentation medium used in the fermentation process to help increase the pace of the process play a vital role as process initiators in an array of applications. Catalysts or process-enhancing chemicals also contribute in the manufacturing cost, chemical reaction time (fermentation time) as well as energy consumption and making the fermentation process more economically attractive. Natural occurring, low cost and productivity enhancing features of fermentation chemicals find applications in a multiple of industries worldwide.

### Market Segmentation

The global fermentation chemicals market is divided on the basis of type of product developed

- Organic Acids
- Alcohols (Ethanol and Other Alcohols)
- Enzymes
- Others (Antibiotics, Vitamins, Xanthan, etc.)

The global fermentation chemicals market is divided on the basis of Industrial application developed,

- Alcohol Industry
- Plastics & Fiber Industry
- Pharmaceutical & Nutritional Industry
- Food & Beverages Industry
- Other Industries

## 1.4 Types of Fermentation

Fermentation is one of the oldest scientific domains which has evolved, refined and been diversified over the centuries. Fermentation processes are mainly classified as either (A) solid state or (B) submerged culture. Solid state fermentation (SSF) differs from submerged fermentation (SmF) with respect to flowing water, which is present in SmF while it is absent or very minor in SSF. In the following sections, we have discussed various aspects of solid-state and submerged fermentation processes.

### 1.4.1 Solid State Fermentation (SSF)

The growth of microorganisms on moist solid substrate particles in the absence or minimum water between the particles is known as solid state fermentation. The moisture content of solid substrate ranges between 12–80%. SSFs are usually used for the fermentation of agricultural products or foods. Some food fermentations involving SSF: Wheat by *Aspergillus*, Soybean by *Rhizopus*, Soybean by *Aspergillus*. During SSF, the substrate may require preparation or pretreatment, like chopping or grinding-reduce particle size, cooking or chemical hydrolysis pasteurization or sterilization-reduce contaminants. Usually a filamentous fungus requiring aerobic condition is used and the inoculum is mixed into substrate for fermentation.

The origin of Solid-state fermentation (SSF) can be traced back to bread-making process in ancient Egypt. SSF processes are those microbiological processes in which growth and product formation takes place on and inside the humidified solid substrate [4]. In this case, processes occur in the absence or near absence of free water. There are four interacting phases present in such processes which include (i) a gas phase, (ii) a solid insoluble support, (iii) a liquid phase containing dissolved substrates and products, and (iv) a biotic phase formed by the microorganisms. Fundamentally, six different types of solid-state fermenters (SSFr) are commonly used, which include – (i) Tray, (ii) Rotary drum, (iii) Packed-bed, (iv) Swing solid state, (v) Stirred vessel, (vi) Air solid fluidized bed bioreactor. Further, based on mixing and aeration, Mitchell [5] divided SSFr into the following four groups:

*Group-I* – the bed is static or mixed occasionally; air is circulated around the bed.
*Group-II* – bed is static or mixed occasionally; air is passed forcefully through the bed.
*Group-III* – bed is continuously mixed; air is circulated around the bed.
*Group IV* – bed is agitated; air is passed forcefully through the bed.

Based on the type of solid substrate used, SSF systems have been divided into: (i) those employing natural material as solid substrate, (ii) those employing inert support impregnated with liquid medium [6]. Different agro-industrial wastes such as cassava bagasse, sugarcane bagasse, sugarbeat pulp/husk, orange bagasse, oil cakes, wheat bran, coir pith, coffee pulp/husk, okaraetc are used as natural solid substrates, selection of which depends on some physical parameters such as particle size, moisture level, intra-particle spacing and nutrient composition within the substrate [6].

*Factors affecting SSF performance:* SSF needs close process parameters monitoring and control, which is very difficult practically. There are various factors which affect the performance of SSF processes, some of which include –

i. *Biological Factors* – These influence the behavior of the microbial species used in SSF. *Selection of a suitable strain* is one of the most important criteria in SSF. Next comes the *selection of substrate* to be used, which depends on the factors related to cost and availability. Then comes the *inoculum size*, and the size of inoculum determines the biomass production. Too low or too high concentration of inoculum is often undesirable and does not give the expected outcome.
ii. *Moisture and water activity* – As compared to submerged fermentation, the low moisture content becomes limiting for growth and metabolism of microbes in solid-state. Water activity ($A_W$) is a useful parameter which is frequently used during SSF to characterize the energetic state of water [7]. Quantitative studies of water relations in SSF have suggested that microbial activity is strongly influenced by water activity of substrates and it determines the type of organism that can grow on given substrate during SSF and, the microbial metabolic production and excretion could be modified by controlling this parameter [7–8].
iii. *Temperature and heat transfer* – Temperature and heat transfer processes in substrate bed has also been reported to greatly influence the microbial growth and metabolite production during SSF processes. A large amount of heat is generated due to metabolic activities of microbes and since the substrates used during SSF have low thermal conductivities, heat removal is decreased resulting in accumulation of heat within the system. Therefore, heat removal to maintain the optimal temperature for growth and metabolite production by microbes becomes the key issue during SSF processes. Additionally, moisture also needs to be controlled, thus coupled control of moisture and temperature becomes an important issue for consideration during SSF [6].
iv. *Off-Gas analysis* – Measuring microbial biomass and growth kinetics during SSF is another major challenge and hence it might affect the repeatability and reproducibility of results. However, this could be addressed by establishing a correlation with $CO_2$ production, which correlates well with microbial metabolism and even low physiological activity could be indicated by this method.
v. *Mass transfer* – SSF rate and efficiency is strongly affected by mass transfer limitations [9–11]. Mass transfer in SSF involves:
   a. Micro-scale phenomenon – this depends on inter and intra particle $O_2$ and $CO_2$diffusion, enzyme, nutrient absorption and metabolite formation.

b. Macro scale phenomenon – this depends on airflow (inflow and outflow), substrate type, design of bioreactor, substrate mixing, interparticle space, variation in particle size and microbes used. Therefore, for maximum mass transfer process the optimally designed SSF bioreactors should be selected.

*Applications of SSF:*
Previously, SSF process was famous for "low volume - high cost" products due to critical technical problems associated with heat and mass transfer for large capacity. But with advancement in SSF technology, this process is inclining towards "high volume - high cost" products. SSF has been successfully applied in various industrial processes where submerged fermentation proves to be challenging. Some of the industrial processes where SSF has successfully been applied include:

i. Enzyme production – this is one of the most important applications of SSF. Some of the advantages of SSF over submerged fermentation for enzyme production include high yield and volumetric productivities along with low cost of and lesser waste generation [6]. Some of the reported enzymes produced through SSF are protease, lipase, cellulose, pectinase, phytase, L-glutaminase, amylase, ligninase, xylanase, etc.
ii. Organic acid production – Some of the organic acids produced via SSF includes lactic acid and gallic acid [12–17].
iii. Secondary metabolite production – There are various secondary metabolites which have been produced using SSF. Some of those products include gibberellic acids [18–20], aroma production [21–23], antibiotics [24–28].
iv. Poly-gamma glutamate production [29]
v. Poly unsaturated fatty acid production [30]
vi. Biocontrol agent production [31–32]

*Advantages of solid state fermentation:*
SSF has several advantages which makes it an attractive technology to be used for production of the above-mentioned industrially important products. Some of these advantages include [33]:

i. Higher product titer
ii. Low capital and recurring expenditure
iii. Low waste water production
iv. Reduced energy requirements
v. Absence of foaming problem
vi. Simple and highly reproducible
vii. Simpler fermentation media
viii. Less space requirement
ix. Easier aeration
x. Economic to use even at smaller scale
xi. Lower cost of downstream processing

However, there are several limitations of SSF, some of which are listed below:

i. Exposure of fungal hyphae to an air phase, which might desiccate them
ii. Rise of temperature well above the optimum level due to inadequate removal of waste metabolic heat resulting in temperature variation during growth in SSF
iii. Poor availability of nutrients to the organisms due to large concentration gradients of nutrients within the particles
iv. Poor oxygen availability to a significant proportion of the biomass
v. Need to use indirect methods for biomass measurement
vi. pH control during SSF is almost impossible
vii. Difficult to control the moisture content of the medium
viii. Microbial types that could be used in SSF are limited
ix. Limited knowledge on engineering and development aspects

To a large extent, the above-mentioned limitations of SSF could be addressed by another mode of fermentation popularly known as submerged fermentation. In the next section, we will briefly discuss various aspects of submerged fermentation.

### 1.4.2 Submerged Fermentation (SmF)

Submerged fermentation is the cultivation of microorganisms in liquid nutrient broth. Industrial enzymes/other products can be produced using this process. This involves growing carefully selected microorganisms (bacteria and fungi) in closed vessels containing a rich broth of nutrients (the fermentation medium) and a high concentration of oxygen. Production medium is required for growth of microorganism as well as production of primary and secondary metabolites. Submerged fermentations have been mainly used to produce flowable formulations.

Contrary to SSF, submerged fermentation (SmF) are those microbiological processes in which growth and product formation take place using substrate present in liquid form. There are three main modes of SmF – (i) Batch mode, (ii) Fed-Batch mode, (iii) Continuous mode. Most industrial fermentation processes, except a few, operate as simple batch or fed-batch process. In order to develop an industrially applicable production process for a desired product, one must understand these three modes of fermenter operation.

*Batch mode* – Batch culture is the simplest mode of operation. Here, initially all the nutrients required for the organism's growth and product formation are added in one vessel at the start of the fermentation process. This is followed by medium sterilization and inoculation of the vessel with desired organism for growth and product formation. This mode of operation operates as closed system and in between nothing is added into the vessel, except air. Finally, fermentation is terminated when either all the nutrient is exhausted or the desired concentration of product is achieved.

*Advantages of batch mode of operation* –

i. Simple to use
ii. Operability and reliability

iii. Remote chances of contamination during process
iv. Can be handled by relatively inexperienced operator

*Fed-batch mode* – This is similar to batch process, except that these do not operate as closed systems. In this mode, one or more substrates, nutrients, and/or inducers are added in between during the process. This mode is normally used to extend the productive phase of the process.

*Advantages of fed-batch mode of operation* –

i. 'Catabolite repression' and 'Crabtree effects' could be controlled by controlling limiting substrate concentration
ii. Organism's growth rate and subsequent oxygen demand could be controlled
iii. High cell density of the cells could be achieved
iv. Increased production of non-growth related metabolites could be achieved
v. This mode of operation could help in reducing the broth viscosity, when needed

*Continuous mode* – In this mode of operation, the organisms are fed with fresh nutrients while removing the spent medium and cells at same rate so as to ensure that factors like culture volume, biomass, product and substrate concentrations along with pH, temperature and dissolved oxygen are constant during the process.

*Advantages of continuous mode of operation* -

i. Feed flow rate could be optimized to improve the productivity and growth rate
ii. Longer period of productivity and shorter down time
iii. High cell density of cells could be achieved by cell recycling
iv. Best system to study culture physiology

All the three modes of operation have their advantages and could be opted depending on the final objective and problems being tried to be solved.

*Major factors influencing submerged fermentation output:* There are several biological as well as physical parameters which can affect the output of submerged fermentation output, and some of them include:

i. Microbial strain
ii. Inoculum stage, age, size
iii. Medium components
iv. pH
v. Temperature
vi. Agitation and aeration
vii. Dissolved oxygen concentration

viii. Broth viscosity
ix. Foaming

### 1.4.3 Solid State (SSF) vs. Submerged (SmF) Fermentation

A comparison between SSF and SmF has been presented in following table:

## 1.5 Classification of Fermentation

Most of the industrial fermentation processes use batch or fed-batch procedures, although continuous fermentation can be more economical, if sterility conditions are properly maintained.

**Batch fermentation:** In a batch process, all the ingredients are combined and the reactions proceed without any further input. Batch fermentation has been used for millennia

**Table 1.1** Comparison between solid state and submerged fermentation.

| Parameter | SSF | Submerged fermentation |
|---|---|---|
| Nature of substrate | Solid | Liquid |
| Volumetric productivity | Relatively low | High |
| Energy requirements | Relatively high | Low |
| Aeration control | Relatively difficult to meet the requirement | Relatively easier to meet the requirement |
| Downstream processing | Relatively difficult | Relatively easier |
| Risk of desiccation of microbe used | High | No |
| Inoculum used | Inoculum sprayed on surface medium | Inoculum is usually in liquid form |
| Temperature control | Difficult to achieve | Not overly difficult |
| pH control | Difficult to provide | Relatively easier to provide |
| Availability of nutrients to the microbes | Cannot be controlled within narrow limits if needed by the process | Can be controlled within narrow limits if needed, through feeding |
| Availability of oxygen to the biomass | Cannot be controlled at a particular level of saturation | Can be controlled reasonably at a particular level of saturation of the medium |
| Contamination problem | Relatively high | Low |
| Power consumption | Relatively low | High |
| Labour required | Relatively high | Low |
| Foaming | No | Yes |

to make bread and alcoholic beverages, and it is still a common method, especially when the process is not well understood. However, it can be expensive because the fermentor must be sterilized using high-pressure steam between batches. Strictly speaking, there is often addition of small quantities of chemicals to control the pH or suppress foaming.

Batch fermentation goes through a series of phases. There is a lag phase in which cells acclimatise to the new environment once the cell adopt to the new environment then exponential growth occurs which is also known as log phase. The growth slows down and becomes non-exponential when many of the nutrients have been consumed, but production of secondary metabolites accelerates. This continues through a stationary phase after most of the nutrients have been consumed, and then the cells die.

**Fed-batch fermentation:** Fed-batch fermentation is a variation of batch fermentation where some of the ingredients are added during the fermentation. Through fed batch process production of the desired product can be controlled as per need during the fermentation process. Addition of a limited quantity of nutrients during the non-exponential growth phase increases the production of secondary metabolites.

**Open:** Open fermentation usually involves the use of a naturally evolved mixed culture. Various open fermentation approaches which can be resistant to contamination may help to combat the high cost of sterilization between batches. This is particularly favoured in wastewater treatment, since mixed populations can adapt to a wide variety of wastes. Thermophiles can produce lactic acid at temperatures of around 50 °C, sufficient to discourage microbial contamination; and ethanol can be produced at a temperature of 70 °C. This is just below its boiling point (78 °C), making it easy to extract. Halophiles can produce bioplastics in hypersaline conditions. Solid-state fermentation adds a small amount of water to a solid substrate; it is widely used in the food industry to produce flavours, enzymes and organic acids.

**Continuous fermentation:** In continuous fermentation, substrates are added and final products are removed continuously. Typically, there are three varieties of continuous fermentation. (i) Chemostats, which keep the nutrient levels constant (ii) turbulostat, which keep cell mass constant and (iii) the plug flow reactors, in which the culture medium flows steadily through a tube and the cells are recycled from the outlet to the inlet. During the process, there is a steady flow of feed as well as effluent and the costs of repeatedly setting up a batch are thus minimized. It may also prolong the exponential growth phase and avoid formation of byproducts. However, it is critical to maintain a steady state without any contamination. Typically the fermentor must run for more than 500 hours to be more economical than batch processors.

## 1.6 Design and Parts of Fermentors

A fermentor or bioreactor is a closed vessel system with inbuilt arrangement for aeration, agitation, temperature and pH control, and drain or overflow vent to remove the waste biomass of cultured microorganisms along with their products. Fermentor is used for the commercial production of bioactive compound in the industries. It is a machine in which low cost substrate is utilized by microbes or enzymes to generate high value products. Fermentors are extensively used for production of antibiotics, food processing, probiotics, waste treatment, etc.

All bioreactors deal with heterogeneous systems dealing with two or more phases, e.g., liquid, gas or solid phase. Therefore, optimal conditions for fermentation requires efficient transfer of heat, mass, and momentum from one phase to the other phase. For the designing and operation of fermentors, chemical engineering principles are employed.

A bioreactor should provide for the following:

i. Agitation (for mixing of cells and medium),
ii. Aeration (aerobic fermentors); for $O_2$ supply,
iii. Regulation of factors like temperature, pH, pressure, aeration, nutrient feeding, liquid level, etc.,
iv. Sterilization and maintenance of sterility, and
v. Withdrawal of cells/medium (for continuous fermentors).

Modern automated and semi-automated fermentors are usually integrated with computers and database for efficient process monitoring and data acquisition. During a fermentation process, about 20-25% of the fermentor volume is left unfilled as head space for splashing, foaming and aeration. The fermentor design mostly depends on the type and the fermentation for which it is used. Bioreactors design should provide best possible growth and biosynthesis for industrially important cultures and allow ease of manipulation for all operations.

**Size of Fermentors**

The size of fermentors ranges from 1-2 L laboratory fementors to 5,00,000 L upto 1.2 million L. The choice for the size of the fermentor depends on the process and operating conditions.

**Construction of Fermentors**

Industrial fermentors can be divided into two major classes, anaerobic and aerobic. Generally, aerobic fermentors require elaborated equipment to ensure adequate mixing and aeration, whereas, an anaerobic fermentor requires specialized equipment except for removal of heat generated during the fermentation process.

A fermentor should be constructed in such a way that it can make provisions for the below activities:

- Sterilization
- Temperature control
- pH control
- Foam control
- Aeration and agitation
- Sampling point
- Inoculation points for microorganisms, media and supplements

- Drainage point for drainage of fermented media
- Harvesting of product
- Cleaning
- Facility of providing hot, cold water and sterile compressed air

Since most industrial fermentation process are aerobic, the construction of a typical aerobic fermentor is the following:

**1. Cooling Jacket**

The material of construction for a large-scale industrial fermentors is mostly stainless steel. A fermentor is a large cylinder closed at the top and bottom connected with various pipes and valves. The exterior of the fermentor vessel is fitted with a cooling jacket through which steam (for sterilization) or cooling water (for cooling) is provided. In order to successfully complete the fermentation in the fermentor, cooling jacket is necessary. Insufficient heat transfer takes place through the jacket in large scale fermentors. Internal coils are provided in the jacket through which steam or cooling water is run to maintain temperature.

**2. Aeration System**

Aeration system is one of the very important part of a fermentor. In a fermentor with a high microbial population density, the culture required high oxygen demand, but due to poor solubility in water, it hardly transfers rapidly throughout the growth medium. It is therefore, important to choose a good aeration system to ensure proper aeration and oxygen availability throughout the culture. two separate aeration devices(sparger and impeller) are used to ensure proper aeration in fermentor. The sparger is typically a series of holes in a metal ring or a nozzle through which filter-sterilized air (or oxygen-enriched air) passes into the fermentor under high pressure. The air enters inside the fermentor as a series of small bubbles through which the oxygen diffused into the liquid culture medium. The impeller (also called agitator) is an agitating device necessary for stirring inside the fermenter.

The stirring accomplishes two things:

i. It helps to mix the gas bubbles through the liquid culture medium and
ii. It helps to mix the microbial cells through the liquid culture medium which ensures the uniform access of microbial cells to the nutrients.

The size and position of the impeller in the fermentor depends upon the size of the fermentor. More than one impeller may be required for adequate aeration and agitation in the lager fermentors. Ideally, the impeller is one third of the fermentors diameter fitted above the base of the fermentor.

**3. Baffles**

The baffles are incorporated into fermentors to prevent a vortex improve aeration in the fermentors of all sizes. They are metal strips of about one-tenth of the fermentors diameter. attached radially to the walls of fermentor.

**4. Controlling Devices for Environmental Factors**

In any microbial fermentation, it is necessary not only to measure growth and product formation but also to control the process by altering environmental parameters as the process proceeds. Various devices for controlling environmental factors such as temperature, oxygen concentration, pH, cells mass, levels of key nutrients, and product concentration are commonly used.

**5. Use of Computer Software in Fermentor**

Computer technology has a remarkable impact in fermentation work and the computers are used to model fermentation processes in industrial fermentors. Integration of computers into fermentation systems is done in such a way, that it can control process monitoring, data acquisition, data storage, and error-detection. The on-line data analysis includes the acquisition measurements, verification of data, filtering, unit conversion, calculations of indirect measurements, differential integration calculations of estimated variables, data reduction, and tabulation of results, graphical presentation of results, process stimulation and storage of data.

**Parts of fermentor and their function**

1. Material used for fermentor
2. Impellers
3. Baffles
4. Inoculation port
5. Sparger
6. Sampling point
7. pH control device
8. Temperature control system
9. Foam control device
10. Bottom drainage system

**1. Material of construction:**

The material used for the construction of a fermentor unit must have these important properties (a) It should not be corrosive (b) It should not add any toxic substances to the fermentation media. (c) It should tolerate steam sterilization process. (d) It should be able to tolerate high pressure and resist pH changes.

The fermentor material to be used depends on type of fermentation process. For example, in case of beer, wine, lactic acid fermentation, the fermentor tanks are made up of wooden material, whereas, iron, copper, glass and stainless steel are used most commonly. Most of the time, 304 and 316 stainless steel is used for designing of a fermentor and these fermentors are mostly coated with epoxy or glass lining.

To develop a successful fermentation process, a fermentor unit should be provided the following control parameters.

**2. Impellers**

Impellers are an agitation device which are mounted on the shaft and introduced in the fermentor through its lid. They are made up of impeller blades attached to a motor on lid. The important function of an impeller is to mix micro-organisms, media and oxygen uniformly. Impeller blades play an important role of reducing the size of air bubbles and distribute them uniformly into the fermentation media. Impellers also help in breaking foam bubbles in the head space of fermentor. The foam formed during fermentation process can be controlled by the use of impellers as high foaming may cause contamination in the batch.

**3. Baffles**

Baffles are mounted on the walls of a fermentor and its main function is to break the vortex formed during agitation process by the impellers. If vortex is not broken, the fermentation medium may spill out of fermentor which may result in contamination. In short, baffles act as a barrier which break the vortex formed.

**4. Inoculation Port**

Inoculation port is the part of the fermenter through which media, substrate and inoculums is added in the fermenter vessel. All inoculations and media transfer should be done aseptically.

**5. Spargers**

A sparger is a system through which sterile air is introduced in the fermentation vessel. Spargers are located at the bottom of the fermentation tank and glass wool filters are used in spargers for sterilization of air and other gases. The sparger pipes contain small holes of about 5-10 mm, through which pressurized air is released in the aqueous fermentation medium.

**6. Sampling point**

During the fermentation process, sampling is required for time to time and sample point helps to withdraw the samples to monitor fermentation process and quality control. This sampling point should provide be steam sterilized and completely aseptic before withdrawal of sample.

**7. pH Control device**

Maintenance of pH to its optimum level is very important for growth of micro-organism to obtain a desired product during fermentation process. The pH probe in the fermentor checks the pH of medium at the given time intervals and adjusts the pH by adding acids or alkalis to maintain the desired pH level with the help of peristaltic pumps.

**8. Temperature control**

Temperature control probe contains a thermometer and jackets/ cooling coil around the fermentor to maintain the desired temperature. During the fermentation process, various reactions take place inside the fermentor and heat is generated & released in the fermentation media. The increase in the temperature may slow down the growth of living cells which may subsequently resulted in the low production of desired molecule or prolong fermentation cycle. This rise in temperature is controlled and maintained by circulating cold water through the coils or jackets around the fermentor.

**9. Foam controlling device**

Foam is usually generated during fermentation. It is necessary to remove or neutralize this foam with the help of anti-foaming agents, otherwise, the media may spill out of fermentor and may lead to contamination. A foam controlling device is therefore placed on the top of fermentor with inlet into fermentor.

**10. Bottom drainage system**

It is an aseptic outlet present at the bottom of the fermentor for removal of fermented media and products formed.

## 1.7 Types of Fermentor

The fermentor (bioreactor) types used extensively in industries are:

i. Continuous stirred tank fermentor
ii. Airlift fermentor
iii. Bubble column fermentor
iv. Fluidized bed fermentor
v. Packed bed fermentor
vi. Photo bioreactor

### 1.7.1 Stirred Tank Fermentor

Stirred tank fermentors consists of a cylindrical vessel with a motor driven central shaft that supports one or more impellers.

A continuous stirred tank bioreactor consists of a cylindrical vessel with motor driven central shaft that supports one or more agitators (impellers). The shaft is fitted at the bottom of the bioreactor. The aspect ratio of a stirred tank bioreactor is usually, 3 to 5 for microbial culture, while for animal cell culture it is less than 2. The diameter of the impeller is usually one third of the vessel diameter and the distance between two impellers is approximately 1.2 times the impeller diameter (Figure 1.1). The distance between two impellers is approximately 1.2 impeller diameter. Different types of impellers (rushton disc, concave bladed, marine propeller etc.) are in use. ln stirred tank bioreactors or in short stirred tank reactors (STRs), the air is added to the culture medium under pressure through a device called sparger.

The sparger along with impellers (agitators) enables better gas distribution system throughout the vessel.

There are many advantages of STRS over other types. These include the efficient gas transfer to growing cells, good mixing of the contents and flexible operating conditions, besides the commercial availability of the bioreactors.

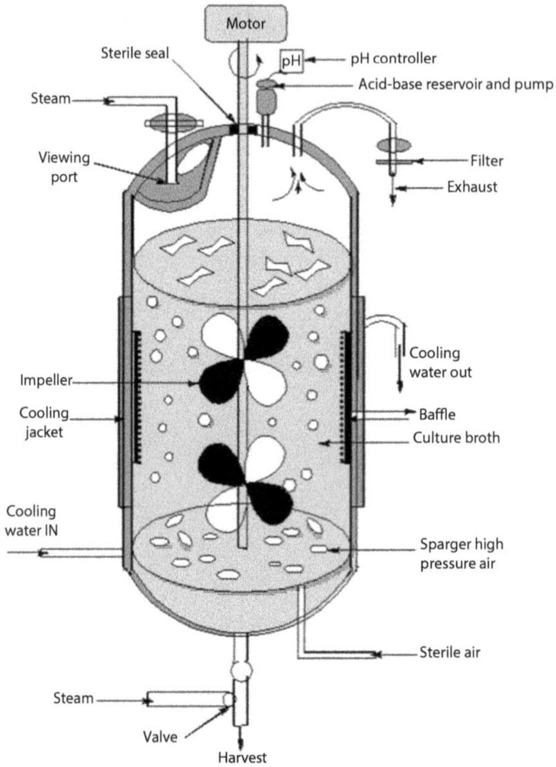


**Figure 1.1** A stirred tank fermentor.

### 1.7.2 Airlift Fermentor

In a airlift fermentor, the liquid culture volume of the vessel is divided into two interconnected zones by means of a baffle or draft tube. Only one of the two zones (riser) is sparged with air or other gas. Down-comer is the other zone that receives no gas. Airlift fermentors are considered to be highly energy-efficient and are often used in large-scale manufacture of biopharmaceutical proteins obtained from fragile animal cells. Heat and mass transfer capabilities of airlift reactors are similar to other systems, and airlift reactors are more effective in suspending solids than are bubble column fermentors. All performance characteristics of airlift fermentor are related ultimately to the gas injection rate and the resulting rate of liquid circulation, and the rate of liquid circulation increases with the square root of the height of the airlift fermentor. Since the liquid circulation is driven by the gas hold-up difference between the riser and the down-comer, circulation is enhanced when there is little or no gas in the down-comer (Figure 1.2).

Airlift bioreactors are used for aerobic bioprocessing technology, where they provide a controlled liquid flow in a recycle system by pumping. The performance of the airlift bioreactors is dependent on the pumping (injection) of air and the liquid circulation. Due to high efficiency, airlift bioreactors are preferred in methanol production, waste water treatment, single-cell protein production.

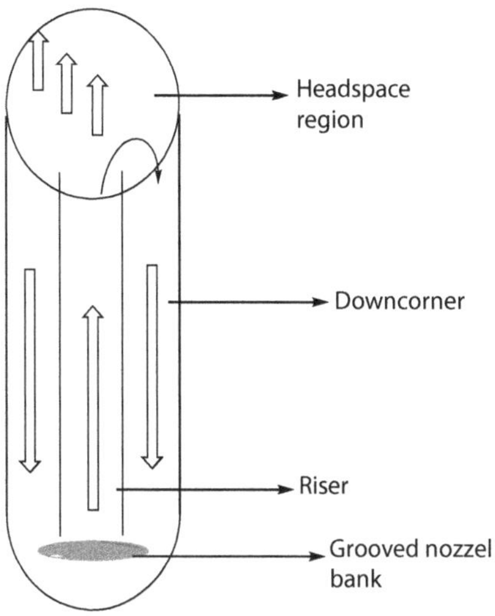


**Figure 1.2** An Airlift fermentor.

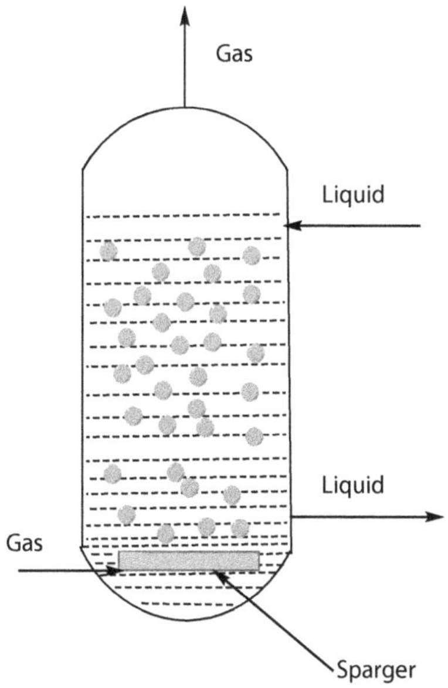


**Figure 1.3** A bubble column fermentor.

### 1.7.3 Bubble Column Fermentor

A bubble column fermentor (Figure 1.3) is usually cylindrical with an aspect (height-to-diameter) ratio of 4–6. Gas is sparged at the base of the column through perforated pipes, perforated plates, or sintered glass or metal micro-porous spargers. $O_2$ transfer, mixing and other performance factors are influenced mainly by the gas flow rate and the rheological properties of the fluid. Internal devices such as horizontal perforated plates, vertical baffles and corrugated sheet packings may be placed in the vessel to improve mass transfer and modify the basic design. At a set gas flow rate, by increasing vessel diameter, the mixing improves. The column diameter does not affect as long as the diameter exceeds 0.1 m. In bubble column reactor, as gas flow rate is increased, mass and heat transfer and the prevailing shear rate also increases.

### 1.7.4 Fluidized Bed Fermentor

Fluidized bed fermentor is comparable to bubble column bioreactor except the top position is expanded to reduce the velocity of the fluid. The design of the fluidized bed fermentor have expanded top and narrow reaction column, such that the solids are retained in the reactor while the liquids flow out (Figure 1.4). These bioreactors are suitable for use of carry out reactions involving fluid suspended biocatalysts such as immobilized enzymes, immobilized cells and microbial flocs.

A suitable gas-liquid-solid fluid bed is created by sparging the gas for an efficient operation of a fluidized bed. It is also important to ensure that the suspended solid particles are not too light or too dense and remain in a good suspended state. It is important to recycle the liquid as it helps to maintain continuous contact between the enzyme and substrate.

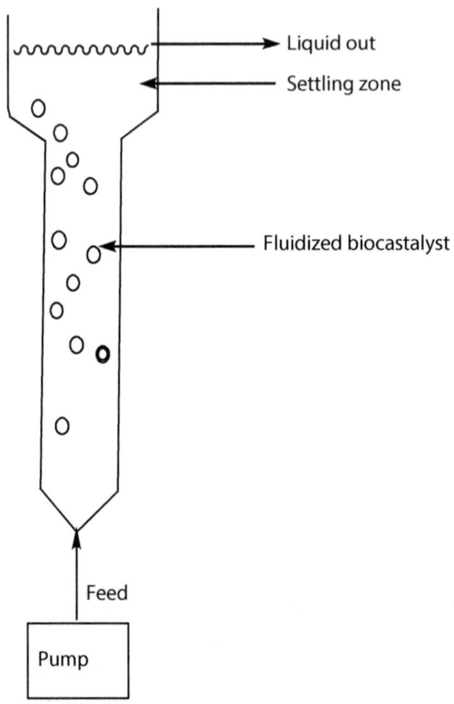


**Figure 1.4** A fludized bed fermentor.

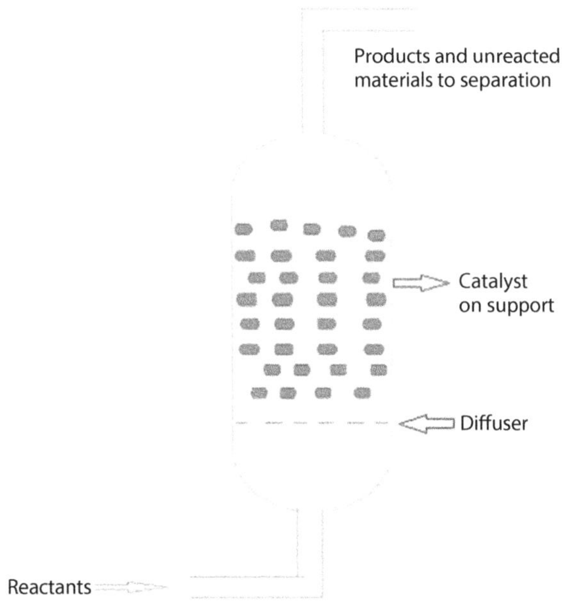


**Figure 1.5** A packed bed fermentor.

### 1.7.5 Packed Bed Fermentor

Packed bed fermentor (Figure 1.5) is a bed of solid particles, having enzymes on or inside the matrix of solids. The solids used may be porous or non-porous gels, and may be compressible or rigid in nature. The medium containing the substrate flows continuously over the immobilised matrix consisting of enzyme. The products formed after the substrate enzyme reaction is released in the medium and collocted. The flow of the medium can be from both side. The concentration of the nutrients (and therefore the products formed) can be increased by increasing the flow rate of the nutrient broth. It is difficult to control the pH of the medium because of poor mixing in the packed bed fermentors. However, these bioreactors are preferred for bioprocessing technology where reactions are inhibited by the product. The packed bed fermentor do not allow accumulation of the desired molecule to any significant extent.

### 1.7.6 Photo Bioreactor

These are the bioreactors specialised for fermentation that can be carried out either by direct sunlight or artificial illumination (Figure 1.6). Since artificial illumination is expensive, only the outdoor photo-bioreactors are preferred. Certain important compounds are produced by employing photo-bioreactors e.g., beta-carotene, asthaxanthin. They are made up of glass or more commonly transparent plastic.

These tubes or flat panels is consist of light receiving systems. With the help of centrifugal pumps or airlift pumps, the medium can be circulated through the solar receivers (Figure 1.7). For higher productivity it is very much needed that the cells

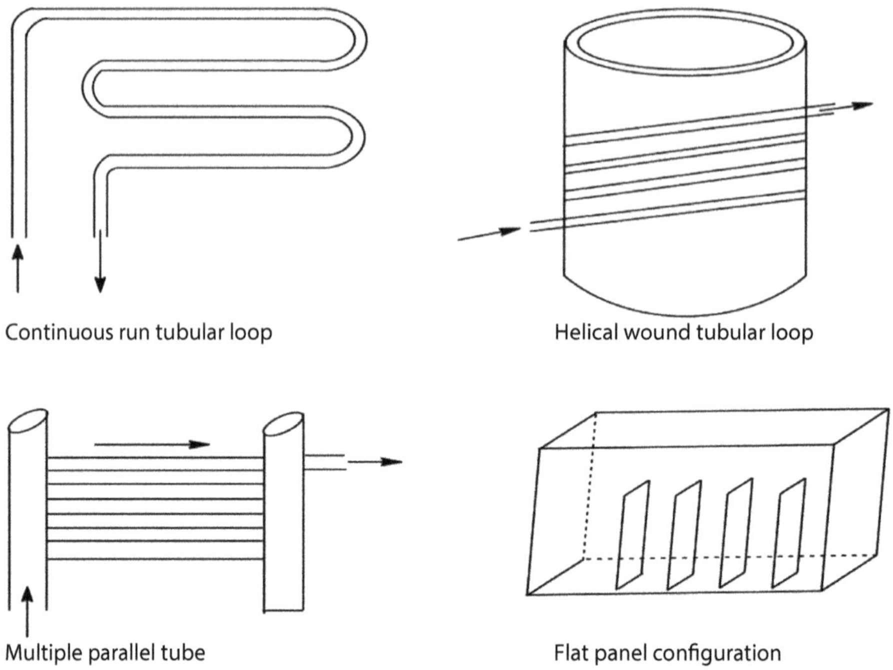


**Figure 1.6** Types of photobioreactors.

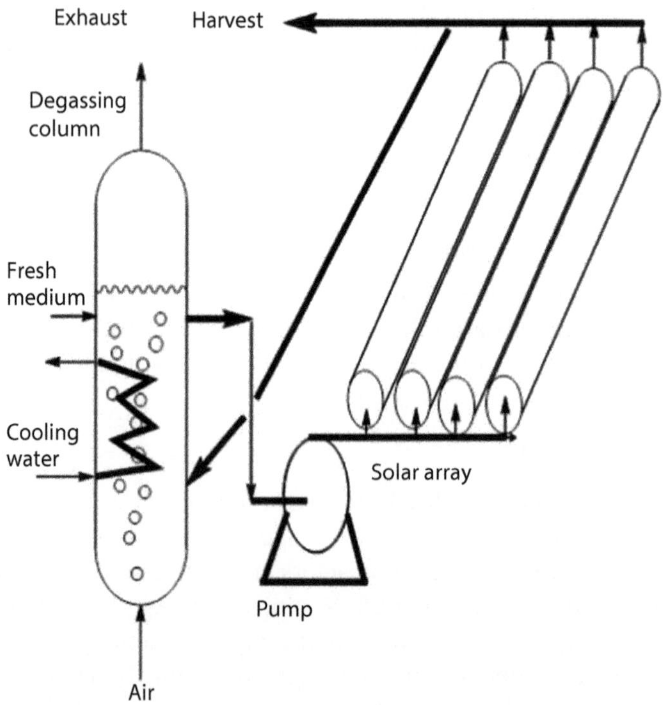


**Figure 1.7** A tubular photobioreactor with parallel run horizontal tubes.

should be in continuous circulation without forming any sediments. Further adequate penetration of sunlight should be maintained. The tubes should also be cooled to prevent rise in temperature. Photo-bioreactors are usually operated in a continuous mode at a temperature in the range of 25–40 °C. Microalgae and cyanobacteria are normally used.

## 1.8 Industrial Applications of Fermentation Technology

Fermentation processes lead to the production of variety of industrially important products including [34–37]:

a. Production of cells (biomass)
b. Production of metabolic products such as amino acids, proteins (including enzymes), vitamins, alcohol, etc., for human and/or animal consumption or industrial use
c. Modification of compounds (through the mediation of elicitors or biotransformation
d. Production of recombinant products such as recombinant proteins

(a) The fermentation process primarily involves the growth of microorganism(s) that may result in the production of a range of metabolites. The required organism must be grown under precise cultural conditions at a particular growth rate for production of desired metabolites. If a microorganism is introduced into a nutrient medium that supports its growth, the inoculated culture will undergo its growth phases as a batch culture. The biomass can be processed according to the required downstream processing for the desired product.

(b) Production of metabolites: Fermentation technology has widely been used for production of both primary as well as secondary metabolites. Primary metabolites are produced during the growth phase of the microorganism. For example, ethanol/ lactic acid are produced during glycolysis, while citric acid is produced in citric acid cycle by some strains of *A.niger* Lysine; threonine, tryptophan and other amino acids are produced by *Corynebacterium* sp.

Secondary metabolites are produced during the stationary phase of the life cycle of the microorganism. The common examples are antibiotics like penicillin, bacteriocins, etc. The secondary metabolites are not generally produced in the presence of rich carbon source which encourage growth, and are released into the medium without rupture of the cell membrane.

(c) Recent studies have shown that the microorganisms have much more potential than that observed by fermentation under laboratory conditions. These cryptic metabolic pathways can be altered in the laboratory using molecular or cultivation-based approaches. Application of small-molecule elicitors to activate secondary metabolite gene clusters may lead to novel metabolites. Botransformation processes using whole cells or enzymes have been the subject of interest among biotechnologists due to high regio, stereo and enantioselectivities.

(d) Recombinant DNA technology has widely been used as this provides the strain that is stable over the course of many fermentation cyles. Moreover, their half-life as novel species may be appreciably less than a conventional fermentation time. Most of the bioprocesses for biopharmaceuticals are mostly therapeutic recombinant proteins, growth hormones or vaccines. Initially, recombinant strains were widely used for production of metabolites by most of the biopharmaceutical companies. Further, more biopharmaceuticals were manufactured in eukaryotic cells, like CHO cells, than in microbes, but used similar bioreactor systems.

The detailed applications of fermentation technology in various industries are discussed in the following chapters of this book. The book is a collection of chapters covering application areas of fermentation technology, placed in respective sections according to their applications. The main sections include antibiotics & immunosuppressants; antibodies, peptides & proteins; sugars & sugar alcohols; metabolic engineering derived products.

## 1.9 Scope and Global Market of Fermentation Technology

According to one of the report, the world market of fermentation based product is valued approximately USD 1,573.15 million in 2017 and is expected that by the end of the year 2023 revenue of fermentation based product will reach to USD 2,244.20 million, growing at a CAGR of around 6.10% between 2017 and 2023. Modern industry has complemented the basic principle of fermentation technology with advances in genetic engineering by extending applications to produce a vast range of products. Moreover, rising petrol prices and depleting reserves of fossil fuels have diversified applications of the microbial fermentation process in the chemical sector to provide products such as alcohols, enzymes, amino acids, alkaloids, and other categories. The rising global demand for these products is likely to play a key role in helping the global microbial fermentation market expand at a promising pace in the in the years to come.

The fermentation technology based products are marketed in different segments on the basis of type of product such as alcoholic beverages, pharmaceutical products, industrial important bulk products, food & feed products. Pharmaceutical based products are further divided into monoclonal antibodies, recombinant proteins, antibiotics, probiotics and other biosimilars. Alcohol beverages segment is expected to grow at the highest CAGR in the global market over the forecast period.

Asia-Pacific is the dominant market in the global microbial fermentation technology during the forecast period. Market growth is particularly expected in China, India, and other emerging countries. Strong growth prospects of these countries are attributed to the vast increase in geriatric population, highly developed chemical industry in China, cutting-edge research in biotechnology and healthcare, the rise in disposable incomes, and changing lifestyles in developing nations. Moreover, improved spending capacity, the high prevalence of chronic ailments such as cancer and diabetes and the consequent rise in demand for new drug candidates will further boost the growth of the regional market in near future.

**Key Trends, Drivers**

In the recent past, a large share of the global market of fermentation products has been contributed by alcohols, followed by enzymes and organic acids. There has been an increase in the manufacturing of antibiotics and steroids worldwide, which in turn involves the use of fermentation products. There has also been a burgeoning demand for fermentation products in the form of enzymes from various manufacturing processes, such as in the manufacturing of paper, personal care products and starch.

It is forecasted that owing to the increased applications in oil and gas well treatment processes in the petroleum industry there will be an increased demand for organic acids as fermentation products in the coming years. Also, the growing market of preserved and nutritional foods involves the consumption of organic acids and hence is expected to further boost demand for the organic acids segment. Fermentation products are also important raw materials in the manufacturing of pet food; hence, the rise in demand for animal feed additives worldwide is expected to further fuel the fermentation product market.

**Market Participants**

Examples of some key market participants in the global fermentation chemicals market are as follows:

- Ajinomoto Company Incorporation
- BASF SE
- The Dow Chemical Company
- Novozymes A/S
- Cargill Inc.
- Evonik Industries
- Du Pont Danisco A/S
- Amano Enzyme Inc
- Archer Daniels Midland
- AB Enzymes

## 1.10 Conclusions

In a relatively short time, modern biotechnology has dramatically passed from a laboratory curiosity to a commercial activity. A century and half after the pioneering work of Louis Pasteur, the science of microbiology has reached its apex. Advances in microbiology and biotechnology have created a favorable niche for the development of enzymes and will continue to facilitate their applications to provide a sustainable environment for mankind and to improve the quality of human life.

## References

1. S.R. Barnum and R. By-Susan, *Biotechnology: An introduction. Modern Biotechnology.* Wadrwoods. pp.1, 2006.
2. S. Herrera, Industrial biotechnology-a chance at redemption. *Nature Biotechnology,* Vol. 22, p. 67, 2004.
3. Bioreactor: History, Design and Its Construction." Biology Discussion. N.p., 16 Sept. 2016. Web. 19 Apr. 2017
4. E. Cannel, M. Moo-Young, *Process Biochemistry*, Vol. 4, p. 2, 1980.
5. D.A. Mitchell, N. Krieger, M. Berovic, *Solid State Fermentation Bioreactors,* p. 1–12, 2006.
6. S. Bhargav, B.P. Panda, M. Ali, S. Javed, *Chemical and Biochemical Engineering Quarterly,* Vol. 22 (1), p. 49, 2008.
7. W. J. Scott, *Australian Journal of Biological Sciences*, Vol. 6, p. 549, 1953.
8. A. Pandey, P. Selvakumar, C.R. Soccol, P. Nigam, *Current Science*, Vol. 77, p. 149, 1999.
9. G. Georgiou, M.L. Shuler, *Biotechnology* and *Bioengineering*, Vol. 28, p. 405, 1986.
10. M. Moo-Young, A.R. Moreira, R.P. Tengerdy, *The Filamentous Fungi*, Vol. 4, p. 117, 1983.
11. J.L. Prosser, *The Growing Fungus,* p. 301, 1994.
12. S.A. Shojaosadati, V. Babaeipour, *Process Biochemistry*, Vol. 37, p. 909, 2002.
13. L.P.S. Vandenberghe, C.R. Soccol, A. Pandey, J.M. Le -bault, *Bioresource Technology*, Vol. 74, p. 175, 2000.
14. S. Miura, T. Arimura, I. Itoda, L. Dwiarti, J.B. Beng, C.H. Bin, M. Okabe, *Journal of Bioscience and Bioengineering,* Vol. 97, p. 153, 2004.
15. R.P. John, K.M. Nampoothiri, A. Pandey, *Process Biochemistry,* Vol. 41, p. 759, 2006.
16. R. Banerjee, G. Mukherjee, K.C. Patra, *Bioresource Technology*, Vol. 96, p. 949, 2005.
17. P.K.R. Kumar, B.K. Lonsane, *Biotechnology and Bioengineering*, Vol. 30, p. 267, 1987.
18. C. Gelmi, R. Pér ez-Correa, M. González, E. Agosin, *Process Biochemistry,* Vol. 35, p. 1227, 2000.
19. L.M. Pastrana, M.P. Gonzalez, J. Pintado, M.A. Murado, *Enzyme and Microbial Technology,* Vol. 17, p. 784, 1995.
20. A.B.P. Medeiros, A. Pandey, P. Christen, P.S.G. Fontoura, R.J.S. Freitas, C.R. Soccol, *World Journal of Microbiology and Biotechnology,* Vol. 17, p. 767, 2001.
21. G. Feron, P. Bonnarame, A. Durand, *Trends in Food Science and Technology*, Vol. 7, p. 285, 1996.
22. M.A. Longo, M.A. Sanromán, *Food Technology and Biotechnology,* Vol. 44, p. 335, 2006.
23. M. Dominguez, A. Mejia, S. Revah, J. Barrios-Gonzalez, *World Journal of Microbiology and Biotechnology,* Vol. 17, p. 751, 2001.
24. C.G. Phae, M. Shoda, H. Kubota, *Journal of Fermentation and Bioengineering,* Vol. 69, p. 1, 1990.
25. C. Sekar, K. Balaraman, *Bioprocess Engineering,* Vol. 18, p. 293, 1998.
26. C. Sekar, V.W. Rajasekar, K. Balaraman, *Bioprocess Engineering,* Vol. 17, p. 257, 1997.
27. A.E. Asagbra, A.I. Sanni, O.B. Oyewole, *World Journal of Microbiology and Biotechnology,* Vol. 21, p. 107, 2005.
28. X. Jian, C. Shouwen, Y. Ziniu, *Process Biochemistry,* Vol. 40, p. 3075, 2005.
29. F.C. Oliveira, D.M.G. Freire, L.R. Castilho, *Biotechnology Letters,* Vol. 26, p. 1851, 2004.
30. T.D. Vrije, N. Antoine, R.M. Buitelaar, S. Bruckner, M. Dissevelt, A. Durand, M. Gerlagh, E.E. Jones, P. Lüth, J. Oostra, W.J. Ravensberg, R. Renaud, A. Rin - zema, F.J. Weber, J.M. Whipps, *Applied Microbiology and Biotechnology,* Vol. 56, p. 58, 2001.

31. I. Larena, R. Torres, A. Cal, M. Li ñán, P. Melgarejo, P. Domenichini, A. Bellini, J.F. Mandrin, J. Lichou, X.O. Eribe, J. Usall, *Biological Control,* Vol. 32, p. 305, 2005.
32. A. Pandey, F. Francis, A. Sabu, C.R. Soccol, *Concise Encyclopedia of Bioresource Technology*, p. 702, 2004.
33. P.F. Stanbury, A. Whiitaker, S.J.Hall, Principles of Fermentation Technology (Second ed.). Butterworth-Heinemann. ISBN 978–0750645010, 1999.
34. P.L.Show, K.O.Oladele, Q.Y.Siew, F.A.A.Zakry, J.C.Wei Lan, T.C.Ling, *Frontiers in Life Science*, Vol.8(3), p. 271, 2015.
35. K. Nakayama, K. Araki, H. Kase, *Advances in Experimental Medicine and Biology*, Vol.105, p.649, 1978.
36. R.K. Pettit, *Microbial Biotechnology, Vol.* 4(4), p. 471, 2011.
37. D. F. Ollis, Philisophical Transactions of the Royal Society B, Vol 298, 1982 (DOI DOI: 10.1098/rstb.1982.0065.

2

# Extraction of Bioactive Molecules through Fermentation and Enzymatic Assisted Technologies

**Ramón Larios-Cruz, Liliana Londoño-Hernández, Ricardo Gómez-García, Ivanoe García-Galindo, Leonardo Sepulveda, Raúl Rodríguez-Herrera and Cristóbal N. Aguilar***

*Group of Bioprocesses and Bioproducts, Food Research Department, School of Chemistry, Universidad Autónoma de Coahuila, Saltillo, Coahuila, (México).*

***Abstract***

Bioactive compounds are a group of compounds that may or may not occur naturally and can have an effect on human health. Polyphenols and polypeptides are examples of such compounds. In order to improve functional properties of foods, several processes have been used to obtain and purify these compounds having high added values. The extraction processes of these compounds include chemical synthesis, particularly used in food industry, distillation, solvent and supercritical fluids extraction. On the other hand, soft chemistry provides new routes for a set of mild chemical operations, which can contemplate reactions such as cationic exchange, dehydration, hydrolysis, redox, among others. These reactions could be found in biological processes. Microorganisms have the ability to synthesize these molecules and the efficiency of enzymes in the degradation and synthesis of compounds. Fermentation and enzyme-assisted extraction form the part of new methodologies to obtain bioactive compounds.

***Keywords:*** Biotransformation, phenolic compounds, peptides, compounds extraction, experimental design, soft chemistry

## 2.1 Introduction

Bioactive compounds are constituents of plants and food substances, which are found in small amounts in them. These compounds are source of antibiotics, secondary metabolites, mycotoxins, alkaloids, pigments, phenolic compounds etc. [1]. In the past years, interest in obtaining phenolic compounds has increased due to the association of foods rich in phenols with human health. It has been reported that phenolic compounds decrease the risk of cardiovascular diseases and cancer conditions. Additionally, these compounds generate stability to the foods by preventing peroxidation [2].

**Corresponding author*: cristobal.aguilar@uadec.edu.mx

Saurabh Saran, Vikash Babu, and Asha Chaubey (eds.) High Value Fermentation Products, Volume 1, (27–60) © 2019 Scrivener Publishing LLC

According to the number of benzene rings and structure of carbon skeleton, the phenolic compounds can be grouped into different groups as phenolic acids, flavonoids, lignans and stilbenes. Flavonoids are the largest group of phenols and have been studied the most [3]. Flavonoids exhibit multiple pharmacological activities including antibacterial and antiviral [4], antiprotozoal [5], antiallergic, anti-inflammatory [6], antidiarrheal [7], vasodilator [8] etc.

As phenolic compounds are beneficial for health and help in the production of stable foods, many methodologies and sources have been evaluated for their production. Traditionally, phenolic compounds have been recovered through processes of solvent extraction, using water or mixtures of solvents [9]. Also, they have been used along with emerging technologies such as microwaves, ultrasound or the use of supercritical fluids. However, for these processes a pre-treatment which is the hydrolysis of materials in order to release the hydroxyl group (O-glycosides) or break to carbon-carbon (C-glycosides) bonds present in the cell wall [3]. In industries, this hydrolysis process is chemically done by making use of acid or alkali, but the obtained products may be toxic, so the process generates environmental pollution [3], [10]. Therefore, there is need of alternative technologies that mimic the natural processes following the principle of soft chemistry, such as biocatalysis chemistry.

Biocatalysis is the process in which enzymes are used to catalyze chemical or biological reactions [11]. This process is highly desirable, since it operates at conditions near to neutral pH, room temperature and atmospheric pressure. Mainly due to the specificity and selectivity of enzymes, using biocatalysis, we can obtain products of higher purity [10]. In recent years, use of biocatalytic process, especially in industries, has been increased to produce a variety of chemicals, mainly pharmaceutical products [12]. Biocatalysis is an alternative with great potential for the development of new bioactive compounds. Among the available technologies, which work on the principle of biocatalysis are fermentation, microbial transformation and extraction assisted by enzymes. These technologies have gained special interest because they are used in the production of new novel compounds. [13].

Studies on biotransformation are important and have gained special interest due to the production of secondary metabolites which are of high added value. Some studies are limited to the availability of the purified compounds in order to identify the degree of biotransformation or the compound produced. Other studies are focused on the source and the enzyme, microorganism or compound used to biotransform. Chávez-González *et al.* [14] studied the enzymatic and microbial biotransformation of tannic acid. It is important to note that they also showed variations in tannic acid, depending on the laboratory conditions. This is an example of the difficulties of the studies in biotransformation due to the differences in the substrate employed. Even if the substrate is the same, by changing an enzyme, the product will change. So, there is a necessity of more studies for identification and purification of many molecules, in order to figure out more biotransformation processes when limitation of high quality equipments for identification of molecules is present.

Considering the benefits of bioactive compounds, alternative technologies available for their production will be described in this chapter. Some of the major bioactive compounds that are available will be defined, technologies that are used traditionally, alternative technologies for obtaining these compounds, the possible metabolic routes used for transformation of biocomposites, the effect of microorganisms or enzymes used to free compounds of the matrices, substrates used and the techniques employed to achieve optimization of the processes.

## 2.2 Definition of Bioactives Compounds

### 2.2.1 Polyphenols and Polypeptides

Bioactive compounds occur naturally in small quantities in plant and food products and are considered as an extra nutritional constituent [15]. Most common bioactive compounds include: secondary metabolites such as, alkaloids, antibiotics pigments, plant growth factors, and phenolic compounds [15–17].

One of the bioactive compounds, i.e., polyphenols is a complex group of organic compounds containing more than one phenolic group ($C_6H_5OH$); they are the second most abundant molecules present in nature and are generally found in the kingdom plantae [18]. The chemical structures of these compounds are well studied, but they are a less understood class of molecules, which are known to have a profound relationship with the biological activities exhibited by them. Their chemical structure impart them acidic nature; it can vary from a simple phenolic structure to highly complex polymeric structures with molecular weight between 300 to 20,000 Da [19]. Phenolic compounds are the most abundant antioxidants present in fruits, plant-derived beverages, citrus peels, pomegranate husk, raspberry and grape waste, which are widely known for their high antioxidant and free radical-scavenging capacity [20–23].

The majority of phenolic compounds are found covalently linked to polysaccharides of the plant cell wall and are present in soluble form in the cytoplasm and cell vacuoles, out of which the most abundant is ferulic acid [24, 25]. Decrease in antioxidant activity implies the reduction of functionality over health, especially in cases when these compounds are consumed as food or as nutraceuticals [2].

Plants produce phenolic compounds, which are not directly related to primary metabolic processes. Because of this reason, these natural compounds appear to function as defensive agents or act as attractants for reproductive advantage and may have antioxidant properties with potential health benefits. These compounds may reduce the risk of cardiovascular disease and cancer [26, 27]. In addition, phenolic compounds can be considered to be added-value product from industrial wastes [23, 28], [29]. Recently, biotechnological methodologies for bioactive compounds production have attracted more attention, Li *et al.* [30]. The mechanism for enzymatic extraction of phenolic compounds from waste (citrus peels) has been reported. In this work, enzymatic degradation of the cell wall was done, making easy extraction of intracellular materials and liberation of phenolic compounds.

### 2.2.2 Importance and Applications of Bioactive Compounds

Degenerative diseases decrease the life quality of human beings, as these diseases advance progressively and cause death. These diseases cause physical or mental impairment to people, causing an imbalance in the mechanisms of cell regeneration. Some of the most common degenerative diseases are: diabetes, obesity, cancer, arthritis, Alzheimer´s and cardiovascular diseases. Some potential causes of these diseases are: poor diet, high consumption of fats and sugars, smoking, alcoholism, pollution or in some cases because they simply are hereditary, such as Alzheimer's disease [31].

The prevention of these diseases can be achieved by: a) A rich diet of natural antioxidants, which are present in fruits (pomegranate, grapes, blueberries, etc.), vegetables, and moderate consumption of red wine, tea and juice, b) A rich and balanced diet based mainly on plant foods, and c) Replacing foods rich in sugars, fats and animal foods. Plant foods in addition provide nutrients that are important sources of energy, vitamins and minerals, containing a number of compounds, which are responsible for giving sensory characteristics, also provide some growth regulators or natural protectors against pest and pathogens [32]. The main importance of consuming these compounds is that some of these have a significant impact on improving human health and on prevention of degenerative diseases. [33]. It is thought that polyphenols help to prevent premature aging and may even have antibiotic effects.

Some polyphenolic compounds have a role as defense chemicals, e.g. in grains, they prevent the loss in premature germination and damage due to microorganisms. Flavonoids and other polyphenols can protect proteins and lipoproteins from degradation; also these compounds can be used as antioxidants due to their activity on free radicals and lipoprotein oxidation with low density [34]. An antioxidant is defined as a natural substance that can neutralize and counter harmful effects of cellular oxidation that are caused by free radicals [20, 35, 36]. Phenolic antioxidants are able to stop the chain reaction of free radical oxidation which is attributed by the ability to donate hydrogen from the phenolic hydroxyl groups, thus forming the stable final product [37]. Some of the most studied polyphenolic compounds are: tannins, flavonoids and phenolic acids, which are abundantly found in many trees such as oak (*Quercus pedunculata*) [38], eucalypt (*Eucalyptus camaldulensis*) [39] and pomegranate (*Punica granatum*) [29], some plants such as creosote bush (*Larrea triedentata*) [40], candelilla (*Euphorbia antysyphilitica Zucc*), fruits such as grapes (*Vitis vinifera*) [41] found mainly in the shell and seeds [19] and in some beverages such as juices, red wine, green and black tea [42], [43]. A group of polyphenol antioxidant compounds present in grapes, green tea, soybeans and wine may lower the risk of different cancers, but the study of these powerful compounds has remained unclear [44].

Secondary metabolites like anthocyanins and flavonoids exhibit a wide range of properties, depending on their particular structures. Yellow, orange, red and blue pigments, as well as various compounds are involved in food flavouring. Interactions with other constituents of the food matrix affects polyphenol properties; for example, it is well known that color intensification is resulting from interactions of anthocyanins with other compounds [45], [46]. Vanillin and eugenol are volatile polyphenols responsible for odor of some plants; these compounds are extremely potent odorants, but the major flavors associated with polyphenols are bitterness and astringency. One polyphenol responsible for astringency perception is eugenol that results from interactions of tannins with salivary proteins. Another example is the formation of turbidity and precipitates in beverages, and for inhibition of enzymes and reduced digestibility of dietary proteins. The presence of different molecules such as proteins and polysaccharides altered tannins astringency [47].

The popularity of soybeans is attributed to its content of isoflavones, which have similar structure to estrogens and can bind to receptors. Apples contain many kinds of polyphenols, and the main components is oligomeric procyanidins [48]. There are reports that show plant-derived polyphenols can slow the growth of cancer cells in mice and curb the spread of cells by triggering a serial of reactions that causes the cell to self-destruct, a process known as apoptosis [49].

Studies using red wine have examined its effects on several cancers, including leukemia, skin, breast and prostate cancer [50]. Scientists are studying resveratrol to learn more about its cancer-preventive activities. Recent evidence from animal studies suggests that anti-inflammatory compounds may be an effective chemo-preventive agent in three stages of the cancer process: initiation, promotion and progression; however, studies of the association between red wine consumption and cancer in humans are in their initial stages [44].

Polyphenols from plants and residual wastes may be used for increasing the shelf life of food by preventing lipid peroxidation and protecting from oxidative damage. Increasing the oxidation stability of vegetable oils is important for industrial practice, and many anti-oxidant tests are based on this ability to retard or inhibit the oil rancidity [51]. Although several *in vitro* studies have yield excellent results using polyphenols from plants, more detailed investigations are still required to extrapolate these results to *in vivo* applications [52].

Wu *et al.* [53] reported the beneficial effects of using wild rice as an antioxidant, particularly after particle size reduction and cooking. Improvement in color stability for different species of rock fish was observed in the presence of antioxidant extracts from shrimp shell waste [54]. Antioxidants have also been proposed for preventing loss or improving the stability of pigments from red beet juice in the food industry [55], as well as for aroma protection and stabilization [56] and for use in oral and topical pharmaceutical and cosmetic compositions [57].

### 2.2.3 Bioactive Peptides

Proteins and peptides have bioactive roles beyond their nutritional impact. Proteins have limited absorption in the gastrointestinal tract; for this reason, proteins showing activities such as enhancing nutrient absorption, inhibiting enzymes and modulating the immune system to defend against pathogens are very important [58]. Examples of these bioactive proteins and peptides are: vitamin- and mineral-binding proteins, proteins with antimicrobial activity, immunosuppressing/-modulatory proteins, proteins with enzyme inhibitory activity and hormones and growth factors. A brief description of each protein type is given next.

**Vitamin binding proteins** can alter the biological activity; one example is Vitamin D-binding protein (DBP) which has been reported affecting the activity of total 25-hydroxyvitamin D (25(OH)D). This issue has relation to the effects of vitamin D on bone mineral density (BMD) and fractures. One example of these kind of proteins is fetuin-A which has been involved in bone metabolism [59].

**Proteins binding to minerals** they have different physiological functions, acting as mineral trappers through specific and non-specific binding sites. Functions of these peptides such as carriers, chelators, of various minerals have been reported. In the human body there have been both positive and negative effects of these kinds of proteins, which are characterized to have a high content of negative charges; in consequence these proteins efficiently bind to divalent cations with the formation of soluble complexes. Different peptides and minerals complexes have been found; for example, peptides bind to Fe, Mg, Mn, Cu and Se. These phosphopeptides are resistant to further hydrolysis because of their high content of negative charges. Some examples of these compounds in foods are peptides from the caseins which have the disadvantage that they may give rise to a bitter taste, in milk are the main mineral binder or chelators of calcium examples are caseins such as 1-casein, as2-casein, b-casein and k-casein. Others reported are whey proteins, b-lactoglobulin, a-lactalbumin and lactoferrin, which are more resistant to enzymatic attack and undergo hydrolysis much more slowly than the caseins these peptides bind to specific minerals like calcium, magnesium, zinc, iron, sodium and potassium. New applications in the food industry of these kind of peptides are mentioned, such as during production of nutraceutical or dietary foods, or during fortification of foods with minerals in a low concentration to overcome mineral deficiency. In addition, these peptides may be used in the pharmaceutical industry [60].

**Antimicrobial peptides** are able to inhibit pathogenic microorganisms or parasites while causing little or no harm to the host cells; these peptides have been reported in different organisms (animals, plants, and microorganisms) and are of great interest to the health, pharmacology, and food industry. Some examples of antimicrobial peptides are the defensins which are found in granules in the polymorphonuclear neutrophils (PMNs) and are responsible for the defense of the organism. Animal defensins, such as dermaseptin, antileukoprotease, and protegrin have been reported because of their activities against bacteria, fungi, and protists. On the other side, examples of the invertebrates defensins are drosomycin and heliomicin, while in plants have been reported types A and B; and in bacteria bacteriocins, e.g., acrocin, marcescin, etc., are found [61]. Bacteriocins from lactic acid bacteria (LAB) are very important in the food industry because they have the QPS status (qualified presumption of safety); humans have consumed bacteriocins from LAB in fermented foods for a very long time without any adverse side effects.

**Immunosuppressing modulatory proteins** are involved in the regulation of the immune system. There are reports about different proteins and peptides with these properties such as MSCs which possess potent immunosuppressive capacity; this protein inhibits T cell proliferation and function via nitric oxide in mice. The inflammatory cytokines IFNγ together with either TNFα or IL-1β the immune regulatory capacity of this protein [62].

Proteins with enzyme inhibitory activity are very important for characterization of eukaryotic and prokaryotic enzymes. In this case, proteins can be used as selective inhibitors to assign functions to enzymes in native biological systems. In addition, these inhibitors have been employed to delineate the biochemical and cellular functions of enzymes, which have an impact on the discovery of metabolic and signaling pathways [63].

Growth factors are proteins that bind to receptors on the cell surface, activating cellular proliferation and/or differentiation. Many growth factors stimulate division of a different cell-type, while others are specific to a particular cell-type. For example, increases in the concentration of blood glucose induce insulin secretion, and then uptake of glucose is promoted by a number of tissues [64].

Bioactive peptides have been reported from different plants. Montoya-Rodríguez *et al.* [58] showed that bioactive peptide sequences from amaranth seed (*Amaranthus hypochondriacus*) proteins may have potential to prevent chronic diseases. Some of them were angiotensin-converting enzyme-inhibitor peptides, dipeptidyl peptidase IV inhibitor, antioxidants, with glucose uptake-stimulating activity and antithrombotic activities. In other research, Sharma *et al.* [65] achieved purification of the peptide "Hispidalin" which showed inhibitory effects against diverse human bacterial and fungal pathogens from seeds of *Benincasa hispida*. This protein is amphipathic and basic with one net positive charge having isoelectric pH 8.1. Bioactive peptides have been also isolated from animals such as *Anemonia sulcata* and bovine hair [66]. In this case, the identified polypeptides showed a strong reducing power and antioxidant capacity [67]. Recently the analysis of genomes, transcriptomes and proteomes reveals the existence of short open reading frames (small ORFs or smORFs) which have roles in cell metabolism, apoptosis and development [68].

## 2.3 Traditional Processes for Obtaining Bioactive Compounds

One of the key steps to produce bioactive compounds is the extraction process. Nowadays, there exist non-conventional, more efficient and environment friendly methods such as: microwave, electric pulse, and ultrasound among others. Nevertheless, traditional processes are commonly used because of the high yield for obtaining bioactive compounds and the low costs of production and recovery process. Note that most traditional methods that are employed in the research are complemented by other technology for better results. Some works for obtaining bioactive compounds using traditional processes and in conjunction with other technologies are analyzed and described.

### 2.3.1 Soxhlet Extraction

Extraction with Soxhlet system is a method that is based on the separation of the components present in a solid matrix through a liquid solvent. The extraction occurs on the solid matrix which is placed within a porous cartridge in Soxhlet equipment. The solvent is heated, evaporated and condensed in the refrigerant and this in turn falls on the solid matrix to dissolve the components. When the solvent reaches the level of the siphon, the solvent mixture and components up by the siphon and return to the container. This process is repeated for a time until all components are dissolved in the solid matrix. Soxhlet extraction system is commonly used to compare with new alternatives for extraction. Citrus peels wastes were used to obtain D-limonene as the main constituent of lemon essential oil. In that study, basic parameters were evaluated in the extraction process as temperature, time and solid-liquid ratio [69]. Soxhlet extraction was used to obtain flavonoid compounds from leaves of mint (*Mentha spicata* L.). The results showed that this conventional methodology obtained higher yields compared to supercritical extraction with carbon dioxide. It was possible to identify compounds such as epicatechin, catechin, rutin, apigenin, among others. These bioactive compounds are of great importance because of the benefits they present to human health [70]. Soxhlet extraction and subcritical extraction with solvents to obtain bioactive compounds from *Arucaria angustifolia* were compared. High yields of fatty acids were obtained with the traditional method; in addition, these compounds have various applications in the food industry, pharmaceuticals and cosmetics [71]. In other research, ultrasound-assisted extraction and extraction with Soxhlet system for the production of ergosterol were compared. This is a bioactive compound of high added value and is the most abundant compound in *Agaricus bisporus* L. Although good yields with Soxhlet system are obtained, a saponification is required for the extraction of bioactive compound [72].

Soxhlet extraction has some advantages: efficient extraction of components due to extensive contact of the solvent with the solid matrix. It does not require filtration. Due to the closed system, the temperature is optimal for obtaining the components. The system is low cost and increases yields [73].

### 2.3.2 Liquid-Liquid and Solid-Liquid Extraction

One of the groups of most studied bioactive compounds are polyphenols, due to the great benefits they present to human health, and in particular research focuses on the parameters for extraction of these compounds. The liquid-liquid extraction is based on the separation of compounds from a liquid using a solvent. For example, the liquid-liquid extraction was evaluated for the identification of bioactive compounds from grapes *Vitis labrusca* and *Vitis vinifera*. They were selected because this method showed the best accuracy and repeatability. Furthermore, this method is very economical, simple and has a short analysis time [74]. The method of liquid-liquid extraction is one of the conventional techniques used due to low production costs and high yield for obtaining bioactive compounds. In other research, extraction of phenolic compounds was evaluated from virgin olive oil. In this work, this method against deep eutectic solvent extraction was compared. They concluded that this new technology had better yields compared to the conventional method [75]. The solid-liquid extraction is another traditional method for obtaining bioactive compounds. This is based on the extraction of components present in a solid material with the aid of a solvent. For example, extraction parameters for obtaining bioactive compounds were optimized from gardenia fruit. The effects of ethanol concentration, extraction time and temperature were investigated. With this method yields can be obtained of 8.41 mg/g, 109 mg/g and 24.87 mg/g of crocin, geniposide and phenolic compounds respectively [76]. In another study, the influence of the kind of solvent (water, ethanol-water (1:1), and absolute ethanol) was evaluated and the type of processing (fresh, lyophilized and cooked) on the extractability of bioactive compounds from elephant foot yam. They concluded that the type of solvent and type of processing was significant in all cases, because the effects caused by each solvent may depend on how the sample is processed [77]. The main drawbacks of the liquid-liquid extraction are: it requires large volumes of solvents in the extraction process. The process and product recovery involves many hours of work. The product concentration can be low. The process can produce undesirable compounds [78]. The disadvantages in the solid-liquid extraction are low product yields and product concentration is low [79].

### 2.3.3 Maceration Extraction

Extraction by maceration defined in the food industry and botanicals for obtaining bioactive compounds of a solid matrix using a liquid extractant may be water, acetic acid or alcohol. However, it has two variants; cold maceration involves immersing the material in a particular solvent extractor for a certain time. Maceration heat is the process in which the same procedure can vary only by the means of maceration and heat is applied to accelerate the process. This process has some disadvantages as high consumption of time; the process is very laborious and must be accompanied by other methods of concentration, identification and separation of compounds [80]. This methodology was used to obtain extracts of *Solanum paludosum* to test their antioxidant activity. They found that maceration extraction was efficient for obtaining bioactive compounds; however, alternative methods are gaining advantages by decreasing time processes and the good yields obtained [81]. Maceration extraction was used to obtain bioactive compounds from chokeberry (*Aronia melanocarpa*). The influence of different solvents, particle size, solid-liquid ratio and extraction time were investigated. With this methodology high yields of phenolic compounds and anthocyanins were achieved [82]. The effect of maceration on the profile of bioactive compounds is effective. They evaluated this process for phenolic compounds and organic acids from new varieties of grape juice like *Vitis labrusca* L. The combination of this process with the use of enzymes for increasing the content of bioactive compounds and antioxidant activity was improved [83].

## 2.4 Fermentation and Enzymatic Technologies for Obtaining Bioactive Compounds

### 2.4.1 Soft Chemistry in Bioactive Compounds

The "era of plastic" of the 1970s and 1980s has passed. When chemists began to redirect their attention towards natural products, they started to recognize raw materials for the production of eco-friendly compounds. They began the synthesis of biopolymers for the production of biodegradable packaging and other products, and began to observe processes achieved by nature. From here comes the new trend of chemistry and bio-inspired products. The challenge then became to elucidate the biological strategies that were used through *in-vitro* tests and with the necessary modifications to implement these natural processes in obtaining new materials [84].

Synthetic chemicals traditionally worked under extreme conditions (high temperatures and low pressures) making them costly processes in terms of energy and using large amounts of organic solvents which are difficult to remove safely after use in the reactions. Following the new trend of chemistry, in 1977 Jacques Livage said, "Nature teaches us that it is possible to carry out chemical reactions at room temperature, untidy and aqueous environments", by ending this principle as "Soft Chemistry" (*Douce Chimie*). Today, this method is used to obtain new materials performing reactions in quasi-physiological conditions, which generate only renewable biodegradable products [84]. Soft chemistry is a type of mild reactive conditions that allows to obtain materials or other compounds which are hard to produce under classical conditions (for example, high temperatures). Reactions involved in soft chemistry include intercalation, deintercalation, cation exchange, dehydration, dihydroxylation, redox, and others [85].

Since that time, scientists have been commissioned to study the characteristics of the processes that could lead to the production of new products. Such is the case of Figlarz [85] who studied the structural and thermodynamic aspects to generate new metastable phases, which cannot be obtained from traditional thermodynamic processes, and Rao and Biswas [86] detailed the routes followed in soft chemistry. Likewise, most of the work using the principle of soft chemistry is focused on the synthesis of polymeric materials.

A level of organic chemistry, the term soft chemistry can also be used since it is possible to carry out reactions following the principles of nature, obtaining a wide variety of biocompounds. The definition of soft chemistry opens the way for the use of biocatalysis, these reactions being carried out by enzymes for the transformation of a compound in natural products of high industrial value.

At this point fermentation processes and enzymatic technologies converge as new alternatives for obtaining bioactive compounds, carried out in a clean environment, free of waste and solvents that can modify the activity of these compounds and enhance the environment.

### 2.4.2 Biotransformation of Bioactive Compounds

In general, biocatalysis can be defined as utilization of enzymes in catalysis of different processes – industrially or not – in artificial conditions (*in vitro*) [11]. On the other hand, biotransformation is defined as the use of biological systems to achieve chemical changes in compounds that do not represent a natural substrate. In the biotransformation, isolated enzymes or microorganisms can be used [87].

Biotransformation can be advantageous over similar chemical methods when each one is compared. Conditions of biological processes are soft and in the majority of cases, they do not require protection for other functional groups. Economically, the biotransformation processes are cheaper and more efficient than its analogues in the chemical industry. Usually, conversions proceed under conditions that are acceptable ecologically [87].

Biotransformation explores two unique properties of biocatalysis, its stereo and regio-specificity and its ability to react at pH and non-extreme temperatures. The biotransformation could be used for the specific conversions of complex substrates from plants, animals, microbial cells or purified enzymes as catalysts. The biotransformation has a great potential to generate new products or known products from an efficient pathway [88]. The reactions involved in the biotransformation of organic compounds by cells of several microorganisms include: oxidation, reduction, hydroxylation, methylation, acetylation, isomerization, glycosylation and esterification [89, 90].

***Hydroxylation***: During the hydrolysis, the rupture of a water bond occurs. The hydroxylation of flavones usually occurs in the *ortho* position of the hydroxyl group from the A ring and over C-4' position at the B ring. In general, microorganisms attack to hydroxyl group of the flavones at the C-4', C-5 and C-6 position; however, in the prenylated flavones the hydroxylation occurs more frequently over the double bond $C_{4\alpha} = C_{5\alpha}$ in the prenyl group (lateral chain from the A ring) [13]. Bustillo *et al.* [91] studied the biotransformation of (±)-1-(4'-chlorophenyl)-2-phenylethanol using fungi *Colletotrichum gloeosporioides* and *Botrytis cinerea.* They indicated that the hydroxylation of (±)-1-(4'-chlorophenyl)-2-phenylethanol occurs in *o*-, *m* and *p*- positions of the aromatic ring without a chloro substituent, producing 2-(2'-acetoxyphenyl)-1-(4"-chlorophenyl)-ethyl acetate, 2-(3'-acetoxyphenyl)-1-(4"-chlorophenyl)ethyl acetate and 2-(4'-acetoxyphenyl)-1-(4"-chlorophenyl)ethyl acetate.

***Dehydroxylation***: The dehydroxylation is based on the breakdown of a substance in $H^+$ and $OH^-$ radicals to form water. During the process, also known as loss of chemical water, the structural transformation occurs that involve the destruction of OH in the structure [92]. According to Lin *et al.* [93] the dehydroxylation of rutin or rutinoside (quercetin-3-rutinoside), a flavonoid found in some plants, and the formation of kaempferol-3-rutinoside occurs.

***O-Methylation***: Methylation is a process in which methyl groups ($-CH_3$) are added into a molecule. There have been reported several processes of methylation in flavonoids during their biotransformation. In particular, the methylation of quercetin and fisetin occurs over the hydroxyl C-3' and C-4', while for luteolin and baicalein occurs in C-6 (OH) [13].

***O-Demethylation***: Contrary of methylation, during demethylation the methyl groups ($-CH_3$) gets eliminated from the molecule. Majority of researches on biotransformation indicated that demethylation of flavonoids occurs in the positions C-3' and C-4'. In particular, studies carried out by Buisson *et al.* [94], indicate that demethylation of tangeretine, a flavonoid polymethylade, occurs to 4'-hidroxytangeretin.

***Glycosylation***: The glycosylation reaction include the biosynthesis of phenolic compounds in which the activate glycosylic donor links to a phenolic aglycone by bonds with hydroxyl groups [3]. Reactions of glycosylation are special due to the conversion of insoluble compounds in water [89]. Studies realized by Hyung Ko *et al.* [95] indicate the use of enzyme DP-glycosyltransferase to form apigenine, genisteine, kaempferol, luteolin, quercetin, naringenin and the glycosylation occurs at the group 3-hydroxyl. Besides, it could transfer a molecule of glucose in the group 7-hydroxyl when the interior is not available.

***Deglycosylation***: It is the inverse reaction of glycosylation. It is a process widely studied, since it is known for the increment in antioxidant capacity and availability of flavonoids [96]. Several studies have been demonstrating structural changes suffered by flavonoids during the deglycosylation. Di Gioia *et al.* [97] indicated that using bifidobacterium daizdin, genistin and glycitin, flavonoids present in commercial milk, are converting to their corresponding aglycones: didzein, genistein and glycitein. Otieno *et al.* [98] used three strains of *Lactobacillus acidophilus*, two of *Lactobacillus casei* and one of *Bifidobacterium* to evaluate β-glucosidase activity and their capacity to transform isoflavone glucosides. The results showed that all strains have the capacity for produce the β-glucosidase enzyme. In addition, during the process occurs a deglycosylation of some compounds such as daizdin and genistin, producing didzein and genistein. Ávila *et al.* [99] evaluated the capacity to transform anthocyanin glycosides of several strains of *Bifidobacteria* and *Lactobacillus*. They identified that *B. lactis* BB-12 strain is capable of deglycosylate delphinidin-3-glucoside to delphinidin aglycon, which in turn is degraded to gallic acid.

***Dehydrogenation***: The dehydrogenation is the loss of a molecule of hydrogen ($H_2$) into an organic compound. There is a limited research of this kind of reaction using phenolic compounds; Werner *et al.* [100] studied the biosynthesis of gallic acid and observed the possible metabolic routes. They suggested that all phenolic oxygen atoms of gallic acid are retained from the carbohydrate-derived precursor 5-dehydroshikimate. Therefore, the dehydrogenation of 5-dehydroshikimate is one of the most probable route of gallic acid formation. Anita *et al.* [101] used a crude peroxidase extract from sawi hijau (*Brassica juncea*) to biotransform eugenol, a natural methoxyphenols. They found that after the reaction a dehydrodimer named as dehydroeugenol was produced. One step during this reaction was dehydrogenation.

***Hydrogenation***: It is a reaction in which a molecule of hydrogen ($H_2$) is added to another compound. There exist a few reports of biotransformation, through these reactions, of flavonoids and other phenolic compounds carried out by enzymes or microorganisms. Stompor *et al.* [90] developed a method of synthesis of dihydrochalcones from flavanone and α,β-unsaturated chalcones using the strain of *Rhodococcus* sp. and *Gordonia* sp. They concluded that *Rhodococcus* sp. DSM 364 strain is capable to biotransform dihydrogen chalconaringenin to phloretin, a dihydrochalcone of naringenin with high productivity.

***Cyclization***: During cyclization reaction, a seal from structures of linear molecules was produced. At the last years, the studies of enzymes obtained by microorganisms with potential to catalyze biotransformation reactions were increased, as cyclization, a reaction that occur in synthesis of flavonoids [102]. The enzyme chalcone isomerase has been studied which catalyzes the cyclization of chalcone, a natural compound with several biological properties [103]. In order to form a flavanone, chalcone isomerase participates in the transformation of chalcone and 6'-deoxychalcone to (2S)-naringenin and (2S)-5-deoxyflavanone [10].

***Carbonyl reduction***: This reaction is based on eliminating a carbonyl group from the molecule. Faramarzi *et al.* [104] studied the transformation of androst-4-en-3,17-dione into six different hydroxyesteriodes by employing the fungus *Mucor racemosus*. In results, they concluded that reactions made by the microorganism and the hydroxylation are in the positions C-6β, C-7α, C-β, C-11α, C-14α and carbonyl reduction in the position C17. Sánchez-Patán *et al.* [105] studied the transformation of flavan-3-ol compounds by *Lactobacillus plantarum*. Among results, they found that the gallic acid was decarboxylated into pyrogallol at 48 h of fermentation.

***Sulfation***: The sulfation reactions consist of transfer sulphate group ($SO_4^-$) to an organic molecule. Some reactions of sulfation are carried out in human organism; it leads flavonoid metabolism [106], [107]. Some researchers demonstrate how biotransformation of flavonoids occurs in sulfation reactions. Araújo *et al.* [108] studied the bioconversion of quecetin and rutin, two flavonoids widely known for their properties. They used 20 filamentous fungi and *Streptomyces* strains for the process. In general, most of the strains tested were able to metabolize quercetin and rutin through different metabolic routes. Particularly, they showed that during the transformation process of rutin by *Cunninghamella echinulate* ATCC 9244 it was produced rutin sulphate as main product of metabolism.

An example of all enzymatic reactions mentioned is shown in Table 2.1. It is presented chemical structure of the reactive and products involved in enzymatic reactions.

### 2.4.3 Enzymatic and Fermentation Technologies

Microorganisms and enzymes have been employed in the biotransformation of several compounds. One of them is tea catechins; tea is a beverage highly consumed in the world; it contains a great quantity of polyphenols as epicatechin gallate, epigallocatechin, epicatechin gallate and epicatechin. From catechins, the epigallocatechin gallate is the majority present in tea. Tannase, polyphenol oxidase and peroxidase are the principal enzymes for biotransformation of catechins. Knowing processes of biotransformation of tea catechins lead to improve some commercial products derived from this [109, 110].

Baik *et al.* [111] showed the application of two enzymes for the biotransformation of catechins from green tea. One enzyme (T1) was produced by solid-state fermentation and the second one (T2) was produced by submerged fermentation. At high concentrations of catechins T2 enzyme were not inhibited and this enzyme was tested on a larger scale for reduction of the polymeric compounds to their basis, gallic acid principally.

The identification of the compound extracted from green tea was performed by Macedo *et al.* [110] and found a predominant mass corresponding to epigallocatechin. Further, antioxidant capacity during tannase biotransformation was evaluated and an enhancement during the process was verified. The antioxidant activity was attributed to the epicallocatechin and gallic acid formed during the degalloylation of gallate. Ni *et al.* [112] studied tannase production by SSF. After that, tannase was applied on catechin biotransformation of tea. The tannase showed hydrolysis activity of catechins, and the epigallocatechin gallate was the compound with a major velocity of biotransformation with this enzyme. Temperature and pH optimal for enzymatic activity were 50 °C and 5.5, respectively. It presents a range of 4–60 °C and 3.0–7.0 of pH.

**Table 2.1** Enzymatic biotransformation of polyphenols.

<table>
<tr><th>Enzymatic reaction</th><th>Refs.</th></tr>
<tr><td>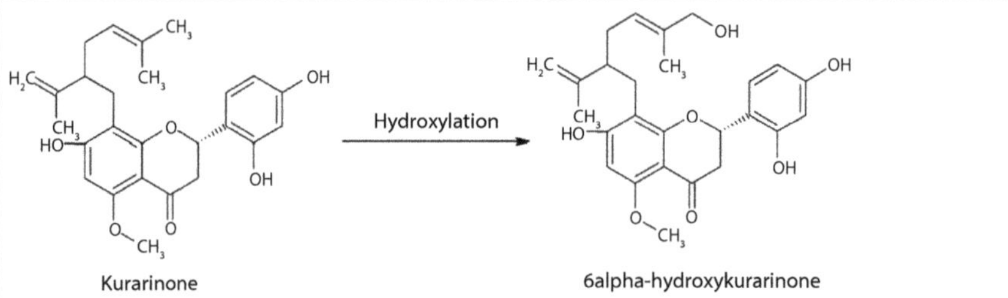
</td><td>[13]</td></tr>
<tr><td>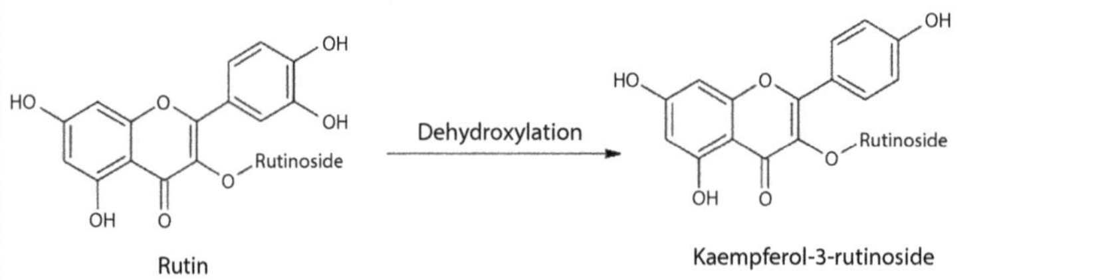
</td><td>[93]</td></tr>
<tr><td>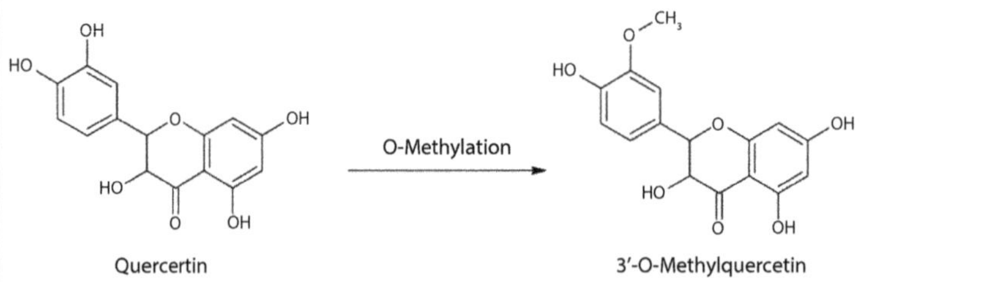
</td><td>[15]</td></tr>
<tr><td>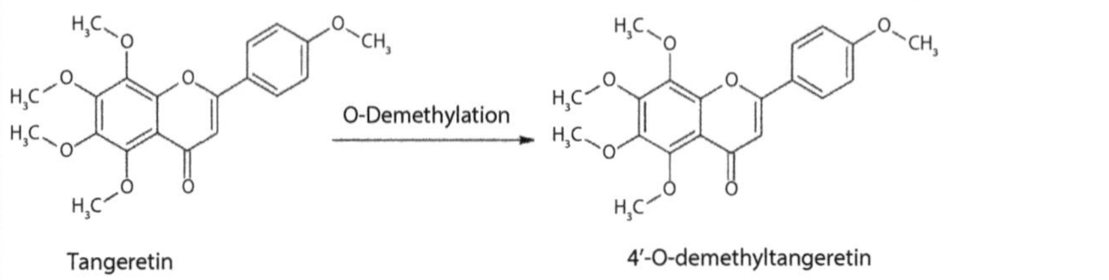
</td><td>[160]</td></tr>
<tr><td>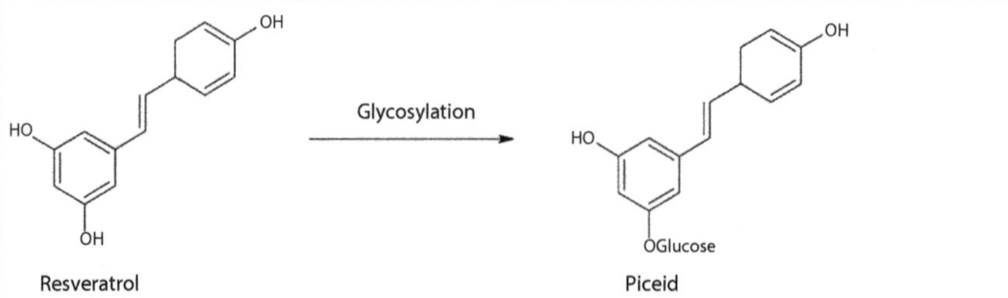
</td><td>[161]</td></tr>
<tr><td>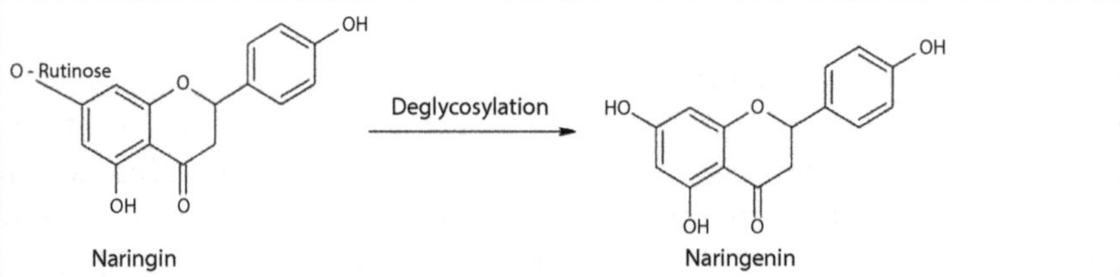
</td><td>[162]</td></tr>
</table>

(*Continued*)

**Table 2.1** Cont.

| Enzymatic reaction | Refs. |
|---|---|
| 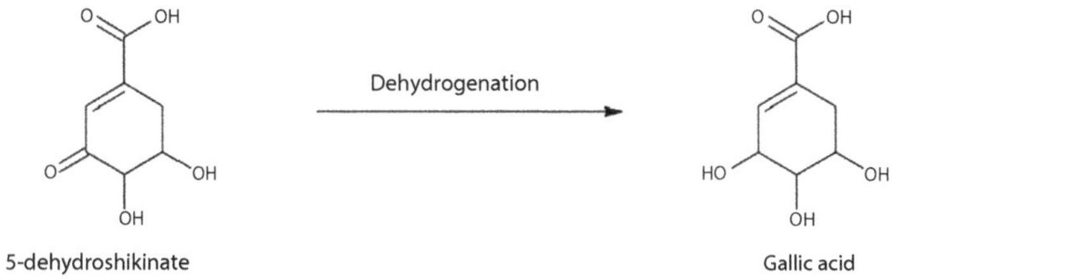  | [100] |
| 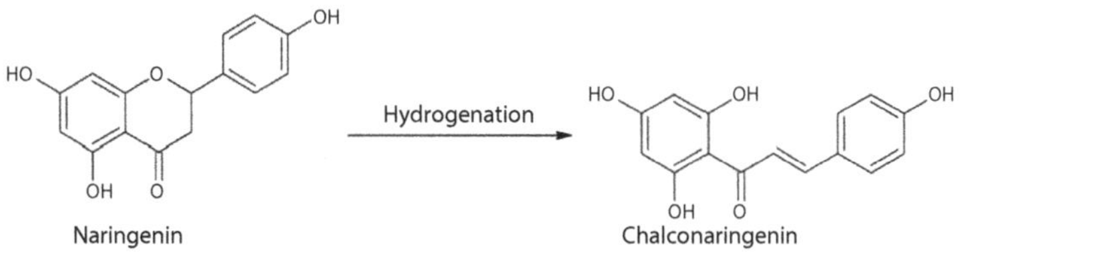  | [90] |
| 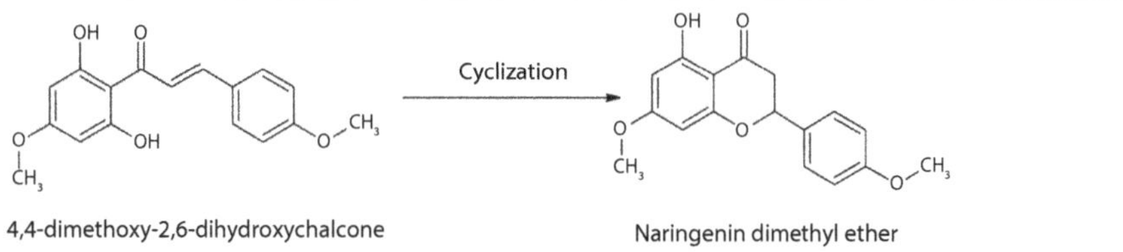  | [163] |
| 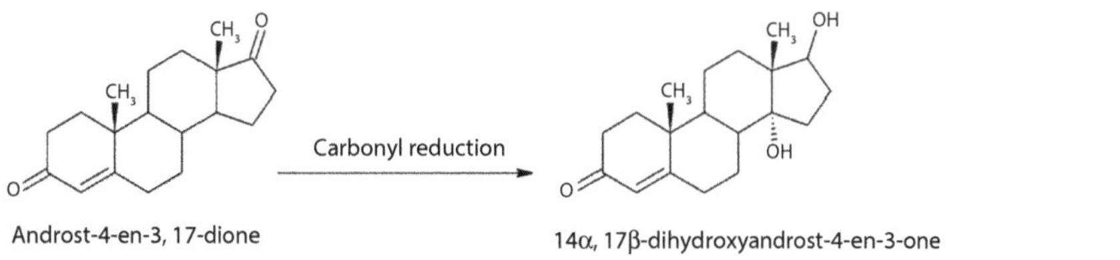  | [104] |
| 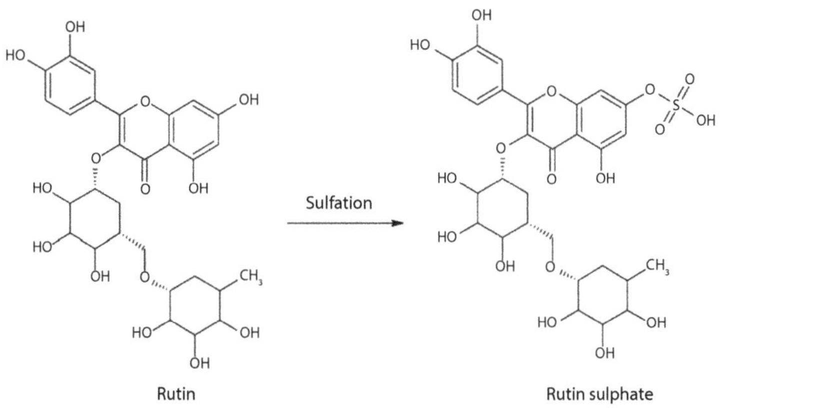  | [108] |

### 2.4.3.1 *Enzymatic Technology*

As was mentioned above, enzymatic activities comprise soft chemistry reactions. These reactions include hydrolysis of polymers such as cellulose carbohydrates. That is why some studies are about the rupture of this kind of bonds. But in some cases, hydrolysis could be followed by a pretreatment in order to improve the enzymatic activity and reach major recovery of products [113]. Other enzymatic reactions are made to break higher molecules that enclosed another compounds, like essential oils, metabolites trapped in cells of some plants. Hydrolysis of some carbohydrates leads to more efficient recovery of essential oil in extraction process. This is achieved in different matrix with enzymatic activity as orange, lemon and grapefruit peel [114]. Some enzymes related to biotransformation of phenolic compounds are tannase, polyphenol oxidase, oxidase among others.

***Tannase.*** Tannin acyl hydrolase, commonly known as tannase (E.C.3.1.1.20), is an enzyme amply characterized. It is recognized by its activity over phenolics and its capacity to hydrolase ester bonds (galloyl ester from an alcohol fraction) and depside bond (galloyl ester of gallic acid) of substrates with high tannin content. The capacity of tannase to catalyze biotransformation of citric polyphenols as hesperidin and naringin has been described. Hesperidin is a molecule of hesperitin bond to a disaccharide in the position C7 (rutinose, 6-O-α-L-ramnosyl-D-glucose). A hydrolysis of this compound leads the formation of hesperitin. Naringin (naringenin-7-O-neohesperidoside) is bond to neohespidoside (2-O-α-L-ramnosyl-D-glucose) through glycosidic bond in the C7 carbon. A break of these bonds is catalyzed by tannase. It was confirmed that tannase can act over substrates rich in polyphenols over alimentary matrices [115].

***Polyphenol oxidase and phenol oxidase.*** There are other enzymes that oxidize phenolic compounds such as mangase peroxidase and polyphenol oxidases. The polyphenol oxidases are divided into tyrosinases, known as polyphenol oxidase or catecolase, and laccases, also known as phenol oxidase. The polyphenol oxidase catalyze the hydroxylation of monophenols with molecular oxygen to form o-diphenols; after that a dehydrogenization occurs to transform it into o-quinones. The laccases catalyzed the oxidation of aromatic compounds and inorganics [116, 117]. Reactions of polyphenol oxidase are irreversible. After the enzymatic activity (o-quinone formation) follows a spontaneous polymerization of quinone [118].

Like other enzymes, their production can be from microorganisms and plants. Recovery of phenol oxidase from tea leaves depends on the collection time of the leaves. The same occurs with the production of this enzyme from microorganisms, in different fermentation time the enzymes are different. The differences are in molecular weight and the in enzymatic activity [119].

***Oxygenase.*** Oxygenase is an enzyme reported to be capable to biotransform polyphenols. Contreras-Domínguez *et al.* [120] found that an oxygenase participates in the biotransformation of a dimmer of procyanidin. It was also showed a possible pathway of biotransformation of procyanidin B2 into ((-)-epicatechin-(4β-8)-(-)-epicatechin lactone) involving an oxygenase (dioxygenase), a hydration and lactonization. Roopesh *et al.* [121] showed a biotransformation of a procyanidin B2 with a dioxygenase produced by *Asperillus fumigatus* MC 8. The enzyme resulted to be a trimer with 117 KDa in non-denaturing conditions. It was obtained two compounds derived of procyanidins called PB2-$X_1$ (PB2-$X_3$) and PB2-$X_2$. The last one resulted to be the same as ((-)-epicatechin-(4β-8)-(-)-epicatechin lactone).

### 2.4.3.2 Fermentation Technology

Currently, filamentous fungi are used as biocatalysts and cause interest in biotechnological processes. Part of this interest is due to their ability to produce large quantities of biomass and an extensive variety of enzymes in a relatively short time, which can biotransform several chemical compounds. Besides, many microorganisms can grow under different conditions and in plenty of substrates [122]. Changes can be done by hydrolysis reactions, oxide-reduction, cross-link C-C, addition-removal and glycosylation, among others. As well as use of natural and non-natural substrates with complex chemical structures.

In the biotransformation of terpenoids, microorganisms such as *Stereum hirsutum*, *Ceripora* sp. ZLY-2010 [123], *Alternaria alternata*, *Corynespora cassiicola* DSM 62474 [124], *Absidia coerulea* AM93, *Absidia glauca* AM254, *Absidia cylindrospora* AM336 [125] and enzymes such as peroxygenase were used [126]. The enzymes used are produced by liquid fermentation or solid-state fermentation generally into inducer substrates in order to produce the enzyme for microbial growth. In case of microorganisms, the use of fungi and bacteria has been highlighted. From enzymes, hydrolases are more used in terpenoid biotransformation. The pH of the medium and the temperature play an important role to improve the enzymatic activities for terpenoids biotransformation. Besides, solvent used in biotransformation is related to the mechanism of biotransformation, and to the compound to biotransform. The solvent used must be capable of solubilizing the enzyme and the compounds, without these suffering changes such as denaturalize or precipitation A mixture between water and a solvent improves the reaction productivity. On the other hand, these reactions must be developed in reactors as agitated tank, airlift, column reactor and in more used erlenmeyer flask in agitation [127].

Studies have reported the biotransformation of compounds using different kinds of microorganisms. Two of the most emphasized compounds are terpenes and polyphenols. The terpenes are compounds that obtain from essential oil of plants and some fruits like pine needles, orange peels, eucalyptus leaves [128]. Within terpene group monotepenoids, sesquiterpenoids, diterpenoids and triterpenoids are some of the interest of compounds for food industry. In the biotransformation process, oxidation, hydroxylation, biotransformation, transformation, metabolization, hydrolyzation, among others are included [129].

The pH of the medium in biotransformation of geraniol, nerol and citral is a parameter of controlling in order to manage the biotransformation. Demyttenaere *et al.* [130] found linalool as the principal identified compound in the biotransformation of geraniol, nerol and citral. An acid pH improves the resulted compounds. It is also highlighted *Aspergillus niger* as the microorganism capable to biotransform compounds. Demyttenaere and De Kimpe [131] showed the biotransformation of geraniol, nerol and citrol (a mixture) into 6-methyl-5-heten-2-ono. In the biotransformation of compounds there were employed two methods of fermentation, one by liquid fermentation and the other by spore suspension. Results showed that in liquid fermentation, there was a biotransformation of compounds near to 73–86%. On the other hand, spores suspension did not present more than 46%. In another study of biotransformation of geraniol by *P. digitatum* ATCC 201167 it was proposed that the three actions of enzymes produced and tested the spores of the fungus. In the biotransformation of geraniol and nerol into methylheptone, there were involved citral dehydrogenase, citral lyase and citral dehydrogenase; also an abiotic isomerization [132].

Enzymes that catalyzed reactions of hydroxylation and oxidation are also considered in obtaining compounds. Chen and Reese [133] used *Aspergillus niger* ATCC 9142 in the biotransformation of terpenes. There were two enzymes involves hydroxylation and oxidation pathway. However, the name of the enzymes is not mentioned because of lack of studies focused on the identification of them. A liquid fermentation was done for the biotransformation.

Another group of compounds studied in biotransformation processes are tannins. Tannins are divided into gallotannins, ellagitannins, complex tannis and condensed tannins based on their structural characteristics [19]. Condensed tannins are complicated to study due to the purification difficulty and availability in the market. However, it was found that the compound is possibly degraded to at least smaller compounds (procyanidins formed by two molecules of catechin). Contreras-Domínguez *et al.* [120] showed the ability of *Aspergillus fumigatus* to degrade procyanidins. Working with the enzymatic extracts of the fungus in presence of procyanidins leads to a proposed biodegradation mechanism that contemplates an oxygenase activity. In addition to resistance to biodegradation, they are capable to form a network with minerals and other macromolecules like proteins. This can generate undesirable situations in the matrices that contain tannins [134]. Chávez-González *et al.* [14] studied the tannic acid degradation by a commercial tannase and by fermentation with *Aspergillus niger* GH1. In the enzymatic degradation, it was possible to figure out different degrees of degradation with commercial tannic acid from one of the substrate respect to the other 3. The same way, they were tested by fungus fermentation; *A. niger* GH1 presented the same specific growth rate in 3 substrates (the same as in enzymatic activity) indicating the capacity of the fungus to degrade tannic acid.

In another study of tannins degradation by *A. niger* GH1 in submerged fermentation and solid-state fermentation was evaluated. Differences between tannase production and tannins degradation in both fermentations were observed. The majority of enzymatic production titles was presented in solid-state fermentation. It was the same system where the major tannins degradation was presented and in minor time of fermentation (over 72 h of fermentation). Tannic acid was the carbon source. Compounds of degradation were gallic acid, digaloylglucose and trigaloylglucose [135].

Ellagitannins are part of the group of hydrolyzable tannins. They are formed by a core of glucose through ester bonds to hexahydroxydiphenic acid (HHDP). When these compounds are hydrolyzed using the fungal enzymes released HHDP group and a molecular rearrangement occurs to form ellagic acid [136]. The importance of this compound is due to the biological properties as anti-inflammatory [137], anticancer [138], antidiabetic [139], anti-conceptive [140], etc. Information on the pathway of biodegradation of these compounds is very low; however, there are some studies on the elucidation of the possible route. In previous studies, polyphenols of pomegranate peel was used as a source of carbon and energy in a solid state culture of *Aspergillus niger* GH1. It was found that the accumulation of ellagic acid in the fermentation kinetics equals the ellagitannase activity, responsible for degradation of the ellagitannins [141]. Subsequently, the production process of ellagitannase using a solid state culture

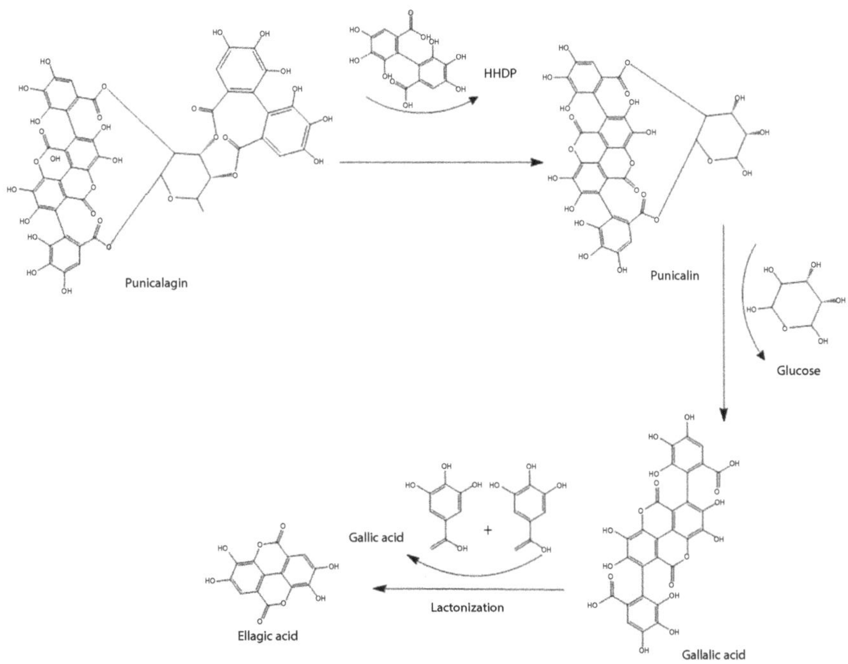


**Figure 2.1** The possible pathway of biodegradation of punicalagin to form ellagic acid.

of *Aspergillus niger* strains was optimized. With these results, the specificity having the fungal ellagitannase to hydrolyze the phenolic compounds and accumulate ellagic acid was evaluated [142]. Finally, the pathway of biodegradation of ellagitannins of pomegranate was established using *Aspergillus niger* GH1 in a solid-state culture with polyurethane foam. In the fermentation kinetics ellagitannins was biodegraded from 6 h to 18 h and coincided with the elagitanasa activity. Fermentation extracts by HPLC/MS was analyzed [143].

Bacteria have the ability to degrade tannic acid. Aguilar-Zárate *et al.* [144] showed the tannic acid degradation by liquid fermentation with *Bacillus subtilis* AM1 and *Lactobacillus plantarum* CIR1. Results showed the presence of gallic acid during the fermentation in both microorganisms, *B. subtilis* AM1 and *L. plantarum* CIR1. Both bacteria have tannase production with maximum activities between 32 and 36 h of fermentation. In addition, other compounds were found after the fermentation process as di-galloyl glucose and tri-galloyl glucose. Both compounds are intermediated of tannic acid degradation and they were observed by HPLC analysis [144].

## 2.5 Use of Agroindustrial Waste in the Fermentation Process

Agroindustrial wastes are commonly used in fermentation processes for the production of bioactive compounds because they are a rich carbon and energy source. These wastes contain high amounts of polysaccharides and other compounds that can be biotransformed microorganisms for the production of bioactive compounds that are of high added value and of great importance in the pharmaceutical, food and cosmetic industries. In agriculture and food industry in Mexico, tons of waste are produced that can be exploited for the production of bioactive compounds as shown in Table 2.2. In this section some research on the use of agro-industrial wastes as substrates and/or supports processes in solid state and submerged fermentation for the production of biotechnological products were analyzed.

### 2.5.1 Cereal Wastes

In the processing of grains and cereals, a lot of waste is generated mainly at the stage of grain separation and screening that can be processed in the textile industry, food, cosmetic and fermentation processes [145]. The filamentous fungi are commonly used in fermentation processes for the production of enzyme, mainly by solid-state fermentation (SSF). Important enzymes are now produced by SSF; the phytase was produced by *Aspergillus oryzae* SBS50 using substrates such as wheat bran and wheat straw were evaluated obtaining that can be increased to 29 times phytase production using wheat bran at 30 °C for 96 hours at an initial ratio of substrate of 1: 2 and a water activity of 0.95 [146]. *Aspergillus awamori* GHRTS was tested for the production of fructosyl transferases and fructo-oligosaccharides in solid-state fermentation with a mixture of substrates compounds; the substrates were wheat bran, rice bran and corncobs. Corncob was the best substrate for the production of secondary metabolites. These compounds are of high value added and they have a potential use in the food and pharmaceutical industry [147]. The effect of 8 factors were evaluated including the relationship between the mass of the substrate and the area of the bed in the fermentation system, the contents of wheat germ, among others, for the production of a prolyl endopeptidase by *Aspergillus niger* ATCC 11414 in solid state fermentation. They found that the interaction between the mixture of the substrate, the moisture content and content of wheat germ reached the highest production of the enzyme; this may be due to the high porosity of the substrate used and the ease of transport of gases in the process fermentation [148]. Wheat bran was used as substrate for the production of polygalacturonase with *Aspergillus sojae* in a solid-state fermentation tray bioreactor. The ratio of the thickness of the bed and the relative humidity in the process was evaluated with statistical tools. It was possible to reach up to 298 U/g of substrate with optimal conditions. The thickness of the substrate is of great importance in static fermentation processes because in this area catabolic processes of the microorganism occur [149]. In another research, lignocellulosic residues were evaluated like wheat bran and distillers dried grains with mixed cultures for the production of volatile fatty acids in submerged fermentation. These substrates evaluated presented high potential for the production of volatile fatty acids due to their chemical content. Volatile fatty acids can be used for the production of biofuels and green chemicals [150].

### 2.5.2 Fruit and Plant Waste

Organic wastes are most commonly used for the production of biotechnological products around the world. Rich in tannins substrates were evaluated as carbon source

**Table 2.2** Potential waste obtained from industrial activity and agriculture in Mexico. (ANH): Area not harvest. (WG): waste generated. (Ha): hectares. (*t*): tons. (SIACON, 2016).

| Product | ANH (Ha) | WG (*t*) |
|---|---|---|
| African palm (*Elaeis guineensis*) | 25,450.68 | 339,692.88 |
| Agave (*Agave tequilana*) | 59,743.33 | 556,4637.90 |
| Alfalfa (*Medicago sativa*) | 1,222.50 | 19,906.96 |
| Apple (*Malus domestica* Var. Golden Delicious) | 2,879.64 | 46,459.72 |
| Avocado (*Persea americana* Var. Hass) | 21,743.80 | 217,015.05 |
| Barley (*Hordeum vulgare*) | 10,643.69 | 191,201.63 |
| Bean (*Phaseolus vulgaris* Var. Flor de Junio) | 14,759.01 | 9,525.22 |
| Bean (*Phaseolus vulgaris* Var. Flor de Mayo) | 24,494.05 | 15,403.05 |
| Bean (*Phaseolus vulgaris* Var. Pinto Saltillo) | 10,081.82 | 7,898.59 |
| Broccoli (*Brassica oleracea*) | 1,030.50 | 15,498.53 |
| Chilli pepper (*Capsicum annuum*) | 1,860.05 | 51,797.53 |
| Coffe (*Coffea arábica*) | 37,982.87 | 63,221.32 |
| Copra (*Cocos nucifera*) | 2,725.50 | 4434.54 |
| Corn (Zea mays Var. Yellow) | 16,683.26 | 102,321.45 |
| Corn (Zea mays Var. White) | 343,970.26 | 1,080,798.55 |
| Feed corn (*Zea mays*) | 26,317.00 | 657,436.86 |
| Grape (*Vitis vinífera*) | 1,560.39 | 23,088.67 |
| Grapefruit (*Citrus x paradisi* Var. Ruby Red) | 943.82 | 27,263.78 |
| Green lemon (*Citrus x latifolia*) | 9,846.23 | 140,752.84 |
| Green tomato (*Physalis ixocarpa*) | 2,279.69 | 34,140.79 |
| Jícama (*Pachyrhizus erosus*) | 1,026.00 | 26,324.67 |
| Lemon (*Citrus x aurantifolia*) | 6,228.70 | 83,850.84 |
| Maguey (*Agave salmiana*) | 7,321.35 | 774,261.47 |
| Mango (*Mangifera indica* Var. Ataulfo) | 2,488.85 | 22,707.44 |
| Mango (*Mangifera indica* Var. Kent) | 3,061.87 | 18,357.01 |
| Mango (*Mangifera indica* Var. Tommy Atkins) | 2,249.90 | 22,178.25 |
| Nopal (*Opuntia ficus-indica*) | 6,812.00 | 100,905.84 |
| Nut (*Caraya illinoinensis*) | 32,624.18 | 544,68.16 |
| Oats (*Avena sativa*) | 4,472.11 | 66,712.36 |
| Olive (*Olea europea* Var. Manzanilla) | 2,397.75 | 6,967.71 |
| Orange (*Citrus x sinensis*) | 3,533.00 | 35,462.23 |
| Orange (*Citrus x sinensis* Var. Valencia) | 9,037.51 | 129,753.07 |
| Papaya (*Carica papaya Var. Maradol*) | 1,460.15 | 85,723.80 |
| Peach (*Prunus persica*) | 3,896.20 | 11,314.31 |
| Pineapple (*Annanas cosmosus* Var. Cayena Lisa) | 15,372.50 | 681,273.11 |

(*Continued*)

**Table 2.2** Cont.

| **Product** | **ANH (Ha)** | **WG (*t*)** |
|---|---|---|
| Pumpkin seed (*Curcúbita* spp.) | 5,531.50 | 2,860.77 |
| Red tomato (*Solanum lycopersicum*) | 843.56 | 27,417.65 |
| Rubber (*Castilla elástica*) | 5,923.91 | 19,192.35 |
| Sesame (*Sesamum indicum*) | 4,297.13 | 2,895.97 |
| Sorghum (*Sorghum* spp.) | 64,588.34 | 269,206.95 |
| Soy (*Glycine max*) | 5,902.00 | 11,118.26 |
| Sugar cane (*Saccharum officinarum*) | 66,775.64 | 4,967,442.85 |
| Tuna (*Opuntia tuna*) | 1,442.00 | 13,789.20 |
| Wheat (*Tririicum turgidum*) | 3,574.80 | 21,696.33 |
| Zacate (*Cenchrus ciliaris*) | 1,418.00 | 18,898.96 |

and support for tannase production by isolate of *Penicillium purpurogenum* PAF6. Tamarind seeds were the most favorable substrate for the production of tannase. In addition, the phosphate source, the nitrogen source and temperature were the parameters which influenced the process of production of the enzyme. Tannase production with optimal conditions reached 6.15 U/g [151]. In another investigation, agro-industrial wastes were used for lipase production by *Candida rugosa* NCIM 3462 in solid state fermentation. The analysis determined that the cake sesame oil was the best substrate for lipase production. At a temperature of 32.36 °C and at 1:3.23 g/mL of substrate moisture are obtained until 22.40U/g [152]. They evaluated the effect of different inducers on bioproduction of cellulase and hemicellulase by *Aspergillus niger* NRRL-567 using apple pomace as substrate in solid state fermentation. In this study, values of 133.68 IU/gds on filter paper cellulase, 60.09 IU/gds β-glucosidase and 1412.58 IU/gds xylanase activity were obtained. The cost of the production of cellulase may be reduced with the use of agro-industrial wastes and inductors inexpensive fermentation processes [153]. *Chlorella protothecoides* was used for the production of biomass and lipids over a heterotrophic culture medium, supplemented with a mixture of waste substrate of brewer fermentation and crude glycerol. The rich mineral content in the substrates used to improve growth of the microorganism and lipid accumulation [154]. The ellagitannase is an enzyme that hydrolyzes the ester bonds in ellagitannins therefore ellagic acid and glucose is released. In this work, *Aspergillus niger* GH1 used different supports like candelilla stalks, sugarcane bagasse, corn cobs and coconut husk for the production of ellagitannase in solid state fermentation. The importance of the support composition and the availability of water in the fermentation system is vital for optimal growth of the microorganism and to achieve high titers of enzyme [155]. In another research, the parameters in submerged fermentation were evaluated for the production of ellagic acid by *Aspergillus niger* GH1. It was found that the filamentous fungus can consume pomegranate husk for the production of ellagic acid; due to the high content of hydrolysable tannins and sugars, that can be used as carbon and energy source [156].

## 2.6 General Parameters in the Optimization of Fermentation Processes

Recently the use of statistical tools such as optimization is of great interest because they analyze the effects presented by some independent factors on the fermentation process. To find the optimum conditions for production of any secondary metabolite is one of the main objectives in a fermentation system. The independent factors most commonly evaluated are: the effect of temperature, carbon and nitrogen source, mineral composition of the culture medium, support type, pH, among others. Some research on aspects evaluated in the optimization of fermentation systems are shown in Table 2.3.

### 2.6.1 Response Surface Methodology

The response surface methodology is a set of math and statistics techniques used to model and analyze problems in which an important factor influences the dependent factor. This methodology has some advantages as much information generating treatments, work plans, is known for the time to develop the work, and provides interaction between the factors, and multiple responses can be evaluated [157].

### 2.6.2 First-Order Model

The parameters that influence the process is not known in detail; the first attempt adjustment is made with a first-order model. The general form of a first-order model with *k* factors, $X_1$, $X_2$...$X_{k,}$ is as shown in equation 1.

$$Y = \beta_0 + i = 1_k \beta_i X_i + \varepsilon$$

**Equation 1**: General first-order model

Where *Y* is response variable, $\beta_0$ is constant regression coefficient, $\beta_i$ is regression coefficient that affect to the independent factor, *Xi* is the level of independent factor, and ε is total error in the treatments.

To solve general first order models, exploratory experimental designs that fit this model are used. Plackett-Burman design uses two levels (+1, -1) and can analyze at least seven factors with 8 treatments in total and up to 127 factors with 128 treatments.

### 2.6.3 Second-Order Model

When you need to optimize a fermentation process and some of the factors that most influence the response variable are recognized, analysis fits a second-order model. The general form of a second-order model is as shown in Equation 2.

$$Y = \beta_0 + i = 1_k \beta_i X_i + i = 1_k \beta_{ii} X_{i2} + i = 1_k - 1_j = 1_j > i_k \beta_{ij} X_i X_j + \varepsilon$$

**Equation 2:** General second-order model.

Where *Y* is response variable, $\beta_0$ is constant regression coefficient, $\beta_i$ is regression coefficient that affect the independent factor, $X_i$ is the level of independent factor, $\beta_{ii}$ is the

**Table 2.3** General aspects evaluated in the optimization fermentation process.

| Culture type | Microorganism | Independent factors | Design statistical | Product | Reference |
|---|---|---|---|---|---|
| Submerged fermentation | *Cordyceps sinensis* UM01 | Initial pH<br>Temperature (°C)<br>Rotation speed (RPM)<br>Inoculums volume (%)<br>Medium capacity (mL) | Orthogonal $L_{25}$ | Bioactive polysaccharides | [164] |
| Submerged fermentation | *Aspergillus niger* | Cashew apple juice (g/L)<br>pH<br>Temperature (°C)<br>Time (day)<br>Methanol (%)<br>$NaNO_3$ (g/L) | Central composite design | Oxalic acid | [165] |
| Solid state fermentation | *Bacillus subtilis* SPB1 | Solid substrate ratio<br>Humidity (%)<br>Inoculum size | Central composite design | Biosurfactant | [166] |
| Solid state fermentation | *Aspergillus niger* GH1 | Temperature (°C)<br>Moisture (%)<br>Inoculum (spores/g)<br>$NaNO_3$ (g/L)<br>KCl<br>$MgSO_4$<br>$KH_2PO_4$ | Plackett-Burman<br>Central composite design | Ellagic acid | [136] |

| | | | | | |
|---|---|---|---|---|---|
| Submerged fermentation | *Aspergillus terreus* | Glucose (g/100mL)<br>Soybean meal<br>Glycerol<br>$KH_2PO_4$<br>$CuCl_2.2H_2O$<br>$CaCl_2.2H_2O$<br>$CaCl_3.H_2O$<br>$FeSO_4.7H_2O$<br>$MgSO_4.7H_2O$ | Plackett-Burman<br>Central composite design | Mevastatin | [167] |
| Solid state fermentation | *Aspergillus tubingensis* CICC 2651 | NH4Cl (%, w/w)<br>α-lactose<br>Humidity (%, v/w)<br>Initial pH<br>Inoculum size (mL)<br>Time (h)<br>Temperature (°C) | Plackett-Burman<br>Central composite design | Tannase | [168] |

regression coefficient that affect the independent factor $(X_{12})$, $X_{12}$ is the level of independent quadratic factor, $\beta_{ij}$ is the regression coefficient that affect the interaction $(X_iX_j)$, $X_iX_j$ is the level factor in the interaction, ε is total error in the treatments.

To solve the second-order models, optimization experimental designs which adjusted to this model, such as the Box-Behnken design, are used. This design has the characteristic analysis from two factors with a total of 9 treatments and up to 10 factors with a total of 243 treatments, depending on the number of selected blocks. The Box-Behnken design are incomplete factorial design with three levels. A sample block corresponding to a two-level factorial design is repeated in different sets of parameters. The parameters that are not included in the factorial design remain at their average level around the block [158].

## 2.7 Final Comments

Fermentation is the process involving the biochemical activity of organisms, during their growth, development, reproduction, even senescence and death. Fermentation technology uses the microorganisms to produce food, pharmaceuticals, alcoholic beverages on a large-scale industrial basis. Here was revised a promissory technology for extraction of bioactive molecules and enrichment of the fermented masses source of such bioactives, under the basic principle involved in the industrial fermentation technology in which organisms are grown under suitable conditions, by providing raw materials meeting all the necessary requirements such as carbon, nitrogen, salts, trace elements and vitamins. The end products formed as a result of their metabolism during their life span, are released into the media, which are extracted by human beings for use and that have a high commercial value. The major products of fermentation technology produced economically on a large-scale industrial basis are wine, beer, cider, vinegar, ethanol, cheese, hormones, antibiotics, complete proteins, enzymes and other useful products. In particular, a great opportunity for extraction of potent bioactive molecules through fermentation or enzymatic technologies associated to traditional extraction methods is open and requires to be focused to exploit all technological potential.

## Acknowledgements

The authors thank the National Council of Science and Technology (CONACYT Mexico) for all financial support and scholarships for Larios Cruz, Londoño-Hernández & García Group of Bioprocesses and Bioproducts in the Postgraduate Program of Food Science and Technology - Universidad Autonoma de Coahuila, Mexico.

## References

1. S. Martins, S. I. Mussatto, G. Martínez-Avila, J. Montañez-Saenz, C. N. Aguilar, and J. A. Teixeira, *Biotechnology. Advances*, Vol. 29, no. 3, p. 365, 2011.

2. L. R. Ferreira, J. A. Macedo, M. L. Ribeiro, and G. A. Macedo, *Food Research. International*, Vol. 51, no. 2, p. 526, 2013.
3. N. T. Huynh, J. Van Camp, G. Smagghe, and K. Raes, *International. Journal of. Molecular Sciences*, Vol. 15, no. 11, p. 19369, 2014.
4. D. D. Orhan, B. Özçelik, S. Özgen, and F. Ergun, *Microbiological. Research*, Vol. 165, no. 6, p. 496, 2010.
5. A. M. M. Nour, S. A. Khalid, M. Kaiser, R. Brun, E. A. Wai'l, and T. J. Schmidt, *Journal of Ethnopharmacology*, Vol. 129, no. 1, p. 127, 2010.
6. S. Shimosaki, Y. Tsurunaga, H. Itamura, and M. Nakamura, *Natural Product Research*, Vol. 25, no. 4, p. 374, 2011.
7. K. Abubakar, M. R. Abubakar, J. C. Ugwah–Oguejiofor, A. A. Muhammad, M. Usman, and H. E. Mshelia, *Advancement in Medicinal Plant Research*, Vol. 3, p. 23, 2015.
8. P. Lagiou, E. Samoli, A. Lagiou, and A. Tzonou, *European Journal of Clinical Nutrition*, Vol. 58, no. 12, p. 1643, 2004.
9. A. Boulila, I. Hassen, L. Haouari, F. Mejri, I. Ben Amor, H. Casabianca, and K. Hosni, *Industrial Crops and Products*, Vol. 74, p. 485, 2015.
10. A. Wang, F. Zhang, L. Huang, X. Yin, H. Li, Q. Wang, Z. Zeng, and T. Xie, *Journal of Medicinal Plants Research*, Vol. 4, no. 10, p. 847, 2010.
11. 11. A. del Monte-Martinez, B. V. C. Avila, J. O. G. Bacerio, D. G. Abradelo, V. F. Espí, and R. C. Vazquez, "Aplicación de enzimas en biocatálisis. Perspectivas de la utilización de nanoarreglos como biocatalizadores/Application of enzymes on biocatalysis. Perspectives of the use of nanoarrays as biocatalysts," Revista Cubana de Ciencias Biológicas, Vol. 2(2), p. 7, 2013.
12. A. Zaks, *Current Opinions in Chemical Biology*, Vol. 5, no. 2, p. 130, 2001.
13. H. Cao, X. Chen, A. R. Jassbi, and J. Xiao, *Biotechnology Advances*, Vol. 33, no. 1, p. 214, 2015.
14. M. Chávez-González, J. Contreras-Esquivel, L. Prado-Barragán, R. Rodríguez, A. Aguilera-Carbo, L. Rodríguez, and C. Aguilar, *Chemical Papers*, Vol. 66, no. 3, p. 171, 2012.
15. P. M. Kris-Etherton, K. D. Hecker, A. Bonanome, S. M. Coval, A. E. Binkoski, K. F. Hilpert, A. E. Griel, and T. D. Etherton, *American Journal of Medicine*, Vol. 113, no. 9, p. 71, 2002.
16. U. Hölker, M. Höfer, and J. Lenz, *Applied Microbiology and Biotechnology*, Vol. 64, no. 2, p. 175, 2004.
17. P. S.-N. Nigam and A. Pandey, *Biotechnology for agro-industrial residues utilisation: utilisation of agro-residues*. Springer Science & Business Media, 2009.
18. H. Baxter, J. B. Harborne, and G. P. Moss, *Phytochemical dictionary: a handbook of bioactive compounds from plants*. CRC Press, 1998.
19. K. Khanbabaee and T. van Ree, *Natural Products Report*, Vol. 18, no. 6, p. 641, 2001.
20. N. G. Baydar, G. Özkan, and S. Yaşar, *Food Control*, Vol. 18, no. 9, p. 1131, 2007.
21. G. K. Jayaprakasha, T. Selvi, and K. K. Sakariah, *Food Research International*, Vol. 36, no. 2, p. 117, 2003.
22. A. Starzyńska-Janiszewska, B. Stodolak, and M. Jamróz, *Food Chemistry*, Vol. 109, no. 2, p. 285, 2008.
23. D. A. Vattem and K. Shetty, *Process Biochemistry*, Vol. 39, no. 3, p. 367, 2003.
24. M. Naczk and F. Shahidi, *Journal of Chromatography A*, Vol. 1054, no. 1, p. 95, 2004.
25. S. Khadem and R. J. Marles, *Molecules*, Vol. 15, no. 11, p. 7985, 2010.
26. F. Conforti, F. Menichini, C. Formisano, D. Rigano, F. Senatore, N. A. Arnold, and F. Piozzi, *Food Chemistry*, Vol. 116, no. 4, p. 898, 2009.
27. G.-N. Kim, J.-G. Shin, and H.-D. Jang, *Food Chemistry*, Vol. 117, no. 1, p. 35, 2009.
28. J. S. Hernández, A. F. Aguilera-Carbó, R. Rodríguez Herrera, J. L. Martínez, and C. N. Aguilar, in *Proceedings of the 10th International Chemical and Biological Engineering Conference*, pp. 1849, 2008.

29. A. Robledo, A. Aguilera-Carbó, R. Rodriguez, J. L. Martinez, Y. Garza, and C. N. Aguilar, *Journal of Industrial Microbiology and Biotechnology*, Vol. 35, no. 6, p. 507, 2008.
30. B. B. Li, B. Smith, and M. M. Hossain, *Separation and Purification Technology*, Vol. 48, no. 2, p. 189, 2006.
31. E. Van Exel, J. Gussekloo, A. J. M. De Craen, A. Bootsma-Van Der Wiel, P. Houx, D. L. Knook, and R. G. J. Westendorp, *Journal of Neurology Neurosurgery Psychiatry*, Vol. 71, no. 1, p. 29, 2001.
32. E. Haslam, *Journal of Natural Products*, Vol. 59, no. 2, p. 205, 1996.
33. B. Bozan, G. Tosun, and D. Özcan, *Food Chemistry*, Vol. 109, no. 2, p. 426, 2008.
34. J. Shi, J. Yu, J. E. Pohorly, and Y. Kakuda, *Journal of Medicinal Food*, Vol. 6, no. 4, p. 291, 2003.
35. M. E. Olsson, J. Ekvall, K.-E. Gustavsson, J. Nilsson, D. Pillai, I. Sjöholm, U. Svensson, B. Åkesson, and M. G. L. Nyman, *Journal of Agricultural and Food Chemistry*, Vol. 52, no. 9, p. 2490, 2004.
36. S. J. S. Flora, *Oxidative Medicine and Cell. Longevity*, Vol. 2, no. 4, p. 191, 2009.
37. K. Shimada, K. Fujikawa, K. Yahara, and T. Nakamura, *Journal of Agricultural Food and Chemistry*, Vol. 40, no. 6, p. 945, 1992.
38. W. Huang, J. Ni, and A. G. L. Borthwick, *Process Biochemistry*, Vol. 40, no. 3, p. 1245, 2005.
39. A. Barra, V. Coroneo, S. Dessi, P. Cabras, and A. Angioni, *Natural Product Communication*, Vol. 5, no. 2, p. 329, 2010.
40. C. N. Aguilar, A. Aguilera-Carbo, A. Robledo, J. Ventura, R. Belmares, D. Martinez, R. Rodríguez-Herrera, and J. Contreras, *Food Technology and Biotechnology*, Vol. 46, no. 2, p. 218, 2008.
41. G. C. Martínez-Ávila, A. F. Aguilera-Carbó, R. Rodríguez-Herrera, and C. N. Aguilar, *Annals of Microbiology*, Vol. 62, no. 3, p. 923, 2012.
42. H. Negishi, J.-W. Xu, K. Ikeda, M. Njelekela, Y. Nara, and Y. Yamori, The *Journal of Nutrition*, Vol. 134, no. 1, p. 38, 2004.
43. N. P. Seeram, M. Aviram, Y. Zhang, S. M. Henning, L. Feng, M. Dreher, and D. Heber, *Journal of Agricultural and Food Chemistry*, Vol. 56, no. 4, p. 1415, 2008.
44. S. He, C. Sun, and Y. Pan, *International Journal of Molecular Sciences*, Vol. 9, no. 5, p. 842, 2008.
45. R. Brouillard, *In The flavonoids*, p. 525,1988.
46. T. Goto and T. Kondo, *Angewandte Chemie International Edition*, Vol. 30, no. 1, p. 17, 1991.
47. V. Cheynier, *American Journal of Clinical Nutrition*, Vol. 81, no. 1, p. 223 S, 2005.
48. R. A. Dixon, *Annual Review of Plant Biology*, Vol. 55, p. 225, 2004.
49. M. R. Campos-Esparza, M. V. Sánchez-Gómez, and C. Matute, *Cell Calcium*, Vol. 45, no. 4, p. 358, 2009.
50. L. B. Valdez, S. Alvarez, T. Zaobornyj, and A. Boveris, *Biological Research*, Vol. 37, no. 2, p. 279, 2004.
51. A. Moure, J. M. Cruz, D. Franco, J. M. Domínguez, J. Sineiro, H. Domínguez, M. J. Núñez, and J. C. Parajó, *Food Chemistry*, Vol. 72, no. 2, p. 145, 2001.
52. M. Friedman, *Journal of Agriculatural and Food Chemistry*, Vol. 45, no. 5, p. 1523, 1997.
53. K. Wu, W. Zhang, P. B. Addis, R. J. Epley, A. M. Salih, and J. Lehrfeld, *Journal of Agriculatural and Food Chemistry*, Vol. 42, no. 1, p. 34, 1994.
54. S. J. Li, T. A. Seymour, A. J. King, and M. T. Morrissey, *Journal of Food Science*, Vol. 63, no. 3, p. 438, 1998.
55. D. Han, S. J. Kim, S. H. Kim, and D. M. Kim, *Jouranal of Food Science*, Vol. 63, no. 1, p. 69, 1998.
56. R. Bertholet, L. Colarow, A. Kusy, and V. Rivier, Google Patents, 03-Feb-1998.

57. H. Wulff, *WO9739632*, 1997.
58. A. Montoya-Rodríguez, M. A. Gómez-Favela, C. Reyes-Moreno, J. Milán-Carrillo, and E. González de Mejía, *Comprehensive Reviews of Food Science and Food Safety*, Vol. 14, no. 2, p. 139, 2015.
59. H. Nimitphong, C. Sritara, L. Chailurkit, S. Chanprasertyothin, W. Ratanachaiwong, P. Sritara, and B. Ongphiphadhanakul, *Nutrition Journal*, Vol. 14, no. 1, p. 29, 2015.
60. G. E. Vegarud, T. Langsrud, and C. Svenning, *British Journal of Nutrition*, Vol. 84, no. S1, p. 91, 2000.
61. H. D. M. Coutinho, K. M. Lobo, D. A. C. Bezerra, and I. Lobo, *Indian Journal of Pharmacology*, Vol. 40, no. 1, p. 3, 2008.
62. C. Xu, G. Ren, G. Cao, Q. Chen, P. Shou, C. Zheng, L. Du, X. Han, M. Jiang, and Q. Yang, *Journal of Biological Chemistry*, Vol. 288, no. 16, p. 11074, 2013.
63. M. J. Niphakis and B. F. Cravatt, *Annual Review of Biochemistry*, Vol. 83, p. 341, 2014.
64. G. Walsh, *Proteins, Second Edition*, pp. 233.
65. S. Sharma, H. N. Verma, and N. K. Sharma, *International Journal of Peptides*, Vol. 2014, 2014.
66. J. Sanchez, T. Bruhn, A. Aneiros, E. Wachter, and L. Béress, *Toxicon*, Vol. 34, no. 11, p. 1361, 1996.
67. W.-C. Zeng, W.-C. Zhang, W.-H. Zhang, and B. Shi, *Reactive. and Functional Polymers*, Vol. 73(3), p. 573, 2013.
68. A. Saghatelian and J. P. Couso, *Nature Chemical Biology*, Vol. 11, no. 12, p. 909, 2015.
69. C. G. Lopresto, F. Petrillo, A. A. Casazza, B. Aliakbarian, P. Perego, and V. Calabrò, *Separation and Purification Technology*, Vol. 137, p. 13, 2014.
70. M. Bimakr, R. A. Rahman, F. S. Taip, A. Ganjloo, L. M. Salleh, J. Selamat, A. Hamid, and I. S. M. Zaidul, *Food and Bioproducts Processing*, Vol. 89, no. 1, p. 67, 2011.
71. C. M. da Silva, A. B. Zanqui, A. H. P. Souza, A. K. Gohara, S. T. M. Gomes, E. A. da Silva, L. Cardozo Filho, and M. Matsushita, *The Journal of Supercritical Fluids*, Vol. 112, p. 14, 2016.
72. S. A. Heleno, P. Diz, M. A. Prieto, L. Barros, A. Rodrigues, M. F. Barreiro, and I. C. F. R. Ferreira, *Food Chemistry*, Vol. 197, p. 1054, 2016.
73. M. D. L. De Castro and F. Priego-Capote, *Journal of Chromatography A*, Vol. 1217, no. 16, p. 2383, 2010.
74. V. M. Burin, N. E. Ferreira-Lima, C. P. Panceri, and M. T. Bordignon-Luiz, *Microchemical Journal*, Vol. 114, p. 155, 2014.
75. A. García, E. Rodríguez-Juan, G. Rodríguez-Gutiérrez, J. J. Rios, and J. Fernández-Bolaños, *Food Chemistry*, Vol. 197, p. 554, 2016.
76. B. Yang, X. Liu, and Y. Gao, *Innovative Food Science and Emerging Technologies*, Vol. 10, no. 4, p. 610, 2009.
77. A. S. G. Costa, J. C. M. Barreira, A. Ruas, A. F. Vinha, F. B. Pimentel, R. C. Alves, I. C. F. R. Ferreira, and M. B. P. P. Oliveira, *Industrial Crops and Products*, Vol. 79, p. 180, 2016.
78. R. E. Clement and C. Hao, "2.03-Liquid–Liquid extraction: basic principles and automation," *Comprehensive. Sampling and Sample Preparationin. Acad. Press. Oxford*, pp. 51, 2012.
79. S. C. Mandal, V. Mandal, and A. K. Das, *Essentials of botanical extraction: principles and applications*. Academic Press, 2015.
80. B. Trusheva, D. Trunkova, and V. Bankova, *Chemistry Central Journal*, Vol. 1, no. 1, p. 13, 2007.
81. S. Siqueira, V. dos Santos Falcão-Silva, M. de Fátima Agra, C. Dariva, J. P. de Siqueira-Júnior, and M. J. V. Fonseca, *Journal of Supercritical Fluids*, Vol. 58, no. 3, p. 391, 2011.

82. N. Ćujić, K. Šavikin, T. Janković, D. Pljevljakušić, G. Zdunić, and S. Ibrić, *Food Chemisty*, Vol. 194, p. 135, 2016.
83. M. dos Santos Lima, M. da C. P. Dutra, I. M. Toaldo, L. C. Corrêa, G. E. Pereira, D. de Oliveira, M. T. Bordignon-Luiz, and J. L. Ninow, *Food Chemistry*, Vol. 188, p. 384, 2015.
84. B. Bensaude-Vicent and J. Simon, *Chemistry, The Impure Science*. London: Imperial College Press, 2008.
85. M. Figlarz, *Materials Science Forum*, Vol. 152, p. 55, 1994.
86. C. N. R. Rao and K. Biswas, in *Essentials of Inorganic Materials Synthesis*, p. 57, 2015.
87. J. Aleu and I. G. Collado, *Journal of Molecular Catalysis B Enzymatic*, Vol. 13, no. 4, p. 77, 2001.
88. A. Giri, K. Osako, A. Okamoto, E. Okazaki, and T. Ohshima, *Food Research International*, Vol. 44, no. 1, p. 317, 2011.
89. A. Giri, V. Dhingra, C. C. Giri, A. Singh, O. P. Ward, and M. L. Narasu, *Biotechnology Advances*, Vol. 19, no. 3, p. 175, 2001.
90. M. Stompor, B. Potaniec, A. Szumny, P. Zieliński, A. K. Żołnierczyk, and M. Anioł, *Journal of Molecular Catalysis B Enzymatic*, Vol. 97, p. 283, 2013.
91. A. J. Bustillo, C. M. García-Pajón, J. Aleu, R. Hernández-Galán, and I. G. Collado, *Tetrahedron: Asymmetry*, Vol. 14, no. 23, p. 3755, 2003.
92. P. R. Palacios, L. De Los Santos Valladares, A. Bustamante, and J. C. González, "Estudio de la deshidroxilación en el óxido férrico hidratado denominado limonita,"*Revista de la Sociedad Química del Perú*, Vol. 78, no. 3, pp. 198, 2012.
93. S. Lin, Q. Zhu, L. Wen, B. Yang, G. Jiang, H. Gao, F. Chen, and Y. Jiang, *Food Chemistry.*, Vol. 145, p. 220, 2014.
94. D. Buisson, J. Quintin, and G. Lewin, *Journal of Natural Products*, Vol. 70, no. 6, p. 1035, 2007.
95. J. Hyung Ko, B. Gyu Kim, and A. Joong-Hoon, *FEMS Microbiology Letters*, Vol. 258, no. 2, p. 263, 2006.
96. K. Németh, G. W. Plumb, J.-G. Berrin, N. Juge, R. Jacob, H. Y. Naim, G. Williamson, D. M. Swallow, and P. A. Kroon, *European Journal of Nutrition*, Vol. 42, no. 1, p. 29, 2003.
97. D. Di Gioia, E. Strahsburger, A. M. L. de Lacey, V. Bregola, I. Marotti, I. Aloisio, B. Biavati, and G. Dinelli, *Journal of Functional Foods*, Vol. 7, p. 671, 2014.
98. D. O. Otieno, J. F. Ashton, and N. P. Shah, *Food Research International*, Vol. 39, no. 4, p. 394, 2006.
99. M. Ávila, M. Hidalgo, C. Sánchez-Moreno, C. Pelaez, T. Requena, and S. de Pascual-Teresa, *Food Research International*, Vol. 42, no. 10, p. 1453, 2009.
100. R. A. Werner, A. Rossmann, C. Schwarz, A. Bacher, H.-L. Schmidt, and W. Eisenreich, *Phytochemistry*, Vol. 65, no. 20, p. 2809, 2004.
101. Y. Anita, A. Sundowo, N. L. P. Dewi, E. Filailla, H. Mulyani, C. Risdian, S. Banjarnahor, M. Hanafi, and E. P. Istyastono, *Procedia Chemisty*, Vol. 16, p. 265, 2015.
102. M. Sanchez-Gonzalez and J. P. N. Rosazza, *Journal of Natural Products*, Vol. 67, no. 4, p. 553, 2004.
103. M. E. Ramírez Escobedo, L. Barajas Bermúdez, C. Pérez Berumen, A. Sáenz Galindo, and S. Y. Silva Belmares, "Síntesis y actividad biológica de chalconas," *Revista mexicana de ciencias farmacéuticas*, Vol. 43, no. 4, 2012
104. M. A. Faramarzi, M. Badiee, M. T. Yazdi, M. Amini, and M. Torshabi, *Journal of Molecular Catalysis B Enzymatic*, Vol. 50, no. 1, p. 7, 2008.
105. F. Sánchez-Patán, R. Tabasco, M. Monagas, T. Requena, C. Peláez, M. V. Moreno-Arribas, and B. Bartolomé, *Journal of Agricultural and Food Chemistry*, Vol. 60, no. 29, p. 7142, 2012.
106. T. G. Pai, M. Suiko, Y. Sakakibara, and M.-C. Liu, *Biochemicaland Biophysical Research Communications*, Vol. 285, no. 5, p. 1175, 2001.

107. J. B. Vaidyanathan and T. Walle, *Drug Metabolism and Disposition*, Vol. 30, no. 8, p. 897, 2002.
108. K. C. F. Araújo, E. M. de MB Costa, F. Pazini, M. C. Valadares, and V. de Oliveira, *Food and Chemical Toxicology*, Vol. 51, p. 93, 2013.
109. S. Sang, J. D. Lambert, C.-T. Ho, and C. S. Yang, *Pharmacological Research*, Vol. 64, no. 2, p. 87, 2011.
110. J. A. Macedo, V. Battestin, M. L. Ribeiro, and G. A. Macedo, *Food Chemistry*, Vol. 126, no. 2, p. 491, 2011.
111. J. H. Baik, H. J. Suh, S. Y. Cho, Y. Park, and H.-S. Choi, *Journal of Bioscience and Bioengineering*, Vol. 118, no. 5, p. 546, 2014.
112. 110. H. Ni, F. Chen, Z. D. Jiang, M. Y. Cai, Y. F. Yang, A. F. Xiao, and H. N. Cai, *LWT-Food Science and Technology*, Vol. 60, no. 2, p. 1206, 2015.
113. H. A. Ruiz, A. A. Vicente, and J. A. Teixeira, *Industrial Crops and Products*, Vol. 36, no. 1, p. 100, 2012.
114. M. L. Chávez-González, L. I. López-López, R. Rodríguez-Herrera, J. C. Contreras-Esquivel, and C. N. Aguilar, *Chemical Papers*, Vol. 70, no. 4, p. 412, 2016.
115. L. Rosas, J. Alves, M. Lima, and G. Alves, *Food Research International*, Vol. 51, p. 526, 2013.
116. N. Duran and E. Esposito, *Applied Catalysis B Environmental*, Vol. 28, no. 2, p. 83, 2000.
117. S. G. Burton, *Trends in Biotechnology*, Vol. 21, no. 12, p. 543, 2003.
118. V. Calabro, S. Curcio, M. G. De Paola, and G. Iorio, *Desalination*, Vol. 236, no. 1–3, p. 30, 2009.
119. G. N. Pruidze, N. I. Mchedlishvili, N. T. Omiadze, L. K. Gulua, and N. G. Pruidze, *Food Research International*, Vol. 36, no. 6, p. 587, 2003.
120. M. Contreras-Domínguez, S. Guyot, N. Marnet, J. Le Petit, I. Perraud-Gaime, S. Roussos, and C. Augur, *Biochimie*, Vol. 88, no. 12, p. 1899, 2006.
121. K. Roopesh, S. Guyot, A. Sabu, M. Haridas, P. G. Isabelle, S. Roussos, and C. Augur, *Process Biochemistry*, Vol. 45, no. 6, p. 904, 2010.
122. L. Thomas, C. Larroche, and A. Pandey, *Biochemical Engineering Journal*, Vol. 81, p. 146, 2013.
123. S.Y. Lee, S.-H. Kim, C.-Y. Hong, H.-Y. Kim, S.-H. Ryu, and I.-G. Choi, *Mycobiology*, Vol. 43, no. 3, p. 297, 2015.
124. G. İşcan, N. Kırımer, F. Demirci, B. Demirci, Y. Noma, and K. Başer, *Chemistry & Biodiversity*, Vol. 9, no. 8, p. 1525, 2012.
125. R. Gniłka, W. Majchrzak, and C. Wawrzeńczyk, *Phytochemistry Letters*, Vol. 13, p. 41, 2015.
126. C. Fuchs and W. Schwab, *Journal of Molecular Catalysis B Enzymatic*, Vol. 96, p. 52, 2013.
127. C. C. C. R. De Carvalho and M. M. R. da Fonseca, *Biotechnology Advances*, Vol. 24, no. 2, p. 134, 2006.
128. G. R. Tortella, O. Rubilar, M. Cea, G. Briceno, A. Quiroz, M. C. Diez, and L. Parra, *International Biodeterioration and Biodegradation*, Vol. 85, p. 8, 2013.
129. I. A. Parshikov and J. B. Sutherland, *Process Biochemistry*, Vol. 49, no. 12, p. 2086, 2014.
130. J. C. R. Demyttenaere, M. del Carmen Herrera, and N. De Kimpe, *Phytochemistry*, Vol. 55, no. 4, p. 363, 2000.
131. J. Demyttenaere and N. De Kimpe, *Journal of Molecular Catalysis B Enzymatic*, Vol. 11, no. 4, p. 265, 2001.
132. W. Wolken and M. Van Der Werf, *Applied Microbiology and Biotechnology*, Vol. 57, no. 5, p. 731, 2001.
133. A. R. M. Chen and P. B. Reese, *Phytochemistry*, Vol. 59, no. 1, p. 57, 2002.
134. M. Chávez-González, L. V Rodríguez-Durán, N. Balagurusamy, A. Prado-Barragán, R. Rodríguez, J. C. Contreras, and C. N. Aguilar, *Food and Bioprocess Technology*, Vol. 5, no. 2, p. 445, 2012.

135. M. L. Chávez-González, S. Guyot, R. Rodríguez-Herrera, A. Prado-Barragán, and C. N. Aguilar, *Process Biochemistry*, Vol. 49, no. 4, p. 541, 2014.
136. L. Sepúlveda, A. Aguilera-Carbó, J. A. Ascacio-Valdés, R. Rodríguez-Herrera, J. L. Martínez-Hernández, and C. N. Aguilar, *Process Biochemistry*, Vol. 47, no. 12, p. 2199, 2012.
137. V. D. Bulani, P. S. Kothavade, H. S. Kundaikar, N. B. Gawali, A. A. Chowdhury, M. S. Degani, and A. R. Juvekar, *Journal of Molecular Structure*, Vol. 1105, p. 308, 2016.
138. D. Kumar, S. Basu, L. Parija, D. Rout, S. Manna, J. Dandapat, and P. R. Debata, *Biomedicine & Pharmacotherapy*, Vol. 81, p. 31, 2016.
139. R. P. Nankar and M. Doble, *Journal of Functional Foods*, Vol. 15, p. 1, 2015.
140. M. T. Mansouri, B. Naghizadeh, and B. Ghorbanzadeh, *Pharmacological Reports*, Vol. 67, no. 3, p. 473, 2015.
141. J. A. Ascacio-Valdés, J. J. Buenrostro, R. De la Cruz, L. Sepúlveda, A. F. Aguilera, A. Prado, J. C. Contreras, R. Rodríguez, and C. N. Aguilar, *Journal of Basic Microbiology*, Vol. 54, no. 1, p. 28, 2014.
142. R. de la Cruz, J. A. Ascacio, J. Buenrostro, L. Sepúlveda, R. Rodríguez, A. Prado-Barragán, J. C. Contreras, A. Aguilera, and C. N. Aguilar, *Preparative Biochemistry and Biotechnology*, Vol. 45, no. 7, p. 617, 2015.
143. J. A. Ascacio-Valdés, A. F. Aguilera-Carbó, J. J. Buenrostro, A. Prado-Barragán, R. Rodríguez-Herrera, and C. N. Aguilar, *Journal of Basic Microbiology*, Vol. 56, no. 4, p. 329, 2016.
144. P. Aguilar-Zárate, M. A. Cruz, J. Montañez, R. Rodríguez-Herrera, J. E. Wong-Paz, R. E. Belmares, and C. N. Aguilar, *Microbial Cell Factories*, Vol. 14, no. 1, p. 209, 2015.
145. C. DU, C. S. K. LIN, and C. WEBB, *Advances in Biorefineries Biomass Waste Supply Chain Exploitation*, p. 303, 2014.
146. B. Sapna, *Applied Biochemistry and Biotechnology*, Vol. 173, p. 1885, 2014.
147. T. Sathish and R. S. Prakasham, *Indian Journal of Microbiology*, Vol. 53, no. 3, p. 337, 2013.
148. Y. Esparza, A. Huaiquil, L. Neira, A. Leyton, M. Rubilar, L. Salazar, and C. Shene, *Food Science and Biotechnology*, Vol. 20, no. 5, p. 1323, 2011.
149. H. Demir and C. Tari, *International Biodeterioration. and Biodegradation*, Vol. 106, p. 60, 2016.
150. I. M. van Aarle, A. Perimenis, J. Lima-Ramos, E. de Hults, I. F. George, and P. A. Gerin, *Biochemical Engineering Journal*, Vol. 103, p. 242, 2015.
151. A. Jana, C. Maity, S. K. Halder, K. C. Mondal, B. R. Pati, and P. K. Das Mohapatra, *Applied Biochemistry and Biotechnology*, Vol. 167, no. 5, p. 1254, 2012.
152. A. Rajendran and V. Thangavelu, *Waste and Biomass Valorization*, Vol. 4, no. 2, p. 347, 2013.
153. G. S. Dhillon, S. Kaur, S. K. Brar, and M. Verma, "Potential of apple pomace as a solid substrate for fungal cellulase and hemicellulase bioproduction through solid-state fermentation," *Industrial Crops and Products*, Vol. 38, p. 6, 2012.
154. X. Feng, T. H. Walker, W. C. Bridges, C. Thornton, and K. Gopalakrishnan, *BioresourceTechnology*, Vol. 166, p. 17, 2014.
155. J. Buenrostro-Figueroa, A. Ascacio-Valdés, L. Sepúlveda, R. De la Cruz, A. Prado-Barragán, M. A. Aguilar-González, R. Rodríguez, and C. N. Aguilar, *Food and Bioproducts Processing*, Vol. 92, no. 4, p. 376, 2014.
156. L. Sepúlveda, J. J. Buenrostro-Figueroa, J. A. Ascacio-Valdés, A. F. Aguilera-Carbó, R. Rodriguez-Herrera, J. C. Contreras-Esquivel, and C. N. Aguilar, *Micologia Aplicada International*, Vol. 26, no. 2, 2014.
157. E. O. Geiger, 1997.

158. M. Cavazzuti, *Optimization methods: From Theory to design scientific and technological aspects in mechanics*. Springer Science & Business Media, 2012.
159. T. Walle, *Free Radical Biology & Medicine*, Vol. 36, no. 7, p. 829, 2004.
160. J. Quintin, D. Buisson, S. Thoret, T. Cresteil, and G. Lewin, *Bioorganic and Medical Chemistry Letters*, Vol. 19, no. 13, p. 3502, 2009.
161. M. N. Krasnow and T. M. Murphy, *Journal of Agricultural and Food Chemistry*, Vol. 52, no. 11, p. 3467, 2004.
162. A. R. Rechner, M. A. Smith, G. Kuhnle, G. R. Gibson, E. S. Debnam, S. K. S. Srai, K. P. Moore, and C. A. Rice-Evans, *Free Radical Biology & Medicine*, Vol. 36, no. 2, p. 212, 2004.
163. C. Dittmer, G. Raabe, and L. Hintermann, *European Journal of Organic Chemistry*, Vol. 2007, no. 35, p. 5886, 2007.
164. L.Y. Wang, K.-L. Cheong, D.-T. Wu, L.-Z. Meng, J. Zhao, and S.-P. Li, *International Journal of Biological Macromolecules*, Vol. 79, p. 180, 2015.
165. E. Betiku, H. A. Emeko, and B. O. Solomon, *Heliyon*, Vol. 2, no. 2, p. e00082, 2016.
166. R. Zouari, S. Ellouze-Chaabouni, and D. Ghribi-Aydi, "Optimization of Bacillus subtilis SPB1 biosurfactant production under solid-state fermentation using by-products of a traditional olive mill factory," *Achievements in the Life Sciences.*, Vol. 8, no. 2, p. 162, 2014.
167. M. B. Syed and M. Rajasimman, *Biotechnology Reports*, Vol. 6, p. 124, 2015.
168. A. Xiao, Y. Huang, H. Ni, H. Cai, and Q. Yang, *Electronic Journal of Biotechnology*, Vol. 18, no. 3, p. 143, 2015.

# 3

# Antibiotics Against Gram Positive Bacteria

**Rahul Vikram Singh, Hitesh Sharma, Anshela Koul and Vikash Babu***

*Fermentation Technology Division, CSIR-Indian Institute of Integrative Medicine, Canal Road, Jammu, 180001 (India)*

***Abstract***
The role of microbial products as a source of remedies for the treatment of various diseases has been accepted since ancient times. Among these microbial products, antibiotics are the potential compounds for the treatment of life-threatening diseases. Due to bacterial drug resistance, demand for new antibiotics has been increasing; therefore, researchers are putting their attention on synthesizing the derivative of drug molecules in the laboratory so that the biological activity can be evaluated. For the treatment of infections caused by Gram positive bacteria, cephalosporins and penicillins are the largest-selling antibiotics globally. Both of these antibiotics contain the"β-lactam" ring which is responsible for inhibiting cell wall synthesis of pathogens. In this chapter, different type of antibiotics, their mode of action, application and fermentation production process are discussed.

***Keywords:*** Gram positive bacteria, Cephalosporins, Penicillins, β-lactam

## 3.1 Introduction

Microorganisms constitute an inexhaustible reservoir of secondary metabolites with pharmacological, physiological, medical and agricultural applications. The most important commercially exploited secondary metabolite is antibiotics, which is being produced by microbes, specifically bacteria, and is used in a wide range. Antibiotics continue to play a crucial role in the screening primarily in biochemistry, molecular biology, microbiology and genetics (including genetic engineering) [1]. Antibiotic is generally defined as a secondary metabolite produced by a microorganism which inhibits the growth of some other microorganisms [2]. Most of the antibiotics used today are from microbes. The term antibiotic means against life or a set of antimicrobial agents that inhibit or kill bacteria. Alexander Fleming, in 1948, was the first to identify the fact that microorganisms could inhibit the growth of each other. He noticed that the growth of the bacterium *Staphylococcus aureus* was inhibited by a mold (fungus) that contaminated his plate. The mold was later identified as *Penicillin notatum* and the antibiotic which was later named penicillin was isolated [3].

**Corresponding author*: vikash@iiim.ac.in

Saurabh Saran, Vikash Babu, and Asha Chaubey (eds.) High Value Fermentation Products, Volume 1, (61–78) © 2019 Scrivener Publishing LLC

Amongst the different classification systems, microbes were categorized as bacteria, fungi and viruses. Bacteria were further categorized as Gram Positive and Gram Negative on the basis of Gram staining, which was discovered by H.C. Gram in 1884. It is an important technique which allows the wide variety of the bacteria to be classified under the category of Gram positive and Gram negative on the basis of staining properties due to difference in cell wall structure [4]. Bacterial cells, having a thick peptidoglycan layer (Figure 3.1) with the teichoic acid and teichouronic acid show positive result for Gram staining solutions, and thus are known as Gram positive bacteria. These are those classes of bacteria which retain the purple color on Gram staining. The cell wall of bacteria plays an important role in the pathogenicity. Figure 3.1 shows representation of the arrangement of components in the cell walls of Gram-positive bacteria. Gram-positive cell walls have an open, hydrophilic structure that retains the cell shape during isolation and purification. The key component of cell wall is peptidoglycan, that covered 50% of the weight of the wall. Linear anionic polymers, termed teichoic or teichuronic acids, are covalently associated to the peptidoglycan, giving the wall a net negative charge. Teichoic acids are linear polymers of repeating units of ribitol or glycerol units linked by phosphodiesters. Teichuronic acids do not contain any phosphate; in its place, they are made up from linear chains of sugar units containing uronic acid residues. Additional key form of teichoic acid found in the Gram-positive cell wall is lipoteichoic acid (LTA). LTA is a glycerolphosphate teichoic acid chain linked covalently to a glycolipid (typically a glycosyl diglyceride) situated on the external face of the cytoplasmic membrane. The glycerophosphate chain extends over the cellwall and is exposed on the cell surface. A number of functionally major proteins are also found both covalently and noncovalently linked to peptidoglycan. These mediate interactions

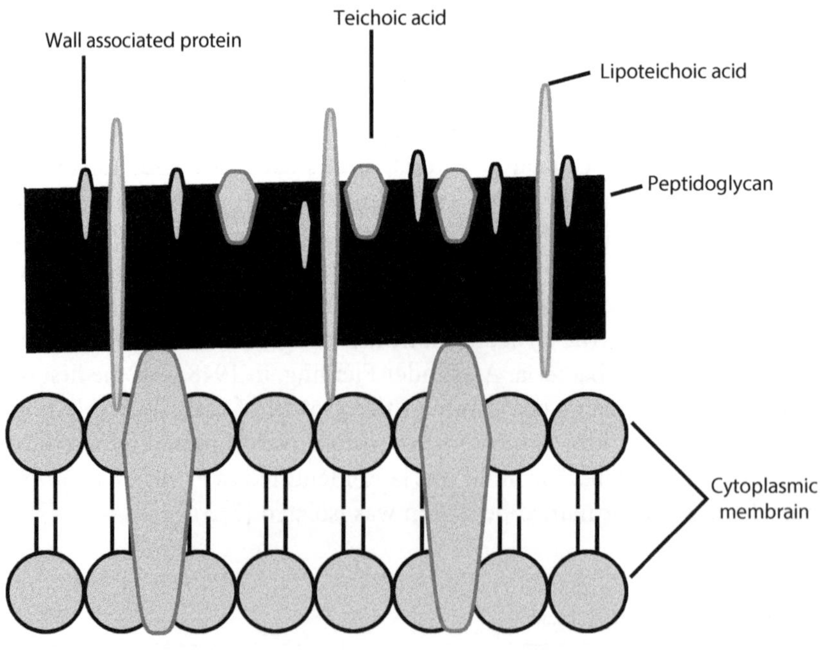


**Figure 3.1** Cell wall structural morphology of Gram Positive bacteria.

between cells and their environment. Many pathogens cooperate precisely with host cells and tissues in infections by creating surface exposed proteins that bind to host proteins. Some Gram positive bacteria generate capsular polysaccharides, that are loosely connected to the cell wall. Capsules form an additional barricade around the cells, protecting against engulfment by predatory cells in natural environments and by host phagocytic cells in infection [5].

*Actinomyces* spp plays a significant role in nature in which some are human pathogens. A few of them are opportunistic pathogens which caused infectious diseases in the human mouth, *i.e.*, periodontitis (inflammation of the gums) and oral abscesses. Genus *Mycobacterium* causes a diverse group of infectious diseases in which *M. tuberculosis* is the causative agent of tuberculosis. Another pathogenic species, *M. leprae*, is the cause of Hansen's disease (leprosy). *C. diphtheria* is the causative agent of diphtheria. Genus *Gardnerella*, contains only one species, *G. vaginalis* which causes vaginosis in women [6].

Although recent global consideration is paying attention to the concern of multidrug resistance (MDR) in Gram-negative bacteria, Gram-positive AMR is also a serious concern. Methicillin-resistant *Staphylococcus aureus* (MRSA) is perhaps the paradigm example, and is of high global importance as a cause of community-acquired and healthcare-associated infection. MRSA is a pathogen of concern due to its inherent resistance to almost all β-lactam antimicrobials (i.e., penicillins, cephalosporins and carbapenems) [7].

There are various species of pathogenic Gram Positive bacteria in nature which cause various diseases in humans and very few drugs are available in the market to cure such diseases. The major drawback associated with the antibiotic use is bacterial drug resistance which has created a chaotic situation in the whole world. Therefore, scientists have a great challenge to discover novel antibiotics which have less tendency to get resisted as well as with high efficiency. Out of the various antibiotics available in the market, few are specific for Gram positive bacteria. They are categorized on the basis of their structure which has been discussed further.

The discovery and development of the β-lactam antibiotics is the most powerful and successful achievement in the area of drug development against Gram negative and Gram positive pathogens. β–Lactam antibiotics have been extensively used for treatment of various bacterial infections for more than half a century [8]. Optimisation of the industrial production for β-lactams have been succesfully carried out by pharmaceutical companies. The discovery story of β-lactam antibiotics was started after the discovery of penicillins in 1921 and after that, researchers were focused on discovering new antibiotics from different microbial sources.

β-lactam antibiotic are broad spectrum class that contain beta-lactam ring, and include penicillin derivatives (penams), cephalosporins (cephems), monobactams and carbapenems as shown in Table 3.1. The major target of β-lactam antibiotics to inhibit cell wall biosynthesis in bacterial organism. β-lactam antibiotics inhibit the important enzymes (transpeptidases, carboxypeptidases) which play an important role in the peptidoglycan biosynthesis. β-Lactams are classified according to their core ring structures, which are described in Table 3.1. In the present time, the total world market for β-lactam antibiotics along with cephalosporin is now more than 15 billion dollars. Commercial production of penicillin-G and V by fermentation process approximately

**Table 3.1** Different classes of β-Lactams antibiotics, according to their chemical structure [9].

| Core ring | Addition group | Name of drug |
|---|---|---|
| **Saturated five-membered ring** | | |
| β-Lactams | Thiazolidine | Penams |
| β-Lactams | Pyrrolidine | Carbapenams |
| β-Lactams | Oxazolidine | Oxapenams or Clavams |
| **Unsaturated five-membered rings** | | |
| β-Lactams | 2,3-dihydrothiazole | Penems. |
| β-Lactams | 2,3-dihydro-1H-pyrrole | Carbapenems. |
| **Unsaturated six-membered rings** | | |
| β-Lactams | 3,6-dihydro-2H-1,3-thiazine | Cephems |
| β-Lactams | 1,2,3,4-tetrahydropyridine | Carbacephems |
| β-Lactams | 3,6-dihydro-2H-1,3-oxazine | Oxacephems |

30,000–100,000 gallons. In 1995 bulk penicillin (33,000 tons) was used for the production of semi-synthetic penicillins and cephalosporins [9, 10].

Other than β-Lactam antibiotics, there are various other classes of antibiotics which are against Gram positive bacteria, the major out of those classes being glycopeptide antibiotics, which are both natural as well as semi-synthetic. Glycopeptide antibiotics (GPAs) are commonly used to treat life-threatening infections caused by multidrug-resistant Gram-positive pathogens, such as *Staphylococcus aureus*, *Enterococcus* spp., *and Clostridium difficile*. These are drugs of last resort against methicillin-resistant *Staphylococcus aureus* (MRSA), which is currently a main cause of community-acquired infections that results in high morbidity and mortality rates in hospital-acquired infections. First-generation GPAs were totally natural products made of glycosylated non-ribosomal heptapeptides produced by a various group of actinomycetes [11]. After that few other antibiotics were discovered for the Gram Positive bacteria.

Resistance to vancomycin in *enterococci* since 1988 and the development of high-level GPA resistance in clinical isolates of MRSA since 2002 have encouraged searching for second-generation drugs belonging to the GPA class. Second-generation GPAs are semi-synthetic derivatives of natural products which are potent for even antibiotic-resistant organisms and also target-specific, which leads to least side effects.

## 3.2 Target of Antibiotics Against Gram Positive Bacteria

Antibiotics have been classified into different categories on the basis of their target site (Figure 3.2) such inhibition of cell-wall biosynthesis enzymes and substrates (β-lactams, vancomycin, and bacitracin), bacterial protein synthesis (chloramphenicol, clindamycin,

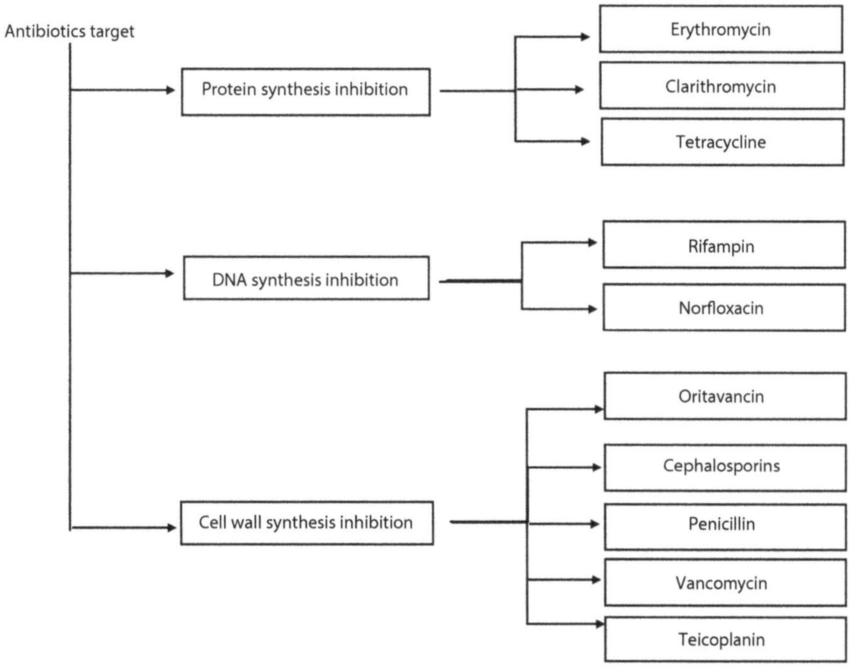


**Figure 3.2** Classification of Antibiotics on the basis of targets.

tetracyclines, macrolides, aminoglycosides, linezolid, mupirocin, and fusidic acid) and bacterial nucleic acid replication. A few of US Food and Drug Administration approved antibiotics, which have a different target site, are described in Figure 3.2 [12–14].

### 3.2.1 Cell Wall Synthesis Inhibition

First target of antibiotics in Gram Positive bacteria is to inhibit the cell wall synthesis. Specific antibiotics inhibit with the synthesis of the cell wall, weakening the peptidoglycan scaffold within the bacterial wall so that the structural integrity ultimately fails. Peptidoglycan construction begins in the cytoplasm with the synthesis of a muramyl peptapeptide precursor containing a terminal D-Ala-D-Ala. Some antibiotics inhibit the synthesis of the basic peptidoglycan building block. For example, D-cycloserine inhibits two enzymes involved in the precursor synthesis, preventing both conversion of L-alanine to D-alanine by racemase, and the construction of D-alanyl-D-alanine by D-Ala-D-Ala ligase. In the cytoplasm, muramyl pentapeptide is anchored via a water-soluble UDP-glucosamine moiety. A few antibiotics such as β-lactam and glycopeptide antibiotics act by inhibiting the synthesis of bacterial cell walls. [15]

The β-lactams incorporate the widely used penicillins and cephalosporins as well as the carbapenems and monobactams. In the second phase of peptidoglycan construction, muramyl pentapeptide N-acetylglucosamine is transferred to a $C_{55}$ undecaprenyl phosphate with the release of UMP to form a Lipid I intermediate. Tunicamycin inhibits the enzymatic conversion of the undecaprenyl phosphate to the lipid I intermediate, stopping the completion of the peptidoglycan structure. An additional glycosylation step completes the peptidoglycan unit, following which it is transported via its $C_{55}$ lipid tail to the external periplasmic surface of the membrane where its peptidyglycan unit becomes integrated into the cell wall matrix. Bacitracin inhibits lipid phosphatase, preventing the release of the finished peptidoglycan from its $C_{55}$ lipid carrier. Other transpeptidases and transglycosylases bond to newly synthesized peptidoglycan structures to the cell wall peptidoglycan matrix. The specificity of β-lactam antibacterials is due to their ability to inhibit transpeptidase enzymes and prevent the assembly of the peptidoglycan layer in both Gram-positive and Gram-negative bacteria. β-Lactam molecules, with their structural similarity to the D-alanyl-D-alanine group within the peptidoglycan structure, compete for the binding sites of transpeptidases. When it was first commercialized, penicillin, a β-lactam antibiotic, was considered a "magic bullet" because of its specificity for bacterial infections without creating any toxicity [16, 17].

### *3.2.1.1 Penicillin*

Penicillin was the first successful chemotherapeutic agent produced by microbes, and it initiated the age of antibiotics [18]. Penicillin is a broad spectrum of antibiotics that exhibit biological activity against Gram Positive and Gram Negative pathogens. It is also known as lactam antibiotics due to containing of β- lactam ring (Figure 3.3A). Penicillin inhibits the cell wall of bacteria by blocking transpeptidase after binding to penicillin binding protein (PBPs) and prevents its synthesis.

Annual bulk production of penicillin is $3 \times 10^7$ kg/year with annual sales of ~$15 billion, and market share is 65% of the total antibiotic market [19]. Large-scale production of penicillin by submerged fermentation was started in 1943 and it was marked as the first bio-based production of a pharmaceutical [20]. With continuous developments in the production of penicillin, including microbial strain improvement, media optimization, physical parameters optimization and chemical engineering methods,

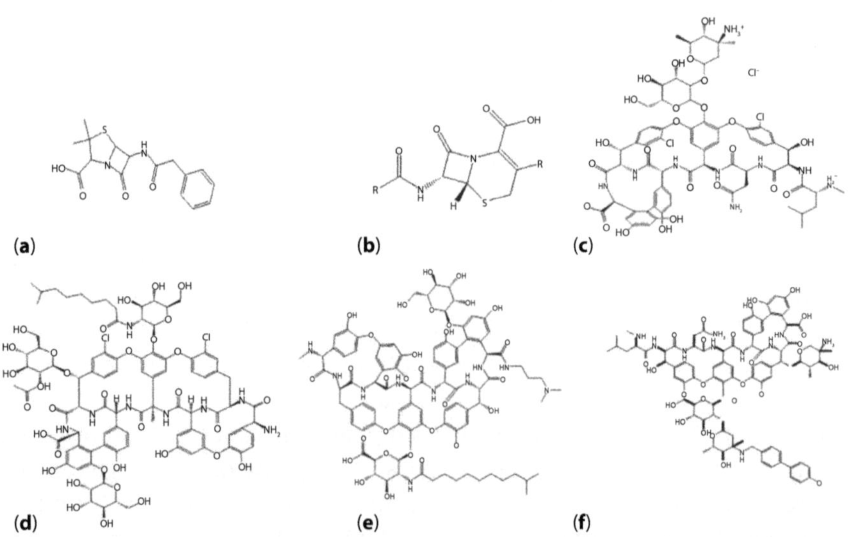


**Figure 3.3** Structures of Cell wall synthesis inhibiting antibiotics: A. Penicillin; B.Cephalosporin; C. Vancomycin; D. Teicoplanin; E. Dalbavancin; F. Oritavancin.

it has been estimated that recent industrial strains could produce 100,000 times more penicillin than Fleming's original strain *Penicillium notatum.* Nowadays, *Penicillium chrysogenum* has proven to be an ideal microbial source for the production of penicillin, and successful approaches i.e., culture medium, pH, temperature, mechanical forces, inoculum size, and dissolved oxygen have been exploited to elevate the production titer [21]. Optimization efforts have led to the increased production of penicillin over 70 g/l [22]. Penicillins were categorized into different groups on the basis of their molecular structure and biological activity against the pathogens, which are described in Table 3.2 [23].

### *3.2.1.2 Cephalosporins*

According to Allied Market Research, the global cephalosporin market was valued at $77,764 million in 2016, and is estimated to reach $1,99,754 million by 2023, growing at a CAGR of 14.4% from 2017 to 2023 [24]. Cephalosporin drugs are also β-lactam antibiotics which target the cell wall of bacteria. It contains a dihydrothiazine ring with D-α-aminoadipic acid (Figure 3.3B). Cephalosporins are less toxic and broad-spectrum antibiotics comparable in action to ampicillin [25]. Today, cephalosporin and its derivatives are widely used in the treatment of a number of infectious diseases caused by bacteria in the respiratory tract, infections of the skin and infections of urinary tract [26]. Initially, Dr. Abraham isolated cephalosporin C from *Cephalosporium acremonium.* Cephalosporin drugs act as similar to penicillin by targeting the necessary enzymes that are required for the synthesis of cell wall of bacteria. This antibiotic is poorly soluble in water; therefore attempts have been made to prepare derivatives of cephalosporin. On the basis of spectrum, generation, chemical structure, resistance to beta lactamases, and clinical pharmacology, it has been categorized into different generations which are described in Table 3.3 [27–30]. Cephalosporins are produced by *Acremonium chrysogenum* (also called *Cephalosporium acremonium*) and *Streptomyces clavuligerus* [31].

**Table 3.2** Classification of Penicillins.

| Class | Drugs | Drug of choice | Toxicity |
|---|---|---|---|
| Penicillin | PenicillinG<br>Aqueous penicillinG<br>Procaine penicillin G<br>Benzathine penicillin G PenicillinV | *Strep. Pyogenes Step. AgalactiaeC. Perfringens* (Bacilli) | Hypersensitivity reaction<br>Hemolytic anemia |
| Penicillinase-resistant-penicillins | Methicillin<br>Nafcillin<br>Oxacillin<br>Cloxacillin<br>Dicloxacillin | PCNase-producing *Staphylococcus aureus* | Interstitial nephritis |
| Antipseudomonal Penicillins | Piperacillin | *Staphylococcus aureus* | - |

**Table 3.3** Different generation of Cephalosporins.

| Generation | Drugs | Targeted pathogen |
|---|---|---|
| **First Generation** | Cefazolin<br>Cephalothin<br>Cephapirin<br>Cephalexin<br>Cefadroxil<br>Cephradine | *MSSA, Streptococci Grp A,B,C, S. viridians, S. Pneumonia, H. Influenza, E. coli, Klebsiella pneumonia, Proteus mirabilis* |
| **Second Generation** | Cefamandole<br>Cefuroxime<br>Cefoxitin<br>Cefotetan<br>Cefmetazole<br>Cefaclor<br>Cefprozil<br>Cefpodoxime<br>Loracarbef | 1st generation coverage plus β-lactamase positive *H. Influenza, Moraxella catarrhali, Neisseria meningitides, E. coli, Klebsiella pneumonia, Proteu, Oral anaerobes,* Cefoxitin & Cefotetan cover *B. fragilis, Staphylococcus aureus* |
| **Third Generation** | Cefotaxime<br>Ceftriaxone<br>Ceftizoxime<br>Ceftazidime<br>Cefoperazone<br>Cefixime | 1st generation coverage plus Expanded Gram-negative coverage, Oral anaerobes, *S. aureus* (OSSA), *S. pneumonia, Strep* Grp A,B,C,G, *S. viridians,* Gram negative rods, *N. Gonorrhoea,* All cover *B. fragilis* except cefotaxime & ceftazidime, *P. aeruginosa* - ceftazidine only |
| **Fourth Generation** | Cefipime | Their spectra is comparable to 3rd generation, Good Gram-positive & Gram-negative coverage, Anti-Pseudomonal (including ceftazidime resistant isolates), Limited anaerobic coverage |
| **Fifth Generation** | Ceftaroline<br>Ceftobiprole | Effective against Gram-positive bacteria, and retains the activity of later-generation cephalosporins with broad-spectrum activity against Gram-negative bacteria |

Enzymatic methods have also employed to produce the key intermediate 7-aminocephalosporanic acid (7-ACA) required to increases the availability of the drug [26].

Many efforts have been carried out by researchers for the the optimization of fermentation conditions, which is of primary importance in the development of any fermentation process owing to their impact on the economy and practical feasibility of the process [26]. A mixture, composed of 40% (v/v) oleic acid and 60% (v/v) linoleic acid, had the strongest stimulatory effect on cephalosporin C (CPC) production from *Acremonium chrysogenum*, due to a synergistic effect of the two fatty acids. the maximum CPC titer (7.44 g/l) was improved about 4.5-fold [31]. For the high yield (20–25 g/l) production of cephalosporin-C at industrial *A. chrysogenum* strain reported for fed-batch fermentations. During the fermentation process, chemical instability of the cephalosporin C molecule that reduces cephalosporin production in compared to penicillin fermentations [10].

#### 3.2.1.3 Vancomycin

Vancomycin was introduced in early 1950, known at first as glycopeptide antibiotic. Vancomycin is a branched tricyclic glycosylated nonribosomal peptide produced by the fermentation of the Actinobacteria species *Amycolatopsis orientalis.* Biosynthesis occurs via unlike nonribosomal protein synthases (NRPSs). Jung *et al.* 2007 reported strain *Amycolatopsis orientalis* KCCM-10836P for the vancomycin production of 11.5 g/l yield [32, 33]. Less toxicity profile as well as availability of less toxic replacements of beta-lactams, considered more against Gram-positive infections. Vancomycin (Figure 3.3C) also prevents the synthesis of peptidoglycan precursors of the bacterial cell wall by blocking the transglycosylation phase and then affecting the transpeptidation step also. Both the transglycosylation and transpeptidation steps are essential for bacterial cell wall cross-linking [34]. In 2009 the Food and Drug Administration (FDA) approved telavancin (Vibativ) for clinical use. Telavancin is a derivative of vancomycin containing addition of a hydrophobic and a hydrophilic group to the vancomycin structure [33, 35]. Nowadays telavancin is manufactured by some companies by semisynthetic route.

#### 3.2.1.4 Teicoplanin

Teicoplanin is another natural glycopeptide that have potential activity against the Gram-positive bacteria, including methicillin-resistant *Staphylococcus aureus* and *Enterococcus faecalis.* It is a semisynthetic glycopeptide antibiotic with similar to vancomycin. It also inhibits bacterial cell wall synthesis. This drug was isolated from strain Actinoplanes teichomyceticus. After semisynthesis teicoplanin was modified that contain a group of five structures as shown in Figure 3.3D [33, 36].

Dalbavancin (Figure 3.3E) is a semi-synthetic drug which is derivative of the teicoplanin-like molecule A40926, that was isolated from the actinomycete *Nonomuraea* sp. ATCC 39727 collected from an Indian soil in the mid-1980s. In contrast with teicoplanin, A40926 lacks the saccharide moiety on the amino acid residue 6 Dalbavancin different from teicoplanin due to the presence of an acylaminoglucuronic acid on amino acid 4 instead of the acylglucosamine. Another structural difference between A40926 and teicoplanin include the terminal methylamino group, the position of one chlorine atom and the length of the fatty acid chain. In 2014, US-FDA approved dalbavancin (Dalvance) for clinical use [33, 37].

#### 3.2.1.5 Oritavancin

Oritavancin is semi-synthetic second-generation GPAs belonging to the vancomycin family that have broad spectrum of both resistant and susceptible against Gram-positive bacteria, including *Staphylococcus aureus, MRSA, Enterococci,* and *Streptococci.* Oritavancin (Orbactiv) (Figure 3.3F) is the *N*-alkyl-*p*-chlorophenyl-benzyl derivative of chloroeremomycin produced by the actinomycete *Amycolatopsis orientalis.* Chloroeremomycin, that differs from vancomycin by the glycosylation pattern on amino acid residues 4 and 6. In 2014 it was approved by the United States FDA for treatment of skin infections. Oritavancin inhibit the transglycosylation and inhibition of transpeptidation that are responsible for the synthesis of cell wall in Gram-positive bacteria. In present time Eli Lilly & Company is the only producer of Oritavancin antibiotics [33].

### 3.2.2 Protein Synthesis Inhibition

Another target site in Gram Positive bacteria is protein synthesis inhibition in which antibiotics binds to the 50S ribosomal sub-unit reversibly and block the binding of tRNA to the acceptor site which prevent translocation of the peptide chain. A few antibiotics are present in the market which act as protein synthesis inhitor such as chloramphenicol, tedizolid, delafloxacin, pleuromutilin [38, 39].

#### *3.2.2.1 Erythromycin*

It was first discovered in 1952 as the metabolic products of *Streptomyces erythreus*. This strain is prominently found in the Philippines. Erythromycin, as shown in Figure 3.4A, is a parent molecule from which Clarithromycin and Azithromycin are synthesized. Erythromycin usually makes the bacterial growth static but at higher doses, it is lethal also. It is most effective against Gram-positive *Cocci* and *Bacilli*. Erythromycin and the other macrolide antibiotics bind to the 50S subunit of sensitive microorganisms which is known to be between the A and the P site, near the binding spot of chloramphenicol. They prohibit the peptide chain from the A to the P site. These drugs are used to treat respiratory and soft tissue infections [40].

Various processes have been developed for the synthesis of erythromycin be it synthetic as well as semi-synthetic. Zou *et al.* 2008 [41] stated that corn steep liquor can regulate oxygen uptake rate to a certain level in the early phase of fermentation, and enhance the metabolic flux of erythromycin biosynthesis. Erythromycin production was successfully scaled up from a laboratory scale (50 l fermenter) to an industrial scale (132 $m^3$ and 372 $m^3$) using oxygen uptake rate as the scale-up parameter.

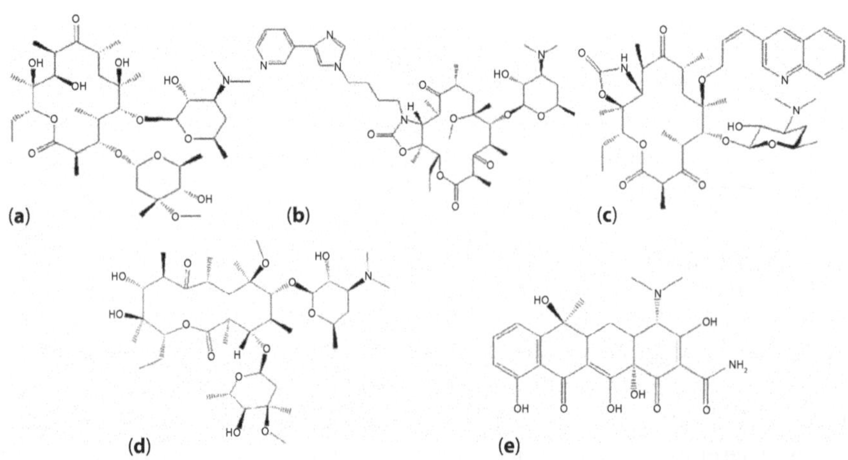

**Figure 3.4** Structures of protein synthesis inhibiting antibiotics: A. Erythromycin; B. Telithromycin; C. Cethromycin; D. Clarithromycin; E. Tetracycline.

Erythromycin derivatives known as Ketolides are another type of antibiotics which are semisynthetic derivatives. They have a 14-membered macrolactone ring, having a keto group instead of an L-cladinose sugar appended at position 3. With addition of hydroxyl groups in positions 11 and 12 that are replaced by cyclic carbamate. For example, telithromycin has an alkylaryl extension that is bound to its cyclic carbamate as shown in Figure 3.4B, whereas ABT-773 (cethromycin) has a quinolylallyl arm at the O-6 position (Figure 3.4C). The crystal structure of telithromycin bound to the large ribosomal subunit of *Deinococcus radiodurans* indicated that telithromycin interacts with domain V (via the 3-keto group and additional hydrophobic interactions) and domain II (via the carbamate extension) of the 23S rRNA.37 Domain V is the peptidyl transferase centre that catalyses peptide bond formation. Telithromycin blocks the ribosomal exit tunnel, thus terminating peptide synthesis. Cethromycin also act alike. While ketolides bind to a similar region of the 50S ribosomal subunit as does erythromycin, they tend to have significantly binding affinity and therefore can still bind to erythromycin resistant ribosomes. The semisynthesis modification of the structural macrolide molecule leads to improved effectiveness against many Gram-positive bacteria, mainly those that have developed resistance to macrolides such as the ketolides have reasonable activity against macrolide-resistant *S. pneumoniae.* Because they have activity against numerous Gram-positive organisms and certain Gram-negative respiratory pathogens, they are often used for respiratory-tract infections, including CAP, acute exacerbations of chronic bronchitis, and sinusitis, as well as *streptococcal pharyngitis*. However they are not suitable in treating MRSA or resistant *Enterococcus* spp. [42].

### *3.2.2.2 Clarithromycin*

Clarithromycin, also known as Biaxin, is an antibiotic used against bacterial infections. It is a macrolide and causes the inhibition of protein synthesis. It is a derivative of erythromycin and is chemically known as 6-O-methylerythromycin as shown in Figure 3.4D [43]. The wholesale cost in the developing world is between 0.13 and 0.79 USD per dose but in the US it costs around 50–100 USD per dose [44].

Clarithromycin is primarily used to treat a number of bacterial infections including pneumonia, *Helicobacter pylori*, in strep throat as an alternative to penicillin, and is also used in cat scratch disease and other infections due to bartonella, cryptosporidiosis, as a second line agent in Lyme disease and toxoplasmosis. As an alternate to penicillin, it may also be used to prevent bacterial endocarditis [45]. It is effective against upper and lower respiratory tract infections, skin and soft tissue infections and *Helicobacter pylori* infections associated with duodenal ulcers [46].

### *3.2.2.3 Tetracyclines*

Tetracyclines are group of antibiotics dicovered by Benjamin Duggar in 1948 from the streptomyces genus. He isolated the natural product aureomycin, or 6-chlorotetracycline (chlorotetracycline), from a bacterial culture [47]. Tetracyclines are broad-spectrum antibiotics of polyketide family of antibiotics and the third most consumed antibiotic, after penicillin and quinolones. Nowadays, tetracyclines are produced by *Streptomyces aureofaciens* or *Streptomyces rimosus.* The different form of tetracyclines are tetracycline, chlortetracycline demeclocycline, oxytetracycline have been discovered and many derivatives of tetracycline such as doxycycline, lymecycline, meclocycline, methacycline, minocycline and rolitetracycline, doxycycline have been synthesized for the treatment of various diseases [48]. The simplest tetracycline to display detectable antibacterial activity is 6-deoxy-6-demethyltetracycline (Figure 3.4E) and so, this structure may be regarded as the minimum pharmacophore [49]. These antibiotics are effective against both Gram-positive and Gram-negative bacteria for the treatment of pneumonias, community acquired pneumonia, rickettsial and chlamydial infections, Lyme disease, cholera, syphilis and periodontal infections [49]. Tetracyclines inhibit protein synthesis by preventing the binding of aminoacyl-tRNA to the A site of the bacterial ribosome [50].

The major objective in any antibiotic production is to reduce the cost either by fermentation optimization or down-streaming processing [51]. Various parameters such as temperature, pH, dissolved oxygen, carbon source, precursor, nitrogen source, aeration, agitation and other components require to be optimized for better antibiotic production [52, 53]. The first large-scale steel product-specific fermenter was constructed with a capacity of 54,000$dm^3$ [54]. The production titer of chlortetracycline and tetracycline has been reached to 30–35 g $L^{-1}$ using strain improvement technologies [55].

### 3.2.3 DNA Synthesis Inhibition

A few antibiotics act as DNA synthesis by inhibiting topoisomerase such as quinolones that inhibit topoisomerase II (DNA gyrase), which are involved in DNA replication. Second-generation quinolones—for example, levofloxacin, norfloxacin, and ciprofloxacin—are potent drugs against Gram-positive bacteria. Enzyme topoisomerases are present in both prokaryotic and eukaryotic cells, but the quinolones are specific inhibitors of bacterial topoisomerase II. A few antibiotics such as rifampicin, nemonoxacin, lipoglycopeptides, etc., block initiation of RNA synthesis by specifically inhibiting bacterial RNA polymerase. It does not interact with mammalian RNA polymerases, making it specific for Gram-positive bacteria. Some antibiotics that interfere with RNA synthesis by inhibiting RNA polymerase, such as doxorubicin and actinomycin D (dactinomycin), are not specific for bacteria and interfere with both bacterial and mammalian systems [56].

Rifampicin (Rifampin) is an antibiotic used to treat some common bacterial infections including tuberculosis, leprosy and Legionnaire's disease. It is used for the prevention of *Haemophilus influenzae* type b and meningococcal disease in isolation; the rest is used in combination with other drugs. Rifampicin (Figure 3.5) was discovered in 1965, marketed in Italy in 1968, and approved in the United States in 1971 [57–60]. It is in the World Health Organization's List of essential Medicines, the most effective and safe medicines needed in a health system [61].

## 3.3 Antibiotics Production Processes

Due to antibiotic resistance, demand for novel antibiotics has been increased, and to overcome this serious problem scientists were approaching new methods of rapid

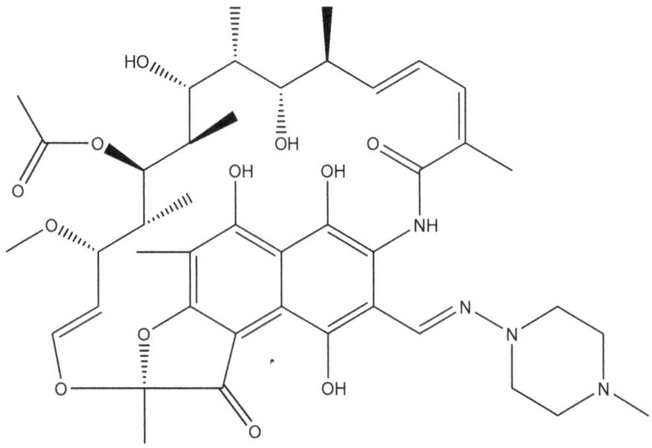


**Figure 3.5** Structure of rifampicin.

screening of potent strains. For the large-scale production of novel antibiotics a few approaches were followed by researchers [62, 63].

- Synthetic route
- Semi-synthetic route
- Microbial source for the production of antibiotics using fermentation Technology.

Synthetic route involves the production of waste water and other eco-hazardous compounds which is a major concern with these processes. On the other hand, in semi-synthetic process, the structural change in a naturally produced antibiotic may or may not occur with a possibility that its derivative may or may not be a potent molecule and also with an influenced specificity. Horizons apart, fermentation processes for antibiotics production involve biodegradable waste generation as well as least waste water generation with a high specificity towards the substrate. But antibiotics produced by microbes are in very small quantity. Also the antibiotic production directly depends upon the microbial biomass, their duplication time, their inherent efficiency to produce these molecules. Therefore, in order to produce these antibiotics on a large scale techniques like recombinant DNA technology are utilized before the fermentation processes. Recombinant DNA technologies were used to transfer that particular targeted gene in suitable host that have less generation time. This will increase the yield with reduction in generation time. Further these cloned gene containing microbes are grown in nutrient medium under fermentation conditions which result in improved yield [64].

The fermentation process has been playing an important role in large-scale production of antibiotics. Industrial fermentation comprises both upstream process and downstream processing stages (Figures 3.6 and 3.7). Upstreaming process included all factors such as isolation, screening and production that include fermentation process. Downstreaming process encompasses all processes such as recovery of a product from cell broth, separation of products from crude and further purification of targeted

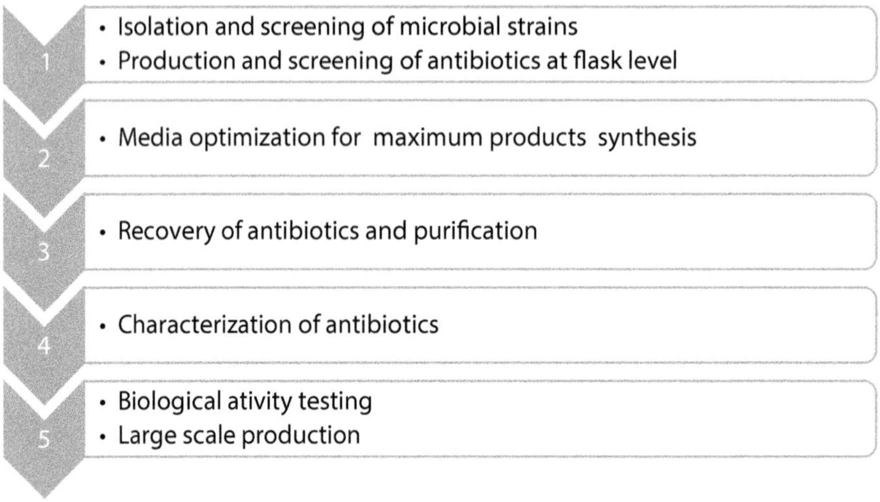


**Figure 3.6** Fermentation process for production of antibiotics.

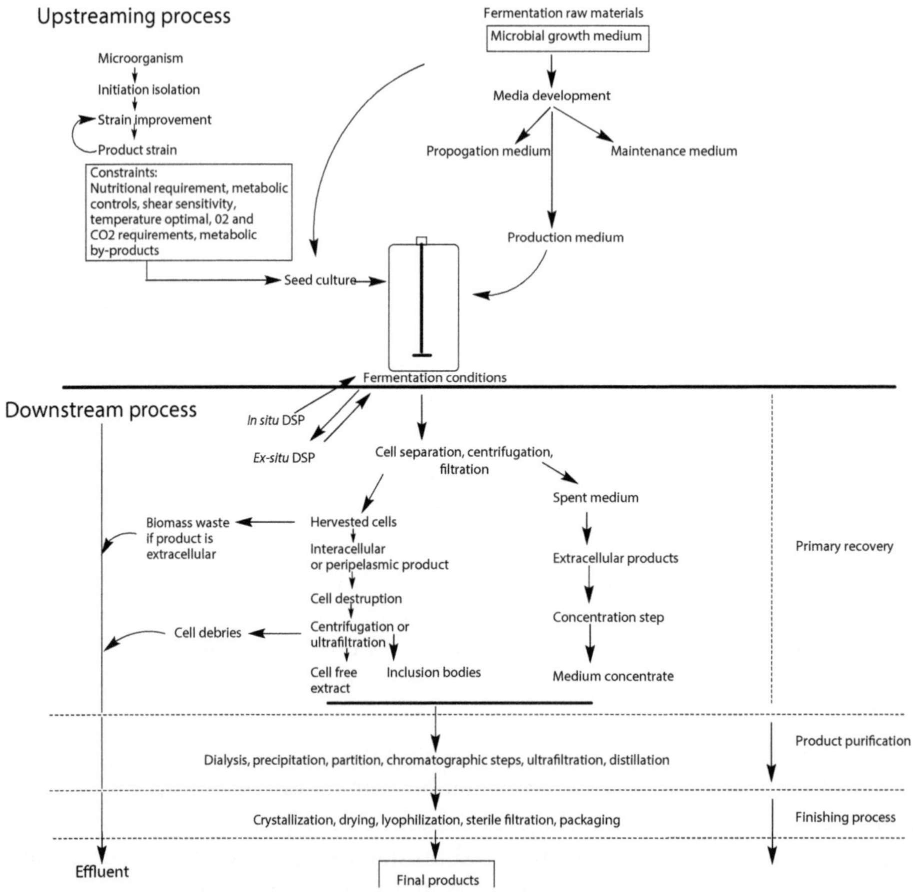


**Figure 3.7** Outline of upstream and downstream processing application.

product [65, 66]. There are several challenges in isolation, screening and production of antibiotics in fermentation conditions such as the following.

- Replacement of old traditional screening methods to new advanced screening methods (e.g., LC-MS, HPLC, HRMS, MALDI-TOF-MS, and GC-MS etc.).
- Up-scaling process for production of antibiotics.
- Fast extraction method for recovery of antibiotics without degrading molecule.
- Ultra-purification of products

The extraction and purification of fermentation products may be difficult and costly. Ideally, one is trying to obtain a high-quality product as quickly as possible at an efficient recovery rate using minimum plant investment operated at minimal costs. Unfortunately, recovery costs of microbial products may vary from as low as 15% to as high as 70% of the total manufacturing costs. Recovery of microbial-fermented products from broth is a challenging task itself; therefore various extraction techniques were developed. After the fermentation process the downstream process starts where fermented product was separated from the cell broth by centrifugation or microfiltration. The next step of the downstream process (Figure 3.7) is extraction of product from the agues broth. It can be separated by solvent extraction in which different solvents such as ethyl acetate, chloroform, dichloroform, hexane, n-butanol, etc., were used and organic layer containing product were dried under the reduced pressure vacuum. After that product purified by column chromatography techniques. In another technique which is generally used in the laboratory for the microbial products, cell biomass is separated from the cell broth by centrifugatation and aqueous phase concentrated into reduced pressure vaccum and then purification process can be done.

## 3.4 Conclusion

There are a number of antibiotics which are effective against Gram Negative bacteria while few antibiotics are available in the market which are effective against the Gram Positive based on their targets. Cephalosporin, penicillin, vancomycin, teicoplanin and oritavancin are the basic antibiotics being used for cell wall synthesis inhibition of Gram positive bacteria but the resistive nature of pathogens has led to development of derivatives of such antibiotics.

In conclusion, increasing multi-drug resistance in pathogenic bacteria has created an emergency of potent antibiotics in the world. Therapeutic approaches for the treatment of bacterial infections have historically relied on the antibiotics that target bacterial protein, DNA, RNA, or cell wall synthesis. However, although antibiotics have been successfully used for decades, the discovery rate of novel antibiotics is unable to keep pace with the emergence of antibiotic-resistant bacteria. Therefore, chemists and biologists are facing a new challenge to discover novel effective antibiotics against the pathogen.

For the large-scale production and up-scaling of antibiotics from a microbial source, it is necessary to understand the ideal fermentation conditions so that the commercialization process of such antibiotics will be easy.

## References


1. K.V. Prabhu, C. Sundaramoorthi, G. Saurabh, K. Karthick and N. Tamilselvi, *International Research Journal of Pharmacy,* Vol. 2, p. 114, 2011.
2. R.W. Jack, J.R. Tagg and B. Ray, *Microbiological Reviews,* Vol. 59, p. 171, 1995.
3. M.S. Abdulkadir and S. Waliyu, *European Journal of Applied Sciences,* Vol. 4, p. 211, 2012.
4. www.columbia.edu/itc/hs/medical/pathophys/id/2009
5. A.P. Lambert Cellular impermeability and uptake of biocides and antibiotics in Gram-positive bacteria and mycobacteria. *Journal of Applied Microbiology,* 92(1). 2002 p. 46S.
6. Gram-Positive Bacteria, https://courses.lumenlearning.com/microbiology/chapter/gram-positive-bacteria.
7. Antimicrobial Resistance Global Report on Surveillance: Summary, WHO, Geneva, Switzerland, 2014.
8. J. Thykaer and J. Nielsen, *Metabolic Engineering,* Vol. 5, p. 56, 2003.
9. M. Barber. *The penicillins business.* Michael Barber and Associates, 18 Croydon Road, Caterham, Surrey, UK. 1996.
10. R.P. Elander, *Applied Microbiology and Biotechnology,* Vol. 61, p.385, 2003.
11. G.M. Rossolini, F. Arena, S. Pollini. Novel infectious diseases and emerging Gram-positive multi-resistant pathogens in hospital and community acquired infections. In *Antimicrobials.* Springer, Berlin, Heidelberg. pp. 11–28. 2014.
12. M.M. Huycke, D.F. Sahm and M.S. Gilmore, *Emerging Infectious Diseases,* Vol. 4, p. 239, 1998.
13. J. Ritchie, J. Lewis, C.M. Nicholls, R. Ormston, *Sage,* Vol. 2, p. 295, 2013.
14. R.E. Hancock, *The Lancet Infectious Diseases,* Vol. 5, p. 209, 2005.
15. Sigma-Aldrich, *Antibiotics for Research Applications, Biofiles,* Vol. 4, p. 7, 2006.
16. L.M. Leung, *Microbial Membrane Glycolipids as Diagnostic Markers During Infection* (Doctoral dissertation, University of Maryland, Baltimore). 2017.
17. D.Y. Aksoy, S. Unal, *Clinical Microbiology and Infection,* 14(5), p.411. 2008.
18. N. Kardos, A.L. Demain. *Applied Microbiology and Biotechnology,* 92(4), p.677. 2011.
19. A. Nandi, S. Pan, R. Potumarthi, M.K. Danquah, and I.P. Sarethy. *Journal of Analytical Methods in Chemistry,* 2014.
20. J.S. Rokem, A.E. Lantz, J. Nielsen. *Natural Product Reports.* 24S (6):1262. 2007.
21. G. Wang, J. Chu, H. Noorman, J. Xia, W. Tang, Y. Zhuang, S. Zhang. *Applied Microbiology and Biotechnology,* 98(6), p.2359. 2014.
22. C. Olano, F. Lombo C. Mendez, J.A. Salas. *Metabolic Engineering.* 10:281–292. 2008.
23. H.F. Chambers, D.H. Deck. *Basic and Clinical Pharmacology,* 10, p.726. 2007
24. https://www.alliedmarketresearch.com/cephalosporin-market.
25. W. Lotfy, *Research Journal of Microbiology,* Vol. 2, p. 1, 2007.
26. N.T. Khan, *Enzyme Engineering.* 6: 159. doi:10.4172/2329-6674.1000159, 2017.
27. A. Dalhoff, N. Janjic, R. Echols, *Biochemical Pharmacology.* 71 (7), 1085. 2006.
28. R. Berkow, M.H. Beers, A.J. Fletcher. *The Merck Manual of Medical Information.* Mark eds. 2nd Home Edition. *Whitehouse Station,* NJ: Merck; 2003.
29. Tumah, H. *Chemotherapy,* 51:80. 2005.

30. L.L. Brunton, B. Chabner, B.C. Knollmann, *The goodman and gilman's manual of pharmacological therapeutics.* McGraw-Hill Professional, 2011.
31. J.C. Kim, S.W. Kang, J.S. Lim, Y.S. Song, and S.W. Kim. *Journal of Microbiology and Biotechnology*, 16(7), p.1120. 2006.
32. H. Jung, S. Kim, H. Moon. *Applied Microbiology and Biotechnology.* 77:789. 2007.
33. P.A. Ashford, S.P. Bew. *Chemical Society Reviews*, 41(3), p.957. 2012.
34. E. Binda, F. Marinelli, G.L. Marcone. *Antibiotics*, 3(4), p.572. 2014.
35. E.C. Nannini, M.E. Stryjewski. *Expert opinion on pharmacotherapy*, 9(12), p. 2197. 2008.
36. H. Wang, X. An, X. Deng, G. Ding. *Electrophoresis*, 38(9–10), p.1374. 2017.
37. B.P. Goldstein, E. Selva, L. Gastaldo, M. Berti, R. Pallanza, F. Ripamonti, P. Ferrari, M. Denaro, V. Arioli, G. Cassani. *Antimicrobial Agents and Chemotherapy.* 31, p.1961. 1987.
38. V. Fines, M. and R. Leclercq, *Journal of Antimicrobial Chemotherapy*, 45(6), p.797. 2000.
39. M.A. Fischbach, C.T. Walsh. *Science*, 325(5944), p.1089. 2009.
40. G.G. Zhanel, E. Hartel, H. Adam, S. Zelenitsky, M.A. Zhanel, A. Golden, F. Schweizer, B. Gorityala, P.R. Lagacé-Wiens, A.J. Walkty A.S. Gin, A.S. *Drugs*, 76(18), p.1737. 2016.
41. X. Zou, H.F. Hang, J. Chu, T.P. Zhuang, S.L. Zhang. *Bioresource Technology*, 100(3), p.1406. 2009.
42. R.E. Hancock. *The Lancet Infectious Diseases*, 5(4), p. 209. 2005.
43. "Clarithromycin". The American Society of Health-System Pharmacists. Archived from the original on September 3, 2015.
44. "Clarithromycin". International Drug Price Indicator Guide. Retrieved 7 September 2015.
45. R.J. Hamilton. *Tarascon Pocket Pharmacopoeia* 2013 Deluxe Lab-Coat Edition. Jones & Bartlett Publishers. 2012.
46. A.H. Kirst, Macrolide Antibiotics (2ed.). *Basel: Birkhäuser Basel.* p. 53. ISBN 9783034881050. 2012.
47. F. Liu, A.G. Myers. *Current Opinion in Chemical Biology*, 32, p.48. 2016.
48. B.M. Vastrad, S.E. Neelagund. *Recent Research in Science and Technology*, 3(2). p.1 2011.
49. F. L. Bahrami, D. Morris, H.M. Pourgholami. *Mini Reviews in Medicinal Chemistry.* 12(1): p. 44. 2012.
50. A.A. Borghi, M.S.A. Palma. *Brazilian Journal of Pharmaceutical Sciences*, 50(1), p.25. 2014.
51. G. Detilley, D.G. Mou and C.L. Cooney. *"Optimization and economics of antibiotics production"* in J.E. Smith, D.R. Berry and B Kristiansen, eds., *The Filamentous Fungi,* Vol. 4, Edward Arnold, London, p. 190, 1983.
52. J.E. Smith, "Concepts of Industrial Antibiotic Production", in D.I. Alani, M. Moo-Young, eds., *Perspectives. in Biotechnology and Applied Microbiology*, Elsevier Applied Science Publishers, pp. 105–142, 1986.
53. M.J. Waites, N.L. Morgan, J.S. Rockey, G. Higton G, I*ndustrial microbiology: an introduction*, Blackwell, Oxford, UK, 2001.
54. J.R. Callahan, *Chemical and Metallurgical Engineering,* Vol. 51, p.94, 1944.
55. J.L. Adrio, A.L. Demain, *FEMS Microbiology Reviews*, 30(2), p.187. 2005
56. D.Y. Aksoy, S. Unal, *Clinical Microbiology and Infection*, 14(5), p.411. 2008
57. Rifampin". The American Society of Health-System Pharmacists. Archived from the original on 2015-09-07. Retrieved Aug 1, 2015.
58. P. Sensi, "History of the development of rifampin". *Reviews of Infectious Diseases.* 5 Suppl 3: S402–6. doi:10.1093/clinids/5.supplement_3.s402. JSTOR 4453138. PMID 6635432. 1983
59. Oxford Handbook of Infectious Diseases and Microbiology. OUP Oxford. 2009. p. 56. ISBN 978-0-19-103962-1. Archived from the original on 2015-11-24.

60. M. Hugh, D. Timothy, Tuberculosis: diagnosis and treatment. Wallingford, Oxfordshire: CAB International. p. 219. ISBN 978–1–84593–807–9. 2011.
61. "WHO Model List of Essential Medicines (19th List)" (PDF). World Health Organization. April 2015. Archived (PDF) from the original on 13 December 2016. Retrieved 8 December 2016.
62. S. Parekh, V.A. Vinci, R.J. Strobel, *Applied Microbiology and Biotechnology*, 54, p. 287. 2000.
63. W. Leuchtenberger, K. Huthmacher, K. Drauz, *Applied Microbiology and Biotechnology*, 69:1. 2005.
64. A.C. Finlay, G.L. Hobby, S.Y. P'an, P. P Regna, J.B. Routien, D.B. Seeley, G.M. Shull, B. A. Sobin, I.A. Solomons, J.W. Vinson and J.H. Kane, *Science*. 111: p. 85, 1950.
65. H. L. Ser, J.W.F. Law, N. Chaiyakunapruk, S.A. Jacob, U.D. Palanisamy, K.G. Chan, B.H. Goh and L.H. Lee, *Frontiers in Microbiology*, Vol. 8. pp. 877. 2017.
66. R.P. Elander, *Applied Microbiology and Biotechnology*, Vol. 61, p. 385, 2003.

# 4

# Antibiotic Against Gram-Negative Bacteria

**Maryam Faiyaz[1], Shikha Gupta[2] and Divya Gupta[3]***

[1]*Department of Bioengineering, Integral University, Lucknow, Dasauli, (India)*
[2]*Gujarat State Biotechnology Mission, Gandhinagar, (India)*
[3]*Department of Life sciences, Uttarakhand Technical University, Dehradun, (India)*

***Abstract***
Already available drugs in the market are unable to keep pace with the rate of production of novel bacterial strains in the environment. Antibiotics, if not used appropriately, give rise to resistant microorganisms. This has become an issue of great concern to scientists as well as medical practitioners, even more so if the bacteria is gram-negative. Efforts from biotechnologists and pharmacist in collaboration with scientists are being made to get large-scale production of antibiotics so that people can benefit from it when it is commercialized. The present study focuses on the steps involved in the fermentation of antibiotics acting against gram-negative bacteria. It also deals with the components and factors involved in the scaling up of antibiotics and the importance of quality control in the whole process. For antibiotic production, solid state fermentation (SSF) has been prioritized over submerged fermentation (SmF) which is discussed in this chapter. Techniques such as protoplast fusion, mutagenesis, recombinant DNA technology, etc., are widely in use to improve strain for better antibiotic yield. Selections of carbon and nitrogen source, activators, minerals, etc., along with the supply of oxygen are a few imperative steps while going for antibiotic fermentation.

***Keywords:*** Submerged fermentation, solid state fermentation, mutagenesis, minerals

## 4.1 Introduction

Gram-negative bacteria, as described by Christian Gram, is any microorganism that appears pink in color after following the staining steps. It differs from Gram-positive microorganism on the basis of its cell wall structure. Gram-negative bacteria are characterized by the presence of protein pores called prions, which have the ability to expel the antibiotics from the bacteria [1]. It has a thin cell wall covering the outer cell membrane [2, 3] which aids in blocking the entry of antibiotics into the microorganism [4].

Due to the peculiar cell wall structure and functioning of the Gram-negative bacteria, they have developed antibiotic resistance and susceptibility to a wide range of antibiotics [2, 4, 5]. The resistance is increasing at an alarming rate and world health leaders

**Corresponding author*: divyagupta.mu@gmail.com

Saurabh Saran, Vikash Babu, and Asha Chaubey (eds.) High Value Fermentation Products, Volume 1, (79–102) © 2019 Scrivener Publishing LLC

have described such antibiotic-resistant organisms as "nightmare bacteria" resulting in a disastrous threat. Gram-negative bacteria have the property of sharing their genetic material resulting in the spread of resistance more quickly and broadly [5].

Moreover, different Gram-negative bacteria offer different mechanistic complexity. For example, in Australia [6] and Pacific [7], *Acinetobacter baumannii* has offered resistance to carbapenem. The emergence of multidrug resistant Gram-negative organisms (MDRGNs) is a subject of major concern to hospitalized patients, as mortality rates have gone from 30% to 70% [8–14]. This condition of MDRGNs may arise due to the ample and improper use of antibiotics belonging to a broad-spectrum. For example, multi-resistant *Klebsiella pneumoniae*, *Acinetobacter baumanni*, and *Pseudomonas aeruginosa* resist the effect of carbapenems which is considered as the most influential and authentic class of drugs [15]. This is creating a problem in coping with the demands for production of new antibiotics against these lethal organisms. The following three approaches can be followed to produce antibiotics at large scale:

i. Natural microbial production using fermentation technology
ii. Semi-synthetic production (post-production modification of natural antibiotics)
iii. Synthetic production of antibiotics

Penicillin, which is now known as Penicillin G, is one of the most eminent antibiotics and the first to be produced on a large scale that is widely recommended even to date [16]. In 1993, its annual production was estimated to be about 33 million pounds, having a market value of around US$344 million [17]. *Penicillium chrysogenum*, a species of mould, has been found more promising in terms of antibiotic production when compared to the traditional *Penicillin notatum* isolated by Fleming. Its antibiotic producing efficiency has been found to be 1,000 times more in contrast to the Fleming's original culture of *P. notatum*. Large-scale production of mold has been further ignited with the introduction of submerged culture techniques [17–19]. Although there are plentiful antibiotics present in the market, there is still the demand for the production of antibiotics, which are effective against bacteria which are known to cause infections and diseases, specifically gram-negative bacterial class.

## 4.2 Gram-Negative Bacteria and Antibiotics

Owing to the health threat imposed by the gram-negative bacteria, there arises an urgent need for the development of novel antimicrobial compounds (Table 4.1). Gram-negative bacteria have adapted various unusual antibiotic biosynthesis pathways leading to the exceptional behavior of this class of microorganism [20]. Development of resistance in the bacteria has become one of the three major problems for human health, according to WHO (World Health Organization) [21]. *Klebsiella pneumoniae, Acinetobacter baumannii, Pseudomonas aeruginosa* and *Enterobacter* species are listed as some of the most frequent MDR micro-organisms. This phenomenon has been described as "ESKAPE" by many researchers [22]. The use of broad-spectrum antibiotics for treatment of infections caused by gram-negative bacteria

**Table 4.1** List of gram-negative bacteria and infections caused by them.

| S. No. | Gram-negative microorganism | Associated infections | References |
|---|---|---|---|
| 1. | *Escherichia coli* | Urinary tract infection (s), enterocolitis | [23] |
| 2. | *Acetobacter baumannii* | Respiratory, gastrointestinal and genitourinary tracts; bloodstream infections (BSI) and pneumonia | [24–26] |
| 3. | *Pseudomonas aeruginosa* | Malignant external otitis, endophthalmitis, endocarditis, meningitis, pneumonia, and septicemia | [27] |
| 4. | *Klebsiella pneumoniae* | Pneumonia, bloodstream infections, wound or surgical site infections, and meningitis | [28] |
| 5. | *Neisseria gonorrhoeae* | Urogenital tract infection, anorectal, conjunctival, pharyngeal, and ovarian/uterine | [29] |

has been effective up to an extent. Recently developed cephalosporins, carbapenems and β-lactamase inhibitors are found to have the ability to act against Gram-negative bacteria [21].

Based on data collected by the National Healthcare Safety Network (NHSN) in 2008, *E. coli* (13%), *Klebsiella* (13%), *P. aeruginosa* (17%), *A. Baumannii* (74%) are found multidrug resistant in intensive-care units (ICU) [30]. Statistical analysis done on infections caused by gram-negative bacteria as of 2002 predicts about 99,000 deaths among 1.7 million, hospital- acquired infections i.e., 4.5 per 100 admissions [31]. With this number, hospital-acquired death stands at $6^{th}$ position in the U.S. [32]. A similar analysis was traced in Europe also [33]. Some broad-spectrum antibiotic has been found promising to treat infections caused by gram-negative bacteria. These include cephalosporins (ceftriaxone-cefotaxime, ceftazidime, and others), fluoroquinolones (ciprofloxacin, levofloxacin), aminoglycosides (gentamicin, amikacin), imipenem, broad-spectrum penicillin with or without β-lactamase inhibitors (amoxicillin-clavulanic acid, piperacillin-tazobactam), and trimethoprim-sulfamethoxazole [34, 35].

Classification of antibiotics acting against gram-negative bacteria is as follows:

### 4.2.1 β-Lactam Drugs

Belonging to the class of broad-spectrum antibiotics, the β-lactam antibiotics consist of beta lactam ring in their molecular structure [36]. They act against the cell wall synthesis in the target bacteria, thereby showing bacteriostatic effect. This is achieved by inhibiting the key enzymes such as transpeptidases, carboxypeptidases, which are required in the biosynthesis of peptidoglycan at the terminal stage [37]. Most of the microbes produce beta-lactamase enzymes, which are species specific and hydrolyze the ring structure of antibiotics, making the drug inactive and leading to the emergence of antibiotic resistance [38, 39]. β-lactam drugs is a branch of broad spectrum antibiotics comprised of antibiotics having beta-lactam ring in their molecular structure. Classification of β-lactam drugs are given below:

#### *4.2.1.1 Cephalosporins*

An example of class cephems, cephalosporins are broad as well as extended-spectrum antibiotics depending on the generation ($1^{st}$–$5^{th}$) containing a sulfur atom in an unsaturated ring of six member (dihydrothiazine ring) [40, 41]. Depending on their activity and mode of action, they are divided into five generations (Table 4.2).

#### *4.2.1.2 Penicillin*

This comes under the class of penems and is a narrow spectrum antibiotic consisting of Penicillin G and Penicillin V having clinical significance [40]. Derived from a fungi *Penicillium*, it may be natural or synthetic in nature. Chemical structure contains a beta-lactam ring with a side chain attached to a thiazolidine ring [45]. A broad classification of penicillin is given in Table 4.3.

#### *4.2.1.3 Carbapenems*

It is one of the most potent wide spectrum antibiotics which are bactericidal in nature [50]. Carbapenems are characterized by the presence of five membered ring structures with a double bond. [40]. Cephalosporin resistant gram-negative bacteria can be treated by this class of beta lactam drug. It includes imipenem, meropenem, doripenem and ertapenem which show activity against both gram-positive and gram-negative bacteria [51].

#### *4.2.1.4 Monobactam*

Just a single beta-lactam core separates it from other drugs of class beta-lactam on the basis of structure [52]. Isolated from *Chromobacterium violaceum* and now prepared as a synthetic drug, Aztreonam is the only available example of this class. It is known as a potent and specific inhibitor of wide-spectrum gram-negative bacteria [53].

### 4.2.2 Macrolide

It contains 14, 15 or 16 membered lactone ring structure which is substituted by two sugar moieties, one of which has aminated functional group. They are known to possess broad spectrum of antimicrobial activity and are used as an alternative to penicillin [54]. For example, Azithromycin is a 15-membered ring structure which has an additional amino group present in the lactone ring, giving rise to a new subclass of this group named "azalides". The first clinically developed macrolide, which is natural in origin is Erythromycin. The rest, more recent molecules, are semi-synthetic derivatives in nature which are acid-stable [40]. Macrolide derivatives such as roxithromycin, clarithromycin or azithromycin are found more or less similar to erythromycin and have shown antibacterial activities. Gram-negative bacteria are found to be affected by the use of these antibiotics [55].

**Table 4.2** Five generations of cephalosporins.

| S. No. | Generation | Examples | Treatment | Activity | References |
|---|---|---|---|---|---|
| 1. | First | cephalexin, cephazolin, cefadroxil | Skin infections | Gram- positive | [42] |
| 2. | Second | Cefoxitin, cefotetan, cefprozil, cefuroxime, cefaclor | Respiratory tract infections | Gram- negative (*Moraxella, Neisseria, Salmonella*, and *Shigella)* | [42] |
| 3. | Third (broad-spectrum) | cefdinir, cefditoren, cefixime, cefotaxime, cefpodoxime, ceftibuten, and ceftriaxone. | Community-acquired respiratory tract infections, resistant infections, and nosocomial infections | *Neisseria* species, *M. catarrhalis, Klebsiella P. aeruginosa* | [42] |
| 4. | Fourth | Cefepime | Intra-abdominal infections, respiratory tract infections, and skin infections | Both gram-positive and negative bacteria, including *P. aeruginosa* and many *Enterobacteriaceae.* | [42] |
| 5. | Fifth | Ceftaroline fosamil | Community-acquired pneumonia infections, skin infections | *E. coli, H.influenzae, Klebsiella, S. aureus* (methicillin-susceptible isolates only), *S. pneumoniae* | [43, 44] |

**Table 4.3** Classification of Penicillin.

| S. No. | Class | Examples | Infections cured | References |
|---|---|---|---|---|
| 1. | Natural penicillins | Penicillin G<br>Penicillin V | Mild localized infections | [45] |
| 2. | Penicillinase-resistant penicillins | Methicillin<br>Nafcillin<br>Dicloxacillin | Hemolytic streptococcal pharyngitis, pneumonia | [46, 47] |
| 3. | Aminopenicillins | Amoxicillin<br>Ampicillin | Prophylaxis against bacterial endocarditis, prophylaxis prior to gastrointestinal and genitourinary procedures | [48] |
| 4. | Extended spectrum penicillin | | Infections caused by *P. aeruginosa* | [49] |

### 4.2.3 Aminoglycosides

One of the highly important and broad-spectrum drugs for treating life absorbing gram-negative bacillary infections is aminoglycosides. Kanamycin, Gentamycin, Tobramycin are a few examples of antibiotics belonging to this class of drug [56]. Amikacin, Dibekacin, Netilmicin comes under semi-synthetic group of aminoglycosides that were produced in concern with the development of resistance in few bacterial strains to the already existing aminoglycoside drug [57]. Aminoglycosides have dibasic cyclitol with one or a few sugars having an amine group connected by a glycosidic bond. Clinically important aminoglycosides has 2-deoxystreptamine whereas streptomycin and fortimicin derivatives have streptidine and fortamine, respectively [58].

Aminoglycosides are highly polar in nature. Uptake of these molecules by the membrane of gram-negative bacteria is automatic. The basic process of inhibiting the microbial action is by getting attached to prokaryotic ribosomes and hence interfering with the synthesis of proteins [59–61].

### 4.2.4 Fluoroquinolones

A quite novel category of synthetic antibiotic showing efficient bactericidal effect is the broad-spectrum quinolones or fluoroquinolones. It has found potential to treat various clinically important pathogens like urinary tract infections (UTI), gastro-intestinal infections [62], respiratory tract infections (RTI), sexually-transmitted disease (STD) and skin infections [63, 64].

The first patented fluoroquinolone is the Flumequine, which was later on followed by norfloxacin, pefloxacin, enoxacin, fleroxacin, ciprofloxacin (most successful and widely used) and ofloxacin. These are all broad-spectrum antibiotic [65]. Regarding their structure, Fluoroquinolones are synthetically derived analogues of nalidixic acid, a 1, 8-naphthyridine with a 4-quinolone nucleus [66]. This class of antibiotics acts by obstructing the bacterial DNA replication and transcription, ultimately, leaving the cell to die [67].

## 4.3 Production of Antibiotics

Production of antibiotics encompasses some basic steps, as described below:

i. Inoculum preparation
ii. Media formulation and optimization
iii. Fermentation
iv. Downstream processing and Purification
v. Quality control

### 4.3.1 Strain Development

Strain development is an important step and finds its application in industries during the production of antimicrobials, enzymes, drugs, beverages, etc. With the main objective of achieving higher microbial yield, inoculum should be in an optimal state and mimic with the native environmental conditions at an appropriate concentration [68, 69]. Several bacteria have the ability to produce the same antibiotic; however, the antibiotic producer strain is chosen on the basis of their ability to produce the selected antibiotic efficiently. The selected strain can be genetically improved using various biotechnological techniques such as recombinant DNA technology, mutagenesis, protoplast fusion, gene transfer, modification of gene expression etc. [70]. Cloning of particular identified genes in the strain is made able to produce the product beyond its natural ability [71, 72]. A few industries opt for strain development program before going to main-scale production to ensure the best product formation [73]. Figure 4.1 represents the key steps involved in strain development program.

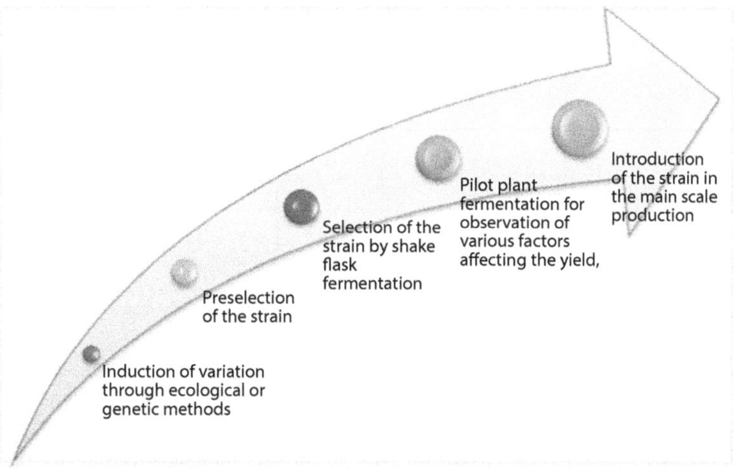


**Figure 4.1** Strain development program.

#### 4.3.1.1 Mutagenesis

Mutagens at low dosage are used such that it favors 10% survival of the strain. It provides a way to control the adverse effects of mutagens [74]. Bio techniques such as nitroso guanidine-induced mutation, induction of mutation during the period of active transcription, or by mutagenesis of cloned genes in-vitro are now available and can be used for inducing mutations directly at the gene of interest [75]. Following is the list of mutagens used in strain improvement (Table 4.4):

#### 4.3.1.2 Protoplast Fusion

Cells having no membrane when fused with the help of chemical fusogens, such as polyethylene glycol, sodium nitrate polyvinyl alcohol etc., are called protoplasts. This technique has opened the gateway for fusion of inter- and intra-specific protoplasts with a high level of recombination frequency (Figure 4.2). This bio technique is helpful to exploit the parental genomes of various cells and combine them into one cell, referred to as heterokaryon, as fusion products are genetically different [71, 85–87]. This method

**Table 4.4** List of mutagens used in strain improvement.

| S No | Bacterial strain | Mutagen | Augmented antibiotic production | Refs |
|---|---|---|---|---|
| 1. | *Penicillium chrysogenum* | Nitrogen mustard, UV radiation, X-radiation | Penicillin | [76] |
| 2. | *Cephalosporium acremonium* | UV radiation, Nitrosoguanidine solution | Cephalosporin C | [77] |
| 3. | *Streptomyces clavuligerus* | UV radiation | Clavulanic acid | [78] |
| 4. | *Streptomyces clavuligerus* | UV radiation | Clavulanic acid | [79] |
| 5. | *Escherichia coli* | UV radiation | Penicillin G acylase (used in the synthesis of beta lactam antibiotic) | [80] |
| 6. | *Saccharopolyspora erythraea* | Transposon mutagenesis | Erythromycin | [81] |
| 7. | *Saccharopolyspora erythraea* | Ethylmethane sulfonate | Erythromycin | [82] |
| 8. | *Streptomyces lividans* | Site directed mutagenesis through rpsL (ribosomal protein S12) mutation | Streptomycin | [83] |
| 9. | *Acremonium chrysogenum* | UV radiation | Cephalosporin | [84] |

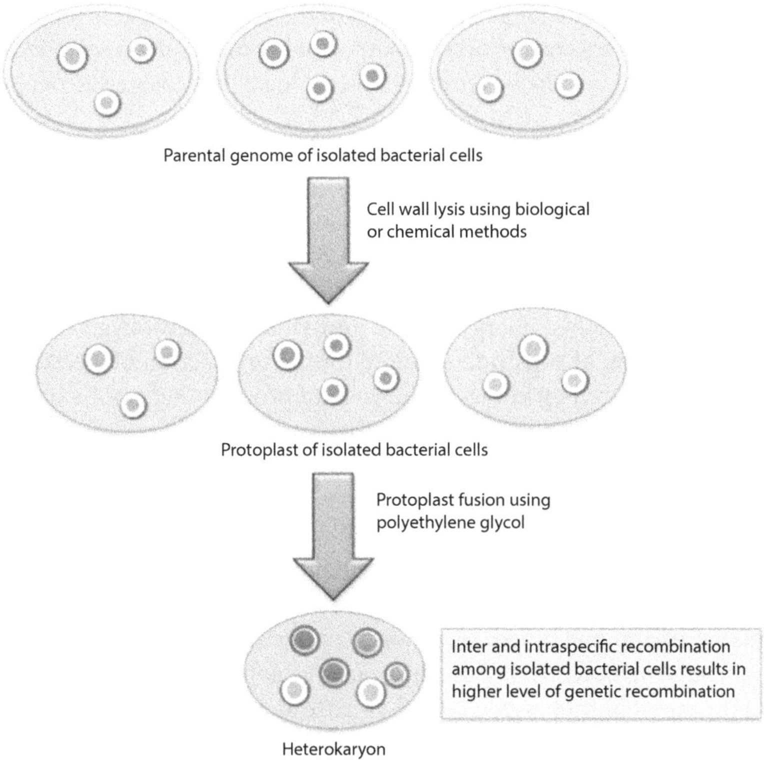


**Figure 4.2** Strain development using protoplast fusion.

of strain development has already been employed by Wesseling and Lago (1981), getting two cell cultures from divergent mutations lines of *Nocardia lactamdurans*, which is a cephamycin C producing microbe. Following protoplast fusion, increment in 10–15% of cephamycin C titer levels have been observed with respect to that of parents in two cell cultures [88]. A similar study was also conducted by Kitano *et al.* with *Streptomyces griseus subsp.cryophilus*, a carbapenem antibiotic producing microorganism revealing a two to four times increase in the production rate after strain development using protoplast fusion method [89]. Improved antibiotic titers were observed with the fusion of different strains of *Cephalosporium acremonium* [90].

### *4.3.1.3 Recombinant DNA*

Production of antibiotics is a highly convoluted process involving regulatory interactions between precursor feedback and various cellular metabolisms. This complexity gets exaggerated because of the scarcity of knowledge about the biosynthesis of antibiotics [91]. Antibiotic biosynthesis involves multistep pathways initiating from precursors (which are often intermediates of primary metabolic cycles) to the component units which attach to form antibiotics. Enzymes specific for each antibiotic participate at each step [92]. It was revealed in *Streptomyces* that actinomycetes contain genes either in the cluster form on a chromosome or on plasmids in some cases, which codes for particular enzymes involved in the production of antibiotic [93–96]. Cloning of genes through recombinant DNA technology will highly depend on the position of the gene responsible for biosynthesis of antibiotic [72]. Taking this factor into account, cloning in *Streptomyces lividans* 66 and S. *coelicolor* A3[2] has been done by creating low and high copy phage vectors [97].

### 4.3.2 Media Formulation and Optimization

Success of the experiment is largely dependent on the media formulation. Media is prepared according to the elemental demands of the microorganism required for production of the desired metabolite. The basic need of microbial growth is water, carbon, nitrogen, mineral elements, energy source and vitamins and oxygen if the culture is aerobic. Concentration of substrate required for the production of a certain amount of product can be calculated by knowing the yield of product that can be obtained from a specific amount of biomass [18]. However, theoretical yields are different as compared to that of experimental yields. This was proved by Cooney in 1979. Based on the calculations from the reaction stoichiometry of substrate and product formation using a biosynthetic pathway, theoretical yield of Penicillin G per gram glucose was 1.1 g. While experimental analysis using fed-batch fermentation was found to be 0.053 g per gram glucose, which is quite less as compared to that of theoretical yields [98].

Mohamed Farid and Ahmed Al Diwany optimized the media for cultivation of *Streptomyces natalensis*, which is a producer of antibiotic Natamycin. For better yield of natamycin production, they used 20.0 g/l glucose as a carbon source, 8 g/l beef extract and 2 g/l of yeast extract. With this optimized media, they got maximum concentration of natamycin production being 1.5 g/l [99]. The main components of media formation are described below:

- **Carbon source:** Production of primary or secondary metabolite is dependent on the rate of metabolizing the carbon source by microbe. Fast growth and hence less antibiotic production are associated with rapidly metabolizing carbon source (Table 4.5). So, it requires providing a carbon source (example, lactose) which is slowly degraded by the microbe in order to get a fair concentration of secondary product, antibiotic [100].

  Choice of substrate is often influenced by the purity of sources of carbon being used in the fermentation process [106–108].
- **Nitrogen Source:** Both organic, as well as inorganic nitrogen sources, are preferred by most microorganisms employed for industrial production of antibiotic. Ammonium gas, ammonium salts or nitrates can be given as a source of inorganic nitrogen [109], whereas amino acid, protein or urea can be added as an organic nitrogen source [18]. According to Aharonowitz, type and concentration of the nitrogen source in the culture medium can affect the antibiotic production by many microorganisms [110]. Fast-depleting nitrogen sources might inhibit antibiotic production. Therefore, complex nitrogen sources are preferred while producing antibiotics on a large scale because they aid in promoting physiological conditions in log phase, which support the formation of antibiotic in idiophase [111].

**Table 4.5** Negative impact of glucose as a carbon source on the yield of various antibiotics.

| S. No. | Microbe | Activity | Antibiotics produced | References |
|---|---|---|---|---|
| 1. | *Penicillium chrysogenum* | Gram-positive | Penicillin | [101] |
| 2. | *Cephalosporium acremonium* | Broad spectrum | Cephalosporin | [102] |
| 3. | *Streptomyces fradiae* | Gram-negative | Neomycin | [103] |
| 4. | *Streptomyces griseus* | Gram-negative | Streptomycin | [104] |
| 5. | *Streptomyces rimosus* | Broad spectrum | Tetracycline | [105] |

Moyer and Coghill considered corn-steep liquor as the best nitrogen source for penicillin production [112]. Similarly, Inskeep *et al.* considered peanut granules as the best nitrogen source for Bacitracin production [113]. Distillers' solubles, pharmamedia soybean meal with ammonium sulfate and soybean meal, etc., have been considered as the main nitrogen sources for the antibiotic production such as Novobiocin, Rifomycin, Polyenes respectively [111, 114, 115]. Kamel and Al-Zahrani evaluated several culture conditions to get the maximum yield of SA-53 antibiotic produced from a strain of *Streptomyces anandii* taken from soil sample of Saudi Arabia [116]. They optimized culture media and found that the strain produced best when supplemented with potassium nitrate along with glucose (20 g/l) as carbon source and ammonium chloride or soybean (280 mg/l) as nitrogen source.

- **Minerals:** Some minerals play an important role in growth and metabolism of all microorganisms [117, 118]. Weinberg (1970), have revealed the regulation of secondary metabolism by the presence of manganese, iron and zinc trace elements [119]. Some other reports have also shown earlier that the concentration of phosphate (obtained from beef extract) affects the target enzymes required for the synthesis of antibiotic. Apart from phosphorous several other metal ions obtained from beef extract such as zinc, iron and magnesium have been shown to have two activities. They either work as activators for enzymes important for secondary metabolite biosynthesis or they can work as binding agents for enzymes which are inhibited by the effect of high phosphate concentration [111, 120, 121].
- **Precursors**: Certain chemical compounds which are added before or simultaneously with the fermentation product, without affecting the property of the molecule are called precursors. They have proven in enhancing the antibiotic yields. This was quoted taking the oldest penicillin production as an example by Moyer and Coghill [112, 122]. The yield of penicillin was found to be increased by adding corn-steep liquor from 20 unit's $cm^{-3}$ to 100 units $cm^{-3}$. This was because of the presence of

phenylethylamine, which worked as a side chain and on attaching to the penicillin molecule yielded benzylpenicillin (Penicillin G). Later on, it was confirmed that side chain work as a limiting factor while considering the activity of penicillin. Moreover, addition of phenylacetic acid as a side chain is being considered as a customary practice [18]. Experiments carried out by Smith and Bide reveal a threefold increase in yield of penicillin as well as enhanced production of benzylpenicillin as a result of directed biosynthesis from penicillin from 0 to 93% as compared with other penicillins [123].

Phenylacetic acid-related compounds and Phenoxy acetic acid have been used as a precursor for the production of Penicillin G and Penicillin V respectively from *P. chrysogenum* [122, 124]

- **Inhibitors:** Inhibitors in the fermentation have a unique role to play. Upon addition, they will either generate a specific product or accumulate the metabolic intermediate, which is usually metabolized during the process.
- **Oxygen:** Metabolite production and growth rate is also influenced by the availability of oxygen in the media during fermentation, especially, if breakdown or consumption of substrate is fast, as in the case of *P. chrysogenum* fungus which is a producer of β-lactam antibiotic penicillin. In the event of this fungus, glucose is utilized more rapidly as compared to sucrose or lactose. When glucose is used as a main carbon source, the specific oxygen uptake rate of the organism is higher [125], thereby requiring more supply of oxygen from some external source.
- **Activators:** Hyun Yong Shing *et al.* used glycerol as a carbon source for the production of cephalosporin. Crude glycerol is one of the byproducts of biodiesel production, which has become an attractive substrate for industrial antibiotic production as the production of biodiesel has significantly increased in the past few years. Due to this reason the glycerol became cheaper as compared to other sources such as glucose, plant oils etc. [126, 127]. Depending on statistical analysis of β-lactam antibiotic production by filamentous fungi, nutritional factors such as novel carbon sources have been optimized in this regard [128, 129].

Hyun Yong Shing *et al.* (2010) experimented with *A. chrysogenum* M35, a UV-Induced mutant strain of *A. chrysogenum* which showed increased production of cephalosporin C in the basal seed medium. They optimized media with the addition of 2–6% (v/v) glycerol and then incubated the culture at 27 °C for 5 days at 300 rpm. With this, cephalosporin production increased by 12 times because of up regulation of the transcription of the gene for isopenicillin synthase (pcbC) and transporter (cefT) in early exponential phase as an effect of glycerol addition to the fermentation media [130].

### 4.3.3 Fermentation

The prime objective of antibiotic production is to limit production cost through the advancement of fermentation and downstream processes [131]. Mechanically agitated and aerated fermenters are used for the production of most of the antibiotics [92]. Temperature, pH, dissolved oxygen, carbon source, precursor, nitrogen source, aeration and other components require to be optimized for better antibiotic production [132]. The need for development of product-specific fermenter led to the construction of the first large-scale plant made of steel with 54,000 $dm^3$ of volume for penicillin production using deep fermentation at Terre Haute in the United States on 15th September 1943 [133]. The upgrade and optimized fermentation technologies with improved strain productivity lead to augmented antibiotic production ultimately reducing antibiotics cost in the market. The harvest titre of penicillin and cephalosporin from major industrial units were recorded to be 40–50 g/l and 20–25 g/l respectively, even in 2003 [134]. The common flow diagram of the major antibiotic fermentation process is shown below in Figure 4.3:

Keeping in mind the availability of oxygen during fermentation, several modifications were made. One of them was the incorporation of modified draught tubes to analyze the transfer rates of oxygen. A solid draught tube was fitted in the internal cooling coil of helical shape in a bubble- column fermenter of 20 $m^3$, initially set on pilot-scale. It was utilized by Carrington *et al.* in their studies for the production of *Streptomyces*

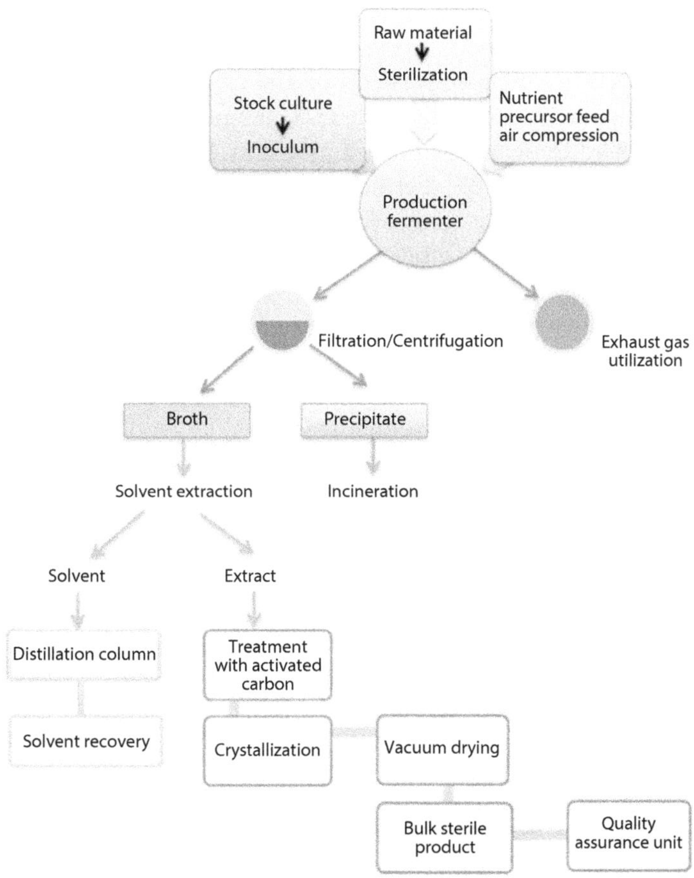


**Figure 4.3** Flow diagram for major antibiotic fermentation process.

antibiotic at a commercial level. Fermentation is done in a complex medium, resulting in the production of a viscous, non-Newtonian broth [135].

Srivastava *et al.* [136] worked on Cephalosporin C, a β-lactam antibiotic working as a starting material for the industrial production of semi-synthetic cephalosporins. Air-lift reactor made up of borosilicate glass for CPC fermentation from *Acremonium chrysogenum* in batch as well as fed-batch modes was utilized by Pradeep *et al.* in their studies. They have developed a relationship between the production of antibiotic and the rate at which feed was supplemented. Significantly higher yield was obtained after using fed-batch mode with a flow rate of feed being low. Increased production is a result of hydrolyzed sucrose, which was added slowly into the feed and caused the reduced catabolic repression. This strategy turns out to be more promising when compared to batch mode in airlift fermenter [136].

Solid state fermentation (SSF) is usually employed for the production of antibiotics. Submerged fermentation (SmF) was traditionally used for the tetracycline production [137]. But SSF has many advantages over SmF, which is described in detail in the previous art [138–141]. SSF plays a crucial role in antibiotic production and is given priority in the processes where crude fermented product is an antibiotic source [142, 143].

Vastrad and Neelagund focused on the factors affecting the fermentation of various strains of *Streptomyces* to get maximum production of tetracycline. For this, they optimized initial moisture content, incubation temperature, initial pH, particle size of substrate and the size of the inoculum used for the scale-up of antibiotic. After 3–7 days of fermentation cycle, they have revealed that the yield of tetracycline reached up to maximum level when 65% moisture level, 35 °C temperature and 5–6.5 pH, substrate particle size measured to be 6 × 4 mm and inoculum size as $1.0 \times 10^8$ spores/ml, were provided during production [137]. Fermenter having multiple inputs for feed has been designed owing to the positive feedback in the antibiotic yield from the addition of several energy sources and many other compounds such as phosphorous, nitrogen etc. [144].

The fermentation process of cephalosporin and penicillin are very much analogous to each other with a key difference involving the microorganism involved, media composition and some purification steps. High-yielding strains of *A. chrysogenum* are used worldwide to produce cephalosporin on a mass level by employing aerated fed-batch reactors maintaining pH levels at 6.2–7.0 and temperature range 24–28 °C. Glucose or sucrose is used as the carbohydrate source and generally added during the growth phase of fermentation, which in turn is replaced by soybean oil or peanut oil in later stages of fermentation. DL-methionine is usually supplied during the early growth phase of fermentation. Soybean in conjunction with cotton seeds, ammonium sulfate and ammonia is employed as good nitrogen source. Corn steep liquor can also be used as a nitrogen source [134]. The inclusion of glycerol during the fermentation process results in the enhanced cephalosporin production by 12 folds [130].The general production process for cephalosporin is shown in Figure 4.4.

### 4.3.4 Downstream Processing and Purification

Every industry is in a race to obtain high-quality product with low cost, per capita investment and in less time with a high degree of purity. But variation in the cost of

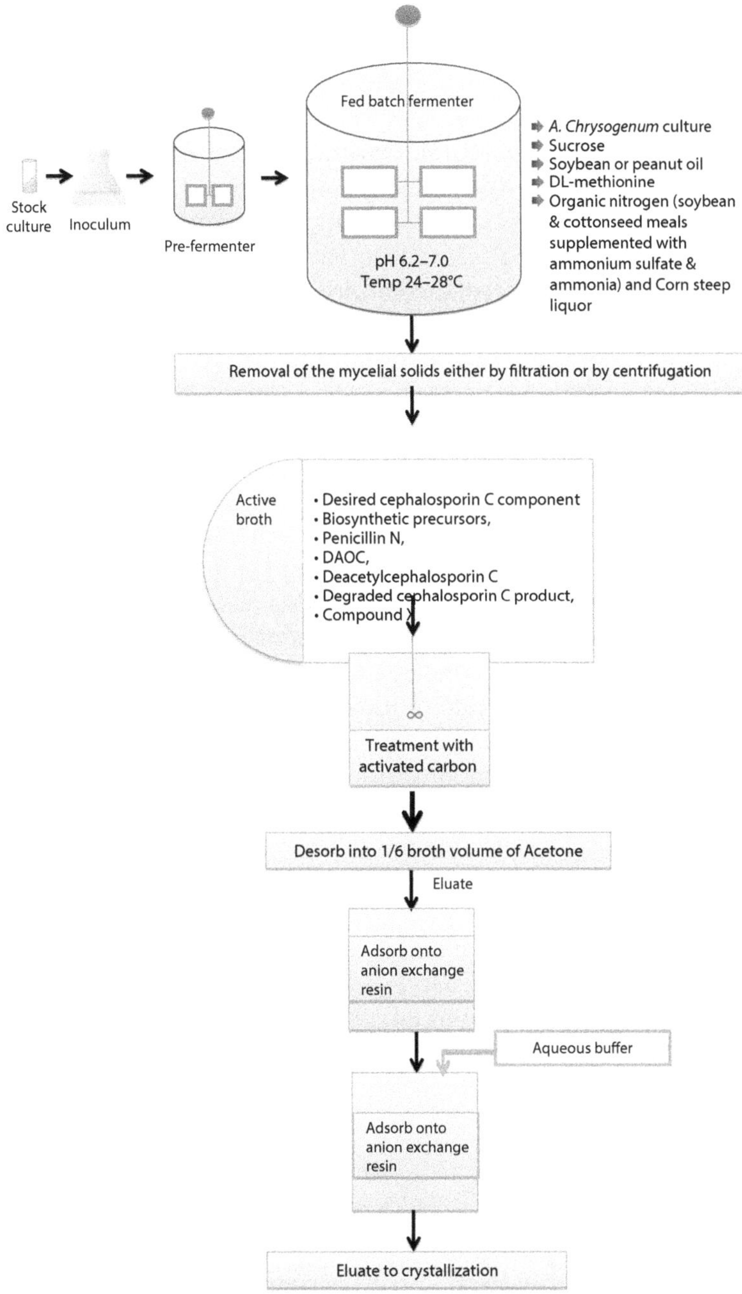


**Figure 4.4** General Production process of Cephalosporin antibiotic.

products to the total manufacturing costs, obtained from microbe has been noticed with the lowest being 15% of maximum of 70% [145–149]. The relative cost of purification depends on the process we choose and obviously on the product of interest. According to Atkinson and Mavituna [150], the purification cost of industrial ethanol is 15% of the total cost while it is 20–30% for bulk Penicillin G, and for enzymes it accounts as much as 70% (because of its high purity demand) [150]. In a few cases, the overall objective of fermentation will be influenced by the cost of downstream processing, for example, high quality and high purity products [18].

Chemical stability and reactivity of the antibiotics and the compounds present in the fermentation broth control the procurement of antibiotic upon downstream processing [92]. Separation and purification of antibiotics pose a great challenge since the concentration to be purified is very less or the antibiotic is present in very dilute form [151]. For complete purification, biological products have to undergo a series of steps that will exclude the impurities at every step, thereby bringing the product approximately near to the final blueprint laid [152].

Different antibiotics are separated by utilizing different means of separation. Not only this but for the purification of single antibiotic various different techniques have been utilized. To identify easy, economical and rapid separation techniques which give larger purification of the product has become a hot research topic nowadays. Chromatography (both HPLC and annular chromatography), supercritical extraction and use of various membranes are new separation techniques being researched as they are believed to reduce the complexity of the process [153]. Modern biotechnology has extended a helping hand in searching and optimizing techniques relevant to this field.

Mathew and Juang have characterized three different categories based on the requirement of purity (Table 4.6) [154]. Amongst the various separation processes, solvent extraction has gained popularity due to its feasibility in scale-up, relatively cheaper, high efficiency and selectivity. It can also be operated in a continuous manner [155]. For the industrial antibiotic purification purpose, liquid-liquid extraction is considered where n-butyl acetate works as a solvent [156]. Alcohol (n-butanol), methyl isobutyl ketone, and pentyl or butyl acetate are a few other examples of popular solvent used in antibiotic extraction. Extraction of antibiotic is influenced by the pH and various other factors.

One of the earliest methods of antibiotic extraction is adsorption, charcoal being used as an adsorbent traditionally. Later on activated carbon, ion-exchange resins, alumina and silica gels were also used as an adsorbent [157]. However, Hu and Gulari found the drawback of using adsorbent in product recovery is that it needs an additional unit of operation for pre-concentration. Also, purification of product takes a

**Table 4.6** Separation techniques used in antibiotic purification from fermentation broth [154].

| S. No. | Range of production | Range of purity | Separation technique (s) used |
|---|---|---|---|
| 1. | High | Low | Liquid-liquid extraction, cell disruption, ultrafiltration |
| 2. | Low | High | affinity separation, ultracentrifugation, electrophoresis |
| 3. | High | High | membrane chromatography, fluidized bed chromatography, monolith column chromatography |

long time and gives a low activity of antibiotic after recovery [158]. Fermentation of industrially important antibiotics of natural origin is aerobic in nature. It was in the 1950s that Bartels *et al.* used carboxylic cation exchange resin for the recovery of streptomycin [159]. Later on, a more simple method of achieving antibiotic purification by ion exchange method was put forward by Cha [160].

### 4.3.5 Quality Control

Quality control being the final step in production of antibiotic is of pragmatic importance. Antibiotics before their implementation undergo a series of clinical tests. Proper drug analysis according to WHO guidelines is done to ensure the quality and safety of the drug. Development of readily available poor-quality drugs in developing countries is one of the main reasons leading to increased rates of mortality and morbidity. Availability of prevalent substandard drugs is one of the dominant factors resulting in the resistance development in antibiotics [161].

Rivera *et al.* in the article "Importance of quality control for the detection of β-lactam antibiotic resistance in Enterobacteriaceae" takes into account that it is essential to interpret susceptibility results of treatment with antibiotics like β-lactams until there is some data available on its clinical efficacy. However, the constant changes in the criteria of interpretation and methodological difficulties let the scientists arrive at a stage where training and external quality control becomes essential [162].

Quality control becomes an important step in antibiotic production due to the liberal and inappropriate use of broad-spectrum antibiotics, which assist in the development of multidrug resistant gram-negative bacteria (MDRGN) [150]. In order to get rid of MDRGN infections, these are supported with additional broad-spectrum antibiotics. This foul cycle created, ends in giving rise to more resistance [15, 163]. The birth of MDRGN organism is a major concern as we have to combat it with the limited number of microbial agents present in the market or in the process of development [164].

## 4.4 Conclusion

The boom in infections and threats caused by Gram-negative bacteria, for example, ESBL or carbapenemase-producing *Enterobacteriacea*, *Acinetobacter baumannii* and *Pseudomonas aeruginosa* is posing a serious problem in hospitals. A larger number of the population is getting affected by these terrifying and life-worsening bacteria. Antibiotic production for these ESKAPE pathogens is in-pipe or in the later stages of drug development. Also, modifications in the already existing antibiotics are done in order to improve its quality and efficacy. Much of the work is being done on the new and potentially active inhibitors of Gram-negative antibiotics such as β-lactamase, carbapenems and cephalosporins.

Selection of carbon and nitrogen source to be used in media for better production has to be made carefully as the choice of substrate varies according to the strains used for large-scale production. Minerals also play an important role in the life cycle of bacteria and hence affect its metabolite production. Environmental conditions such as pH, temperature, salt, oxygen demand by microbe, etc., is optimized. Mode of fermentation, whether batch or fed-batch highly affect the antibiotic yield. For example, in case of production of Cephalosporin C, fed-batch fermentation was chosen over batch mode of fermentation as they gave higher yields of antibiotic.

## References


1. S. Murakami, R. Nakashima, E. Yamashita, T. Matsumoto and A. Yamaguchi, *Nature*, Vol. 443, p. 173, 2006.
2. H.C. Neu and T.D. Gootz, "Antimicrobial Chemotherapy", in S. Baron, eds., *Medical Microbiology,* 4th edition, Galveston (TX): University of Texas Medical Branch at Galveston, 1996.
3. J.L. Gould, W.T. Keeton, *Biological Science*, 6th edition, W.W. Norton & Company Inc, London, 1996.
4. Rang HP, *Pharmacology*, 5th edition, Edinburgh, New York: Churchill Livingstone, 2003.
5. Centers for Disease Control and Prevention (CDC), http://www.cdc.gov/drugresistance/pdf/ar-threats-2013-508.pdf, 2013.
6. E.G. Playford, J.C. Craig, J. Simpson, J.R. Iredell, *Journal of Hospital Infection*, Vol. 65(3), p. 204, 2007.
7. T. Naas, M. Levy, C. Hirschauer, H. Marchandin and P. Nordmann, *Journal of Clinical Microbiology*, Vol. 43(9), p. 4826, 2005.
8. S. Bratu, D. Landman, R. Haag, R. Recco, A. Eramo, M. Alam and J. Quale, *Archives of Internal Medicine*, Vol. 165(12), p. 1430, 2005.
9. C.S. Bryan, K.L. Reynolds and E.R. Brenner, *Reviews of Infectious Diseases*, Vol. 5(4), p.629, 1983.
10. H.F. Geerdes, D. Ziegler, H. Lode, M. Hund, A. Loehr, W. Fangmann and J. Wagner, *Clinical Infectious Diseases*, Vol.15(6), p. 991, 1992.
11. C. Gudiol, F. Tubau, L. Calatayud, C. Garcia-Vidal, M. Cisnal, I. Sanchez-Ortega, R. Duarte, M. Calvo, and J. Carratala, *Journal of Antimicrobial Chemotherapy*, Vol. 66(3), p. 657, 2010.
12. L. Leibovici, Z. Samra, H. Konigsberger, M. Drucker, S. Ashkenazi, S. and S.D. Pitlik, *The Journal of American Medical Association*, Vol. 274(10), p. 807, 1995.
13. E. Mouloudi, E., Protonotariou, A. Zagorianou, E. Iosifidis, A. Karapanagiotou, T. Giasnetsova, A. Tsioka, E. Roilides, D. Sofianou and N. Gritsi-Gerogianni, *Infection Control & Hospital Epidemiology*, Vol. 31(12), p. 1250, 2010.
14. T.M. Perl, L. Dvorak, T. Hwang and R.P. Wenzel, *The Journal of American Medical Association*, Vol. 274, p. 338, 1995.
15. L.B. Rice, *Current Opinion in Microbiology*, Vol. 12, p. 476, 2009.
16. D.Voet, and J.G. Voet, *Biochemistry*, 3rd edition, John Wiley, New York, 2004.
17. M.J. Pelczar, E.C.S.Chan, and N.R. Krieg, *Microbiology*, 6th edition, McGraw-Hill, New York, 1993.
18. P.F. Stanbury and A. Whitaker, *Principles of Fermentation Technology*, Pergamon Press, New York, 1987.
19. A.L. Demain and N.A. Solomon, *Scientific American,* Vol.245, p. 66, 1981.
20. J. Masschelein, M. Jenner and G.L. Challis, *Natural Product Reports*, Vol. 34, p. 712, 2017.
21. M. Bassetti, F. Ginocchio and M. Mikulska, *Critical Care*, Vol. 15, p. 215, 2011.
22. L.B. Rice, *J Infect Dis*, Vol. 197, p. 1079, 2008.
23. http://www.microbiology5.org/microbiology5/book/637–648.pdf
24. L.S. Munoz Price and R.A. Weinstein, *N. Engl. J. Med,* Vol. 358, p. 1271, 2008.
25. J.Gilad and Y. Carmeli, *Drugs*, Vol. 68, p. 165, 2008.

26. M. Paul, M. Weinberger, Y. Siegman-Igra, T. Lazarovitch, I. Ostfeld, I. Boldur, Z. Samra, H. Shula, Y. Carmeli, B. Rubinovitch, and S. Pitlik, *Journal of Hospital Infections*, Vol. 60, p. 256, 2005.
27. G.P. Bodey , R. Bolivar , V. Fainstein , and L. Jadeja , *Review of infectious diseases*, Vol. 5(2), p. 279, 1983.
28. Centers for Disease Control and Prevention (CDC), https://www.cdc.gov/hai/organisms/klebsiella/klebsiella.html,2012.
29. K.E. Miller, *American Family Physician*, Vol. 73(10), p. 1779, 2006.
30. Centers for Disease Control and Prevention (CDC), https://www.cdc.gov/hai/organisms/gram-negative-bacteria.html, 2011.
31. R.M. Klevens, J.R.Edwards, C.L. Richards Jr, T.C. Horan, R.P. Gaynes, D.A. Pollockand D.M. Cardo. *Public Health Rep.*, Vol. 122, p.160, 2007.
32. H.C. Kung, D.L. Hoyert, J. Xu, and S.L. Murphy, *Natl Vital Stat Rep.*, Vol. 56, p. 1, 2008.
33. I. Chopra, C. Schofield, M. Everett, A. O'Neill, K. Miller, M. Wilcox, J.M. Frère, M. Dawson, L. Czaplewski, U. Urleband P. Courvalin, *Lancet Infectious Diseases*, Vol. 8, p. 133, 2008.
34. R.P. Wenzel, D.F. Sahm, C. Thornsberry, D.C. Draghi, M.E. Jones, M.E. and J.A. Karlowsky, *Antimicrobial Agents and Chemotherapy*, Vol. 47(10), p. 3089, 2003.
35. NEJM Journal watchhttp: //www.jwatch.org /id200312120000007 /2003/12/12/ gram-negative-bacteria -and-broad-spectrum, 2003.
36. K.B. Holten, and E.M. Onusko, *American Family Physician*. Vol. 62 (3), p. 611, 2000.
37. R.Williamson, E. Collatz and L.Gutmann, *Presse Medicale*, Vol. 15 (46), p. 2282, 1986.
38. E.P. Abraham and E. Chain, *Nature*, Vol.146, p. 837, 1940.
39. R.B. Sykes, and M. Matthew, *Journal of Antimicrobial Chemother*, Vol. 2, p.115, 1976.
40. F. Van Bambeke, Y. Glupczynski and P.M. Tulkens, *Infectious Diseases*, Vol. *7*, p.128, 2010.
41. M.S. Masoud, A.E. Ali, and N.M. Nasr, *Journal of Chemical and Pharmaceutical Research*, Vol. 6(11), p. 28, 2014.
42. H.C. Neu, *American Journal of Medicine*, Vol. 88(4A), p. 3S, 1990.
43. D.E. Low, T.M.J. File, P.B. Eckburg, G.H. Talbot, H.D. Fredland, J. Lee, L. Llorens, I.A. Critchley and D.A. Thye, *Journal of Antimicrobial Chemotherapy*, Vol. 66 (3), p. 33–44, 2011.
44. L.D. Saravolatz, G.E. Stein and L.B. Johnson, *Clinical Infectious Disease*, Vol. 52(9), p. 1156, 2011.
45. M.J. Zaworotko, H.H. Hammud, I. Abbas and V.C. Kravtsov, *Coordination Chemistry Reviewes*, Vol. 59(1), p. 65, 2006.
46. M.W. Edmunds, and M.S. Mayhew, *Pharmacology for the primary care provider*, St. Louis, Mo: Mosby, 2000.
47. W.H. Hindle, *Clinical Obstetrics and Gynecology*, Vol. 37, p. 916, 1994.
48. KB. Holton, and E.M. Onsusko, *American Family Physician*, Vol. 62(3), p. 611, 2000.
49. J.S. Tan and T.M. File, *Medical Clinics of North America*, Vol. 79, p. 679, 1995.
50. B.G. Spratt, V. Jobanputra and W. Zimmermann, *Antimicrobial Agents and Chemotherapy*, Vol. 12, p. 406, 1977.
51. R.N. Jones, *The American Journal of Medicine*, Vol.78 (6A), p. 22, 1985.
52. K.G. Naber, G.A. Dette, F. Kees, H. Knothe and H.Grobecker, *Antimicrobial Agents and Chemotherapy*, Vol.17(4), p. 517, 1986.
53. M. Xia, T. Hang, F. Zhang, X. Li and X. Xu, *Journal of Pharmaceutical and Biomedical Analysis*, Vol. 49(4), p. 937, 2009.
54. H.A. Kirst, "Introduction to the macrolide antibiotics," in W. Schönfeld and H.A. Kirst, eds., *Macrolide Antibiotics. Milestones in Drug Therapy MDT*, Birkhäuser, Basel, pp. 1–13, 2002.
55. H. Hof, *Immunitat und Infektion*, Vo. *22*(2), p. 66. 1994.

56. D. N.Gilbert, "Aminoglycosides," in G. L. Mandell, J. E. Bennett and R. Dolin, eds., *Principles and practice of infectious diseases*, Churchill Livingstone, New York, N.Y., pp. 279, 1995.
57. M. P. Mingeot-Leclercq, and P. M. Tulkens, *Antimicrobial Agents and Chemotherapy*, Vol. 43(5), p. 1003, 1999.
58. M.P. Mingeot-Leclercq, Y. Glupczynski, and P.M. Tulkens, *Antimicrobial agents and chemotherapy*, Vol. 43(5), p. 727, 1999.
59. R.E.W. Hancock, *Journal of Antimicrobial Chemotherapy*, Vol.8, p. 249, 1981.
60. R. E. Hancock, S. W. Farmer, Z. S. Li, and K. Poole, *Antimicrob. Agents Chemother.*35, p. 1309, 1991.
61. M.Vaara, *Microbiology Reviews*, Vol. 56, p. 395, 1992.
62. D.J. Abraham, *Burgerís Medicinal Chemistry Drug Discovery*, Wiley, 2003.
63. O. Guneysel, O. Onur, M. Erdede, A. Dehizbasi, *The Journal of Emergency Medicine*, Vol. 36(4), p. 338, 2009.
64. J.M.Blondeau, *Clinical Therapeutics*, Vol. 21(1), p. 3, 1999.
65. P.C. Appelbaum, P.A. Hunter, *International Journal of Antimicrobial Agents*, Vol.16(1), p. 5, 2000.
66. M.Moshirfar, J.Chew, L. Werner, J.J. Meyer, B. Hunter, S. Stevens, M. Jensen, G. Kleinmann, N. Mamalis, *Graefe's Archive for Clinical and Experimental Ophthalmology*, Vol. 246(10), p. 1455, 2008.
67. F.J. Schmitz, B. Hofmann, B. Hansen, S. Scheuring, M. Lückefahr, M. Klootwijk, J. Verhoef, A. Fluit, H.P. Heinz, K. Köhrer, and M.E. Jones, *The Journal of Antimicrobial Chemotherapy*, Vol. *41*(4), p. 481, 1998.
68. S. Sood, S. Bhat, R. Singhal and A. Kumar, "Inoculum Preparation", in M.Y. Murray, eds., *Comprehensive Biotechnology (Second Edition),* Elsevier, pp.151–164, 2011.
69. F.L. Mould, K.E. Kliem, R. Morgan and R.M. Mauricio, *Animal Feed Science and Technology*, Vol. 123, p. 31, 2005.
70. S.T. Williams and E.M.H. Wellington, "Principles and problems of selective isolation of microbes", in J.D. Bu'LocK, L.J. Nisbet and D.J. Winstanley, eds., *Bioactive Microbial Products: Search and Discovery,* Academic Press, London, pp. 9–26, 1982.
71. C. Ball, "Protoplast fusion in commercially important microorganisms", in L.J. Nisbet, and D.J. Winstanley, eds., *Bioactive Microbial Products* 2: *Development and Production,* Academic Press, London, pp. 19–32, 1983.
72. J.F. Martín, and J.A. Gil, *Nature Biotechnology*, Vol. 2(1), p. 63, 1984.
73. G.J.M. Hersbach, C.P. Van Der Beek and P.W.M. Van Duck, "The penicillins: properties, biosynthesis and fermentation", in E. Vandamme, eds., *Biotechnology of Industrial Antibiotics,* Marcel Dekker Inc., New York and Basel, pp. 45–140, 1984.
74. R.T. Rowlands, "Industrial fungal genetics and strain selection", in J.E. Smith, D.R. Berryand B. Kristiansen, eds., *Filamentous Fungi*, Vol. 4, pp. 346–372, 1983.
75. R.T. Rowlands, and I.D. Normansell, "Current strategies in industrial strain selection", in L.J.Nisbetand D.J. Winstanley, eds., *Bioactive Microbial Products 2: Development and Production*, Academic Press, London, pp. 1–18, 1983.
76. R.P.Elander, *Journal of Industrial Microbiology and Biotechnology*, Vol. 22, p. 241, 1999.
77. P. Ellaiah, K. Adinarayana, G. M. Chand, G. S. Subramanyam, and B. Srinivasulu, *Pharmazie*, Vol. 57(7), p. 489, 2002.
78. S.D. Lee, S.W. Park, K.K. Oh, S.I. Hong, and S.W. Kim, *Letters in Applied Microbiology*, Vol. 34(5), p. 370, 2002.
79. H. Korbekandi, P. Darkhal, Z. Hojati, D. Abedi, J. Hamedi, and M. Pourhosein, *Iranian Journal of Pharmaceutical Research: IJPR*, Vol. 9(2), p. 177, 2010.

80. R. Arshad, S. Farooq, and S. S. Ali, *Brazilian Journal of Microbiology*, Vol. 41(4), p. 1133, 2010.
81. J.M. Weber, A. Reeves, W.H. Cernota, and R.K. Wesley, *In Vitro Mutagenesis: Methods and Protocols*, p. 257, 2017.
82. C. Subathra Devi, *Malaya Journal of biosciences*. Vol. 1, p. 248, 2014.
83. Y. Okamoto-Hosoya, S. Okamoto, and K. Ochi, *Applied and Environmental Microbiology*, Vol. 69(7), p. 4256, 2003.
84. S. Saxena, in *Applied Microbiology*, Springer Indias, p. 155, 2015.
85. D.A. Hopwood, *Annual Review of Microbiology*, Vol. 35(1), p. 237, 1981.
86. L. Ferenczy, *Symposium Society General Microbiology*, Vol. 31, p. 1, 1981.
87. J.F. Peberdy, "Genetic recombination in fungi following protoplast fusion and transformation", in J.E. Smith, eds., *Fungal Differentiation, Mycology series*, Marcel Dekker Inc., New York, Vol. 4, pp. 559–581, 1983.
88. A.C. Wesseling, and B.D. Lago, *Developments in Industrial Microbiology*, Vol. 22, p. 641, 1981.
89. K. Kitano, Y. Nozaki and A. Imada, Strain improvement in carbapenem antibiotic production by *Streptomyces griseus* subsp. *cryophilus, Abstracts* of *the Fourth International Symposium on the Genetics* of *Industrial Microorganisms* (Kyoto), p. 66, 1982.
90. P.F. Hamlyn and C. Ball, "Recombination Studies with Cephalosporium", in O.K. Sebek, and A.I. Laskin, eds., *Genetics of Industrial Microorganisms*, American Society for Microbiology, Washington, USA, pp. 185–191, 1979.
91. R. Hutter, T. Leisinger, J. Nuesch, and W. Wehrli, Antibiotics and Other Secondary Metabolites - Biosynthesis and Production, Academic Press, London, 1978.
92. J.E. Smith, "Concepts of Industrial Antibiotic Production", in D.I. Alani, M. Moo-Young, eds., *Perspectives in Biotechnology and Applied Microbiology*, Elsevier Applied Science Publishers, pp. 105–142, 1986.
93. D.A. Hopwood, and M.J. Merrick, *Bacteriological Reviews*, Vol. 41, p. 595, 1977.
94. D.A. Hopwood, "Actinomycetes genetics and antibiotic production", i*n* L.C.Vining, eds., *Biochemistry and Genetic Regulation of Commercially Important Antibiotics*, Addison-Wesley, Reading, MA, pp. 1–24, 1983.
95. P.M. Rhodes, N. Winskill, E.J. Friend and M. Warren, *Journal* of *General Microbiology*, Vol. 124, p. 329, 1981.
96. J.S. Feitelson, and D.A. Hopwood, *Molecular and General Genetics*, Vol. 190, p. 394, 1983.
97. M.J. Bibb, K.K. Chater and D.A. Hopwood, "Developments in Streptomyces cloning", in M. Inoye, eds., *Experimental Manipulation of Gene Expression, Academic Press*, New York, pp. 54–82, 1983.
98. C.L. Cooney, *Process Biochem*istry, Vol. 14(1), p. 31, 1979.
99. M.A. Farid, H.A. El-Enshasy, A.I. El-Diwany and E.S.A. El-Sayed, *Journal of Basic Microbiology*, Vol. 40(3), p. 157, 2000.
100. M. J. Johnson, Recent advances in penicillin fermentations. *Bulletin of the World Health Organization*, Vol. 6(1–2), p. 99, 1952.
101. S. J. Pirt, and R. C. Righelato, *Applied Microbiology*, Vol. 15 (6), p. 1284, 1967.
102. M. Matsumura, T. Imanaka, T. Yoshida and H. Taguchi, *Journal of Fermentation Technology*, Vol.56(4), p. 345, 1978.
103. M.K. Majumdar and S.K. Majumdar, *Applied Microbiology*, Vol.13(2), p. 190, 1965.
104. E. Inamine, B.D. Lago and A.L. Demain, in. Perlman, eds., *Fermentation Advances*. Academic Press, New York, pp. 199–221, 1969.
105. N. Singh, and V. Rai, *International Journal of Pharmacy and Pharmaceutical Sciences*, Vol. 4, p. 94, 2012.

106. E.O. Karrow and S.A. Waksman, *Ind. Eng. Chem.*, Vol.39, p. 821, 1947.
107. J.C. Woodward, R. L. Snell and R.S. Nicholls, Conditioning molasses and the like for the production of citric acid by fermentation, US Patent 2492673 A, assigned to Miles Lab, December 27, 1949.
108. J.E. Smith, K. Nowakowska-WaszczuK and J.G. Anderson, "Organic acid production by mycelial fungi", in B. Spencers, eds., *Industrial Aspects of Biochemist!* Part A, North Holland, Amsterdam, pp. 297–317, 1974.
109. S.H. Hutner, *Annual Review of Microbiology*, Vol. 26, p. 313, 1972.
110. Y. Aharonowitz, *Annual Review of Microbiology*, Vol.34, p. 209, 1980.
111. J.F. Martin and L. E. McDaniel, *Advances in Applied Microbiology*, Vol.21, p. 1, 1977.
112. A.J. Moyer and R. D. Coghill, *Journal of Bacteriology*, Vol. 51, p. 79, 1946.
113. G. C. Inskeep, R. E. Benett, J.F. Dudley and M.W. Shepard, *Industrial and Engineering Chemistry*, Vol. 43, p. 1488, 1951.
114. H. Hoeksema and C. G. Smith, *Progress in Industrial Microbiology*, Vol. 3, p. 91, 1961.
115. P. Sensi and J. E. Thiemann, *Progress in Industrial Microbiology*, Vol. 6, p. 21, 1967.
116. Z. Kamel and S.H. Al-Zahrani, in *Perspectives in Biotechnology and Applied Microbiology*, p. 143, 1986.
117. M.N. Hughes and R.K. Poole, "Metal Toxicity", in M.N. Hughes and R.K. Poole, eds., *Metals and Microorganisms*, Chapman & Hall, New York, pp.252–302, 1989.
118. M.N. Hughes, and R. K. Poole, *Journal of General Microbiology*, Vol.137(4), p. 725, 1991.
119. E. D. Weinberg, *Advances in Microbial Physiology*, Vol.4, p. 1, 1970.
120. P. Liras, J.A. Asturias and J.F. Martin, *Trends in biotechnology*, Vol. 8, p. 184, 1990.
121. J.A. Gil, P. Liras and J.F. Martin, *Drugs and the Pharmaceutical Sciences*, p. 513, 1984.
122. A.J. Moyerand R.D. Coghill, *Journal of Bacteriology*, Vol. 53, p. 329, 1947.
123. E. L. Smith and A.E. Bide, *Biochemical journal*, Vol. 42, p. xvii, 1948.
124. Q.F. Soper, C.W. Whitehead, O.K. Behrens, J.J. Corse and R.G. Jones, *Journal of the American Chemical Society*, Vol. 70(9), p. 2849, 1948.
125. M.J. Johnson, *Annals of the New York Academy of Sciences*, Vol. 48, p. 57, 1946.
126. M.I. Galan, J. Bonet, R. Sire, J.M. Reneaume, A.E. Plesu, *Bioresource Technology*, Vol. 100(15), p. 3775, 2009.
127. D. Peters, *Advances in biochemical engineering/biotechnology*, Vol. 105, p. 1, 2007.
128. P.S. Saudagar, R.S. Singhal, *Bioresource Technology*, Vol. 98, p. 2010, 2007.
129. W.A. Lotfy, *Bioresource Technology*, Vol. 98, p. 3491, 2007.
130. H.Y. Shin, J.Y.Lee, Y.R. Jung and S.W. Kim, *Bioresource Technology*, Vol. 101, p. 4549, 2010.
131. G. Detilley, D.G. Mou and C.L. Cooney, "Optimization and economics of antibiotics production" in J.E. Smith, D.R. Berry and B Kristiansen, eds., *The Filamentous Fungi*, Vol. 4, Edward Arnold, London, pp. 190–209, 1983.
132. M.J. Waites, N.L. Morgan, J.S. Rockey, G. Higton G, *Industrial Microbiology: An Introduction*, Blackwell, Oxford, UK, 2001.
133. J.R. Callahan, *Chem. Metal. Eng.*, Vol. 51, p. 94, 1944.
134. R.P. Elander, *Applied Microbiology and Biotechnology*, Vol. 61(5–6), p. 385, 2003.
135. R. Carrington, K. Dixon, A.J. Harrop and G. Macaloney, "Oxygen transfer in industrial air agitated fermentations", in M.R. Ladisch and A. BOSE, eds., *Harnessing Biotechnology for the 21st Century*, American Chemical Society, Washington, DC, pp. 183–188, 1992.
136. P. Srivastava, P. Mishra and S. Kundu, *Journal of Scientific and Industrial Research*, Vol. 65, p. 599, 2006.
137. B. M. Vastrad and S.E. Neelagund, *Recent Research in Science and Technology*, Vol. 3(2), p. 1, 2011.
138. M.R.Johns, *Solid substrate cultivation*, Vol.17, p. 341, 1992.

139. K.Balakrshana, A.Panday, *Journal of Scientific & Industrial Research*, Vol. 55, p. 365, 1996.
140. J. Barrios-Gonzalez, T.E. Castillo, A. Mejia, *Biotechnology Advances*, Vol. 11, p. 525, 1993.
141. A. Ohno, T. Ano, M. Shoda, *Biotechnology Letters*, Vol. 14, p. 817, 1992.
142. T.Robinson, D.Singh, P.Nigam, *Applied microbiology and biotechnology*, Vol. 55, p. 284, 2001.
143. J.Barrios-Gonzalez, A. Tomassini, G.Viniegra-Gonzalez, L. Lopez, *Biotechnology Letters* 10, p.793, 1988.
144. M.T. Kuenzi and J.AL. Auden, "Design and control of fermentation processes", inL.J. Nisbet and Winstanley, eds., *Bioactive Microbial Products* 2: *Development and Production*, Academic Press, London, pp. 91–110, 1983.
145. S. Alba, A.E. Humphrey and N.F. Millis, "Recovery of fermentation products", *Biochemical Engineering* 2nd edition, Academic Press, New York, pp. 346, 1973.
146. R.R. Swartz, *Annual reports on fermentation processes*, Vol.3, p. 75, 1979.
147. G.W. Pace and I.H. Smith, "Recovery of microbial polysaccharides", *Abstracts of Communications. Second European Congress of Biotechnology, Eastbourne*, Society of Chemical Industry, London, pp. 36, 1981.
148. B. Atkinson, and P. Sainter, *Journal of Chemical Technology & Biotechnology*, Vol. 32, p. 100, 1982.
149. R. Data, *Process Biochemistry*, Vol. 21(1), p. 19, 1986.
150. B. Atkinson and F. Mavituna, *Biochemical Engineering and Biotechnology Handbook* (2nd edition), Chapters 16 and 20. Macmillan, London, 1991.
151. E.W. Leserand J.A. Asenjo, "Process Integration in Biotechnology", in M.S. Verrall, eds., *Downstream Processing of Natural Products: A Practical Handbook*, John Wiley, New York, pp. 123–138, 1996.
152. E.Hochuli, *Pure and applied chemistry*, Vol. 64, p. 169, 1992.
153. P.R. Bienkowski, D.D. Leeand C.H. Byers, *Applied biochemistry and biotechnology*, Vol.18, p. 261, 1988.
154. D.S. Mathew, and R.S. Juang, *Separation and purification technology*, Vol. 53, p. 199, 2007.
155. K. Tonova, and Z. Lazarova, *Biotechnology advances*, Vol. 26, p. 516, 2008.
156. P. Y.Bi, H.R. Dong, and Q.Z. Guo, *Separation and purification technology*, Vol. 65, p. 228, 2009.
157. S.H. Mohd-Setapar, S.N. Mohamad-Aziz, C.S. Chuong, M.A. Che Yunus, M.A. Ahmad Zaini and M.J. Kamaruddin, *Chemical Engineering Communications*, Vol. 201(12), p. 1664, 2014.
158. Z. Hu, and E. Gulari, *Journal of Chemical Technology and Biotechnology*, Vol. 65, p. 45, 1996.
159. C.R. Bartels, G. Kleiman, J.N. Korzun, and D.B. Irish, *Chemical Engineering Progress*, Vol. 54, p. 49, 1958.
160. D.Y. Cha, *Reactive Polymers*, Vol. 5, p. 269, 1987.
161. T. Prazuck, I. Falconi, G. Morineau, V. Bricard-Pacaud, A. Lecomte, F. Ballereau, *Sexually Transmitted Diseases*, Vol. 29(11), p. 624, 2002.
162. A. Rivera, N. Larrosa, B. Mirelisand F. Navarro, *Enfermedades infecciosas y microbiologia clinica*, Vol. *32*, p. 30, 2014.
163. P.D. Tamma, C.K. Lee, *The Pediatric Infectious Disease Journal*, Vol. 28, p. 534, 2009.
164. K. Bush, *Current Opinion in Microbiology*, Vol. 13, p. 558s, 2010.

# 5

# Role of Antifungal Drugs in Combating Invasive Fungal Diseases

**Kakoli Dutt**

*Department of Bioscience and Biotechnology, Banasthali Vidyapith, Rajasthan*

***Abstract***

In the era of antibiotics, antifungal drugs have gained prominence due to the invasive fungal infections affecting millions globally. These drugs are created either to be fungistatic or fungicidal and are generally targeted to three categories of fungal diseases like dermatophytic, systemic and opportunistic infections. Azoles, polyenes and allylamines/thiocarbonates have been the classical families of antifungal agents. Currently, several new drug families like echinocandins and flucytosine have also entered the market with increasing prominence. The mode of action for antifungal drugs is either to inhibit the cell wall biosynthesis or to interfere with cellular division. However, despite new drugs being discovered, resistance to antifungal agents makes the search for newer, better and faster-acting drugs a necessity.

***Keywords:*** Antifungal agents, azole, polyene, echinocandins, ergosterol inhibitor

## 5.1 Introduction

Globally, over a billion people are impacted with fungal diseases. It is estimated that these fungi cause at least 1.4 million deaths worldwide per year [1, 2]. Out of approximately 30 fungal species currently identified as human pathogens, *Aspergillus*, *Candida*, *Cryptococcus*, *Pneumocystis jirovecii*, endemic dimorphic fungi including *Histoplasma capsulatum* and *Mucormycetes* are most prominent. According to GAFFI, in descending order, the prevalence of fungal infections listed are: cutaneous (skin, hair and nails) fungal diseases, mucosal candidiasis, allergic fungal conditions usually complicating asthma, chronic skin, lung sinus and bone infections and the acute life or sight-threatening fungal infections [3]. Among them, again, *Candida allbicans* for mucosal diseases, *Aspergillus fumigatus* for allergic fungal diseases and *Trichophyton* spp. especially *T. rubrum* for skin infections are most notable. Compared to other microbial pathogens causing bloodstream infections, *Candida* spp. are ranked fourth after other common bacterial pathogens [4]. *Aspergillus* infections are the most common microbial infections in hematopoietic stem cell transplant (HSCT) recipients [5]. About 30–50% of invasive aspergillosis patients still

*Corresponding author*: kakoli_dutt@rediffmail.com

Saurabh Saran, Vikash Babu, and Asha Chaubey (eds.) High Value Fermentation Products, Volume 1, (103–144) © 2019 Scrivener Publishing LLC

die, and the mortality from candidemia remains high at ~50% [6]. It has been established that fungal infections predominantly affect six main groups of patients a) cancer leukemia, transplantation and AIDS; b) critical care involving premature babies, ICU and major surgery; c) lung diseases including severe asthma, TB, chronic obstructive pulmonary disease (COPD) and cystic fibrosis; d) injury of eye, burns, trauma and skin especially in the tropical countries; e) skin, hair and nails infection; f) recurrent thrush, especially in pregnant women [3]. Chronic pulmonary aspergillosis, allergic bronchopulmonary aspergillosis worsening asthma, cystic fibrosis and *Aspergillus* bronchitis which complicates cystic fibrosis and cryptococcal meningistis in AIDS and recurrent vulvovaginal candidiasis incidences are cosmopolitan [7–10]. However, a precise estimate of global prevalence and incidence for each fungal infection is yet not clearly known due to scanty data, especially in the developing world. The lacuna is due to many reasons like lack of regular national surveillance system, no obligatory reporting of fungal diseases, poor diagnosis outside specialization units complemented with poor diagnostic tests with few well-designed units. This has led to the recent recognition of some fungal diseases (Table 5.1) [6, 11, 12].

**Table 5.1** Most prevalent fungal diseases in humans.

| Infection | Disease | Causative organism |
|---|---|---|
| **Dermophytic** | Tinea corporis (ringworm) | *Microsporum canis, Trichophyton mentagrophytes* |
| | Tinea pedis (athlete's foot) | *T. rubrum, T. mentagrophytes, Epidermophyton floccosum* |
| | Tinea cruis (jock itch) | *T. rubrum, T. mentagrophytes, E. floccosum* |
| | Tinea capis (scalp) | *M. canis T. tonsurans* |
| | Tinea barbae (beard/hair) | *T. rubrum, T. mentagrophytes* |
| | Tinea unguium (nails) | *T. rubrum, T. mentagrophytes, E. floccosum* |
| **Systemic** | Coccidioidomycosis | *Cocidioides immitis* |
| | Histoplasmosis | *Histoplasma capsulatum* |
| | Brazilian Blastomycosis | *Paracoccidioides brasiliensis* |
| | Blastomycosis | *Blastomyces dermatitidis* |
| **Opportunistic** | Candidaisis, Thrush, Vulvovaginitis | *Candida albicans* |
| | Cryptococcal meningitis | *Cryptococcus neoformans* |
| | Aspergillosis | *Aspergillus* sp. |
| | Mucormycosis | *Mucor* sp. |
| | Pneumocystis carinii pneumonia | *Pneumocystis carinii* |

## 5.2 Antifungal Agents

Prior to the 1970s, the market for antifungal drugs was highly limited as fungal diseases were considered to be treatable. Thus, fungal chemotherapy was dependent on potassium iodide for treating sporotrichosis and two polyenes drugs – nystatin and amphotericin B, which were introduced in the 1950s. Interestingly, little development took place in fungal chemotherapy research in the next two decades with only one major drug flucytosine introduced in 1964. The development of azole drugs marks the gaining momentum of antifungal therapy. Still, there has been a limitation on the diversity of antifungal drugs as many of them are primarily polyenes and azoles and more recently, cancidas.

The escalating global incidences of both dermal and invasive fungal diseases accompanied with increasing mortality is forcing the search for newer and better alternatives of existing therapy. Invasive fungal diseases are increasing in immuncompromised patients along with nosocomial infections, particularly in low-income group countries. In Table 5.2 and Table 5.3, the mortality rates of some prominent fungal diseases are cited.

In Table 5.4, a compilation of currently approved drugs is cited from GAFFI [3]. A major reason for lower turnout for newer drugs is that 80% of the antifungal targets are false positives with little potential to be developed as target-based inhibitors. As both fungi and humans have the same eukaryotic machinery, it is tough to identify and develop as target with few or no complications in host protein and cellular machinery [14]. Fungal therapy (Table 5.5; Figure 5.1) today includes drugs from polyenes, azoles and triazoles, allylamines, morpholines, echinocandins, flucytosine and some others [14, 15]. The reasons for the increasing antifungal drug use are manifold. Among hospitalized patients, the empiric use of antifungals in both hematology-oncology as well as intensive care patients is now common. Often, treatment is initiated based on preliminary microbiology results, and definite diagnosis of invasive infection versus colonization may be difficult [16–18]. New antifungal drugs such as itraconazole, caspofungin, and voriconazole have become available and broadened therapeutic options [19]. In some cases, an increasing incidence of invasive fungal infections and the emergence of infections due to rare and atypical organisms have been observed, and this changing epidemiology has contributed to more intense use of antifungal drugs [20]. In the ambulatory care setting, there was a shift from prescribing intravaginal antifungal preparations to fluconazole, raising concern about the possible development of azole drug resistance [21, 22].

Scarcity in the range and variety of drugs is imposing an increasing burden on the existing healthcare systems [23, 24]. Unfortunately, the approval rates for any new antimicrobial drug are very low [25, 26] with increasing tougher regulatory criteria to be met for any new drug class licensed for use [27]. Elaborate protocols have been established to provide clear data with respect to the side effects, combination therapy. Also, the increased evidences of resistance towards antifungals are generating another set of problems. [27]. Generally, clinical trials are larger and more complex in their design and thus, take longer to recruit for and to perform. Once time and cost are considered together for anti-infective drug development programmes, it can lead to drug discovery and development companies steering clear from investing in anti-infectives

**Table 5.2** Global estimates and related deaths due to fungal infections per year in AIDS patients [3, 13].

| Fungal infection | Fungal species | Annual burden | Case fatality rate with treatment | Estimated annual deaths | Clinical outlook |
|---|---|---|---|---|---|
| Cryptococcal meningitis | *Cryptococcus neoformans*<br>*C. gattii* | 371,700 – 957,900 | 15–20% USA >50% in LMICs | 125,000 – 624,700 | • Even with ART, there are still many new cases<br>• No new therapies in more than 25 years<br>• Cryptococcal antigen levels are not used to monitor outcome<br>• Differences in outcome correlate with health-care resource availability<br>• IRIS and increased ICP have been associated with this infection and need to be managed |
| *Pneumocystis* pneumonia | | >400,000 | 15% with best treatment | >200,000 | |
| Disseminated histoplasmosis | | >100,000 | 15–30%, if diagnosed and treated | >80,000 | |
| Chronic pulmonary aspergillosis | | >185,000 | 15–40% mortality in HIC | >100,000 | |
| Invasive aspergillosis | *Asprgillus fumigatus*<br>*A. flavus*<br>*A. terreus*<br>*A. calidoutus* | >45,000 | 30% mortality if treated in HIC | >30,000 | • Increasing resistance to azoles<br>• Early treatment or prevention is essential |

| | | | | | |
|---|---|---|---|---|---|
| Invasive candidiasis | *Candida albicans*<br>*C. tropicalis*<br>*C. glabrata*<br>*C. parapsilosis*<br>*C. krusei*<br>*C.auris* | | | | • High mortality attributable to infections<br>• Drug resistance develops in pathogens<br>• Requires early-stage treatment |
| *T. marneffei* infection | | >8,000 | 18–33% | >2,000 | |
| Mucosal and skin fungal infection | | >10,120,000 | <1% | <1,000 | |
| Mucormycosis | *Rhizopus*<br>*Mucor.*<br>*Cunninghamella bertholletiae* | | | | • Control underlying diseases, initiate polyene therapy and consider surgery<br>• Poor prognosis |
| Dimorphic mycoses | *Blastomyces dermatitidis*<br>*Coccidioides immitis*<br>*Coccidioides posadasii*<br>*Histoplasma capsulatum*<br>*Sporothrix.* | | | | • Geographically restricted<br>• Infect both immunocompetent and immunosuppressed individuals |

**Table 5.3** Global incidences of some fungal diseases [3].

| Fungal infection | Number affected | Case fatality rate | Estimated deaths | Comments |
|---|---|---|---|---|
| *Cryptococcal* meningitis | 370,000 in AIDS | 15–20% USA >50% developing world | 250,000 in AIDS | CDC estimate |
| *Pneumocystis* pneumonia | >400,000 in AIDS<br>>100,000 in non-AIDS | ~15% in AIDS with best reatment<br>~50% in non-AIDS | >200,000 in AIDS<br>>50,000 non-AIDS | Most cases in Africa not diagnosed and 100% mortality |
| Disseminated Histoplasmosis | ~100,000 | 15–30%, if diagnosed and treated | treated<br>>80,000 | Most common in the Americas |
| Invasive aspergillosis | >300,000 | ~30% mortality if treated in HIC –in AIDS~50% non-AIDS, in HIC | >30,000 in AIDS<br>>125,000 in non-AIDS | Many missed diagnoses globally |
| Invasive candidiasis | >750,000 | ~40% mortality treated | >350,000 | |
| Chronic pulmonary aspergillosis | >3,000,000 | ~15–40% mortality in HIC<br>~15% mortality in the developed world | >450,000 in non hospitalised populations | Under-diagnosed and mistaken for tuberculosis |
| Severe asthma with fungal sensitisation (SAFS) | >6,500,000 | <1% but no good estimates. | 350,000 – 489,000 asthma deaths<br>~50% related to SAFS | Uncertain |
| Total | ~13,000,000 | | 1,600,000 | Probably a significant underestimate |

(HIC- high income group country)

**Table 5.4** Status and activity of current and approved oral and intravenous antifungals.

| **Disease/status** | **Intravenous only** | | **Intravenous and oral (azoles and flucytosine)** | | | | | | **Oral only** | |
|---|---|---|---|---|---|---|---|---|---|---|
| | **AmB** | **Candins*** | **Flu** | **Itra** | **Vori** | **Posa** | **Isavu** | **5FC** | **Terbinafine** | **Griseofulvin** |
| On the WHO EML? | Y | N | Y | N | N | N | N | Y | N | Y |
| Generics available? | Y | N | Y | Y | Y | N | N | Y | Y | Y |
| Cryptococcosis | +++ | | ++ | ++ | ++ | ++ | ? | +++ | | |
| Candidiasis | +++ | +++ | ++ | ++ | ++ | ++ | ++ | ++ | | |
| Aspergillosis | + | + | | + | ++ | ++ | ++ | ? | | |
| Histoplasmosis | +++ | | + | +++ | ++ | +++ | ? | | | |
| *Pneumocystis* | | | | | | | | | | |
| Mucormycosis | ++ | | | | | ++ | ++ | | | |
| *T. marneffei* infection | +++ | | + | +++ | ++ | ? | ? | ? | | |
| Skin, hair nails | | | ++ | ++ | ? | ? | ? | | +++ | ++ |
| Mycetoma | | | | | | | | | | |
| Chromoblastomycosis | | | | + | ? | ++ | ? | | | |

Y yes; N no;? not studied; + some activity; ++ good activity but not maximally effective; +++ highly efficacious

* micafungin, caspofungin, anidulafungin azoles and flucytosine

*Note*: AmB = amphotericin B; azoles = fluconazole, itraconazole, voriconazole, posaconazole, isavuconazole.

(Sourced from Gaffi Road Map 2017)

**Table 5.5** Antifungal agents used in fungal therapy.

| Class | Drug | Target | Adverse effect |
|---|---|---|---|
| Azoles | Saperconazole | Ergosterol biosynthesis | Gl disturbance; rare hepatotoxicity |
| Fluorinated pyrimidines | Flucytosine | RNA biosynthesis | Bone marrow suppression; hepatotoxicity; Gl disturbance |
| Imidazoles | Miconazole | Inhibit the ATPase system in the cell membrane, inhibit the activity of *C. albicans* plasma membrane glucan synthase, chitin synthase, adenylcyclase and 5-nucleotidase enzymes | Headache; pruritus; thrombophlebitis; hepatotoxicity; autoinduction of hepatic degrading enzymes |
| | Ketoconazole | inhibit the ATPase system in the cell membrane, inhibit the activity of *C. albicans*plasma membrane glucan synthase, chitin synthase, adenylcyclase and 5-nucleotidase enzymes; affect the transformation of *C. albicans* from the budding form to the pseudomycelial form | Gl disturbance; hepatotoxicity |
| | Fluconazole Itraconazole | reduction of obtusifolione to obtusifoliol, which results in the accumulation of methylated sterol precursors | Headache; dizziness; drowsiness; abdominal pain; diarrhea; loss of appetite; allergic reactions including skin inflammation, itching, rash; unpleasant taste in mouth. |
| Polyenes | Amphotericin B | Ergosterol synthesis | Nephrotoxicity; fever, chills; phlebitis; anaemia; |
| | Amphotericin B (deoxycholate) | | Gl disturbance |
| | Amphotericin B (lipid formulation) | | Reduced azotaemia |
| Allylamines | Naftifine | Non-competitive inhibitors of squalene epoxidase, an enzyme involved in the conversion of squalene to lanosterol, which is an essential step in the synthesis of the fungal cell membrane | Headache; dizziness; application site reactions like burning, stinging, irritation, redness, dry skin, or itching |
| | Terbinafine | | GI disturbance; nausea; mild skin rash; itching |

| | | | |
|---|---|---|---|
| Thiocarbamates | Tolnaftate | inhibit squalene epoxidase, an important enzyme in the biosynthetic pathway of ergosterol (a key component of the fungal membrane) in a similar way to allylamines | |
| Alkylpyridinylidine–octanamine derivatives | Octenidine Pirtenidine | Alters membrane lipid and sterol contents Affect ergosterol biosynthesis by inhibition of 14α-demethylase | burning sensations; reddening; itching; and heat sensations in the treated areas.<br>Dizziness; nausea; vomiting; decreased or loss of appetite; weight loss; stomach pain or upset; tiredness |
| Synthetic phenylmorpholine derivative (Morpholines) | Amorolofine | It acts on the ergosterol pathway, inhibiting the Erg24P (D14-reductase) reaction and the Erg2P (D8–D7 isomerase enzyme) reaction. | Nail becoming discoloured, loose or starting to separate from the nail bed; Burning sensation or allergic skin reaction (contact dermatitis). |
| Histatins | | Inhibits mitochondrial ATP synthesis by loss in transmembrane potential. | |
| Polyoxins, Nikkomycin, Demethylallosamidin | | Chitin synthase inhibitor; chitinase inhibitor | |
| Echinocadins | Caspofungin<br>Micafungin<br>Anidulafungin<br>Papulacandins | Beta-(1,3)-glucan synthase inhibitor | Fever; headache; nausea; vomiting; phlebitis; abnormal liver function tests; histamine-related symptoms have been noted (rash, pruritus, facial flushing). |
| | Pradimicin A; Benanomycin A | Calcium – dependent complexing with the saccharide of mannoprotein, thereby disrupting the cytoplamic membrane, causing leakage of intracellular potassium | |
| | Sphingofungins | Inhibition of serine palmitoyltrasferase | Serious gastrointestinal side effects |
| | Floimycin (concanamycin A) | Inhibitor of V-type H+-ATPase | |
| | Hydroxypyridones | Inhibit ATP synthesis and cellular uptake of essential component | |

(Continued)

**Table 5.5** Cont.

| Class | Drug | Target | Adverse effect |
|---|---|---|---|
| Azoles | Saperconazole | Ergosterol biosynthesis | Gl disturbance; rare hepatotoxicity |
| | Cispentacin | Interfering with amino-acid synthesis | |
| | RI-331 | Inhibit homoserine dehydrogenase | |
| | Eflorinithine (difluoromethyl ornithine) | Inhibitor of ornithine decarboxylase | |
| Hydroxypyridones | Ciclopiroxolamine, ciclopirox and rilopirox | Inhibition of cellular uptake of essential compounds; at high concentrations, they can alter membrane permeability | unusual or severe itching, redness, burning, dryness, or irritation of treated skin; or discoloration or other changes in the nails |
| Sordarins | | Inhibit protein synthesis by blocking the function of fungal translation elongation factor 2 (EF2). | |
| | Cyclosporine, tacrolimus or rapamycin | mTOR or calcineurin | |
| | Rifampin | RNA polymerase | |
| | Verapamil | Calcium channel | |
| | Sertraline | Vesicle transport, serotonin reuptake | |

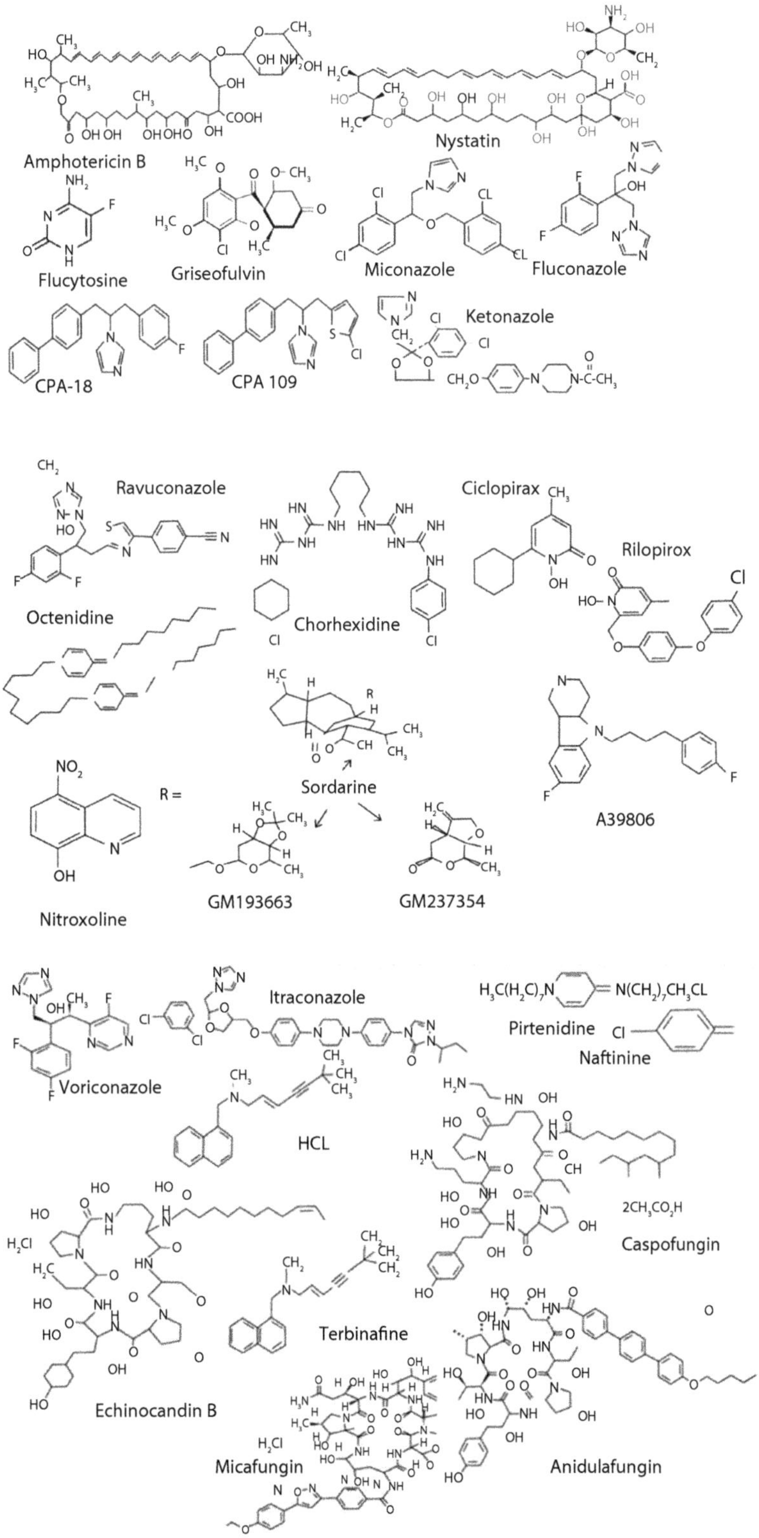


**Figure 5.1** Structures of various antifungal drugs.

research programmes, and particularly more challenging antifungal programmes on a risk-return basis.

### 5.2.1 Azoles

One of the earliest antifungal drug groups, azoles was represented by imidazole, and some triazole compounds. The first azole was synthesized in 1944 by Woolley [28], but it was not until 1958 that the scientific community began to consider azoles as potential antifungal agents. Their ring structure basically contains a short aliphatic chain in which the second carbon is linked to a halogenated phenyl group. Considerable variability exists in structure between the two groups. Imidazoles are five membered ring structures containing two nitrogen atoms with a complex sidechain attached to one of the nitrogen atoms. For antifungal activity, the unsubstituted imidazole and the N–C covalent linkage between the imidazole and the rest of the molecule are relevant. In late 1960s, clotrimazole, econazole, and miconazole became available for treatment [29]. However, their use was restricted to external application due to their high toxicity when administered orally [30, 31]. In 1968, miconazole became the first antifungal available for parenteral injection, but due to its toxicity and relatively limited range among fungal species [32], its use decreased until it was no longer commercialized. In 1981, the Food and Drug Administration (FDA) approved a new antifungal, ketoconazole [33] with several drawbacks. The drug exhibited poor absorbtion on oral administration and no formulation was made for intravenous injection. Also, it cannot cross the cerebrospinal barrier and is less active in immunosuppressed patients [29, 34, 35], while causing some severe side effects such as decrease in testosterone or glucocorticoids production, along with liver and gastrointestinal complications [36–38] and numerous interactions with other drugs.

These drawbacks led to the development of triazoles. The structure of triazoles is similar to imidazole but they contain three nitrogen atoms in the rings. Triazole compounds currently approved for clinical use are itraconazole, fluconazole, voriconazole, lanconazole, ravuconazole and posaconazole. Imidazoles in current clinical use are clotrimazole, miconazole, econazole and ketoconazole [39]. Fluconazole became available for use by clinicians in 1990 and provided many advantages over the use of imidazoles. It is highly hydrosoluble, easily injected intravenously, almost completely absorbed through the gastrointestinal tract, and it diffuses throughout the whole body, including cerebrospinal fluid [40, 41]. Fluconazole is suitable for the treatment of superficial candidiasis (oropharyngal, esophageal, or vaginal), disseminated candidiasis, cryptococcal meningitis, coccidioidomycosis, and cutaneous candidiasis. Unfortunately, the over-prescription of this drug by physicians for prophylaxis or treatment led to an increase in resistance against azole drugs. Moreover, fluconazole is almost ineffective against most molds.

Itraconazole was approved and made available by the FDA in 1992. This triazole possesses a broad spectrum of activity across fungal species comparable to this of ketoconazole and wider than fluconazole. Moreover, it is less toxic than ketoconazole and replaced it for treatment of histoplasmosis, blastomycosis, and paracoccidioidomycosis. Contrary to fluconazole, it is also used for the treatment of infections due to species belonging to the genera *Aspergillus* and *Sporothrix* [42]. However, itraconazole is hydrophobic and is more toxic than fluconazole, thus only indicated for the treatment of onychomycosis, superficial infections, and in some cases for systemic aspergillosis [43]. A new itraconazole formulation with an enhanced absorption and a decreased toxicity was approved by FDA in 1997 [44]. An injectable formulation of itraconazole was made available in 2001 [45].

Fluconazole and itraconazole show some nonnegligible drug interactions with such drugs that are used in chemotherapy or with AIDS treatment. These interactions can result in a decrease in azole concentration or even an increase in toxicity [46]. Furthermore, both these drugs are ineffective against some pathogens like *Scedosporium*, *Fusarium*, and Mucorales [47].

Voriconazole and posaconazole were approved by FDA in 2002 and 2006, respectively. Ravuconazole is currently under clinical trial phase of drug development. These three drugs possess a wide range of activity against *Candida*, *Aspergillus*, *Fusarium*, *Penicillium*, *Scedosporium*, *Acremonium*, and *Trichosporon*, and dimorphic fungi, dermatophytes, and *Cryptococcus neoformans* [48, 49]. While new generation triazoles as against classical triazoles exhibited higher efficacy against *Candida* and *Aspergillus* [49], their side effects and drug interactions were similar to those of fluconazole and itraconazole [50].

Overall, azoles exhibit a broad spectrum antifungal activity towards candidiasis, crytococcosis, blastomycosis and histoplasmolysis. These compounds inhibit ergosterol synthesis, which is an integral component of smooth endoplasmic reticulum and the inner mitochondrial membrane [51, 52]. This is due to inhibition of sterol 14α-demethylase [cytochrome P450 51 or CYP51] which catalyses the last step of ergosterol biosynthesis. Apart from inhibiting cytochrome C oxidative and peroxidative enzymes, azoles cause leakage of proteins and amino acids from the cell, inhibit catalase systems while decreasing fungal adherence and inhibiting germ tube and mycelia formation [52, 53].

### 5.2.2 Polyenes

These are characterized by cyclic amphiphilic organic molecules known as macrolides. Generally, they consist of a 20 to 40 carbons macrolactone ring conjugated with d-mycosamine group. Hydroxyl groups [6–14] are distributed along the macrolide ring on alternate carbon atoms. The amine group present in some of the polyenes is associated with an amine sugar connected to the macrolide ring through a glycosidic linkage. The carbohydrate moiety in amphotericin B and nystatin is the mycosamine ($C_6H_{13}O_4N$; 3-amino-3, 6-dideoxymannose) sugar. Polyenes show limited solubility in water and non-polar organic solvents but are easily dissolved in polar organic solvents such as dimethyl sulphoxide or dimethyl formamide [39, 53]. Their amphiphilic properties are due to the presence of several conjugated double bonds on the hydrophobic side of the macrolactone ring, and the presence of several hydroxyl residues on the opposite hydrophilic side [54].

This group of antifungal compounds increases the permeability of the cell membrane causing leakage of various cellular constituents like amino acids, sugars, other metabolites, etc., leading to cell lysis and death. Inhibition of respiration is also observed, which may be due to the leaky membrane. Pore formation promotes plasma membrane destabilization, and channels allow leakage of intracellular components such as $K^+$ ions, responsible for cell lysis [54]. Thus, the fungistatic effect of polyene is primarily due to the binding of the compound to the cell membrane. The binding is selective for ergosterol in the fungal cells. For polyene susceptibility sterols must be present in the outer membrane. The toxicity is dependent on the fatty acyl composition of the membrane phospholipids. Changes in the ratio of various phospholipids have a direct relationship to polyene sensitivity [39]. Their amphiphilic nature allows them to bind to the lipid bilayer and form pores. Magnetic resonance data shows that 8 AmB molecules bind to an equal number of ergosterol through their hydrophobic ends while the central channel of 70–100nm diameter is formed by the hydrophilic ends. It is also suggested that polyene drugs are able to induce an oxidative stress (particularly in *C. albicans* [53, 55] but their activity seems to be reduced in hypoxic conditions [56].

Polyenes possess a lower but non-negligible affinity for cholesterol, which explains the high toxicity associated with antifungals and is responsible for several side effects [54]. Among the polyene drugs, AmB is given systemically, while nystatin and natamycin are only used locally or orally. The latter molecules fortunately possess limited systemic activity, since they exhibit minimalistic absorption through gastrointestinal mucosa [57, 58]. Thus, AmB is the most favored polyene antifungal for treating systemic infections [54]. However, its use is accompanied with adverse effects, mostly at the level of kidneys and liver. New AmB formulations, such as liposomal AmB or lipid AmB complexes, minimize such side effects [59]. AmB is recommended for the treatment of infections caused by *Candida*, *Aspergillus*, *Fusarium*, *Mucor*, *Rhizopus*, *Scedosporium*, *Trichosporon*, *Cryptococcus*, etc. It is also widely used to treat parasitic infections such as leishmaniasis and amebiasis [54]. Natamycin and nystatin are active against *Cryptococcus*, *Candida*, *Aspergillus*, and *Fusarium*. Nystatin is frequently used for the treatment of cutaneous, vaginal, and esophageal candidiasis, and natamycin can be used for the treatment of fungal keratosis or corneal infections [58].

More than 200 polyene molecules have been identified as antifungals, most of them being produced by *Streptomyces* bacteria. However, only three possess a toxicity allowing their use in clinical practice: amphotericin B (AmB), nystatin, and natamycine. *Streptomyces* synthesize polyenes through a phylogenetically conserved gene cluster within these species. This cluster contains genes coding for several polyketide synthases, ABC (ATP-binding cassette) transporters, cytochrome P450-dependent enzymes, and enzymes responsible for the synthesis and the binding of the mycosamine group [60]. Although it is possible to synthesize polyenes chemically, they are still produced from *Streptomyces* cultures for economic reasons.

### 5.2.3 Allylamine/Thiocarbonates

These synthetic compounds have a chemical structure similar to naphthalene ring which is substituted at the 1-position with an aliphatic chain. Both allylamines and thiocarbamates function as non-competitive inhibitors of squalene epoxidase, an enzyme involved in the conversion of squalene to lanosterol, which is an essential step in the synthesis of the fungal cell membrane. Cell death is dependent on the accumulation of squalene rather than ergosterol deficiency, as high levels of squalene increase membrane permeability, leading to disruption of cellular organization [39, 53]. Tolnaftate and tolciclate are used as topical treatment against dermatophytes [61].

### 5.2.4 Other Antifungal Agents

i. Flucytosine or 5-fluorocytosine (5-FC): A synthetic drug, white, odourless and crystalline oral antimycotic agent, 5-fluorocytosine (flucytosine) is a fluoridated pyrimidine analog, which inhibits DNA and RNA synthesis by being incorporated into growing nucleic acid chain, preventing further extension that eventually leads to cellular defects in protein biosynthesis and cell division. FC partially deaminates to form 5-FC at higher temperatures or on cellular uptake (Figure 5.2). Among fluoropyrimidines, only 5-fluorocytosine (5-FC) and 5-fluorouracil (5-FU) are synthetic structural analogs of the DNA nucleotide cytosine (Figure 5.2). 5-FC was synthesized in 1957 by Duschinsky *et al.*, initially, as an antitumor agent [62]. In 1963, Grunberg and coworkers discovered its antifungal property [63]. Several years later, 5-FC was successfully used for the treatment of systemic candidiasis and of cryptococcal meningitis [64]. However, 5-FC due to cytostatic effects and high rates of emerging resistance is not very useful for monotherapy [65–67].

The drug rapidly enters the fungal cell through specific transporters, such as cytosine permeases or pyrimidine transporters [68], before it is converted into 5-FU by the cytosine deaminase [65]. 5-FU itself is converted into 5-fluorouracil monophosphate (5-FUMP) by another

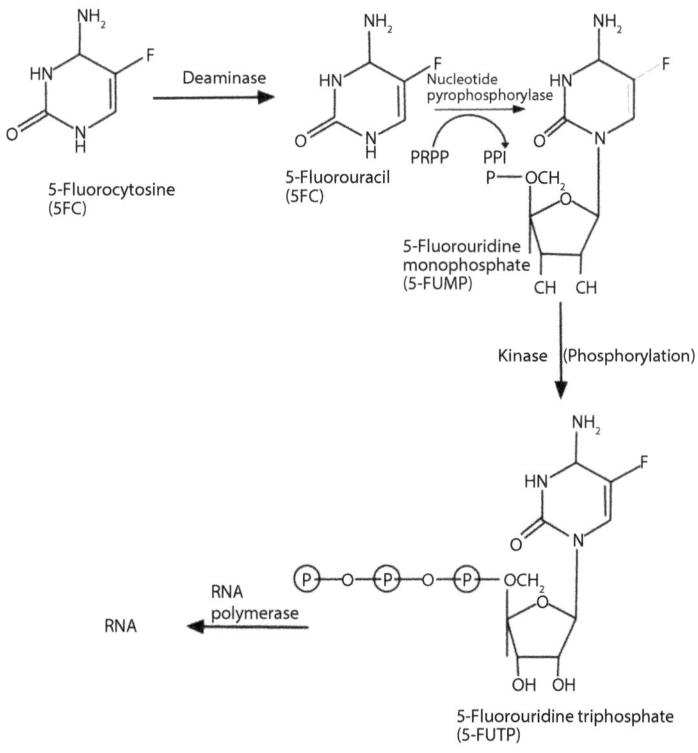


**Figure 5.2** Mode of action of 5-FC [6].

enzyme, uridine phosphoribosyltransferase (UPRT). 5-FUMP can then be either converted into 5-fluorouracil triphosphate, which incorporates into RNA in place of UTP and inhibits protein synthesis, or converted into 5-fluorodeoxyuridine monophosphate, which inhibits a key enzyme of DNA synthesis, the thymidylate synthase, thus inhibiting cell replication [65, 68–70].

It has been successfully used against *Candida*, *Torulopsis*, *Cryptococcus* and *Geotrichum*. Moderate success has been recorded against *Aspergillus* and *Chromomycosis* causing dematiaceous fungi [39, 71–73].

This drug exhibits limited activity on *Phialophora*, *Cladosporium*, and *Aspergillus*. 5-FC is also active against protozoans *Leishmania* and *Acanthamoeba* [74].

Due to its high hydrosolubility and small size, it diffuses rapidly throughout body even when orally administered [75]. Generally, it produces negligible side effects, despite some severe adverse effects, such as hepatotoxicity or bone marrow lesions [77–79], occurring in rare cases [65]. Surprisingly, these side effects are identical to those observed with 5-FU treatment, despite the fact that humans do not possess a cytosine deaminase enzyme that is responsible for the conversion from 5-FC into 5-FU in fungal cells [80]. Some data suggest that the gut microbes may be responsible for the 5-FU production and the observed side effects [81, 82].

The use of 5-FC in clinical practice is decreasing because of the frequent occurrence of innate or acquired resistance to this drug in fungal pathogens. Thus, with few exceptions [66], it is never used as monotherapy but always in combination with another antifungal, such as amphotericin B [83, 84]. However, the elevated renal and liver toxicities of amphotericin B, that further increase 5-FC hepatotoxicity, has led to combination therapy of 5-FC with azole drugs.

ii. Echinocandins – The first echinocandin-type antimycotic, echinocandin B, was isolated independently by the researchers of Ciba-Geigy, Sandoz and Eli Lilli from the fermentation broth of "*Aspergillus nidulans* var. *echinolatus*", "*Aspergillus nidulans* var. *roseus*" and *Aspergillus rugulosus* in the 1970s in random screening of the available strain collections [85–87]. Echinocandin B was followed by the isolation and characterization of more than 20 natural echinocandins. All these secondary metabolites are produced by ascomycota fungi (Table 5.6). They possess cyclic lipo-hexapeptide N-acylated structure with an aliphatic chain of different length [88, 89]. The various echinocandins differ in having different substituents in the hexapeptide ring or a distinct fatty acid chain. This latest class of antifungals is currently represented by three drugs – caspofungin, micafungin and anidulafungin. These drugs affect cell wall biosynthesis by noncompetitive inhibition of β-1,3-glucan synthase [90–92]. Echinocandins show a high heterogeneity in the number and position of hydroxyl-groups on L-Orn, L-homoTyr, L-Gln and (mehyl)-L-Pro and/or the presence of the methyl-group on L-Pro even within the

**Table 5.6** Natural producers of Echinocandins [93].

| Echinocandins | Producer | Echinocandins | Producer |
|---|---|---|---|
| Echinocandin B-D | "*Aspergillus nidulans var. echinulatus*<br>*A. nidulans var.roseus*<br>*A. rugulosus* | Pneumocandin A-E | *Glarea lozoyensisc*<br>*Pezicula carpinea*<br>*Cryptosporiopsis sp.* |
| Aculeacin A-G | *A. aculeatus*<br>"*A. japonicus var. aculeatus*" | Sporiofungin A-C | *Cryptosporiopsis sp.* |
| Mulundocandin deoxymulundocandin | *Aspergillus sydowi* | Catechol-sulfate" echinocandins (FR901379, FR901381–82, FR190293, FR209602–4, FR220897, FR220899, FR227673) | *Coleophoma empetri*<br>*Coleophoma Crateriformis*<br>*Chalara sp.*<br>*Tolypocladium parasiticum* |
| Cryptocandin | *Cryptosporiopsis quercina* | | |

main groups. Because of high-complex chemical structure echinocandins possess their industrial-scale production is based on fermentation. Among the natural echinocandins, echinocandin B, pneumocandin B0 and FR901379 are produced for commercial purposes. Since the fermentation and purification costs of natural echinocandins are the primary reason for the high costs of semisynthetic derivatives, the optimization of the fermentation process is crucial to make a competitive product. Thus, little information has been published in this field so far [93].

Echinocandins are mostly used for treating for invasive candidiasis and aspergillosis [94]. These have low host toxicity, fewer drug interactions and due to large molecular size are available only in IV formulations. All three semisynthetic echinocandins used in human therapy, caspofungin, micafungin and anidulafungin, are approved for the treatment of oesophageal candidiasis and invasive candidiasis in adults and in children (caspofungin) over three months of age. Micafungin has been FDA approved for antifungal prophylaxis in hematopoietic cell transplantation and caspofungin has been approved for empirical therapy of febrile neutropenia and for salvage and primary therapy of invasive aspergillosis. All of them have limited oral bioavailability and are therefore administered by intravenous infusion [93].

The FDA-approved echinocandins show good fungicidal or fungistatic activity against the human pathogenic fungi, *Candida* and *Aspergillus*, in vitro at clinically attainable concentrations [95–100]. Echinocandins show concentration-dependent fungicidal activity either in RPMI-1640 or antibiotic medium 3 against many *Candida* species including *C. albicans, C. glabrata, C tropicalis, C. dubliniensis, C krusei, C. lustitaniae, C. parapsilosis* and *C. guilliermondii* [95, 97, 98, 101]. The majority of these species (e.g., *C. albicans* and *C. tropicalis*) are innately susceptible to echinocandins with minimum inhibitory concentrations (MICs) in the range of 0.015- 0.25 mg/L. Other species (e.g., *C. parapsilosis, C. lusitaniae* and *C. guilliermondii*) show higher MIC values to echinocandins in RPMI-1640 as test medium which is explained by their naturally occurring Fks1 polymorphisms [95, 97–100, 102]. Semisynthetic echinocandins show a strong fungistatic effect against *Aspergillus* including *A. fumigatus, A. terreus, A. flavus* and *A. niger* [95, 103–106]. Since echinocandins are unable to completely inhibit the growth of these species, clear determination of MIC is difficult. Therefore, an alternative method—determination of the minimum effective concentration (MEC)—was introduced to describe the activity of echinocandins against *Aspergillus* strains.

iii. Octenidine and pirtenidine – these drugs are alkylpyridinylidine-octanamine derivatives. They alter the membrane lipid and sterol contents causing extensive cell leakage and damage. These cause gross morphological and ultrastructural changes in the cells. Octenidine treated cells show increased zymosterol and obtusifoliol contents which affects ergosterol biosynthesis. On the other hand, pirtenidine treated cells show increased concentrations of squalene and 4,14- dimethylzymosterol [39].

iv. Morpholines – discovered in the 1980s, morpholines are synthetic phenylmorpholine derivatives. They act on the ergosterol pathway (Figure 5.3), inhibiting the Erg24P($\Delta$14-reductase) reaction and the Erg2P($\Delta$8-$\Delta$7 isomerase enzyme) reaction. This group is represented by amorolofine and used as topical treatment of superficial mycoses [39].

## 5.3 Targets of Antifungal Agents

### 5.3.1 Cell Wall Biosynthesis Inhibitors

Fungal cell wall is generally made up of three polymeric components i.e., glucan, chitin and mannoproteins. These serve as targets for the antifungal agents and various drugs are identified to act as inhibitors for these three cell wall components (Table 5.7)

a. Glucan synthesis inhibitors –
Glucan, a polysaccharide constituted by glucose monomers linked by (1,3)-β or (1,6)-β bonds, is an essential component of the cell wall [107]. Echinocandins are reported to be potent glucan synthesis inhibitors. The success of the lipopeptide class of glucan synthesis inhibitors has prompted the search for other antifungal agents with improved features over the echinocandins (lack of oral absorption). Other cyclic peptides

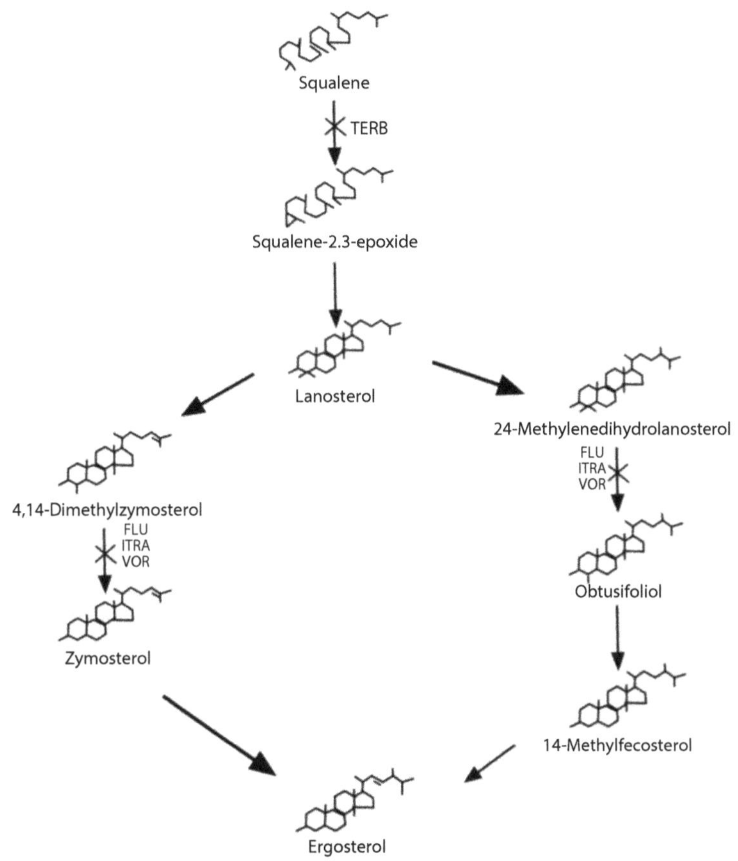


**Figure 5.3** Steps showing the inhibitory activity of antifungal agents during ergosterol synthesis. TERB-terbinafine; FLU- flucanozole; ITRA- itraconazole; VOR-voriconazole [53].

like arborcandins have been described as as glucan synthesis inhibitors. These contain a 10-amino-acid ring and two lipophilic tails [108] like FR901469 (macrocyclic lipopeptidolactone) composed of 12 amino acids and a 3-hydroxypalmitoyl moiety [109]. However, besides cyclic lipopeptides, only two other types of glucan synthesis inhibitors are known, the papulacandins and related compounds, and the acidic triterpenes. The papulacandins are glycolipids discovered in the late 1970s [110, 111]. Despite medicinal chemistry efforts, neither papulacandins nor any of their relatives have been developed as drugs, basically due to their limited potency in animal models [112]. The most recently discovered compound class of glucan synthesis inhibitors are triterpenes containing a polar (acidic) moiety [113]. This polar moiety can be a glycoside (in enfumafungin and ascosteroside), a succinate (in arundifungin) or a sulphate-derivative amino acid (in ergokonin A).

The main effect of echinocandins is glucan inhibition of cell wall but a secondary effect is obtained by reduction of ergosterol (Figure 5.3) and lanosterol content with increased chitin content. This produces

**Table 5.7** Naturally produced fungal cell wall inhibitors [89].

| Compound | Producing species | Compound | Producing species |
|---|---|---|---|
| **Sphingolipid biosynthesis** | | **Protein synthesis** | |
| Sphingofungins | *Aspergillus fumigatus Paecilomyces variotii* | Sordarin | *Sordaria araneosa* |
| Lipoxamycin | *Streptomyces sp.* | Zofimarin | *Zopfiella marina* |
| Viridiofungins | *Trichoderma viride* | BE31405 | *Penicillium minioluteum* |
| Myriocin | *Isaria sinclairii* | SCH57404 | *Unidentified sterile fungus* |
| Fumonisin B1 | *Fusarium moniliforme* | Xylarin | *Xylaria sp.* |
| Australifungin | *Sporormiella australis* | Hypoxysordarin | *Hypoxylon croceum* |
| Aureobasidin A | *Aureobasidium pullulans* | GR135402 | *Graphium putredinis* |
| Rustmicin | *Micromonospora chalcea Streptomyces galbus Micromonospora sp.* | **Lipopeptides** | |
| Galbonolide B | *Micromonospora sp.* | Echinocandin B | *Aspergillus nidulans A. rugulosus* |
| Minimoidin | *Sporomiella minimoides* | Aculeacin | *Aspergillus aculeatus* |
| **Acidic terpenoids** | | Mulundocandin | *Aspergillus sydowii* |
| Efumafungin | *Hormonema sp.* | Sporiofungins | *Penicillium arenicola Cryptosporiopsis sp.* |
| Arundifungin | *Arthrinium arundinis A. phaeospermum Leotiales anamorphs Coelomycete undetermined* | Pneumocandins. | *Glarea lozoyensis Pezicula sp. Cryptosporiopsis sp* |
| Ascoteroside | *Ascotricha amphitricha Mycoleptodiscus atromaculans* | Cryptocandin | *Cryptosporiopsis quercina* |
| Ergokonin A | *Trichoderma longibrachiatum T. koningii T. viride* | WF11899 and related sulfate-derivatives | *Coleophoma empetri Coleophoma crateriformis Tolypocladium parasiticum Chalara sp.* |
| | | Clavariopsins | *Clavariopsis aquatica* |
| | | Papulacandins | *Papularia sphaerosperma* |

cytological and ultrastructural changes, such as growth of pseudohyphae, thickened cell wall, buds failing to separate from mother cells, cells becoming osmotically sensitive and restriction of lysis to the growing tips of budding cells [114, 115].

b. Chitin synthesis inhibitors

Chitin is an insoluble polysaccharide made of β-(1,4)-linked N-acetylglucosamine units. This biological polymer is one of the structural microfibrillar components of the fungal cell wall structure which maintains the morphological shape of the cells and plays an essential role in fungal morphogenesis [116]. In yeasts, chitin accounts for 1% of the cell wall and is distributed differently from glucan. Chitin also links covalently to the cellular glucan, thereby strengthening the wall. There are at least three different chitin synthases; chitin synthase I, involved in repair function during cytokinesis; chitin synthase II, playing a role in primary septum formation between mother and daughter cells; and chitin synthase III, synthesizing lateral chitin in the cell wall [117]. Nikkomycins and polyoxins (peptide-nucleoside antimycotic agents) are well-known inhibitors of chitin synthesis [118].

c. Mannoprotein synthesis inhibitors
Mannoproteins form the outer layer of the cell wall and contain about 50% carbohydrate. The majority of the cell wall mannoproteins are anchored by β-(1,6)- and β-(1,3)-glucan [119] and play several roles in the function of fungal membranes. Inhibitors of the mannoproteins function are the pradimicin/benanomycinfamily, whose chemical structure possesses a benzo [a] naphthacenequinone skeleton [120, 121]. The free carboxyl group of these compounds interacts with the saccharide portion of cell-surface mannoprotein, which disrupts the plasma membrane and causes leakage of intracellular potassium.

### 5.3.2 Sphingolipid Synthesis Inhibitors

Sphingolipids are essential for cellular functions but are present in relatively smaller proportion in the fungal cytoplasmic membrane [122]. Inhibition of sphingolipid synthesis results in growth inhibition and cell death [123, 124]. Three key enzymes in the sphingolipid synthesis pathway have been targeted to search for novel antifungals: serine palmitoyltransferase, ceramide synthase and inositol phosphoceramide (IPC) synthase; the latter lacks a mammalian counterpart and, inhibitors to all three have been discovered from natural sources. Sphingofungins [123, 125], lipoxamycin S.M. [126] and viridiofungins [127] inhibit serine palmitoyltransferase. Fumonisin B1 [128, 129] and australifungin [128] inhibit ceramide synthase, and aureobasidins [130], khafrefungin [131], and rustmicin [132] inhibit IPC synthase (Figure 5.4).

a. Serine palmitoyltranferase inhibitors-
Sphingofungins A to F constitute a family of novel chemical structures that resemble the long-chain base intermediates in the sphingolipid pathway. Lipoxamycin was discovered in 1970 as an antifungal compound of unknown mechanism of action [133]. Lipoxamycin and hydroxylipoxamycin, an analogue co-produced in the fermentation, have a long alkyl chain and an amino-containing polar head group, but otherwise do not resemble the sphingoid bases as closely as the sphingofungins do [126]. Viridiofungins A, B and C comprised a novel family of amino alkyl citrates that have potent, broad spectrum antifungal activity [134]. It has been demonstrated by a variety of biological and biochemical means that sphingofungins, lipoxamycin, hydroxylipoxamycin and viridiofungins

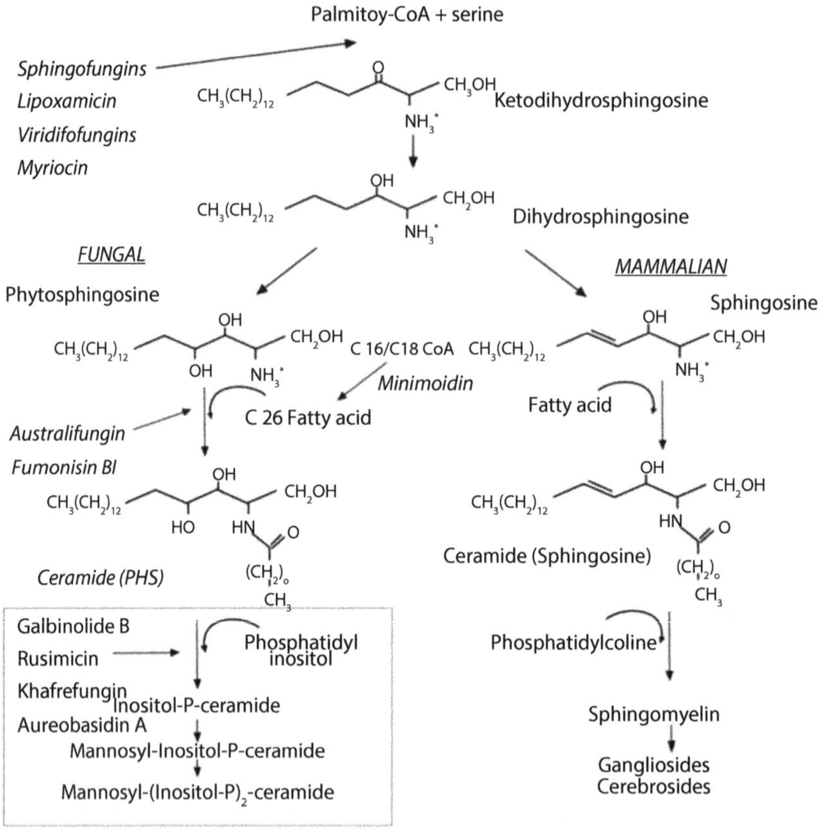


**Figure 5.4** Sphingolipid biosynthesis pathway showing different inhibitors [89].

are specific inhibitors of serine palmitoyltransferase in fungi at nanomolar concentrations [123, 125, 127].

b. Ceramide synthase inhibitors –
The fumonisins are mycotoxins initially characterized as tumour-promoting agents associated with severe toxicological effects in animals [129]. Although fumonisin B1 inhibits fungal ceramide synthase in vitro [128], fumonisins have very poor activity against whole cell fungal sphingolipid synthesis or growth and its limited penetration could account for their poor antifungal activity. Australifungin is a highly potent, broad-spectrum antifungal compound containing a unique combination of α-diketone and β-ketoaldehyde functional groups. It was the first non-sphingosine-based inhibitor described for the sphingolipid biosynthetic pathway. Australifungin inhibits ceramide synthase in vitro at nanomolar concentrations and the enzyme inhibition accounts for the arrest of sphingolipid synthesis [127].

c. IPC synthase inhibitors –
Aureobasidins A to R, are cyclic depsipeptides described as antifungal agents with high in-vitro activity, particularly against *C. albicans*. Aureobasidins A, B, C and E are reported as most potent [135]. Their mechanism of action remained unknown until the inhibition of IPC

synthase by aureobasidin A was observed [130]. Khafrefungin is a novel 22-carbon linear polyketide acid esterified to an aldonic acid that shows a broad antifungal spectrum, with *C. albicans* being the most susceptible organism in vitro. The compound causes the accumulation of ceramide and inhibits the IPC synthase of *S. cerevisiae* and pathogenic fungi at picomolar to nanomolar concentrations [131]. Rustmicin (also named galbonolide A) is a macrolide antifungal agent which inhibits of sphingolipid synthesis at the IPC synthase level [132]. The rustmicin-related macrolide galbonolide B was also reported to inhibit IPC synthase, but with less potency [136].

d. Fatty acid elongation inhibitors
Minimoidin is a novel compound that indirectly inhibits sphingolipid synthesis by blocking the fatty acid longation pathway, thus depriving the ceramide synthase of substrate [130]. Also, it inhibited the incorporation of 14C-malonyl-CoA into long-chain fatty acids.

e. Inhibition of Membrane-Associated Lipids
N'-(3-bromo-4-hydroxybenzylidene)-2-methylbenzo hydrazide (BHBM) and 3-bromo-N'-(3- bromo-4-hydroxybenzylidene) benzohydrazide (D0) represents a new class of antifungal compounds, termed "hydrazycins". These agents were identified in a screen of synthetic compounds inhibiting sphingolipid biosynthesis in *Cryptococcus neoformans*, a process demonstrated to be required for fungal growth in vivo [137]. These compounds inhibit vesicle trafficking of precursor lipids, such as ceramides, to the cell surface, thereby inhibiting glucosylceramide biosynthesis and cell division. Additionally, these compounds specifically inhibit fungal, but not human, glucosylceramide synthesis, suggesting fungal-specific cellular inhibition.

BHBM and D0 were identified in a screen for antifungal activity against *Cryptococcus neoformans* at alkaline pH (138). BHBM showed promising in vitro inhibitory activity against multiple isolates of two *Cryptococcus* species (C. *neoformans* and *C. gattii*), as well as *Histoplasma capsulatum*, *Blastomyces dermatitidis*, *Pneumocystis murinum* and *Pneumocystis jirovecii*. Notably, the strains tested included fluconazole-resistant *Cryptococcus* strains. In vivo activity for BHBM and D0 was assessed in murine models of infection due to *Cryptococcus neoformans*, *Candida albicans* and *Pneumocystis murinum*, resulting in significant increases in survival times compared to controls. Additionally, these drugs showed synergy with both fluconazole and amphotericin B, suggesting potential for combinatorial therapy. These hydrazycins were well tolerated in animal models of invasive fungal infection, though some interaction was observed with the immunosuppressive corticosteroid dexamethasone, which resulted in a decreased compound half-life in vivo [137].

### 5.3.3 Ergosterol Synthesis Inhibitors

The mode of action of azole derivatives and allylamine antifungal agents is based on the ergosterol biosynthesis pathway inhibition at different steps (Figure 5.3). Ergosterol is the major component of the fungal cell membrane which functions as a bioregulator of membrane fluidity, asymmetry and integrity. Earlier imidazole derivatives had a complex mode of action inhibiting several membrane-bound enzymes as well as membrane lipid biosynthesis [53, 55, 139]. Cytochrome P-450 lanosterol 14-alpha-demethylase, encoded by the *ERG11* gene for Erg11p, is the point of action of fluconazole, voriconazole, itraconazole and posaconazole [140–142] and mutations in this gene result in resistance against fluconazole. The pathway, at lanosterol step, is not blocked when the enzyme is inhibited by azole derivatives. The 14-methyl group of lanosterol is important to generate 14-methylated intermediates; one of these is toxic (14-methylergosta-8,24(28)-dien-3,6-diol) and responsible for fungal inhibition. Some authors found that the fungal inhibition was based in accumulation of these toxic intermediates rather than in lack of ergosterol permeabilizing the plasma membrane (union with phospholipid). C14-sterol reductase ends the C-14 modification reactions initiated by cytochrome P-450 lanosterol 14-alphademethylase. Attack at this target by fenpropimorph in *S. cerevisiae* produces an accumulation of ignosterol (ergosta-8,14-dienol) which is toxic. It perturbs membrane and inhibits uptake of glucose and pyridines [143].

Delta 5,6-desaturase, encoded by the *ERG3* gene, transforms 14-methyl intermediates into the toxic compound 14-methylergosta-8,24 [28] -dien-3,6-diol [144.]. *ERG3* gene inactivation confers azole resistance in *C. albicans* and *S. cerevisiae* but not in *C. glabrata* and other species. Previous reactions in the ergosterol biosynthesis pathway can be exploited by other antifungal families. Betahydroxymethylglutarate reductase, encoded by the *HMG1*gene in *C. albicans*, is the target of some statins (lovastatin, zocor) that can synergically act with fluconazole. Allylamine antifungals (terbinafine and naftifine) are also ergosterol biosynthesis pathway inhibitors at squalene epoxidase; this reaction is encoded by the *ERG1 gene*. Their mode of action is achieved by inhibiting earlier but different steps of ergosterol biosynthesis pathway than azole derivatives. The inhibition site is located at squalene-epoxidation, producing an accumulation of the sterol precursor squalene and the absence of any other sterol intermediate. This effect produces fungal death rather than ergosterol deficiency through an ergosterol depletion and accumulation of squalene [145.]. An overexpression of *ERG1* has been demonstrated in cells exposed to terbinafine (*C. albicans, S. cerevisiae, Aspergillus fumigatus*) [143].

### 5.3.4 Protein Synthesis Inhibitors

Among the reported antifungal agents, sordarin is the only reported protein synthesis inhibitor exhibiting specificity for only fungal translation elongation factor [139]. Several sordarin derivatives with different spectra of susceptible species have been reported and they show promise for future use. Interestingly, early antifungal screens in the 1970s excluded sordarin, but two decades later, renewed appreciation for it arose as a consequence of its potent *in vitro* inhibition of protein synthesis in *C. albicans.* Sordarin [1] was isolated in 1969 from the fungus *Sordaria araneosa* by scientists at the Sandoz Co., in Switzerland (Table 5.8), and it was patented as SL 2266 [146, 147]. Upon degradation of sordarin with concentrated HCl in acetone a diterpenoid aglycone called sordaricin [2] was obtained. Sordarin production by fermentation was optimized to simplify purification [148] and to increase the yield [149].

**Table 5.8** Natural Sordarin producing strains [157].

| Fungal species (strain code) | Strain origin | Compound reported |
|---|---|---|
| *Sordaria araneosa* Cain (ATCC36386) | Unknown | Sordarin |
| *Sordaria araneosa* Cain (ATCC36386) | Unknown | Neosordarin, hydroxysordarin |
| *Hypoxylon croceum* (M97–25) | Driftwood, mangrove estuary, Everglades (Florida, US) | Hypoxysordarin, sordarin |
| *Graphium putredinis* (F13302/F13310) | Soil, excavated site, Leeds (UK) | GR135402 (and steroisomers), zofimarin, acetyl-sordarin, 6-hydroxy-GR135402 |
| *Penicillium minioluteum* (F31405) | Soil, Saitama prefecture (Japan) | BE-31405 |
| *Podospora pleiospora* (D01035) | Rabbit dung, Braunton Burrows, Devon (UK) | Sordarin, sordarin B, hydroxysordarin, sordaricin |
| *Xylaria* sp. (PSU-D14) | *Garcinia dulcis* leaves, Songkhla Prov. (Thailand) | Sordaricin |
| Unidentified fungus (SCF1082A) | Unknown | SCH57404 |
| *Xylaria* sp. (A19–91) | Wood, Lescun (France) | Xylarin (¼SCH57404) |
| *Zopfiella marina* (SANK 21274, CBS 155.77) | Marine mud, Chinese Sea (Taiwan) | Zofimarin |
| *Morinia pestalozzioides* (MF6856) | *Sedum sediforme*, Sierra Alhamilla, Almerı´a (Spain) | Moriniafungin |

Unlike traditional antifungal agents, which target only the integrity of the cell membrane through binding of ergosterol or inhibition of its biosynthesis [139, 150], sordarin acts on elongation factor 2 (EF2). The target selected is often EF2 and EF3 (soluble elongation factors) which are necessary for protein synthesis [151]. This enzyme catalyzes the translocation of the ribosome along mRNA during elongation of the emerging polypeptide chain [152]. Sordarin inhibits this translocation by stabilizing the EF2/ribosome complex. Strong activity against *Saccharomyces cerevisiae* [51, 52, 153–156], and a number of pathogenic fungi make sordarin a promising antimycotic agent.

### 5.3.5 Novel Targets

a. ***Calcineurin.*** The calcineurin pathway is important in eukaryotes and is potentially a target of selective inhibitors that could be used as antifungal [158, 159]. The serine/threonine phosphatase calcineurin (also known as protein phosphatase 3; a heterodimer composed of the subunits calcineurin A and calcineurin B) is the target for the most commonly used transplant anti-rejection drugs: tacrolimus and cyclosporine. Cyclosporine was initially discovered in a screen as an anti- *Candida* molecule. Calcineurin is central to the stress responses of many pathogenic fungi, for instance, it has been shown to be essential for the virulence composite of the major fungal pathogens (*Aspergillus fumigatus*, *Candida albicans* and *Cryptococcus neoformans*) and thus for fungal survival and fitness in the host [160]. Furthermore, calcineurin can be linked to other well-studied stress signalling pathways, including the heat shock protein 90 (Hsp90) pathway [161]. Hsp90 (Table 5.9) has also become an excellent target for antifungal inhibitors, and an antibody against Hsp90 (efungumab; also known as Mycograb; NeuTec Pharma/Novartis) in combination with amphotericin B was used for a substantial, comparative clinical trial where it demonstrated therapeutic efficacy in invasive candidiasis [162]. The potential to make calcineurin a viable target for antifungal drug discovery has been contingent on finding specific inhibitors having the differential ability to block only fungal calcineurin. In this respect, the inhibitors must be fungicidal with minimal or no immunosuppressive activity. As the structures of both fungal and human homologues of calcineurin are known, a boost has been given to the progress in discovering potent antifungal, anti-calcineurin molecules with reduced immunosuppressive activity. Some are currently being tested in animal models [159, 163].
b. ***Trehalose.*** The trehalose pathway comprises several enzymes that are connected to glycolysis and creates the regulatory molecule trehalose-6-phosphate and the disaccharide sugar trehalose [164]. Relative to compounds targeting the RAS and calcineurin pathways, molecules targeting this pathway are at a much earlier developmental stage, and fewer inhibitors have been discovered thus far. Despite this, many targets within the trehalose pathway possess attractive features of antifungal agents. Unlike many of the fungal stress pathways that are currently

**Table 5.9** Targets identified for novel antifungal drugs under development.

| Antifungal drug | Target | Antifungal drug | Target |
|---|---|---|---|
| VT-1124/VT-1598<br>BHBM | Plasma membrane | Tamoxifen | Stress response (Oestrogen receptor) |
| F901318 | DNA/RNA biosynthesis (Inhibition of pyrimidine biosynthesis by blocking dihydroorotate dehydrogenase activity, preventing nucleotide biosynthesis) | CD1010=/ biafungin<br>SCY-078<br>Nikkomycin Z | Cell Wall (Chitin synthase) |
| T-2307<br>Ilicicolin-H | Mitochondria | MGCD290 | Glycosyl phosphatidylinositol synthesis (prevents attachment of mannose proteins to the outer cell wall).<br>Probably blocks fungal acetyl-CoA synthetase 1<br>Increases host immune response by downregulating host chaperone proteins |
| Sampangine | Heme Biosysnthesis | APX001 | Unknown, but taken up by Sit1 |
| AR-12 | Fungal carbon Metabolism specifically acetyl-Co A synthase | AR-12 | Inositol phosphorylceramide synthase |
| E1210/1211 | GPI Anchor Biosynthesis through inhibition of fungal Gwt1 protein; inhibition of germ tube formation, biofilm formation and adherence to plastic | ASP2397 | Binds to granulocytes and thereby concentrates the antifungal at the infection |
| Efungumab (also known as Mycograb)<br>Geldanamycin-like agents | Hos2 | Aureobasidin A | Hsp90 |
| Cloudbreak molecules (bispecific antibodies) | Dihydroorotate dehydrogenase | | |

being studied, this pathway is present in fungi, bacteria, plants and invertebrates, but not in mammals. Therefore, minimal potential for inhibitor toxicity will be seen, unless there are off-target issues. Again, this pathway contains two primary synthesizing enzymes (trehalose-6-phosphate synthase (Tps1) and trehalose-6-phosphate phosphatase (Tps2). Both have been validated as essential to tolerating stress in all of the major fungal pathogens [164]. Genetically blocking this pathway kills pathogenic fungi, which clearly demonstrates that this pathway is a fungicidal target [165, 166]. Recently, the crystal structures of the *C. albicans* proteins Tps1 and Tps2 have been solved [167], so molecules that specifically bind to active catalytic components to inhibit the enzymes could be found computationally or in compound libraries [167].

## 5.4 Development of Resistance towards Antifungal Agents

One of the biggest problems with some of the established drugs like amphotericin B was nephrotoxicity [168], and emerging resistance to azole drugs [169]. Lately, lipid formulations of low toxicity polyenes along with new triazoles like voriconazole, rovuconazole and pasaconazole with wider action spectrum have been introduced [170]. Thus, three major groups of antifungal agents i.e., azoles, polyenes and allylamine/thiocarbonates were established which work by inhibiting synthesis of ergosterol which is a major component of the fungal cell wall or directly interact with it to destroy cellular integrity.

### 5.4.1 Minimum Inhibitory Concentration

Antifungal-drug resistance is usually quantified using the minimum inhibitory concentration (MIC), in which growth in the presence of a range of drug concentrations is measured over a defined time period according to a standard protocol. The lowest drug concentration resulting in significant reduction of growth (usually either 50% or 90% reduction of growth compared with growth in the absence of the drug) is the MIC.

However, the limitations of MIC are as follows:

a. If the fungistatic or fungicidal activity is clearly seen, estimating MIC is easy but when the growth is not inhibited clearly, as is the case for many strains that show strong trailing, the MIC can be difficult to assess. In this case, fitness measurements at specified drug concentrations might provide a better measure of resistance.
b. MIC always does not predict the clinical outcome of antifungal therapy [171] because disease is the result of the complex interactions between the pathogen and host [172].
c. MIC has a limitation as a predictor of the progression of disease caused by a fungal pathogen because it can be inaccurate in predicting the fitness of the infecting microorganism, owing to strong environment/genotype

interactions [173, 174]. Whether fungal fitness measures, such as population doubling time during the exponential growth phase or assessing the number of doublings required to attain maximum cell density over a range of drug concentrations, might be better predictors of therapeutic outcome has not been evaluated. Also, in some cases, tolerance might be important to the outcome of treatment. Tolerance is measured in time-kill assays and tests of minimum lethal concentration of drugs [175]. Thus, drug tolerance and drug resistance are not always positively correlated.

### 5.4.2 Antifungal-Drug-Resistance Mechanisms

Multiple resistance mechanisms have been hypothesized (Figure 5.5). Unlike bacteria which show plasmid or transposoon mediated mechanism for antibiotic resistance, fungi possess no such tool. Here the mechanisms to incur resistance are quite well understood. For example, the resistance to polyenes is acquired by fungi which possess ergosterol membranes. The fungicidal activity towards polenes decreases with prolonged exposure [176]. On the other hand, azoles exhibit multiple mechanisms for the development of both primary and acquired resistance, including mutations

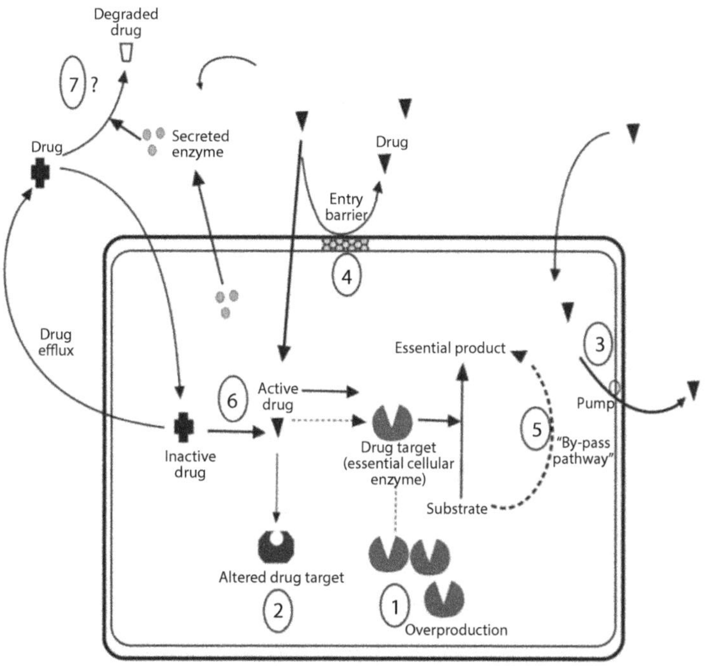


**Figure 5.5** Mechanism for development of resistance. 1- overproduction of target enzyme which prevents inhibition of total biochemical activity; 2- alteration of drug target does not allow binding to take place; 3: uptake and transportation; 3- uptake and translocation of drug via efflux pump; 4- cellular entry of drug is prevented at membrane or cell wall; 5- cell compensation for loss of function inhibition by an alternative pathway; 6- enzyme mediated conversion of inactive drug to active form is inhibited; 7- degradation of drug via the secretion of some extracellular enzyme [53].

in the gene encoding lanosterol 14α-demethylase [*ERG11*]. This makes the catalytic activity of fungi free of the blocking activity of azole, and the amplification of *ERG11*, for overwhelming the inhibitory capacity of azole drugs [171]. Some fungal species have amplified or induced efflux pumps to remove azoles from the fungal cell further reducing its efficacy [177]. In the past five to six years, with the increasing use of echinocandins, an increase is resistance to the drugs particularly in yeasts has been reported. It is suggested that development of mutations in *FKS1* encoding the 1,3-β-d-glucan enzyme that helps in the formation of the fungal cell wall is responsible for the resistance. This echinocandin resistance mechanism has been most prominently observed in the haploid yeast, *Candida glabrata* [178]. Recently, mutations in MSH*1*, a mismatch repair gene, have also been observed to cause multiple drug-resistant phenotypes (to azoles and echinocandins) in yeast strains [179]. Drug-resistant mutations in the pyrimidine pathway occur at the rate of 1 in every $10^6$–$10^7$ yeast cells, so 5-flucytosine is very vulnerable to the development of drug resistance if used as a monotherapy in high fungal burden infections [180]. In principle, these mechanisms fall into three distinct categories (Table 5.10):

1. Decrease of effective drug concentration – The drug intracellular concentrations can be decreased by active efflux. It is known that drug resistance can be mediated by the activity of several efflux transport systems, including ATP-binding cassette (ABC) transporters and transporters of the major facilitator superfamily (MFS). The analysis of fungal pathogen genomes have identified varying numbers of ABC transporters and MFS transporters with different topologies. ABC transporters are arranged in different subfamilies; however, they all contain membrane spanning domains and use ATP hydrolysis for drug transport. MFS transporters are transmembrane proteins, which use the electrochemical proton-motive force to mediate drug efflux. MFS are involved in multidrug resistance (MDR) (MFS– MDR transporters) function as proton antiporters and are classified into two groups: the drug: H+ antiporter-1 DHA1 family and the drug:H+ antiporter-2 DHA2 family [181, 182]. Fungal ABC transporters have been arranged into several classes; however, only ABC transporters of the pleiotropic drug resistance (PDR) class are relevant for antifungal drug resistance. The drug target is overexpressed. By increasing the number of drug targets, the effective drug concentration needs to be also increased to saturate all target molecules, which results in drug resistance [183].
2. Alteration of target enzyme – Changes in target protein either prevent binding of the antifungal drug or prevent the allosteric inactivation of the target after the inhibitor binds. This is a relatively small mutational target, as only a few specific amino-acid changes confer resistance. Alternatively, the target protein might be overexpressed, resulting in sufficient activity in the presence of the drug. Drug targ*et al*terations have been reported for at least two classes of antifungal agents, including azoles and echinocandins. The targets of these two drugs are a 14α-lanosterol demethylase and a β-1,3 glucan synthase, respectively.

**Table 5.10** Summary of mode and mechanism of Antifungal drugs.

| Antifungal | Mode of action | Mechanism of action | Mechanism of resistance |
|---|---|---|---|
| Polyenes<br>Amphotericin B<br>Nystatin | Fungicidal | Binding to sterols in cell membrane forming aqueous pores | Reduction of ergosterol concentration ablating drug-target binding. Alteration in POL gene family |
| Azoles<br>Fluconazole<br>Itracanazole<br>Ketaconazole<br>Posaconazole<br>Voriconazole<br>Miconazole | Fungistatic | Affects ergosterol biosynthesis by blocking the key enzyme, lanisterol 14α-demethylase (Erg11p) | Efflux of drugs by multidrug transporters; ABC gene family.<br>Amino acid substitution to Erg 11p affecting drug-target binding<br>Over expression of Erg 11p minimizing effect of drug<br>Change in toxic sterol concentration due to mutation in Erg 3 alleles |
| Echinocandins<br>Capsofungin<br>Micafungin<br>Anidulafungin | Fungistatic or Fungicidal | Binding to β-(1,3)-D-glucan syntahse and inhibition of β-(1,3)-D-glucan, a cell wall component | Mutation in Fks 1 and Fks 2 binding units |
| Allylamines<br>Terbinafine<br>Naftifine | Fungistatic | Interferes with ergosterol biosynthesis by inhibiting squalene epoxidase (Erg 1p) | Interference from multidrug transporters |
| Pyrimidine analog<br>5-Flucytosine | Fungicidal | Inhibits cellular DNA and RNA synthesis affecting molecular machinery | Mutations in cytosine permease and deaminase |

Lanosterol demethylase is encoded by *ERG11* in *C. albicans* and *Cyp51A* and *Cyp51B* in *A. fumigatus*. Mutations in *ERG11* resulting in non-synonymous amino acid substitutions that are present in azole-resistant *C. albicans* isolates are numerous and were shown to decrease the affinity of the target to azoles [184]. The effects of *ERG11* mutations have different outcomes on azole MICs that depend on structural features of azole drugs. Although most known mutations decrease affinity to fluconazole, they have only a moderate effect on posaconazole affinity [185]. In many cases, simultaneous *ERG11* mutations can be present on the same *ERG11* allele and be accompanied by drug transport modifications, thus resulting in azole-resistant isolates with high MIC values against azoles (for example, fluconazole MIC > 128 μg/ml) [186].

3. Alteration of metabolism – Metabolic bypasses occur when given metabolic pathways are perturbed by loss or strong decrease of specific functions. These alterations can be compared to compensatory mechanisms in which cells divert the toxic effects exerted by some antifungal agents.
4. Loss of enzyme activity prevents the accumulation of a toxic product in the presence of the drug. This is a relatively broad mutational target, as myriad amino-acid changes result in loss of function and a resistant phenotype. In another strategy, the drug is sequestered in extra- or intracellular compartments. Fungal pathogens have the ability to sequester drugs within extracellular compartments. Several fungal pathogens, including *Candida* and *Aspergillus* spp., are able to form biofilms in specific growth conditions [187]. Biofilms are multicellular structures in which cells form a dense network that is covered by the so-called matrix. The matrix is composed of different elements in *C. albicans* biofilms, including several cell wall polymers [188]. Biofilm formation is known to be associated with resistance to several drugs, including azoles, polyenes, and pyrimidine analogs [189]. Interestingly, recent data showed that the matrix participates to this process by its capacity to sequester antifungal agents. This process has been clearly documented for fluconazole [188, 190] as suggested for AmB in *C. albicans* [191].

## 5.5 Market and Drug Development

It is more challenging to develop a safe and non-toxic antifungal drug as against antibacterial. This is predominantly due to the similarities seen in effective drug targets between fungi and closely related human genes/gene products. This is why mostly the target for antifungals is the fungal cell membrane components such as ergosterol which differ from the human counterparts [139]. Antifungal drug targets which are closely related to human genes or their products have the potential to interact with host tissue as well as the target pathogen, leading to variable level of cytotoxicity within the therapeutic dose range. Thus, for such molecules, the therapeutic indices will be too low to be taken forward as potential new medicinal entities. Despite countless highly potent antifungal molecules having been discovered, only a few of them have worked out to be viable therapeutic candidates [192, 193], because proving their safety for human use is much more difficult.

As discussed in the earlier sections, there is great need of new antifungal therapies for serious/life-threatening fungal infections. An increasing need is also there to have new solutions to superficial/mucosal fungal infections which, whilst not life-threatening, have a major impact on the quality of life of those who suffer from seborrhoea, vulvo-vaginal candidiasis, dermatophytoses etc. [194–198]. These chronic conditions provide a substantial market opportunity [199] as 'lower-hanging fruit' for a commercial path to develop new antifungal remedies. The major factors driving the market are increasing awareness of fungal infections with the growing incidence of antifungal infections all over the world, increasing government and corporate funding in the industry. Further, the development of public-private partnership in the pharmaceutical industry and the rising popularity of over-the-counter antifungal drugs for dermal infections is responsible for the growing antifungal drugs market size. Also, the rise in population with weak immune systems due to cancer chemotherapy or acquired immunodeficiency syndrome, in whom the occurrence of fungal infections is much more than average, will drive the market. High penetration of conventional drugs used for the treatment of fungal infections is restraining the market growth. The key challenges faced by the market are cost of drugs and their effectiveness in treatments.

A number of hurdles to antifungal drug development must be overcome to determine the clinical utility of these compounds and the development path [13]:

a. The value of limited- versus broad-spectrum antifungal activity must be balanced.
b. The toxicity data will always be crucial for eukaryotic targets
c. Biomarkers should be accepted as validated primary end points instead of relying exclusively on mortality as these patents may have other comorbidities. So efficacy signals will require large number of patients.
d. Although there are substantial numbers of invasive fungal infections, many patients need to be screened to get a single patient enrolled in clinical trials. This is particularly true for candidaemia studies.
e. The use of compounds for prophylaxis or for treatment should be considered as an initial focus for the development of safe and broad-spectrum compounds.
f. Although more rapid fungicidal activity and shorter treatment courses are needed, initial studies will probably require prolonged treatment duration.
g. Invasive fungal infections classically occur in patients with the most severe illnesses, so both underlying diseases and tolerability can affect the outcome assessment of the antifungal drug.
h. Many of the scoring systems use radiographic results which is imprecise.
i. Success should include disease control, rather than disease eradication.
j. The clinical studies are particularly expensive. These patients are fragile and require close monitoring as well as detailed records. Furthermore, there may be interruptions of assessments due to decisions by the clinical care team.

The global antifungal drugs market is segmented based on drug type into echinocandins, azoles, polyenes, allylamines and other drugs. Also, the market is segmented by type of therapeutic indications into aspergillosis, dermatophytosis, candidiasis and other therapeutic indications. And the market is also divided by types of formulations into powders, ointments, drugs and pastes. While triazoles (fluconazole, itraconazole, posaconazole and voriconazole) and imidazoles (ketoconazole) are systemically acting, echinocandins (anidulafungin, caspofungin and micafungin) are lipopeptides derived from natural fungal fermentation products and polyenes (amphotericin B deoxycholate and liposomal amphotericin B Corifungin) are systemically acting and disrupt eukaryotic cellular membranes. There has been a major increase in the prescription of antifungal drugs after the introduction of fluconazole into the market in the late 1980s, and again in the late 1990s. The systemic antifungal market has continued to experience growth since 1999, increasing in value from $2.1 billion to $3.3 billion in 2003. The azoles dominate the systemic antifungal market, accounting for 52% of total sales in 2003 followed by the echinocandins antifungal drug market [200].

On the basis of geography, the global market is analysed under regions, namely North America, Europe, Asia-Pacific, Latin America, and Middle East & Africa. North America and Europe together contribute to the major share of the global antifungal drugs market, while the market in Asia-Pacific is the fastest growing, followed by Latin America. Some of the major companies dominating the market include Alternaria, Pfizer, Novartis, Sanofi-Aventis, Merck & Co., Asperqillus, Kramer Laboratories, Bayer Healthcare, Enzon Pharmaceuticals, Glaxosmithkline Gilead and Abbott Laboratories. The Global Antifungal Drugs Market was worth $12.15 billion in 2016 and estimated to be growing at a CAGR of 3.68%, to reach $14.56 billion by 2021. The market is expected to grow enormously in the next few years, with the growing global population.

## 5.6 Conclusions

Human exploration brings mankind into contact with new reservoirs of pathogens which is accelerated with the evolution and natural transformation of pathogenic organisms. New patterns of pathogenesis are emerging which are often proving resistant to the conventional therapeutics. Modelling with existing outcome data shows that mortality associated with these priority fungal diseases concurrently with the Joint United Nations Program on HIV/AIDS (UNAIDS) 90–90–90 campaign could save over 1.6 million lives of persons living with HIV globally over the next five years [11].

Rapid acquisition of resistance towards the antifungal drugs is making it evident that a massive effort needs to be taken up to synthesise new and novel drugs. It is also important to discover new drug targets to facilitate the fungicidal and fungistatic activity. Over 80% of patients could be saved from dying with universal availability of fungal diagnostics and potent antifungals agents, based on well-documented treatment response rates. However, the early recognition and management of serious fungal infections is always a challenge, but especially in resource-limited settings as many conventional diagnostics tests are slow, antifungal treatment can be expensive and/or toxic and is not equally available in all countries. Though many new drugs and natural products are being identified, still concerted efforts are needed to develop a comprehensive and aggressive strategy for commercialization and application of fungal chemotherapy. Other factors governing better outcomes include patient compliance with long-term treatment, drug-drug interactions, limited clinical experience of excellent care in many settings and co-morbidities reducing the potential for survival and cure [3].

## Acknowledgement

The author acknowledges the support of the facilities extended by Banasthali Vidyapith towards the completion of this chapter.

## References

1. G. D. Brown, D. W. Denning, N. A.R. Gow, S. M. Levitz, M. G. Netea, and T. C. White, *Science Translational Medicine,* Vol. 4, p.165, 2012.
2. J. Kriengkauykiat, J.I Ito, and S.S. Dadwal, *Clinical Epidemiology*, Vol. 3, p. 175, 2011.
3. Global Action Fund for Fungal Infections (GAFFI). http://www.gaffi.org/media/fact-sheets/
4. H. Wisplinghoff, T. Bischoff, S. M. Tallent, H. Seifert, R. P. Wenzel, and M. B. Edmond, *Clinical Infectious Diseases,* Vol. 39, p. 309, 2004.
5. D.P. Kontoyiannis, K. A. Marr, B. J. Park, B. D. Alexander, E. J. Anaissie, and T.J. Walsh, *Clinical Infectious Diseases,* Vol. 50, p. 1091, 2010.
6. D.W. Denning, *Thorax*, Vol. 70, p. 613, 2015.
7. R. Rajasingham, M.S. Rachel, J.P. Benjamin, N.J. Joseph, P.G. Nelesh, M.C.Tom, D.W. Denning, A.Loyse, and D.R. Boulware, *Lancet Infectious Diseases*, Vol.17, p. 873, 2017.
8. D.W. Denning, A. Pleuvry, and D.C. Cole, *European Respiratory Society*, Vol. 41, p. 621, 2013.
9. D.W. Denning, *Mycoses*, Vol. 56, p. 13, 2013.
10. D.W. Denning, *European Journal of Clinical Microbiology and Infectious Diseases*, Vol 36, p. 923, 2017.
11. D.W. Denning, *Philosophical Transactions of the Royal Society B: Biological Science*, Vol. 371, p. 20150468, 2016.
12. M. Pegorie, and D.W. Denning, *Journal of Infection*, Vol. 74, p. 60, 2017.
13. J.R.Perfect, *Nature Reviews Drug Discovery,* Vol. 16 (9), p. 603, 2017.
14. T.M.Arnold, E. Dotson, G.A. Sarosi, and C.A. Hage, *Proceedings of the American Thoracic Society*, Vol. 7, p. 222, 2010.
15. W.J. Steinbach, P.R. Juvvadi, J.R Fortwendel, and L.E Rogg, *Medical Mycology*, Vol. 49 (Suppl 1), S77, 2011.
16. Current Index of Medical Specialties, July-October 2016. UBM Medica India Pvt. Ltd. 2016.
17. K. S. Rao, M. Nundy, and A. S. Dua, National Commission on Macroeconomics and Health. Financing and Delivery of Health Care Services in India. New Delhi: Ministry of Health & Family Welfare, Government of India, Delivery of Health Services in the Private Sector, p. 89, 2005.
18. G.M. Allan, J. Lexchin, and N. Wiebe, *PLoS Medicine,* Vol. 4, p. 283, 2007.
19. D. R. King, and P. Kanavos, *Croatian Medical Journal,* Vol.43, p. 462, 2002.
20. A. S., Kesselheim, A. S. Misono, J. L. Lee, M. R. Stedman, M. A. Brookhart, N. K. Choudhry, and W.H. Shrank, *JAMA,* Vol. 300, p. 2514, 2008.

21. B.M. Davit, P.E. Nwakama, G.J. Buehler, D.P. Conner, S.H. Haidar, D.T. Patel, Y. Yang, L.X. Yu, and J. Woodcock, *Annals of Pharmacotherapy*, Vol. 43, p. 1583, 2009.
22. W. H. Shrank, E.R. Cox, M.A. Fischer, J. Mehta, and N.K. Choudhry, *Health Affairs (Millwood)*, Vol. 28, p. 546, 2009.
23. J. Menzin, J.L. Meyers, M. Friedman, J.R. Korn, J.R. Perfect, A.A. Langston, R.P. Danna, and G. Papadopoulos, *American Journal of Infection Control*, Vol. 39, p.15, 2011.
24. C.R. Hennen, *Current Medical Research and Opinion*, Vol. 25, 1751. 2009.
25. B.Spellberg, J.H. Powers, E.P. Brass, L.G. Miller, and J.E. Edwards Jr., *Cinical Infectious Diseases*. Vol. 38, 1279, 2004.
26. J. Conly, and B. Johnston, *Canadian Journal of Infectious Diseases*, Vol. 16, p. 159, 2005.
27. FDA Document: Guidance for Industry - Developing Antimicrobial Drugs - General Considerations for Clinical Trials. 1998.
28. D. W. Woolley, *Science*, Vol. 100 (2609), p. 579, 1944.
29. R. A. Fromtling, *Clinical Microbiology Reviews*, Vol. 1(2), p. 187, 1988.
30. M. A. Burgess, and G. P. Bodey, *Antimicrobial Agents Chemotherapy*, Vol. 2 (6), p. 423, 1972.
31. D. Tettenborn, *Postgraduate Medical Journal*, Vol. 50 (S 1), p. 17, 1974.
32. R. C. Heel, R. N. Brogden, and G. E. Pakes, *Drugs*, Vol.19 (1), p. 7, 1980.
33. J. Heeres, L. J. J. Backx, J. H. Mostmans, and J. Van Cutsem, *Journal of Medicinal Chemistry*, Vol.22 (8), p. 1003, 1979.
34. J. W. Van Der Meer, J. J. Keuning, and H. W. Scheijgrond, *J. Antimicrobial Chemotherapy*, Vol. 6 (4), p. 552, 1980.
35. J. R. Perfect, D. T. Durack, J. D. Hamilton, and H. A. Gallis, *Journal of the American Medical Association*, Vol. 247 (24), p. 3349, 1982.
36. A. Pont, P. L.Williams, and D. S. Loose, *Annals of Internal Medicine*, Vol. 97 (3), p. 370, 1982.
37. J. H. Lewis, H. J. Zimmerman, G. D. Benson, and K. G. Ishak, *Gastroenterology*, Vol. 86 (3), pp. 503, 1984.
38. W. E. Dismukes, G. Cloud, and C. Bowles, *Annals of Internal Medicine*, Vol. 103 (6), Part 1, p. 861, 1985.
39. K.H. Abu-Elteen and M. Hamad, "Antifungal agents for use in human therapy" in K. Kavanaugh ed. *Fungi: Biology and Applications*. Wiley Publishing House, pp. 191–217, 2005.
40. C. A. Arndt, T. J. Walsh, C. L. McCully, F. M. Balis, P. A. Pizzo, and D. G. Poplack, *Journal of Infectious Diseases*, Vol. 157 (1), p. 178, 1988.
41. K. W. Brammer, P. R. Farrow, and J. K. Faulkner, *Review of Infectious Diseases*, Vol. 12 (S 3), p. S318, 1990.
42. A. Espinel-Ingroff, S. Shadomy, and R. J. Gebhart, *Antimicrobial Agents in Chemotherapy*, Vol. 26 (1), p. 5, 1984.
43. C. L. Terrell, *Mayo Clinic Proceedings*, Vol. 74 (1), p. 78, 1999.
44. J. A. Barone, B. L. Moskovitz, J. Guarnieri, A. E. Hassell, J. L. Colaizzi, R. H. Bierman, and L. Jessen, *Antimicrobial Agents and Chemotherapy*, Vol. 42 (7), p. 862, 1998.
45. M. A. Boogaerts, J. Maertens, R. Van der Geest, A. Bosly, J.-L. Michaux, A. Van Hoof, M. Cleeren, R. Wostenborghs, and K. De Beule, *Antimicrobial Agents in Chemotherapy*, Vol.45 (3), p. 981, 2001.
46. E. Albengres, H. Le Loüet, and J.P. Tillement, *Drug Safety*, Vol. 18 (2), p. 83, 1998.
47. D. W. Denning, K. Venkateswarlu, K. L. Oakley, M.J. Anderson, N.J. Manning , D.A. Stevens , D.W. Warnock , and S.L. Kelly, *Antimicrobial Agents in Chemotherapy*, Vol. 41(6), p. 1364, 1997.

48. J. A. Sabo, and S. M. Abdel-Rahman, *Annals of Pharmacotherapy*, Vol. 34 (9), pp. 1032, 2000.
49. C. Chiou, A. Groll, and T. Walsh, *Oncologist*, Vol. 5 (2), p. 120, 2000.
50. B. A. Potoski, and J. Brown, *Clinical Infectious Diseases*, Vol. 35 (10), p. 1273, 2002.
51. K. M. Pianalto, and J. A. Alspaugh, *Journal of Fungi*, Vol. *2*(4), p. 26, 2016.
52. B.H. Van den, G. Willemsens, W. Cools, P. Marichal, and W. Lauwers, *Biochemical Society Transactions*, Vol 11, p. 665, 1983.
53. M. A. Ghannoum, and L. B. Rice, *Clinical Microbiology Rev*iew, Vol.12(4), p. 501, 1999.
54. A. Lemke, A. F. Kiderlen, and O. Kayser, *Applied Microbiology and Biotechnology*, Vol. 68 (2), p. 151, 2005.
55. R. A. Akins, *Medical Mycology*, Vol. 43(4), p. 285, 2005.
56. P. A. Warn, A. Sharp, J. Guinea, and D. W. Denning, *Journal of Antimicrobial Chemotherapy*, Vol. 53 (5), p. 743, 2004.
57. E. Cevher, D. Sensoy, M. A. M. Taha, and A. Araman, *AAPS PharmSciTech*, Vol.(3), p. 953, 2008.
58. S. B. Zotchev, *Current Medicinal Chemistry*, Vol. 10 (3), p. 211, 2003.
59. J. P. Barrett, K. A. Vardulaki, C. Conlon, J. Cooke, P. Daza-Ramirez, E.G. Evans, P.M. Hawkey, R. Herbrecht, D.I. Marks , J.M. Moraleda, G.R. Park , S.J. Senn, and C. Viscoli, *Clinical Therapeutics*, Vol. 25(5), p. 1295, 2003.
60. P. Caffrey, S. Lynch, E. Flood, S. Finnan, and M. Oliynyk, *Chemistry and Biology*, Vol. 8 (7), p. 713, 2001.
61. N.S. Ryder, I. Frank, and M-C. Dupont, *Antimicrobial Agents Chemother*apy, Vol. 29(5), p. 858, 1986.
62. R. Duschinsky, E. Pleven, and W. Oberhansli, *Acta—Unio Internationalis Contra Cancrum*, Vol. 16, p. 599, 1960.
63. E. Grunberg, E. Titsworth, and M. Bennett, *Antimicrobial Agents Chemotherapy*, Vol. 161, p. 566, 1963.
64. D. Tassel, and M. A. Madoff, *Journal of American Medical Association*, Vol. 206 (4), p. 830, 1968.
65. A. Polak, and H. J. Scholer, *Chemotherapy*, Vol. 21 (3–4), p. 113, 1975.
66. J. M. Benson, and M. C. Nahata, *Clinical Pharmacy*, Vol. 7 (6), p. 424, 1988.
67. D. R. Hospenthal, and J. E. Bennett, *Clinical Infectious Diseases*, Vol. 27 (2), p. 260, 1998.
68. A. Polak, and M. Grenson, *European Journal of Biochemistry*, Vol. 32 (2), p. 276, 1973.
69. A. R. Waldorf, and A. Polak, *Antimicrobial Agents and Chemotherapy*, Vol. 23(1), p. 79, 1983.
70. J. E. Bennett, "Antifungal agents," in J. G. Hardman *et al*., eds., *Goodman and Gilman's The Pharmacological Basis of Therapeutics*, McGraw-Hill, New York, NY, USA, pp. 1175–1190, 1996.
71. F. Dromer, C. Bernede-Bauduin, D. Guillemot, and O. Lortholary, *PLoS ONE,* Vol 3, p. 2870, 2008.
72. T. Bicanic, R. Wood, G. Meintjes, K. Rebe, A. Brouwer, A. Loyse, L.-G. Bekker, S. Jaffar, and T. Harrison, *Clinical Infectious Diseases*, Vol 47, p. 123, 2008.
73. J.R. Perfect, W.E. Dismukes, F. Dromer, D.L. Goldman, J.R. Graybill, R.J. Hamill, T.S. Harrison, R.A. Larsen, O. Lortholary, and M.-H. Nguyen, *Clinical Infectious Diseases*, Vol. 50, p. 291, 2010.
74. R. L. Stiller, J. E. Bennett, and H. J. Scholer, *Journal of Infectious Diseases*, Vol. 147(6), p. 1070, 1983.
75. T. K. Daneshmend and D. W. Warnock, *Clinical Pharmacokinetics*, Vol. 8 (1), p. 17, 1983.

76. A. Vermes, H. J. Guchelaar, and J. Dankert, *Journal of Antimicrobial Chemotherapy*, Vol. 46(2), p. 171, 2000.
77. A. M. Stamm, R. B. Diasio, W. E. Dismukes, S. Shadomy, G.A. Cloud, C.A. Bowles, G. H. Karam, and A. Espinel-Ingroff, *American Journal of Medicine*, Vol.83 (2), p. 236, 1987.
78. P. Francis, and T. J. Walsh, *Clinical Infectious Diseases*, Vol. 15 (6), p. 1003, 1992.
79. C. A. Kauffman, and P. T. Frame, *Antimicrobial Agents and Chemotherapy*, Vol. 11 (2), p. 244, 1977.
80. R. Diasio, D. Lakings, and J. Bennett, *Antimicrobial Agents and Chemotherapy*, Vol. 14 (6), p. 903, 1978.
81. K. M. Williams, A. M. Duffield, R. K. Christopher, and P. J. Finlayson, *Biomedical Mass Spectrometry*, Vol. 8 (4), p. 179, 1981.
82. B. E. Harris, B. W. Manning, T. W. Federle, and R. B. Diasio, *Antimicrobial Agents and Chemotherapy*, Vol. 29(1), p. 44, 1986.
83. J.E.Bennett, W. E. Dismukes, and R. J. Duma, *New England Journal of Medicine*, Vol. 301(3), p. 126, 1979.
84. R. Patel, *Mayo Clinic Proceedings*, Vol. l. 73 (12), p. 1205, 1998.
85. F. Benz, F. Knüsel, J. Nüesch, H. Treichler, W.Voser, R. Nyfeler, and W. Keller- Schierlein, *Helvetica Chimica Acta*, Vol. 57, p. 2459, 1974.
86. C. Keller-Juslén, M. Kuhn, H.R. Loosli, T.J. Petcher, H.P. Weber, and A. Von Wartburg, *Tetrahedron Letters*, Vol.17, p. 4147, 1976.
87. R. Nyfeler, and S.W. Keller, *Helvetica Chimica Acta*, Vol. 57, p. 2459, 1974.
88. W.W. Turner and W. Current, "Echinocandin antifungal agents" in W.R. Strohl, ed. *Biotechnology of Antibiotics*.New York: Marcel-Dekker Inc., pp. 315–334, 1997
89. M. F. Vicente, A. Basilio, A. Cabello, and F. Pelàez, *Clinical Microbiology and Infection*, Vol. 9, p. 15, 2003.
90. K. M. Pianalto, and J. A. Alspaugh, *Journal of Fungi*, Vol. 2(4) p. 26, 2016.
91. C.S Taft, T.Stark, and C.P. Selitrennikoff, *Antimicrob. Agents Chemotherapy*, 32(12), p.1901, 1988.
92. M.A. Maligie, and C.P. Selitrennikoff, *Antimicrobial* Agents *Chemothe*rapy, Vol. 49, p. 2851, 2005.
93. T. Emri, L. Majoros, V.Tóth, and I.Pócsi, *Applied Microbiology and Biotechnology*, Vol. 97, p. 3267, 2013.
94. D.W. Denning, *Lancet*, Vol. 362 p.1142, 2003.
95. S.C. Chen, M.A. Slavin, and T.C. Sorrell, *Drugs* Vol. 71, p.11, 2011.
96. P.G. Pappas, C.A. Kauffman, D. Andes, D.K. Benjamin Jr, T.F. Calandra, J.E. Edwards Jr, S.G. Filler, J.F. Fisher, B.J. Kullberg, L. Ostrosky- Zeichner, A.C. Reboli, J. H. Rex, T.J. Walsh, and J. D. Sobel, *Clinical Infectious Diseases*, Vol. 48, p.503, 2009.
97. D.S. Perlin, *Drug Resistance Update* Vol. 10, p.121, 2007.
98. D.S. Perlin, *Future Microbiology*, Vol. 6, p. 441, 2011.
99. M.A. Pfaller, and D.J. Diekema, *Clinical Microbiology Reviews*, Vol. 20 p.133, 2009.
100. M.A. Pfaller, D.J. Diekema, D. Andes, M.C. Arendrup, S.D. Brown, S.R. Lockhart, M. Motyl, and D.S. Perlin, *Drug Resistance Update*, Vol. 14, p. 164, 2011b.
101. G. Sóczó, G. Kardos, I. Varga, B. Kelentey, R. Gesztelyi, and L. Majoros, *Antimicrobial Agents in Chemotherapy*, Vol. 51, p. 4474, 2007.
102. G. Garcia-Effron, S.K. Katiyar, S. Park, T.D. Edlind, and D.S. Perlin, *Antimicrobial Agents in Chemotherapy*, Vol. 52, p. 2305, 2008.
103. E. Calvo, F.J. Pastor, E. Mayayo, and J. Guarro, *International Journal of Antimicrobial Agents*, Vol. 38, p.360, 2011.
104. E. Calvo, F.J. Pastor, V. Salas, E. Mayayo, and J.Guarro, *Mycopathologia*, Vol.173, p.251, 2012.

105. P.H. Chandrasekar, and J.D. Sobel, *Clinical Infectious Diseases*, Vol.42, p. 1171, 2006.
106. P.K. Mukherjee, D. Sheehan, L. Puzniak, H. Schlamm, and M.A.Ghannoum, *Journal of Chemotherapy*, Vol. 23, p. 319, 2011.
107. G.H. Fleet "Cell walls" in: A.H. Rose and J.S. Harrison, eds. *The Yeasts*. London: Academic Press, pp. 199–277,1991.
108. T. Ohyama, Y. Kurihara, Y. Ono, T. Ishikawa, S. Miyakoshi, K. Hamano, M. Arai, T. Suzuki, H. Igari, Y. Suzuki, and M. Inukai, *Journal of Antibiotics*. Vol. 53, p. 1108, 2000.
109. A. Fujie, T. Iwamoto, H. Muramatsu, T. Okudaira, I. Sato, T. Furuta, Y. Tsurumi, Y. Hori, M. Hino, S. Hashimoto, and M. Okuhara, *Journal of Antibiotics*, Vol. 53, p. 920, 2000.
110. K. Kaida, R. Fudou, T. Kameyama, K. Tubaki, Y. Suzuki, M. Ojika, and Y. Sakagami, *Journal of Antibiotics*, Vol. 54, p.17, 2001.
111. P. Traxler, J. Gruner, and A. Auden, *Journal of Antibiotics*, Vol 30, p. 289, 1977.
112. N.H. Georgopapadakou, and J.S. Tkacz, *Trends in Microbiology*, Vol. 3, p. 98, 1995.
113. J. Onishi, M. Meinz, J. Thompson, J. Curotto, S. Dreikorn, M. Rosenbach, C. Douglas, G. Abruzzo, A. Flattery, L. Kong, A. Cabello, F. Vicente, F. Pelaez, M.T. Diez, I. Martin, G. Bills , R. Giacobbe , A. Dombrowski, R. Schwartz, S. Morris, G. Harris, A. Tsipouras, K. Wilson, and M.B. Kurtz, *Antimicrobial Agents in Chemotherapy*, Vol. 44, p. 368, 2000.
114. A.J. Carrillo-Muñoz, G. Giusiano, P.A. Ezkurra, and G. Quindós; Vicente, M.F., Basilio, A., Cabello and A., Peláez, F. *Clinical Microbiology and Infection*, Vol. 9, p. 15, 2003.
115. J.A. Vázquez, *Clinical Therapeutics*, Vol. 27, p. 657, 2005.
116. J. Ruiz-Herrera, R. Sentandreu and J.P. Martinez, "Chitin biosynthesis in fungi" in: D.K. Arora, R.P. Elander and K.G. Mukerji, eds. *Handbook of Applied Mycology*. New York: Marcel Dekker, Vol. 4, pp. 281–312, 1992.
117. J.A. Shaw, P. C. Mol, B. Bowers, S. J. Silverman, M. H. Valdivieso, A. Durán, and E. Cabib, *Journal of Cell Biology*, Vol. 114, p. 111, 1991.
118. E. Gabib, *Antimicrobial Agents in Chemotherapy*, Vol. 35, p. 70, 1991.
119. J.M. Van Der Vaart, F.A.Van Schagen, and A.T. Mooren, *Biochimica et Biophysica Acta*, Vol. 1291, p. 206, 1996.
120. T. Oki , M. Konishi , K. Tomatsu , K. Tomita , K. Saitoh , M. Tsunakawa , M. Nishio , T. Miyaki, and H. Kawaguchi, *Journal of Antibiotics*, Vol. 41, p. 1701, 1988.
121. T. Takeuchi, T. Hara, H. Naganawa, M. Okada, M. Hamada, H. Umezawa, S. Gomi, M. Sezaki, and S. Kondo, *Journal of Antibiotics* Vol. 41, p. 807, 1988.
122. G. B. Wells, and R.L. Lester, *Journal of Biological Chemistry*, Vol. 258, p. 10200,1983.
123. M. M. Zweerink, A. M. Edison, G. B. Wells, W. Pinto, and R. L. Lester. *Journal of Biological Chemistry*, Vol. 267, p. 25032, 1992
124. S.M. Mandala, R.A. Thornton, B.R.Frommer, J.E. Curotto, W. Rozdilsky, M.B.Kurtz, R.A. Giacobbe, G.F. Bills, M.A., Cabello, I. Martin, F. Peleaz, and G.H. Harris, *Journal of Antibiotics*, Vol. 48, p. 349, 1995.
125. W. S. Horn, J. L. Smith , G. F. Bills , S. L. Raghoobar , G. L. Helms , M. B. Kurtz , J. A. Marrinan , B. R. Frommer , R. A. Thornton, and S. M. Mandala, *Journal of Antibiotics,* Vol. 47, p. 376, 1992.
126. S.M. Mandala, R. A. Thornton, B. R. Frommer, S. Dreikorn, and M. B. Kurtz, *Journal of Antibiotics*, Vol. 47 p. 376, 1994.
127. S. M. Mandala, R.A. Thornton, B. R. Frommer, S. Dreikorn, and M. B. Kurtz. *Journal of Antibiotics*, Vol. 50, p 339, 1997.
128. W.I. Wu , V.M. McDonough , J.T. Jr Nickels , J. Ko , A.S. Fischl , T.R. Vales , A.H. Jr Merrill, and G.M. Carman, *Journal of Biological Chemistry*, Vol. 270, p13171, 1995.
129. E. Wang, W. P. Norred, C. W. Bacon, R. T. Riley, and A. H. Jr Merrill, *Journal of Biological Chemistry*, Vol. 266, p.14486, 1991.

130. M. M. Nagiec, E. E. Nagiec, J. A. Baltisberger, G. B. Wells, R. L. Lester, and R. C. Dickson, *Journal of Biological Chemistry*, Vol. 272, p. 9809, 1997.
131. S. M. Mandala, R. A. Thornton, M. Rosenbach, James Milligan, Margarita Garcia-Calvo, Herbert G. Bull, and M.B. Kurtz, *Journal of Biological Chemistry*, Vol. 272, p. 32709, 1997.
132. S.M. Mandala, R.A. Thornton, J. Milligan, M. Rosenbach, M. Garcia-Calvo, H.G. Bull, G. Harris, G. K. Abruzzo, A.M. Flattery, C.J. Gill, K. Bartizal, S.Dreikorn, and M.B. Kurtz, *Journal of Biological Chemistry*, Vol. 273, p. 14942, 1998.
133. H. A. Whaley, O. K. Sebek, and C. Lewis, *Antimicrobial Agents in Chemotherapy*, Vol. 10, p.455, 1970.
134. J. C. Onishi, J. A. Milligan, A. Basilio, J. Bergstrom, J. Curotto, L. Huang, M. Meinz, M. Nallin-Omstead, F. Pelaez, D. Rew, M. Salvatore, J. Thompson, F. Vicente, and M.B. Kurtz, *Journal of Antibiotics*, Vol. 50, p. 334, 1997.
135. K. Takesako, K. Ikai, F. Haruna, M. Endo, K. Shimanaka, E. Sono, T. Nakamura, I. Kato, and H. Yamaguchi, *Journal of Antibiotics*, Vol. 44, p. 919, 1991.
136. G. H. Harris, A. Shafiee, M. A. Cabello, J. E. Curotto, O. Genilloud, K.E. Goklen, M.B. Kurtz, M. Rosenbach, P. M. Salmon, R. A. Thornton, D. L. Zink, and S.M. Mandala, *Journal of Antibiotics*, Vol. 51(9) p. 837, 1998.
137. V.Mor, A. Rella, A.M. Farnoud, A. Singh, M. Munshi, A. Bryan, S. Naseem, J.B. Konopka. I. Ojima, E. Bullesbach, A. Ashbaugh, M.J. Linke, M. Cushion, M. Collins, H. K. Ananthula, L. Sallans, P.B. Desai, N. P. Wiederhold, A.W. Fothergill, W.R. Kirkpatrick, T. Patterson, L. H. Wong, S. Sinha, G. Giaever, C. Nislow, P. Flaherty, X. Pan, G. V. Cesar, P. de Melo Tavares, S. Frases, K. Miranda, M. L. Rodrigues, C. Luberto, L. Nimrichter, and M. Del Poeta, *mBio*, Vol. 6, p.1, 2015.
138. P.C. Rittershaus, T.B. Kechichian, J.C. Allegood, A.H. Merrill Jr, M. Hennig, C. Luberto, and M. Del Poeta, *Journal of Clinical Investigation*, Vol.116, p. 1651, 2006.
139. F.C. Odds, A.J Brown, and N.A. Gow, *Trends in Microbiology*, Vol. 11, p. 272, 2003.
140. F. M. Garibotto, A. D. Garro, M. F. Masman. A. M.Rodriguez, P. G. Luiten, M. M. Raimondi, S.A. Zacchino, C. Somlai, B. Penke, and R. D. Enriz, *Bioorganic and Medicinal Chemistry*, Vol. 18, p.158, 2010.
141. A.J.Carrillo-Muñoz, G. Quindós, and J.L.López-Ribot, *Current Medicinal Chemistry-Anti-Infective Agents*, Vol. 3, p.297, 2004.
142. S. Arikan, and J.H. Rex, *Expert Opinion on Emerging Drugs,* Vol. 7, p.3, 2002.
143. R.A. Akins, *Medical Mycology*, Vol. 42, p. 285, 2005.
144. N.P. Wiederhold, R.E. Lewis, N.P. Wiederhold, and R.E. Lewis, *Expert Opinion of Investigational Drugs*, Vol. 12, p. 1313, 2003.
145. R., Leber, S, Fuchsbichler, V. Klobucniková, N. Schweighofer, E. Pitters, K. Wohlfarter, M. Lederer, K. Landl, C. Ruckenstuhl, I. Hapala, and F. Turnowsky, *Antimicrobial Agents and Chemotherapy*, Vol. 47, p. 3890, 2003.
146. H. P. Sigg and C. Stoll, Antibiotic SL 2266. U.S. Patent 3,432,598, assigned to Sandoz Ltd, March 11, 1969.
147. D. Hauser, and H. P. Sigg, *Helvetica Chimica Acta*, Vol.*54*, p.1178, 1971.
148. G. H. Harris, and P. Dai, *Journal of Liquid Chromatography and Related Technologies*, Vol. 24, p. 1775, 2001.
149. T. P. Tully, J. S. Bergum, S. R. Schwarz, S. C. Durand, J. M. Howell, R.N. Patel, P. M. Cino, *Journal of Industrial Microbiology and Biotechnology*, Vol. 34, p. 193, 2007.
150. M. Liu, M.D. Healy, B.A. Dougherty, K.M. Esposito, T.C. Maurice, C.E. Mazzucco, R.E. Bruccoleri, D. B. Davison, M. Frosco, J. F. Barrett, and Y.-K. Wang, *Eukaryotic Cell*, Vol. 5, p. 638, 2006.
151. G. P. Belfield, N. J. Ross-Smith, and M. F.Tuite, *Journal of Molecular Evolution*, Vol. 41, p. 367, 1995.

152. G. R. Anderson, P. Nissen, and J. Nyborg, *Trends in Biochemical Science*, Vol. 28, p. 434, 2003.
153. M. C. Justice, M.-J. Hsu, B. Tse, T. Ku, J. Balkovec, D. Schmatz, and J.Nielsen, *Journal of Biological Chemistry*, Vol. 273, p. 3148, 1998.
154. M.Shastry, J. Nielsen, T. Ku, M.-J. Hsu, P. Liberator, J. Anderson, D. Schmatz, and M. C. Justice, *Microbiology* (Reading, U. K.), Vol. 147, p. 383, 2001.
155. J. M. Domínguez, and J. J Martín, *Antimicrobial Agents in Chemotherapy*, Vol. 42, p. 2279, 1998.
156. J. M. Domínguez, M. G. Gómez-Lorenzo, and J. J.Martín, *Journal of Biological Chemistry*, Vol. 274, p. 22423, 1999.
157. F. Vicente, A. Basilio, G. Platas, J. Collado, G.F.Bills, A.Gonza´lez Del Val, J. Martin, J. R. Tormo, G. H. Harris, D. L. Zink, M. Justice, J. Nielsen Kahn, and F. Pela´ez, *Mycological Research*, Vol. 1 (13), p. 754, 2009.
158. F. Lamoth, P. R. Juvvadi and W. J. Steinbach, Editorial: Advances in *Aspergillus fumigatus* pathobiology. *Frontiers in Microbiology*, Vol. **7**, p. 43, 2016.
159. P. R. Juvvadi, S.C. Lee, J. Heitman, and W.J. Steinbach, *Virulence* Vol. 8, 186–197, 2017.
160. J. R. Blankenship, W. J. Steinbach, J. R. Perfect, and J.Heitman, *Current Opinion in Investigational Drugs*, Vol. 4, p. 192, 2003.
161. L. E. Cowen, S.D. Singh, J. R. Köhler, C. Collins, A. Zaas, W. A. Schell, H. Aziz, E. Mylonakis, J. R. Perfect, L. Whitesell, and S. Lindquist, *Proceedings in National Academy of Sci USA*, Vol. 106, p. 2818, 2009.
162. J. Pachl, P. Svoboda, F. Jacobs, K. Vandewoude, Be van der Hoven P. Spronk, G. Masterson, M. Malbrain, M. Aoun, J. Garbino, J. Takala, L. Drgona, J, Burnie, R Matthews, and Mycograb Invasive Candidiasis Study Group, *Clinical Infectious Diseases*, Vol. 42, p. 1404, 2006.
163. M. Nambu, J. A. Covel, M. Kapoor, X.Li, M. K. Moloney, M. M. Numac, Q.A. Soltow, M.Trzoss, P. Webb, R.R. Webb, and M. Mutz, *Bioorganic and Medicinal Chemistry Letters*, Vol 27, p. 2465, 2017.
164. J. R. Perfect, J. L. Tenor, Y. Miao, and R. G. Brennan, *Virulence* Vol. 8, p. 143, 2016.
165. E. W. Petzold, U. Himmelreich, E. Mylonakis, T. Rude, D. Toffaletti, G. M. Cox, J. L. Miller, and J. R. Perfect, *Infection and Immunity*, Vol. 74, p. 5877, 2006.
166. P. Ngamskulrungroj, U. Himmelreich, J. A. Breger, C. Wilson, M. Chayakulkeeree, M. B. Krockenberger, R. Malik, H-M. Danie, D. Toffaletti, J. T. Djordjevic, E. Mylonakis, W. Meyer, and J. R. Perfect, *Infection and Immunity*, Vol. 77, p. 4584, 2009.
167. Y. Miao, J. L. Tenor, D. L. Toffaletti, E. J. Washington, J. Liu, W. R. Shadrick, M. A. Schumacher, R. E. Lee, J. R. Perfect, and R. G. Brennan, *Proceedings of the National Acaemy of Sciences USA*, Vol. 113, p. 7148, 2016.
168. N.H Georgopapadakou, and T. J.Walsh, *Science*, Vol 264, p. 371, 1994.
169. J.D. Carledge, J. Midgley, and B.G. Gazzard. *AIDS*, Vol 11, p.1839, 1997.
170. F. Granier, *La Presse Médicale*, Vol 29, p. 2051, 2000.
171. D. Sanglard, and F. C. Odds, *The Lancet Infectious Diseases*, Vol 2, p. 73, 2002.
172. A. Casadevall, and L. A. Pirofski, *Nature Reviews Microbiology*, Vol 1, p. 17, 2003.
173. L. E., Cowen, L. M. Kohn, and J. B. Anderson, *Journal of Bacteriology*, Vol. 183, p. 2971, 2001.
174. J. B Anderson, C. Sirjusingh, A. B. Parsons, C. Boone, C. Wickens, L. E. Cowen, and L. M. Kohn, *Genetics* Vol. 163, p. 1287, 2003.
175. M. A. Pfaller, D. J. Sheehan, and J. H. Rex, *Clinical Microbiological Reviews*, Vol.17, p. 268, 2004.
176. L. Rodero, S. Córdoba, P. Cahn, F. Hochenfellner, G. Davel, C. Canteros, S. Kaufman, and L. Guelfand *Journal of Antimicrobial Chemotherapy*, Vol.45, p. 239, 2000.

177. P. E. Verweij, A. Chowdhary, W. J. Melchers, and J. F. Meis, *Clinical Infectious Diseases*, Vol. 62, p. 362, 2016.
178. B. D. Alexander, M.D. Johnson, C.D. Pfeiffer, C. Jiménez-Ortigosa, J. Catania, R. Booker, M. Castanheira, S.A. Messer, D.S. Perlin, and M.A. Pfaller, *Clinical Infectious Diseases,* Vol. 56, p. 1724, 2013.
179. K. R. Healey, Y. Zhao, W. B. Perez, S. R. Lockhart, J. D. Sobel, D. Farmakiotis, D. P. Kontoyiannis, D. Sanglard, S. J. Taj-Aldeen, B. D. Alexander, C.Jimenez-Ortigosa, E. Shor, and D. S. Perlin, *Nature Communications,* Vol. 7, p. 11128, 2016.
180. T. D. Edlind, and S. K. Katiyar, *Antimicrobial Agents in Chemotherapy,* Vol. 54, p. 4733, 2010.
181. M. Gaur, N. Puri, R. Manoharlal, V. Rai, G. Mukhopadhayay, D. Choudhury, and R. Prasad, *BMC Genomics,* Vol. 9, p. 579, 2008.
182. C. Costa, P.J. Dias, I. Sá-Correia, and M.C. Teixeira, *Frontiers in Physiology,* Vol. 5, p. 197, 2014.
183. D. Sanglard, *Frontiers in Medicine*, Vol. 3(11), p.1, 2016.
184. D. Lamb, D. Kelly, T. White, and S. Kelly, *Antimicrobial Agents in Chemotherapy,* Vol. 44, p. 63, 2000.
185. D. Sanglard, and A.T. Coste, *Antimicrobial Agents in Chemotherapy*, Vol. 60, p. 229, 2015.
186. F. Morio, C. Loge, B. Besse, C. Hennequin, and P. Le Pape, *Diagnostic Microbiology and Infectious Diseases,* Vol. 66, p. 373, 2010.
187. G. Ramage, S.P. Saville, D.P. Thomas, and J.L. Lopez-Ribot, *Eukaryotic Cell*, Vol. 4, p.633, 2005.
188. K. F. Mitchell, R. Zarnowski, H. Sanchez, J. A. Edward, E. L. Reinicke, J. E. Nett, A.P. Mitchell, and D.R. Andes, *Proceedings of the National Acaemy of Sciences U S A,* Vol. 112(13), p. 4092, 2015.
189. J. V. Desai, A.P. Mitchell, and D. R. Andes, *Cold Spring Harbour Perspectives in Medicine,* Vol. 4, p.19729, 2014.
190. J. Bonhomm, and C. d'Enfert, *Current Opinion in Microbiology*, Vol. 16 p. 398, 2013.
191. G. Vediyappan, T. Rossignol, and C.d'Enfert, *Antimicrobial Agents Chemotherapy,* Vol. 54, p. 2096, 2010.
192. R.E. Lewis, *Mayo Clinic Proceedings*, Vol. 86, p. 805, 2011.
193. V. M. S. Duncan, and D. A. O'Neil, *Fungal Biology Reviews*, Vol. 26, p. 156, 2013.
194. R. Aly, and T.Berger, *Clinical Infectious Diseases,* 22 (Suppl 2), S128, 1996.
195. M.D. Andrews, and M. Burns, *American Family Physician*, Vol. 22, p. 1415, 2008.
196. S. Asticcioli, L. Sacco, R. Daturi, C. Matti E. Nucleo, F. Zara, and L. Pagani, *New Microbiology,* Vol. 32, p. 199, 2009.
197. B. Havlickova, V.A. Czaika, and M. Friedrich, *Mycoses,* 51 (Suppl 4), p. 2, 2008.
198. A. Weinstein, and B. Berman, *American Family Physician*, Vol. 65, p. 2095, 2002.
199. Health Protection Agency, Report of a working group of the HPA Advisory Committee for Fungal Infection and Superficial Parasites London, 2006.
200. S.C. Das, M. Mandal, and S.C. Mandal, *Drug Today* Vol. 93 (1), July-September Lorina Publications (India) Inc., 2007
201. A. Tiwari, P. Reddy, and C. Goyal, *Indian Journal of Pharmacy and Pharmacology,* Vol. 3(4), p. 192, 2016.

# 6

# Current Update on Rapamycin Production and its Potential Clinical Implications

Girijesh K. Patel[1,2,*], Ruchika Goyal[1] and Syed M. Waheed[1]

[1]*Department of Biotechnology, Graphic Era University, Dehradun, India*
[2]*Department of Oncologic Sciences, University of South Alabama, Mobile, USA*

***Abstract***

Rapamycin is a nitrogen containing triene macrolide, produced by a soil bacterium *Streptomyces hygroscopicus*. Besides the initial application as antibiotics, it is widely used as a potent immunosuppressant during organs- and bone marrow- transplants and also possesses impressive anti-aging, anti-inflammatory-, neuro-protective- and anti-cancer activities along with other medicinal applications in Parkinson's disease, rheumatoid arthritis, etc. Because of lower production efficiency of wild strain and multifaceted activities in medicine, attempts have been made from time to time to improve production and functional efficacy of rapamycin by optimizing different culture conditions, media components, temperature, pH, agitation speed and random- as well as site-directed mutagenesis in rapamycin biosynthetic pathways-related genes. Several analogs of rapamycin (rapalogs) have also been synthesized to improve its efficacy. In addition to its therapeutic values, extended use of rapamycin increases the risk of diabetes by developing resistant to insulin and affects glucose homeostasis.

***Keywords:*** Rapamycin; cancer; immunosuppressant, mTOR, *S.hygroscopicus*

## 6.1 Introduction

Rapamycin, also known as Sirolimus (USAN-assigned generic name) or Rapamune (trade name), is a macrocyclic lactone isolated from a soil bacterium *Streptomyces hygroscopicus* found on Easter Island (Rapa Nui) [1]. Initially, it was characterized as an antifungal agent against ten different strains of *Candida albicans* with a minimum inhibitory concentration ranges from 0.02–0.2 μg/ml [1, 2], while found inactive against many Gram-positive and Gram-negative bacteria. Later on, different potential pharmacological properties of rapamycin were identified, including immunosuppression [3–5], anti-inflammatory, anti-cancerous [6–9], neuroprotective [10], anti-aging [11–13], positive effect on AIDS patients [14], rheumatoid arthritis [15, 16], etc. In 1999, FDA approved rapamycin as a potential drug for clinical use and presently, it is marketed under the tradename of Rapamune by Pfizer.

**Corresponding author*: patelgirijesh@gmail.com

Saurabh Saran, Vikash Babu, and Asha Chaubey (eds.) High Value Fermentation Products, Volume 1, (145–164) © 2019 Scrivener Publishing LLC

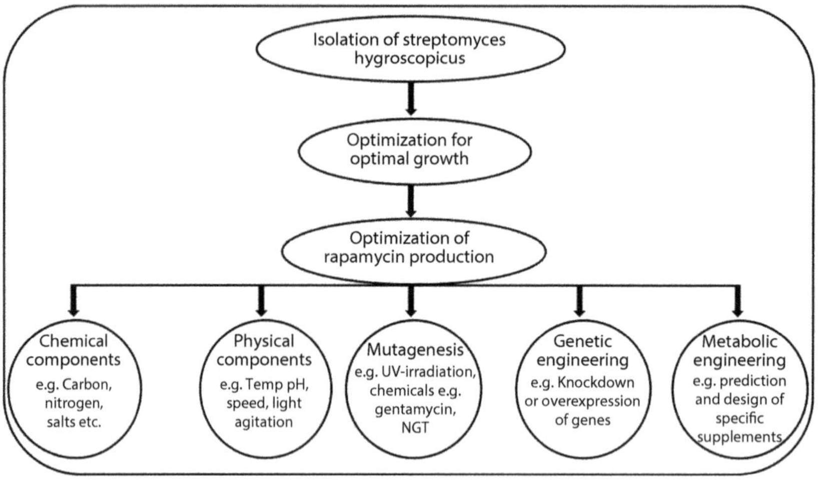


**Figure 6.1** Different strategies to increase rapamycin production.

The low yield of rapamycin is a great challenge for industrial-scale production and commercialization. So, to enhance the rapamycin yield and its functional efficacy, different strategies have been applied (Figure 6.1), including optimization of carbon, nitrogen and salts [17], physical and chemical mutagenesis [18, 19], genetic engineering [20, 21], protoplast fusion [22] and fermentation methods [23–25]. Moreover, the biochemical and metabolomics studies showed different molecules, such as propionate, methionine, acetate, shikimate and pipecolate, serve as precursor molecules in rapamycin biosynthesis [26–28]. The alteration in such genes expression helped to generate high rapamycin yielding strains [19, 21, 28]. Various attempts have been made for the production of the different derivatives of rapamycin to improve the different properties, including solubility, functional efficacy, etc. For this, several methods were employed, including chemical synthesis [29], semisynthetic methods [30], precursor-directed biosynthesis [31] or by manipulation of the biosynthetic gene clusters [32]. These compounds are called as rapamycin analogs or rapalogs.

Although, rapamycin has potential use in human health, but prolonged clinical administration substantially impairs glucose tolerance and insulin resistance [33–36]. In the present chapter, we have summarized the various attempts that have been made to improve the efficacy and yield of rapamycin at industrial scale. Moreover, we also described the potential applications of rapamycin in several diseases and its limitations.

## 6.2 Biosynthesis of Rapamycin

Rapamycin is produced by a soil actinobacterium *Streptomyces hygroscopicus.* It is a white crystalline solid compound possessing molecular weight of 914.187 with molecular formula $C_{51}H_{79}NO_{13}$. It is insoluble in water and readily soluble in some organic solvents (DMSO, methanol, chloroform, ether, etc.) and has melting point range of 183 °C–185 °C. Structurally, rapamycin was characterized by two dimensional nuclear

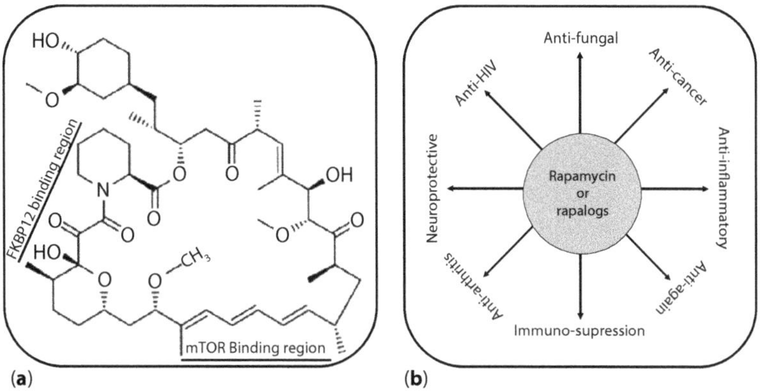


**Figure 6.2** Structure and function of rapamycin.

magnetic resonance (2D-NMR) studies (Figure 6.2). Chemically, rapamycin is a macrocyclic compound, and belongs to polyketide class of natural products which are secondary metabolites produced by some living organisms. The reported biological half-life of rapamycin is 57–63 h. Different steps of the rapamycin biosynthesis in detail are described in the following sections.

### 6.2.1 Microbial Strain

The mycelia of *S. hygroscopicus* (ATCC-29253) radiates out from its periphery and grows optimally at 25–28 °C. For rapamycin production, first spores slants are prepared from isolated colonies on solid media. The spore's suspension is inoculated into seed medium to prepare enough inoculum for fermentation called as seed culture which is inoculated for large-scale fermentation to produce appreciable yield of rapamycin. The composition of seed- and fermentation-medium varies according to the type of study taken into account for the production of rapamycin.

### 6.2.2 Optimization of Carbon, Nitrogen Sources and Salts

Sehgal and co-workers [1] used various types of media and found that tomato paste-oatmeal agar medium was the best medium for growth and sporulation of this bacterium. A detailed study for the carbon source optimization was done by Kojima and coworkers [17]. They used to test 35 different carbon sources in the semi-defined media and found 2% fructose and 0.5% mannose are the best combination for maximum yield from *S. hygroscopicus* strain (AY-B1206; C9). Subsequently, Cheng and colleagues [37] optimized the salt concentration for growth and rapamycin production phase. It was shown that phosphate (100 mM), ammonium (25 mM) and magnesium (256 mg/l) inhibited the rapamycin production which are required at higher concentration for optimal growth, while the less concentration of phosphate (5 mM), ammonium (<5 mM) and magnesium salts (2.5 mg/l) supported higher rapamycin production. In addition, $Fe^{2+}$ stimulated rapamycin production at higher concentration (0.36 mM) while trace amount is required for optimal growth [37, 38]. Fang and Domain [38] reported that exogenous addition of shikimic acid (57 mM) enhanced the production of rapamycin [38]. Moreover, Cheng *et al.* [39] tested 18 different amino acids and reported that the rapamycin production was increased by 150% by the addition of L-lysine which in turn converted into pipecolic acid and serves as rapamycin precursor in the chemically defined media at pH 6.0, while addition of phenylalanine or methionine significantly suppressed the rapamycin production [39]. Later on, Kim and co-worker [40] demonstrated that methionine represses either demethyl rapamycin methyltransferase and / or S-adenosyl methionin esynthetase enzyme and thus reduced rapamycin yield. They further tried to shorten the lag phase by 2–3 days using casein acid hydrolysate (3 g/l) in the medium [40]. In addition to use of amino acids as a nitrogen source, Lee and colleagues evaluated six non-amino acids nitrogenous compounds (ammonium sulfate, ammonium nitrate, ammonium chloride, ammonium citrate, urea, and potassium nitrate) as main nitrogen source for optimum growth and rapamycin production. They reported that except the ammonium citrate, all nitrogenous compounds supported good growth, while ammonium sulfate (40 mM) was found as the best nitrogen source for the optimal rapamycin production [41].

Different culture conditions as well as fermentation methods optimized by Dutta and co-worker [23] have reported that *S. hygroscopicus* MTCC 4003 yielded highest concentration of rapamycin (1316 mg/l) in batch continuous-stirred tank reactor. Kim and colleagues found that glycerol is a potentially effective and economical feedstock for fermentation. It was consumed more quickly after addition of ammonium sulphate during rapamycin biosynthesis phase, and enhanced the rate of rapamycin biosynthesis [42]. Different optimization methods and comparative yields are listed in Table 6.1.

### 6.2.3 Strain Manipulation to Improve Rapamycin Production

The wild strain *S. hygroscopicus* was found to be low yielding and does not meet the need of rapamycin at industrial scale. Different research groups have tried to improve the classical strain by applying different ways, such as chemical and physical mutagenesis, genetic engineering, protoplast fusion followed by high throughput screenings which helped significantly to increase the rapamycin production.

#### *6.2.3.1 Random Mutagenesis*

UV-irradiation is one of the classical methods for random mutagenesis. Zhu and co-workers used UV-irradiation (15 W, 30 cm, 40 s) to generate the random mutagenesis followed by Selection of the high rapamycin yielding strain (*S. hygroscopicus* HD-04-S), using different media supplements which was found to yield 450 mg/l of rapamycin in shaking flask culture [24]. The fermentations were further scaled up using 120 and 20,000 liter fed-batch fermenters, yielding 812.5 and 783 mg/l, respectively at pH 4.8 [24].

In another study, Jung *et al.* [28] exposed spores suspension of *S. hygroscopicus* (ATCC 29253) to UV-radiation (40 W, 30 cm, 70 s) for random mutagenesis and these irradiated spores were grown for 3–4 days and separated colonies were monitored for the rapamycin production. Individual colonies showing increased diameter of growth inhibition zone were selected, assuming the higher production of rapamycin

**Table 6.1** Optimization of different methods for rapamycin production.

| Microbial strain | Method optimized | Yield (mg/l) | References |
|---|---|---|---|
| *S. hygroscopicus* strain (ATCC 29253) | tomato paste-oatmeal agar medium | 87 | [1] |
| *S. hygroscopicus AY-B1206* | Shake batch Flask | 40 | [17] |
| *S. hygroscopicus strain* C9 | Shikimate | 87 | [38] |
| *S. hygroscopicus* strain C9 from AY-B1206 | 2% fructose and 0.5% mannose | 97 | [37] |
| *S. hygroscopicus* strain C9 from AY-B1206 | L-lysine | 134 | [39] |
| *S. hygroscopicus strain* C9 | Ammonium sulfate | 147 | [41] |
| *S. hygroscopicus FC904* | *FC904*<br>*C-14 (gentamicin)*<br>*C14–1(Gentamicin)*<br>*C14–2 (NTG+Gentamicin)* | 87<br>118<br>139<br>195 | [43] |
| *S. hygroscopicus* D7–804 and *S. erythreus* ZJU325 | Protoplast fusion<br>Genome shuffling | 345<br>445 | [22] |
| *S. hygroscopicus R060107* | Fed-batch | 500 | [44] |
| *S. hygroscopicus* | Batch high dissolved $O_2$ | 780 | [45] |
| *S. hygroscopicus HD-04-S* | Shake flask in batch<br>120 L Fed batch<br>20000 L Fed batch | 450<br>812.5<br>783.0 | [24] |
| *S.hygroscopicus* | Mutagenesis and glycerol | 220.7 | [42] |
| *S. hygroscopicus ATCC 29253* | *pfk* knockout and *rapK*, *dahP overexpressed* | 250.8 | [21] |
| *S. hygroscopicus MTCC 4003* | UV-30–17<br>NTG-30–27 | 127.73<br>315.58 | [19] |

and analysed by high performance liquid chromatography (HPLC). Selected colonies with higher rapamycin production were further cultivated in tryptic soy broth (TSB) production medium and second (UV2) and third round of UV-exposure (UV3) were given. They found the yields, 12.3, 14.1, and 11.8 mg/l of rapamycin for UV1, UV2 and UV3, respectively [28]. The UV2 irradiated strains showed relatively high rapamycin production compared to wild strain (7.3 mg/l). The higher yielding UV2-strain was selected and further exposed to UV and after screening selected strain UV2–2 displayed 3.2-fold improvement (23.6 mg/l). These colonies were also evaluated for gene copy number related to rapamycin biosynthetic pathway using quantitative real time PCR (qRT-PCR). The expression of *rapA* almost suppressed at day 4 while the polyketide synthase (*PKS*) and Non-ribosomal peptide synthase (*NRPS*) genes were overexpressed in the UV2–2 strain, suggesting the role of these genes in rapamycin production pathway. This concludes that propionyl-CoA carboxylase pathway is involved in increasing the malonyl-CoA pool in UV2–2 strain, which is ultimately responsible for the enhanced rapamycin production (42.8mg/l). Thus, it revealed that the combined approach involving random mutagenesis and metabolic engineering helped to enhance the production of rapamycin.

#### *6.2.3.2 Chemical Mutagenesis*

To enhance the rapamycin production, different attempts have been made by many research groups. Cheng *et al.* [43] demonstrated that *S.hygroscopicus* strain F904 (87μg/l) when treated with gentamicin (1μg/ml for 2 h) give rise to a higher rapamycin yielding mutant (C-14) strain (118 μg/l). The protoplast of C-14 mutant exposed with gentamycin resulted in C14–1 strain which showed 60% higher rapamycin production (139 mg/l). Another strain (C14–2) with enhanced rapamycin production (195 mg/l) was also obtained when protoplast of mutant C-14 was treated with both a chemical mutagen, N-methyl-N-nitro-N-nitro soguanidine (NTG) 2mg/ml and gentamicin (1μg/ml for 2 h) [43]. Similarly, Dutta *et al.* [19] reported that NTG treatment to *S. hygroscopicus*strain MTCC-4003 resulted in a high-yielding strain which produced 67% higher rapamycin (238.45mg/l) than the parent strain (79.23 mg/l). Production was further improved up to 315.58 mg/l by optimization of temperature (23 °C), agitation speed (120 rpm) and pH (7.6) using Taguchi methodology [19].

#### *6.2.3.3 Metabolic Engineering Combined with Genetic Engineering to Improve Rapamycin Production*

An attempt of genetic manipulation in *S. hygroscopicus* strain ATCC 29253 was done in order to increased rapamycin production but the deletion of polyketide synthase gene (*PKS*) resulted in the loss of rapamycin production, indicating it as an important gene in rapamycin biosynthesis pathway [46]. Another group showed that introduction of a frameshift mutation into *rapL* gene, encoding a specific L-lysine cyclodeaminase, using ϕC31 phage-mediated gene alteration, resulted in a mutant *S.hygroscopicus* NRRL 5491 with decreased level ofrapamycin production but the exotic addition of L-pipecolate restored the wild form of rapamycin production. Moreover, the addition of proline derivatives in *rapL* mutant strain culture word produced a novel form of rapamycin analog by addition of pipecolate incorporation in cyclic chain of rapamycin [32]. A detailed study of metabolomics was performed by Zhao *et al.* [47] and has identified 91 compounds important in rapamycin synthesis. Based on their primary finding, they designed a M3 media and exogenously added pipecolate, which helped to improve rapamycin synthesis from 196 mg/l to 307 mg/l in fed-batch fermentation [47].

Recently, a computer-guided metabolic engineering approach was proposed to identify several potential genetic targets to improve the yield of rapamycin [21] They demonstrated that simultaneous knockdown of 6-phosphofructokinase gene (*PFK*) and overexpression of shikimate pathway's genes namely; 3-deoxyd-arabino-heptulosonate-7-phosphate (*DAHP*) synthase [48] and chorismatase (*RAPK)* in *S. hygroscopicus* (ATCC 29253) improve the rapamycin yield up to 250.8 mg/l compared to the parent strain (103.5 mg/l).

### 6.2.3.4 *Protoplasts Fusion*

The protoplast fusion is another method adapted to increase the rapamycin production. Chen *et al.* [22] demonstrated that protoplasts fusion of interspecies, namely *S. hygroscopicus* D7–804 and erythromycin producing strain *S.erythreus* ZJU325, resulted in a high-yielding rapamycin producer (345 mg/l). The protoplast-based one round of genome shuffling in *S. hygroscopicus* produced another high-yielding mutant that showed enhanced (445 mg/l) yield of rapamycin [22].

### 6.2.3.5 *Optimization of Fermentation Methods*

Fermentation is defined as a chemical process that consumes different media components to yield desired product in anaerobic conditions. Industrial-scale fermentation is a large-scale microbial process that is not restricted to absence of oxygen and it has been widely used for production of different products. Several fermentation methods are optimized for small- to industrial-scale production of rapamycin, including batch, fed-batch and continuous-stirrer tank reactor, which are briefly discussed in the following section.

Batch fermentation mode involves reactor feeding and product extraction only once per cycle of fermentation. In this, once the reactor is fed with medium and fermentation started, it is left undisturbed till completion of fermentation. Then, biomass and media are separated for further downstream processing and reactor is again set up for the next round of fermentation. The product formation is generally stopped at a particular cell density (turbid static) or at a particular chemical concentration (chemostatic) and monitored during process. The laboratory and small-scale optimization was done in a batch culture [1, 17, 24, 45].

As we know that *S. hygroscopicus* needs different components for the optimum growth and rapamycin production, in that regard the fed-batch is a quite common method used for rapamycin production. Different groups have used the fed-batch culture optimization for rapamycin production [44, 49]. It is shown that mutagenesis along with Feb-batch process optimization by using different media components, such as sugar, amino acids, shikimic acids glycerol (250 g/l) and $K_2HPO_4$(10 g/l), revealed the maximum yield up to 812 mg/l [24]. The Plackett–Burman experimental design (PBD) was used to evaluate the different media components for the rapamycin production [50]. They also applied the response surface methodology, represented by central composite design (CCD) enhanced the rapamycin production up to 95 mg/l compared to normal production. In addition, the (Plackett–Burman design PBD) method, artificial neural networks (ANN), and genetic algorithms (GA) improved the yield of rapamycin [51]. The PBD was used to screen mannose, soybean meal, and L-lysine concentrations for rapamycin production. ANN was used to construct rapamycin production model. These strategies have been resulted in a significant enhancement in rapamycin production (320.89 mg/l) compared to the parental strain (57 mg/l). A novel two-stage feeding

**Table 6.2** Optimization of different fermentation methods for rapamycin production.

| Microbial strains | Method | Yield (mg/l) | Refs |
|---|---|---|---|
| *S. hygroscopicus AY-B1206* | Batch | 40 | [17] |
| *S. hygroscopicus NBS-9746* | Fed-Batch | 110 | [49] |
| *S. hygroscopicus R060107* | Fed-batch | 500 | [52] |
| *S. hygroscopicus DSMZ 41530* | CCD | 95 | [50] |
| *S. hygroscopicus* HD-04-S | Batch and Fed-Batch | 812 | [24] |
| *S. hygroscopicus* MTCC 4003 | Continuous-stirred tank | 1316 | [23] |
| *S. hygroscopicus* FMT11 | Box-Behnkendesign | 860.6 | [13] |
| *S. hygroscopicus* ATCC 29253 | ANN-GA method | 320.9 | [51] |

strategy was developed by Zou and Li (2013) to increase the rapamycin biosynthesis. Using statistical optimization and Box-Behnken design helped to optimize amino acid components, which were fed in the initial phase of fermentation, and then switched to feed 2 g/l ammonium sulphate at 72 h in the production phage and the maximum rapamycin production achieved upto 860.6 mg/l [13].

In case of continuous fermentation mode, there is continuous addition of fresh media from inlet and removal of bioreactor fluid from outlet at the same flow rate which is optimized in such a way that maximum yield can be obtained. This ensures the long run of fermentation operations without shutting down the bioreactors. Different culture conditions as well as fermentation methods were optimized by Dutta *et al.* [23] and reported that *S. hygroscopicus* MTCC 4003 yielded highest concentration of rapamycin (1316 mg/l) in batch continuous-stirred tank reactor [23]. The different fermentation method optimization for yield of rapamycin is listed in Table 6.2.

## 6.3 Organic Synthesis of Rapamycin

Besides the biosynthesis using different microbial strains, different research groups have chemically synthesized the rapamycin [53–55]. Due to the challenging structural feature of rapamycin, a series of complex strategies were developed that includes iron-carbonyl chemistry, addition of allylsilanes to oxonium ions, butane-2,3-diacetal-controlled aldol condensation, and a highly-efficient catechol-templated intramolecular macrocycle construction which finally resulted in the total synthesis of rapamycin. The complete detail of rapamycin total synthesis is described by Ley and colleagues [55].

## 6.4 Extraction and Quantification of Rapamycin

The culture media is subjected to centrifugation at low speed (5000×g) for 10–15 min to separate supernatant and biomass, which were subsequently used to extract the rapamycin by methanol extraction method. The quantification of rapamycin is done

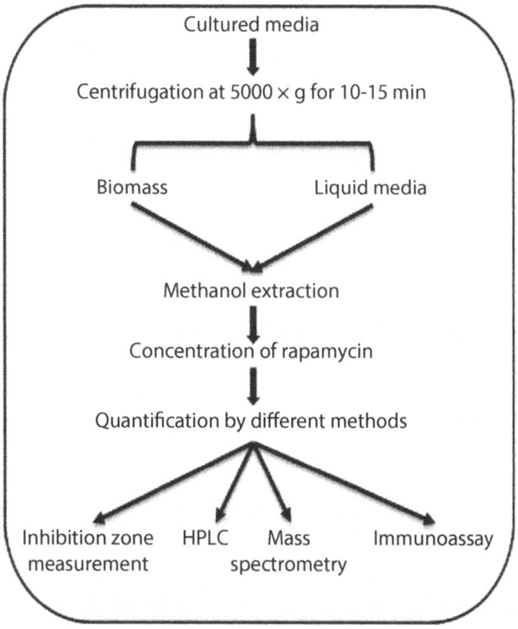


**Figure 6.3** Extraction and quantification of rapamycin.

using different methods, such as HPLC, mass spectroscopy, immunoassay or agar disc diffusion method [17, 18, 23, 41]. Different steps of rapamycin isolation and quantification are illustrated in Figure 6.3.

## 6.5 Physiological Factors Affecting Rapamycin Biosynthesis

The yield of rapamycin depend on various parameters, including media components, pH, metabolite concentration, agitation speed, viscosity, dissolved oxygen (DO), $CO_2$ respiratory quotient (RQ), concentration of carbon sources, turbidity, etc., which are described in brief in subsequent sections.

### 6.5.1 Effect of Media Components

Extensive studies on rapamycin production by *S.hygroscopicus,* have indicated that nutritional and environmental factors highly affect rapamycin production. Several research groups have optimized the carbon and nitrogen sources on the yield of rapamycin production. The detail of the media components are discussed in detail in earlier section 6.2.2.

### 6.5.2 Effect of pH on Rapamycin Production

In any fermentation process, media pH is one among the most deterministic factor for the final product yield. It has been observed that the biosynthesis of rapamycin is affected by the initial pH of the production medium. The comparable yields have been obtained at pH range from 5.5–6.5. The higher pH up to 7.0 adversely affects the yield of rapamycin and remains half of that recorded at pH 6.5. It has been shown that below the optimum pH, rapamycin production significantly suppressed and eventually ceased at pH 4 [17, 37, 38, 41, 56].

### 6.5.3 Effect of Physical Gravity

Apart from other physiochemical factors, the role of gravity on rapamycin production was examined by Fang *et al.* [25]. They studied the effect of simulated microgravity (SMG) environment in rotating wall bioreactor results in pelleted biomass with low dry weight and cell-bound rapamycin. The addition of Teflon beads under simulated microgravity increase the extracellular rapamycin production [25].

### 6.5.4 Effect of Morphological Changes

Qi *et al.* [57] have reported that there is a direct correlation between *S. hygroscopicus* morphology and rapamycin production [57]. They found that long filaments and high branching in hyphal growth in the periphery of *S. hygroscopicus* pellets led to increase in rapamycin production. There is a change of pellet morphology with change in fermentation media, thereby affecting rapamycin production. The bacterial pellets are heterogeneous cell-mass with maximal activity at peripheral region and minimal at the core due to limited supply of nutrients and oxygen in the core. This revealed that the change of intracellular metabolite profile with change in morphology also affects rapamycin production [57].

### 6.5.5 Effect of Dissolved Oxygen (DO) and Carbon Dioxide ($DCO_2$)

The dissolved oxygen (DO) and carbon dioxide ($DCO_2$) are dynamic parameters which levels keep on changing with the fermentation proceed from growth to production phase. The level of both, DO and $DCO_2$ in the fermentation media are very important for the optimum production of rapamycin, as DO and CO2 affect the morphology of *S. hygroscopicus* and thus affect the production. Yen and Hsiao (2013) have demonstrated the effect of dissolved oxygen (DO) on rapamycin production in the pellet form of *S. hygroscopicus.* They reported that higher DO level promotes rapamycin production at speed <200 rpm [45]. Chen *et al.* (2008b) investigated the relationship between rapamycin production and the evolution of $DCO_2$. A high limit of DCO2 (hDCO2) was used to control medium feed which ranged from 3.0% at the beginning to 5.3% at the end of the fed-batch run [52].

## 6.6 Production of Rapamycin Analogs

The literature survey revealed that various rapamycin analogs are available resulted either through addition of specific media supplements, mutations at gene level or chemical modifications. Khaw *et al.* [32] have reported the production of propylrapamycin analog after growing *S. hygroscopicus* in proline supplemented medium [32]. Wong and colleagues [58] demonstrated that the antifungal activity of three different analogs, namely, prolylrapamycin, 32-desmethylrapamycin, and 32-desmethoxyrapamycin and further reported that rapamycin is more effective antifungal agent than these three analogs, in liquid medium [58], while in the disc diffusion method, 32-desmethylrapamycin was found equally effective as rapamycin. Graziani *et al.* [59] demonstrated the precursor-directed biosynthesis of Sulphur containing immunosuppressive rapalogs, such as 20-thiarapamycin and 15-deoxo-19-sulfoxylrapamycin, when the *S. hygroscopicus* cultures were supplemented with L-nipecotic acid and either (S)-1,3-thiazane-4-carboxylic acid or (S)-1,4-thiazane-3-carboxylic acid, respectively [59]. Similarly, Lowden *et al.* [60] reported precursor-directed biosynthesis of rapamycin derivatives [60]. Various rapamycin analogs were obtained when carboxylic acids such as cyclohexane carboxylic acid, cycloheptane carboxylic acid, were added instead of 4,5-dihydroxycyclohex-1-enecarboxylic acid (DHCHC) as pseudo starters [61]. In another study, sulphur containing pipecolate analogs like 1,4-thiazane-3-carboxylic acid and 1,3-thiazane-4-carboxylic acid in rich medium led to the production of rapamycin analogs like 4-thiarapamycin and 9-deoxo-5-sulfoxyrapamycin, respectively [31]. Mita *et al.* [62] demonstrated that Deferolimus (AP23573), a rapalog, showed antitumor activity to a range of malignancies when administered intravenously [62, 63]. The different analogs of rapamycin are listed in Table 6.3.

## 6.7 Mechanism of Action of Rapamycin

The mammalian target of rapamycin (mTOR) is a 289 kDa intracellular serine/threonine kinase. It has been known to exist in two functionally distinct complex forms in the cytoplasm, namely, mTORC1 and mTORC2. The mTORC1 complex resulted after interaction and binding of different proteins including, mTOR, regulatory-associated protein of TOR (raptor), mammalian lethal with SEC-13 protein 8 (mLST8 also known as GβL) and the non-core proteins namely, DEP-domain containing mTOR-interacting protein (DEPTOR) and a proline-rich Akt substrate 40 kDa (PRAS40). The mTOR pathway is activated through various stimulus such as insulin, insulin-like growth factors (IGF-1 or IGF-2), and amino acids along with nutrient, oxygen, and energy levels. This complex is responsible for various cellular functions such as proliferation, cell survival, transcription, protein synthesis, autophagy, metabolism, etc. (Li *et al.*, 2014). The mTORC2 consists of mTOR, mLST8, rapamycin insensitive companion of mTOR (rictor) and stress-activated protein kinase-interacting protein 1 (mSIN1), along with rictor 1/2 (protor 1/2) and DEPTOR, the regulatory components. The mTORC2 complex has been shown to regulate the actin cytoskeleton, through its stimulation of F-actin stress fibers, small GTPase, protein kinase C α (PKCα) and paxillin (Sarbassov *et al.*, 2004) and regulate metabolism and survival by phosphorylating the serine/threonine protein kinase Akt/PKB at Ser473 position [65].

Deregulation of mTOR pathway results in different disease conditions, including arthritis, tuberous sclerosis complex, diabetes, cancer, etc. [35, 66] and found to be potentially regulated by rapamycin or rapalogs. After the entry into the cell, rapamycin

**Table 6.3** List of Rapamycin analogs /derivatives and their activities.

| Analog name | Production strategies | Medicinal use | Target molecule | References |
|---|---|---|---|---|
| 4-hydroxyprolyl-rapamycin | *rapL* gene deletion | Growth inhibition | mTOR | [32] |
| 32-desmethylrapamycin | C32-derivative | Antifungal | mTOR | [58] |
| 32-desmethoxyrapamycin | C32-derivative | Antifungal | mTOR | [58] |
| 9-deoxo-5-sulfoxy-rapamycin | C5-sulfur-containing | Antifungal | mTOR | [31] |
| 20-thiarapamycin | C20- sulfur-containing | Immuno-suppressive | mTOR | [59] |
| 15-deoxo-19-sulfoxylrapamycin | C19-sulfur-containing | Immuno-suppressive | mTOR | [59] |
| | | | | [61] |
| Temsirolimus (CCI-779) | C40-derivatives | Antitumor | Cyclin D1, p70s6k and c-Myc | [63] |
| Everolimus (RAD001) | C40-derivatives | Antitumor | Cyclin D1, p70s6k and c-Myc | [63] |
| Deferolimus (AP23573) | C40-derivatives | Antitumor | mTOR | [62] |
| prolylrapamycin | Proline replacementbypipecolate | Antifungal | mTOR | [64] |

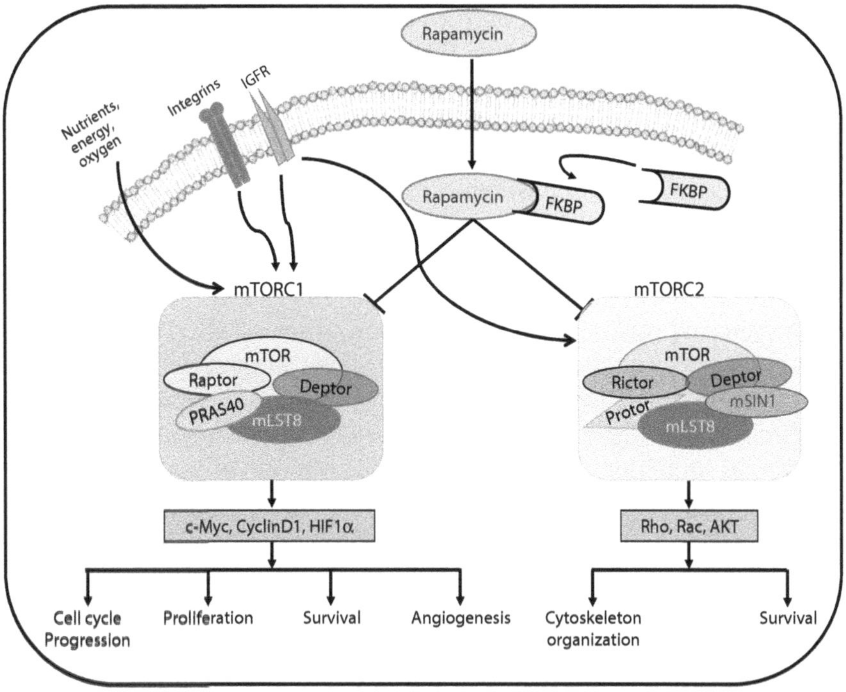


**Figure 6.4** Mechanism of action of rapamycin: mTOR forms two complexes namely, mTORC1 and mTORC2. Based on stimulation through different factors, including oxygen, energy levels, nutrients (amino acid) and growth factors, mTORC1 induces the synthesis of protein, lipids and nucleotide for cell growth and proliferation, while the mTORC2 stimulated only through the growth factors and regulates cytoskeleton organization and cell survival. The rapamycin mainly inhibit the mTORC1 but the extended use inhibits the mTORC2 also.

or rapalogs binds with a 12 kDa cytoplasmic FK506-binding protein. This complex (Ramycin-FKBP12) acts as an allosteric inhibitor of mTOR complexes and inhibits their activities, leading inhibition of various processes such as protein and nucleotide synthesis, glycolysis, cell survival, actin organization, immunogenic reactions, etc. [35, 67]. The mechanism of rapamycin-mediated inhibition of mTOR complexes are depicted in Figure 6.4.

## 6.8 Use of Rapamycin in Medicine

Rapamycin showed multifaceted activities and is useful in various disease conditions. Initially, it was identified as potent anti-*Candida* agent, subsequently, its different properties such as immune-suppressive, anti-tumor, anti-aging and others, were identified, which are briefly described below.

### 6.8.1 Anti-Fungal Agent

*Candida albicans* is one of the major infectious fungal pathogens in humans. Rapamycin is an antifungal agent which effectively inhibit the growth of different

*Candida sp.*at very low concentration with minimum inhibitory concentration (MIC) range (0.02–0.2 μg/ml), compared to the other known anti-fungal agents [2]. It is shown that rapamycin inhibit the mTORC1 activity, an essential regulator of metabolism that hampers morphogenesis and nitrogen acquisition in *C. albicans*. It has been reported that alteration in TOR1 or FKBP12 genes, led to rapamycin resistance in *C.albicans*, indicating their important role in mTOR pathway [68].

### 6.8.2 Immunosuppression

Several immunosuppressive agents (e.g., cyclosporine or FK506, azathioprine, etc.) have been used to prevent the rejection reaction during organ transplant but they pose potential threats to cancer development. The rapamycin is the effective drug candidate which not only inhibits the organ transplantation rejection, but also possesses anticancer properties by inhibiting the mTOR signalling pathway and suppresses the synthesis of IL2, required for activation of the lymphocytes, thus acting as an immunosuppressive agent [69, 70].

### 6.8.3 Anti-Cancer Agent

The enhanced activity of the mTORC1is observed in several malignancies which promotes glycolysis by upregulating the expression of hypoxia-inducible factor alpha (HIF1α) and c-Myc which promotes lipid biosynthesis and the pentose phosphate pathway by sterol regulatory element binding protein 1 (SREBP-1) (7), and positively controls glutamine metabolism by SIRT4 repression [8]. Thus, rapamycin and its analogs play an important role in selective targeting of mTORC1 complex. Also, it is shown that rapamycin suppresses the metastatic tumor growth and angiogenesis in mouse model by inhibiting the expression of vascular endothelial growth factor (VEGF), a known angiogenesis promoting factor [71]. It is reported that the rapamycin and rapalogs are a more effective antitumor agent in tumors possessing defective PTEN [72].

### 6.8.4 Anti-Aging Agent

It is shown that rapamycin and rapalogs have potential implication in increasing the life span of different animal models. It has been demonstrated that the median and maximum life span of male and female mice were increased when treated with rapamycin [11]. Moreover, Wilkinson *et al.* (2012) showed that the treatment of rapamycin decreased the age-related phenotypes in mice [73]. In another study, it was shown that genetically altered or inhibition of mTOR signalling promoted the life span of yeast, nematodes, and fruit flies [36].

### 6.8.5 Role in HIV Treatment

Rapamycin is cytostatic to T-cells and inhibit the signalling mediated by IL2 [74]. It is shown that rapamycin down-regulate C-C chemokine receptor type 5 (CCR5) and promotes the accumulation of anti-HIV β-chemokines [75]. In another study, treatment with rapamycin inhibits the replication of CCR5-utilizing strain of human immunodeficiency virus (HIV) in mice homozygous for the *scid* mutation (SCID mice) [14]. It is observed that the treatment with rapamycin prevented the decrease in $CD4^+$ cell population and thus increased $CD4^+/CD8^+$ ratio compared to the control group in a dose-dependent manner. Moreover, it is reported that rapamycin possesses anti-viral activity and is safe in treatment of HIV-infected individuals with Kaposi sarcoma [76].

### 6.8.6 Rheumatoid Arthritis

Rheumatoid arthritis is an immune-mediated inflammatory disease where various analgesics, glucocorticoids, non-steroidal anti-inflammatory drugs (NSAIDs), and immunosuppressive agents, such as cyclophosphamide and cyclosporine are used to inhibit the inflammation and control the disease progression. Rapamycin are known to inhibit the effect of cytokines mediators, resulting in reduced inflammation. It is demonstrated that rapamycin can potentially control synovial fibroblast proliferation in rheumatoid arthritis by inhibition of platelet derived growth factor (PDGF) and thus, is helpful in treatment of inflammatory disease [16, 77].

## 6.9 Side Effects of Long-Term Use of Rapamycin

Sirolimus are used in varieties of therapeutic intervention. However, its use has been reported with important issues of pulmonary toxicity including interstitial pneumonitis, bronchiolitis obliterans, organizing pneumonia, and alveolar proteinosis during liver transplants[78]. Recent evidences suggest that prolonged use of rapamycin impairs glucose homeostasis and leads to glucose intolerance in mice [33], insulin resistance and obesity in rats [34, 79] and humans [80, 81]. It is still not clear that how rapamycin play both, positive and negative role. Lamming *et al.* (2012) reported that reduced activity of mTORC1 complex increases life span of animal and maintains normal glucose homeostasis, while disruption of mTORC2 activity results in insulin resistance [36].

## 6.10 Conclusions

In the last several years, considerable advancement has been made toward rapamycin yield and its higher efficiency by optimizing different culture conditions, genetic alteration and producing multiple analogs. The structure-function studies of rapamycin enhanced our knowledge for its potential clinical implications in different physiological and pathophysiological conditions. In cancer, rapamycin and its analogs are well studied as mTOR inhibitors with significant outcomes. However, the inhibition of mTORC1 with such drugs leads to the activation of Akt pathway, resulting in more aggressive phenotypes. Hence, the use of rapamycin with inhibitor of Akt or PI3K (PI-I03) serves as a remedy to overcome this problem and showed increased overall survival, revealing the importance of combined therapy. So, it underscores the role and importance of rapamycin in combination therapy for the treatment of different malignancies. Moreover, rapamycin has also shown an anti-aging effect and therapeutic value against several other diseases such as HIV, rheumatoid arthritis, etc.

## Acknowledgements

The authors would like to thankfully acknowledge the editors of this book for providing us the opportunity to contribute this book chapter.

## References

1. S. N. Sehgal, H. Baker, and C. Vezina, *Journal of Antibiotics (Tokyo),* Vol. 28, p. 727, 1975.
2. C. Vezina, A. Kudelski, and S. N. Sehgal, *Journal of Antibiotics (Tokyo),* Vol. 28, p. 721, 1975.
3. R. E. T. R. Morris, *Transplantation Reviews,* Vol. 16, p. 39, 1992.
4. R. Y. Calne, D. S. Collier, S. Lim, S. G. Pollard, A. Samaan, D. J. White, *Lancet,* Vol. 2, p. 227, 1989.
5. T. Weber, D. Abendroth, and H. Schelzig, *Transplant International,* Vol. 18, p. 151, 2005.
6. J. Douros and M. Suffness, *Cancer Treatment Reviews,* Vol. 8, p. 63, 1981.
7. J. L. Yecies and B. D. Manning, *Cancer Research,* Vol. 71, p. 2815, 2011.
8. A. Csibi, S. M. Fendt, C. Li, G. Poulogiannis, A. Y. Choo, D. J. Chapski, *Cell,* Vol. 153, p. 840, 2013.
9. I. Amodeo, M. Colnaghi, G. Raffaeli, G. Cavallaro, F. Ciralli, S. Gangi, *et al., Medicine (Baltimore),* Vol. 96, p. e8871, 2017.
10. B. D. Kahan and J. S. Camardo, *Transplantation,* Vol. 72, p. 1181, 2001.
11. D. E. Harrison, R. Strong, Z. D. Sharp, J. F. Nelson, C. M. Astle, K. Flurkey, *Nature,* Vol. 460, p. 392, 2009.
12. I. Bjedov, J. M. Toivonen, F. Kerr, C. Slack, J. Jacobson, A. Foley, *Cell Metabolism,* Vol. 11, p. 35, 2010.
13. X. L. Zou, J., *Annals of Microbiology,* Vol. 63, p. 1371, 2013.
14. F. Nicoletti, C. Lapenta, S. Donati, M. Spada, A. Ranazzi, B. Cacopardo, *Clinical and Experimental Immunology,* Vol. 155, p. 28, 2009.
15. R. R. Martel, J. Klicius, and S. Galet, *Canadian Journal of Physiology and Pharmacology,* Vol. 55, p. 48, 1977.
16. B. Foroncewicz, K. Mucha, L. Paczek, A. Chmura, and W. Rowinski, *Transplant International,* Vol. 18, p. 366, 2005.
17. I. Kojima, Y. R. Cheng, V. Mohan, and A. L. Demain, *Journal of Industrial Microbiology,* Vol. 14, p. 436, 1995.
18. Z. N. Xu, W. H. Shen, X. Y. Chen, J. P. Lin, and P. L. Cen, *Biotechnology Letter,* Vol. 27, p. 1135, 2005.
19. S. Dutta, B. Basak, B. Bhunia, A. Sinha, and A. Dey, *World Journal of Microbiology and Biotechnology,* Vol. 33, p. 90, 2017.
20. E. Kuscer, N. Coates, I. Challis, M. Gregory, B. Wilkinson, R. Sheridan, *et al., Journal of Bacteriology,* Vol. 189, p. 4756, 2007.
21. L. Dang, J. Liu, C. Wang, H. Liu, and J. Wen, *Journal of Industrial Microbiology and Biotechnology,* Vol. 44, p. 259, 2017.
22. X. Chen, P. Wei, L. Fan, D. Yang, X. Zhu, W. Shen, *Applied Microbiology and Biotechnology,* Vol. 83, p. 507, 2009.
23. S. Dutta, B. Basak, B. Bhunia, S. Chakraborty, and A. Dey, *3 Biotech,* Vol. 4, p. 523, 2014.
24. X. Zhu, W. Zhang, X. Chen, H. Wu, Y. Duan, and Z. Xu, *Biotechnology and Bioengineering,* Vol. 107, p. 506, 2010.
25. A. Fang, D. L. Pierson, S. K. Mishra, and A. L. Demain, *Applied Microbiology and Biotechnology,* Vol. 54, p. 33, 2000.

26. N. L. Paiva, A. L. Demain, and M. F. Roberts, *Journal of Natural Products*, Vol. 54, p. 167, 1991.
27. N. L. Paiva, A. L. Demain and M. F. Roberts, *Enzyme and Microbial Technology*, Vol. 15, p. 581, 1993.
28. W. S. Jung, Y. J. Yoo, J. W. Park, S. R. Park, A. R. Han, Y. H. Ban, *Applied Microbiology and Biotechnology*, Vol. 91, p. 1389, 2011.
29. F. C. Nelson, S. J. Stachel, C. P. Eng, and S. N. Sehgal, *Bioorganic and Medicinal Chemistry Letters*, Vol. 9, p. 295, 1999.
30. R. Sedrani, L. H. Jones, A. M. Jutzi-Eme, W. Schuler, and S. Cottens, *Bioorganic and Medicinal Chemistry Letters*, Vol. 9, p. 459, 1999.
31. F. V. Ritacco, E. I. Graziani, M. Y. Summers, T. M. Zabriskie, K. Yu, V. S. Bernan, *Applied and Environmental Microbiology*, Vol. 71, p. 1971, 2005.
32. L. E. Khaw, G. A. Bohm, S. Metcalfe, J. Staunton, and P. F. Leadlay, *Journal of Bacteriology*, Vol. 180, p. 809, 1998.
33. J. T. Cunningham, J. T. Rodgers, D. H. Arlow, F. Vazquez, V. K. Mootha, and P. Puigserver, *Nature*, Vol. 450, p. 736, 2007.
34. M. Fraenkel, M. Ketzinel-Gilad, Y. Ariav, O. Pappo, M. Karaca, J. Castel, *et al.*, *Diabetes*, Vol. 57, p. 945, 2008.
35. J. Li, S. G. Kim, and J. Blenis, *Cell Metabolism*, Vol. 19, p. 373, 2014.
36. D. W. Lamming, L. Ye, P. Katajisto, M. D. Goncalves, M. Saitoh, D. M. Stevens, *Science*, Vol. 335, p. 1638, 2012.
37. Y. R. Cheng, L. Hauck, and A. L. Demain, *Journal of Industrial Microbiology*, Vol. 14, p. 424, 1995.
38. A. Fang and A. L. Demain, *Folia Microbiologica (Praha)*, Vol. 40, p. 607, 1995.
39. Y. R. Cheng, A. Fang, and A. L. Demain, *Applied Microbiology and Biotechnology*, Vol. 43, p. 1096, 1995.
40. W. S. Kim, Davis, S., Wong, G., & A. L. Demain, *Journal of Microbiology and Biotechnology*, Vol. 13, p. 560, 2003.
41. M. S. Lee, I. Kojima, and A. L. Demain, *Journal of Industrial Microbiology and Biotechnology*, Vol. 19, p. 83, 1997.
42. Y. H. Kim, B. S. Park, S. K. Bhatia, H. M. Seo, J. M. Jeon, H. J. Kim, *et al.*, *Journal of Microbiology and Biotechnolog*, Vol. 24, p. 1319, 2014.
43. Y. R. Cheng, J. Huang, H. Qiang, W. L. Lin, and A. L. Demain, *Journal of Antibiotics (Tokyo)*, Vol. 54, p. 967, 2001.
44. Y. Chen, Krol, J., Huang, W., Cino, J.P., Vyas, R., Mirro, R. and Vaillancourt, B., *Process Biochemistry*, Vol. 43, p. 351, 2008.
45. H. W. Yen and H. P. Hsiao, *Journal of Bioscience and Bioengineering*, Vol. 116, p. 366, 2013.
46. N. Lomovskaya, L. Fonstein, X. Ruan, D. Stassi, L. Katz, and C. R. Hutchinson, *Microbiology*, Vol. 143, p. 875, 1997.
47. S. Zhao, D. Huang, H. Qi, J. Wen, and X. Jia, *Applied Microbioogy andl Biotechnology*, Vol. 97, p. 5329, 2013.
48. S. Tapas, G. K. Patel, S. Dhindwal, and S. Tomar, *Journal of Molecular Modeling*, Vol. 17, p. 621, 2011.
49. Y. Chen, Krol, J., Sterkin, V., Fan, W., Yan, X., Huang, W., ... and Julien, C., *Process Biochemistry*, Vol. 34, p. 383, 1999.
50. Y. R. Abdel-Fattah, *Journal of Microbiology*, Vol. 3, p. 405, 2008.
51. R. S. Sinha, S., P. Srivastava, *Bioprocess and Biosystems Engineering*, Vol. 37 p. 829, 2014.
52. Y. Chen, Krol, J., Huang, W., Cino, J.P., Vyas, R., Mirro, R. and Vaillancourt, B., *Process Biochemistry*, Vol. 43, p. 351, 2008.

53. K. C. Nicolaou, Chakraborty, T.K., Piscopio, A.D., Minowa, N. and Bertinato, P *Journal of the American Chemical Society,* Vol. 115, p. 4419, 1993.
54. M. L. Maddess, Tackett, M. N., Watanabe, H., Brennan, P. E., Spilling, C. D., Scott, J. S., and S. V. Ley, *Angewandte Chemie International Edition,* Vol. 46, p. 591, 2007.
55. S. V. Ley, M. N. Tackett, M. L. Maddess, J. C. Anderson, P. E. Brennan, M. W. Cappi, *Chemistry,* Vol. 15, p. 2874, 2009.
56. L. A. R. Sallam, El-Refai, A.F., Osman, M.E., Hamdy, A.A., Ahmed, E.M. and Mohamed, M.A., *Journal of American Science,* Vol. 6, p. 188, 2010.
57. H. Qi, Zhao, S., Fu, H., Wen, J., Jia, X., *Biochemical Engineering Journal,* Vol. 91, p. 186, 2014.
58. G. K. Wong, S. Griffith, I. Kojima, and A. L. Demain, *Journal of Antibiotics (Tokyo),* Vol. 51, p. 487, 1998.
59. E. I. Graziani, F. V. Ritacco, M. Y. Summers, T. M. Zabriskie, K. Yu, V. S. Bernan, *et al.*, *Organic Letters,* Vol. 5, p. 2385, 2003.
60. P. A. Lowden, G. A. Bohm, S. Metcalfe, J. Staunton, and P. F. Leadlay, *Chembiochem,* Vol. 5, p. 535, 2004.
61. M. A. Gregory, H. Petkovic, R. E. Lill, S. J. Moss, B. Wilkinson, S. Gaisser, *Angewandte Chemie International Edition in English,* Vol. 44, p. 4757, 2005.
62. M. M. Mita, A. C. Mita, Q. S. Chu, E. K. Rowinsky, G. J. Fetterly, M. Goldston, *Journal of Clinical Oncology,* Vol. 26, p. 361, 2008.
63. B. K. Law, *Critical Reviews in Oncology/hematology,* Vol. 56, p. 47, 2005.
64. S. R. Park, Y. J. Yoo, Y. H. Ban, and Y. J. Yoon, *Journal of Antibiotics (Tokyo),* Vol. 63, p. 434, 2010.
65. C. Betz, D. Stracka, C. Prescianotto-Baschong, M. Frieden, N. Demaurex, and M. N. Hall, *Proceedings of the National Academy of Sciences of the United States of America,* Vol. 110, p. 12526, 2013.
66. T. H. Sasongko, N. F. Ismail, and Z. Zabidi-Hussin, *The Cochrane Database of Systematic Reviews,* Vol. 7, p. CD011272, 2016.
67. M. Waldner, D. Fantus, M. Solari, and A. W. Thomson, *British Journal of Clinical Pharmacology,* Vol. 82, p. 1158, 2016.
68. M. C. Cruz, L. M. Cavallo, J. M. Gorlach, G. Cox, J. R. Perfect, M. E. Cardenas, *et al.*, *Molecular and Cellular Biology,* Vol. 19, p. 4101, 1999.
69. B. D. Kahan, *Lancet,* Vol. 356, p. 194, 2000.
70. A. M. Shapiro, J. R. Lakey, E. A. Ryan, G. S. Korbutt, E. Toth, G. L. Warnock, *et al.*, *The New England Journal of Medicine,* Vol. 343, p. 230, 2000.
71. M. Guba, P. von Breitenbuch, M. Steinbauer, G. Koehl, S. Flegel, M. Hornung, *Nature Medicine,* Vol. 8, p. 128, 2002.
72. M. S. Neshat, I. K. Mellinghoff, C. Tran, B. Stiles, G. Thomas, R. Petersen, *Proceedings of the National Academy of Sciences of the United States of America,* Vol. 98, p. 10314, 2001.
73. J. E. Wilkinson, L. Burmeister, S. V. Brooks, C. C. Chan, S. Friedline, D. E. Harrison, *Aging Cell,* Vol. 11, p. 675, 2012.
74. S. N. Sehgal, *Clinical Biochemistry,* Vol. 31, p. 335, 1998.
75. A. Heredia, O. Latinovic, R. C. Gallo, G. Melikyan, M. Reitz, N. Le, *et al.*, *Proceedings of the National Academy of Sciences of the United States of America,* Vol. 105, p. 20476, 2008.
76. S. E. Krown, D. Roy, J. Y. Lee, B. J. Dezube, E. G. Reid, R. Venkataramanan, *Journal of Acquired Immune Deficiency Syndromes,* Vol. 59, p. 447, 2012.
77. K. Migita, K. Eguchi, T. aoyagi, T. Tsukada, M. Tsuboi, Y. Kawabe, *Clinical and Experimental Immunology,* Vol. 104, p. 86, 1996.
78. M. J. Perez, R. O. Martin, D. M. Garcia, J. M. Rey, J. de la Cruz Lombardo, and J. M. Rodrigo Lopez, *Transplantation Proceedings,* Vol. 39, p. 3498, 2007.

79. V. P. Houde, S. Brule, W. T. Festuccia, P. G. Blanchard, K. Bellmann, Y. Deshaies, *et al.*, *Diabetes,* Vol. 59, p. 1338, 2010.
80. A. Teutonico, P. F. Schena, and S. Di Paolo, *Journal of the American Society of Nephrology,* Vol. 16, p. 3128, 2005.
81. O. Johnston, C. L. Rose, A. C. Webster, and J. S. Gill, *Journal of the American Society of Nephrology,* Vol. 19, p. 1411, 2008.

# 7

# Advances in Production of Therapeutic Monoclonal Antibodies

**Richi V Mahajan, Subhash Chand, Mahendra Pal Singh, Apurwa Kestwal and Surinder Singh***

*National Institute of Biologicals (Ministry of Health & Family Welfare) Government of India Plot No.A-32, Sector-62 Institutional Area, NOIDA, (U.P.), (India)*

***Abstract***
Monoclonal antibodies (mAbs) are target-specific protein therapeutics used by patients suffering from cancer, communicable and autoimmune diseases etc. mAb-based therapeutics have grown exponentially with the advent of mammalian cell culture processes and formulation technology. The mAbtherapeutics market is expected to be US$245.8 billion by 2024. A large number of mAbs have been patented for therapeutic use by the US Food and Drug Administration (FDA) or the European Union health authorities. In the last three decades, various powerful techniques such as hybridoma technology, phage display and B-cell amplification with the help of PCR have been developed for the construction of numerous types of antibodies and antibody fragments with widespread structural modifications. The current review chapter highlights the structure and nomenclature of mAbs, upstream strategies and techniques for cost-effective production of mAbs and recent advances in antibody technology.

***Keywords:*** Monoclonal antibodies, production, hybridoma, phage-display

## 7.1 Introduction

Monoclonal antibodies (mAbs) are the fast-emerging class of biomolecules, for their potential as therapeutics in the treatment of various human diseases like cancer, leukaemia, asthma, arthritis, psoriasis and transplant rejection, etc. [1, 2]. Since the inception of muromonab as a therapeutic to reduce acute rejection in patients with organ transplants in 1986, to date more than 68mAbs have been approved by FDA and more than 470 molecules are under clinical trials [3, 4].

Therapeutic mAbs, like other therapeutic proteins, are enormously complex drugs typically produced by mammalian tissue culture cells through recombinant DNA technology, hybridoma technology, B lymphocyte immortalisation or other technologies (phage display technology, genetically engineered cells, etc.) [5]. As a result of

**Corresponding author*: surinder2560@gmail.com

Saurabh Saran, Vikash Babu, and Asha Chaubey (eds.) High Value Fermentation Products, Volume 1, (165–192) © 2019 Scrivener Publishing LLC

naturally-occurring molecular heterogeneity, imperfect cellular processing, chemical and enzymatic alterations during manufacturing and additional modifications upon storage, antibody drugs display a wide variety of minor chemical changes, collectively termed micro heterogeneity [6]. Common examples include glycan structural differences, deamidation, oxidation etc.

Therapeutic mAbs were generated originally in mice. Chimeric antibodies were developed to reduce the immunogenicity of mouse antibodies before introduction of humanized antibodies to minimize anti-idio type antibody generation. Today, almost all antibody products currently in development are humanized or fully human [7].

Manufacturing process for protein involves multiple factors like ready availability of raw materials, production vessel, production conditions, robustness and purification of the final drug products [8]. The focus of the industrial process should be on long-term supply to ascertain reproducibility and scalability. Current industrial production of commercial mAbsis is based on the overproduction of therapeutic proteins of interest by genetically engineered mammalian cells [8]. However, for the production of small fragments bacterial or yeast based process may be more economical and viable [9].

## 7.2 Discovery and Clinical Development

The history of antibody research goes back over a hundred years to Paul Ehrlich, who received the Nobel Prize for Physiology or Medicine in 1908. Ehrlich gave the idea of a "magic bullet", the selective drugs like antitoxins aimed specifically at their corresponding targets having an affinity for specific pathogens, without having much effect on host cells [10]. However, these specific targeted therapies made a breakthrough in the late nineteenth century when Behring and Kitasa were the first to use blood serum therapy to fight against infectious diseases [11].

Klinman in 1969 reported a way of isolating single B lymphocytes which secreted a single type of antibody [12]. By early 1975 Gerhard successfully generated single specificity antibodies with known precision against influenza viruses using the technique [13]. However, the cells that produced the antibodies survived for a maximum of only three months. In 1975, Köhler and Milstein created an immortal cell line capable of producing an endless supply of identical antibodies with known specificity. The method came to be known as "hybridoma technology" and the antibodies it produced "monoclonal antibodies" to signify the fact that they were derived from a single hybrid cell [13].

The first licenced monoclonal antibody (mAb) to be approved by US FDA was Orthoclone OKT3 (muromonab-CD3) in 1986 for the prevention of kidney transplant rejection [14]. However, murine hybridoma antibodies were of little success since they were recognized as foreign by the human immune system and in response induced human anti-mouse antibody (HAMA), resulting in short half-life, reduced efficacy, and increased toxicity to patients [15]. In the mid-1980s, several researches were carried out to improve the characteristics of murinem Abs, and the invention of recombinant DNA technology resulted in development of chimeric and humanized mAbs that contained a combination of rodent- and human-derived sequences. Further, technologies to generate fully human antibodies, such as phage display libraries and transgenic mice were established in the early 1990s. The first chimeric and humanized antibodies were approved by the FDA for human use in 1993 and 1997 respectively [16]. The first fully human mAb to be approved was adalimumab in 2002 [17]. Since then mAb therapeutics has been of great interest in the biopharmaceutical industry.

## 7.3 Structure and Classification

All antibody molecules have a "Y" shaped structure having 4 protein chains, *viz.* two heavy chains and two light chains, linked by disulphide bonds (Figure 7.1) [18]. There are two domains associated with light chain and four with heavy chains. Each light chain consists of one variable domain (VL) and one constant domain (CL). Each heavy chain consists of one variable domain (VH) and three constant domains (CH1, CH2, CH3). The hyper-variable regions in these variable domains (VL & VH) are called complementarity-determining regions (CDRs) [18]. The difference in their length and sequences marks the specificity (recognition) and affinity (binding) of the antibodies to their target markers. Papain digestion of an antibody molecule produces two Fab fragments and one Fc fragment. Fab fragments are the antigen-binding sites consisting of the entire light chain and part of the heavy chain. The Fc region comprises part of the heavy chains and determines the mechanism to destroy the antigen [18].

The first approved murine antibody faced a heavy challenge of a short life span under physiological conditions, limited penetration of solid tumours and inadequacy in mobilising the host effector functions due to dissimilarity between murine and human immune systems (Figure 7.2A) [19]. These shortcomings were overcome by the development of recombinant technology. Chimeric antibodies were developed wherein variable regions of murine antibody were fused onto constant regions of human antibody (Figure 7.2B) [19]. Specifically, human gene

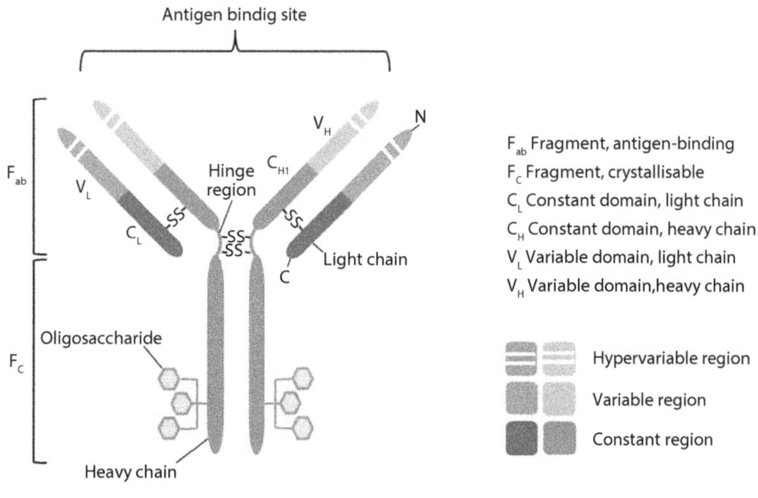


**Figure 7.1** The Structure of IgG Antibody.

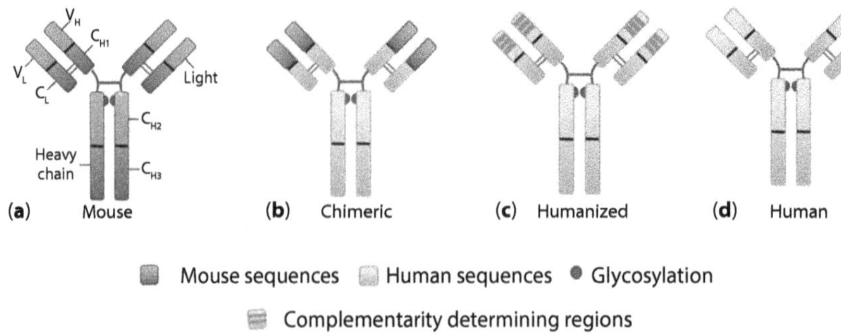


**Figure 7.2** Graphic representation of mAbs indicating their source of origin.

sequences from the kappa light chain and the IgG1 heavy chain were fused to result in antibodies that are approximately two-thirds human. These antibodies had a reduced risk of immunogenicity, and increased serum half-life. However, the presence of residual murine components, still limited the efficacy of these mAb products. These limitations were beaten by the development of mAbs that were predominately humanized or were entirely of human origin. Humanized antibodies are one step enhancements, produced by grafting the murine hypervariable region onto the fully human Immunoglobulin protein (Figure 7.2C) [20]. Fully human antibodies are the ones which are devoid of any murine sequence and are chiefly produced by either of the two sources i.e., phage display technologies or transgenic mice (Figure 7.2D) [21].

The transgenic approach involves the engineered strains of mice to produce human instead of murine antibodies [22]. Immunization of these transgenic mice leads to production of human antibodies and further conventional hybridoma techniques are used to modify these immunoglobulin for desired properties. The approach has an advantage since the need of affinity maturation is circumvented due to in-vivo production of mature antibodies. Furthermore, developed full length IgG presents the preferred way of therapy. However, the technique is ineffective in cases wherein inducing immunogen is toxic to the murine or when the inducing antigen shares close homology to its murine ortholog [22]. A list of few mAbs along with their source of production is presented in Table 7.1.

## 7.4 Nomenclature of Monoclonal Antibodies

Monoclonal antibodies require an International Non-proprietary Name (INN) allocated by the World Health Organization (WHO) for regulatory authorizations [26]. Currently, INNs for mAbs are allocated by the suffix, -mab, preceded by two sub-stems infix: a) target of antibody e.g. -b(a)-bacterial, -c(i)-cardiovascular, -t(u)-/-ta for tumor and b) source from which the antibody is derived e.g.-o-(murine), -xi-(chimeric), -zu-(humanized) or -u-(human) chimeric antibody [27]. A random prefix is added to the substem A of the infix. Table 7.2 presents the nomenclature scheme of mAbs as decided by WHO.

**Table 7.1** List of Therapeutic Monoclonal Antibodies along with their source of origin and their Clinical Indications [23–25].

| S. No | Name of MAb | Source of origin | Clinical indications |
|---|---|---|---|
| 1. | Muromonab-DC3 | Murine | Transplantation rejection |
| 2. | Abciximab | Chimeric | High risk angioplasty(prevention of blood clots) |
| 3. | Rituximab | Chimeric | Non-Hodgkin's lymphoma, Chronic lymphocytic leukemia, Rheumatoid arthritis |
| 4. | Basiliximab | Chimeric | Reversal of transplantation rejection |
| 5. | Infliximab | Chimeric | Crohn's disease, Ulcerative colitis, Rheumatoid arthritis, Ankylosing spondylitis, Psoriatic arthritis |
| 6. | Trastuzumab | Humanized | Breast cancer; Metastatic gastric or gastroesophageal junction adenocarcinoma |
| 7. | Adalimumab | Human | Rheumatoid arthritis, Juvenile idiopathic arthritis, Psoriatic arthritis, Ankylosing spondylitis, Crohn's disease |
| 8. | Omalizumab | Humanized | Asthama |
| 9. | Bevacizumab | Humanized | Metastatic colorectal cancer, Non-small cell lung cancer, Metastatic breast & renal cell carcinoma |
| 10. | Cetuximab | Chimeric | Head and neck cancer, Colorectal cancer |
| 11. | Natalizumab | Humanized | Multiple sclerosis (relapsing), Crohn's disease |
| 12. | Ranibizumab | Humanized | Neovascular (wet) age-related macular degeneration, Macular edema following retinalvein occlusion |
| 13. | Denosumab | Human | Reversal of transplantation rejection |
| 14. | Golimumab | Human | Rheumatoid arthritis, Psoriatic arthritis, Ankylosing spondylitis |
| 15. | Ofatumumab | Human | Chronic lymphocytic leukemia |
| 16. | Tocilizumab | Humanized | Rheumatoid arthritis |
| 17. | Ipilimumab | Human | Melanoma |
| 18. | Pertuzumab | Humanized | Breast cancer |
| 19. | Raxibacumab | Human | Prevention and treatment of inhalation anthrax |
| 20. | Obinutuzumab | Humanized | Chronic lymphocytic leukemia |

**Table 7.2** Nomenclature scheme of mAbs [26, 27].

| Prefix | Infix (Substem A) (Before 2017 revision) | Infix (Substem A) (2017 revision) | Infix (Substem B) | Suffix |
|---|---|---|---|---|
| Random | -b(a)- bacterial | -ba- bacterial | -a- rat | -mab |
| | -am(i)- serum amyloid protein | -ami- serum amyloid protein | -axo- rat-mouse | |
| | -c(i)-cardiovascular | -ci-cardiovascular | -e- hamster | |
| | -f(u)-fungal | -fung-fungal | -i- primate | |
| | -v(i)-viral | -vi-viral | -o- mouse | |
| | -tox(a)-toxin | -toxa-toxin | -u- human | |
| | -n(e)-neural | -ne-neural | -xi- chimeric | |
| | -k (i)-interleukin | -ki-interleukin | -xizu- chimeric-Humanized | |
| | -t(u)-tumour | -t(u)-tumour | -zu- humanized | |
| | -s(o)- bone | -os- bone | | |

## 7.5 Production of Monoclonal Antibodies

### 7.5.1 Hybridoma Technology

Georges Köhler, César Milstein and Niels Jerne were awarded the Nobel Prize for physiology or medicine in 1984 for the discovery of Hybridoma technology [28]. This technology depends on the fact that B-lymphocytes (plasma cells) of mammals produce antibodies or immunoglobulin protein molecules as the part of the defence mechanism in the body against the invading foreign antigens [29]. Each antigen has specific epitopes on its surface. The complementary determining regions (CDRs) present on the antibody are primarily responsible for its specificity. B-Lymphocytes produce many different types of antibodies against the same antigens. These are termed as polyclonal antibodies. However, due to lack of specificity and variability in response they have limited use as diagnostics or therapeutics [30].

In contrast, mAbs are a single type of antibodies targeted against the specific epitope of an antigen [30]. Early approach for production of mAbs included the immunization of laboratory animals with specific epitope followed by isolation and culturing of B-lymphocytes under *in-vitro* conditions. However, the technique was not successful due to challenges of culturing B-Lymphocytes and variability in antibody production owing to the loss of strain [31]. However, if these normal mAbs producing B-cells are fused with the myeloma cells they become immortal thus leading to continuous production of mAbs. These immortalized B-lymphocytes are called Hybridomas [31].

A brief description of hybridoma technology is presented in Figure 7.3 [32]. Designing and creation of hybridomas followed by production of mAbs starts with the immunization of a laboratory animal (usually mouse) with specific antigen at the

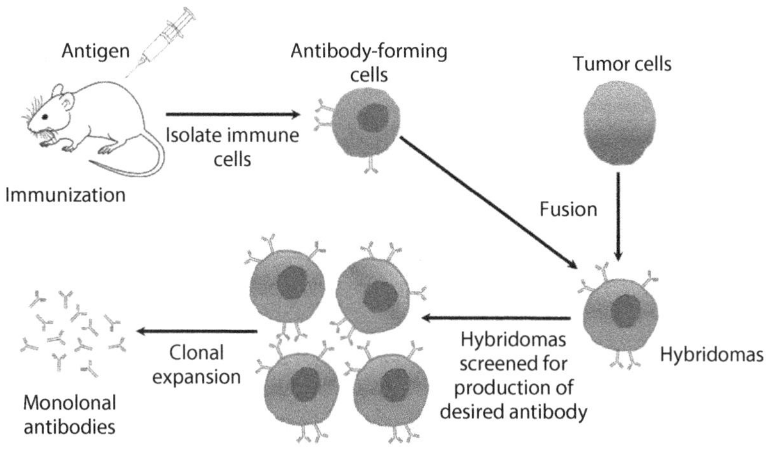


**Figure 7.3** Graphic outline of Hybridoma Technology.

multiple sites for amplified stimulation of responding B-lymphocytes. Before sacrificing the animal, a final dose is given intravenously for enhanced immune-stimulation. Once the animal is sacrificed, spleen is removed aseptically and lymphocytes are separated by density gradient centrifugation. Myeloma cells are screened for two criteria, *viz.* non-production of antibodies and lack of hypoxanthine guanine phosphoribosyl-transferase (HGPRT) gene responsible for production of a key enzyme in *salvage pathway*, making them sensitive to thehypoxanthine-aminopterin-thymidine (HAT) medium. Isolated B cells are fused with immortalized myeloma cells lacking the HGPRT enzyme - using polyethyleneglycol (PEG) or the Sendai virus. Fused cells are cultured in the HAT medium for seven to ten days allowing only hybridoma cells to survive. Myeloma cells do not survive as they are unable to produce nucleotides by de-novo or salvage pathway. Unfused B lymphocytes have a short life span hence phases out in subsequent subculturing steps. The surviving hybridomas are diluted to such an extent that while seeding a multiwell plate each well contains a single cell. The suspension from individual cell is then grown on normal growth media for production of the desired antibody. The supernatant of every hybridoma culture is intermittently tested for specificity of antibody using enzyme-linked immunosorbent assay (ELISA) and radioimmunoassay (RIA). Thus, hybridoma cells generating the desired antibodies are screened and these antibodies are referred to as mAbs. These mAbs are characterized for their biochemical and biophysical properties to access their specificity. It is also important to determine the stability of created hybridomas and their ability for cryopreservation and revivals.

*In vitro* methods for production of mAbs through hybridoma technology faced significant failures. Such conditions can be used for initial selection of hybridoma clones; however, continued production of antibody may have adverse effect on their growth and ability to secrete the desired proteins. In such cases the ascites method is used for production of antibodies. In this method hybridomas are grown ascites in the peritoneal cavity of laboratory animal (usually mouse) [33]. There may be a 10–100 fold increase in production using this technique, hence it is considered better than *in vitro* cultivation [34] but there is associated risk of murine pathogenic contaminations, hence it is not a preferred method [35].

Moreover, the yield of mAbs obtained in culture flasks or bottles is significantly low; therefore, it is not viable for commercial needs. The yield can be increased substantially by increasing the density of hybridomas in culture. This is done by encapsulation of hybridomacells in alginate gels and coating these beads with poly-lysine solution[36]. The yield of mAbs may enhance ten times through this technique. Hybridoma cell technology is used for commercial production of mAbs by Damon Biotech Company and Cell-Tech. These companies are using 100-liter bioreactors to get a yield of about 100 g mAbs in a period of 14 days [37].

### 7.5.2 Epstein-Barr Virus Technology

Production of mAbsusing murine myeloma cells possesses the challenge of immunogenicity. Hence, an alternate technique used for immortalization of B-lymphocytes for specific antibody production is transfecting the cells with Epstein–Barr virus (EBV) which is a lymphotropicherpes virus that converts normal human B lymphocytes into continuous cell line [38]. Selection of specific immortalized cell line is carried out using the plate dilution method as explained for hybridoma technology (Section 4.1). However, production of mAbs is not economically viable due to low yields hence; this method is not common amongst the commercial producers.

### 7.5.3 Phage Display Technology

This technology is again used mostly for the production of human monoclonal antibodies. The method was initially developed by George P. Smith in 1985 using filamentous phage display for short peptide fragments [39]. Adalimumab was the first phage display-derived fully human mAb [40]. The phage display process involves the preparation of antibody-library, followed by insertion of desired antibody genes into a phage display vector. The basic idea behind this display technology is that a repertoire of antibodies of potential interest is created, of which a phage displaying the desired antibody on its surface can be isolated using a specific target ligand to which it binds amongst the billion phages displaying different antibodies (Figure 7.4) [41].

The library is created by inserting variable heavy (VH) and variable light (VL) chain genes into a vector encoding phage protein p3 [42]. The final phage assembly thus, will have p3 protein attached to a functional antibody fragment. V genes repertoires are trapped using RT-PCR amplification of tissue RNAs from lymphoid cells encoding the entire sequence of antibody variable (V) gene fragments [41]. This can also be done de novo by gene synthesis method. The display libraries thus created are potential source of antibodies against multiple disease targets even against those which are poorly antigenic in animals. After the creation of library, antigen specific phage antibodies are screened using *phage panning* technology in which immobilized/labelled antigens are used [41]. Briefly, the desired antigen is immobilized on microplate wells or magnetic beads and the phage constructs are added for specific binding with appropriate antibodies. Subsequently, plate is washed thoroughly for removal of nonspecific constructs; the phage with related antibody will be retained. The screening procedure is repeated

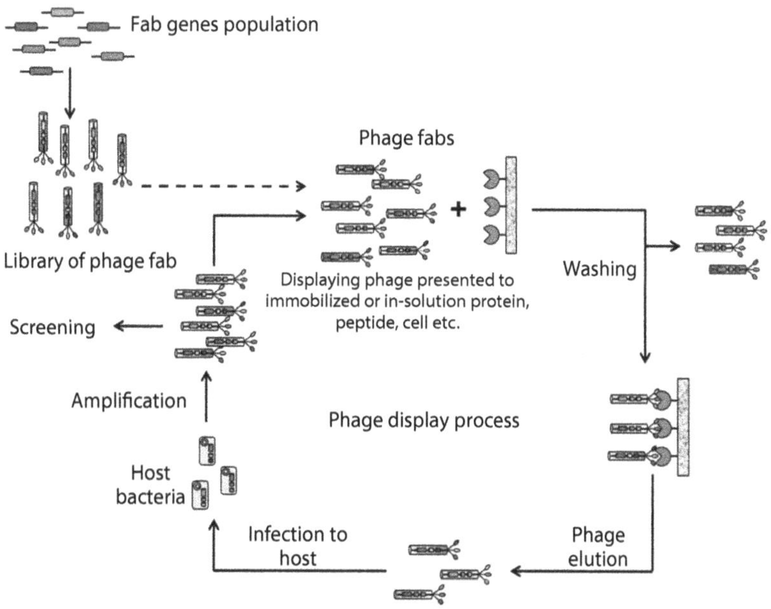


**Figure 7.4** Graphic outline of Phage Display Technology.

for several cycles resulting in a high percentage of anticipated antibodies expressing phages that bind to the antigen of interest. The screened phages are then added to the bacterial culture and grown for further expansion of antibody production and binding analysis [41].

Finally, desired antibodies are identified and characterized using ELISA or FACS analysis, within a number of random clones. Furthermore, antibody genes can be directly identified through sequencing by genetic engineering techniques [41].

### 7.5.4 Cell Line Based Production Techniques

Mammalian cell based expression systems are currently the preferred platform for production of biopharmaceutical products, as they are able to produce large and complex biomolecules with post-translational modifications (mostly glycosylation) similar to those produced under physiological conditions [43–45]. Another advantage of using mammalian cell lines is that during the production process proteins are secreted into the media hence purification process is relatively simple. Therapeutic antibodies are mainly produced in mammalian host cell lines including NS0 murine myeloma cells, PER.C6® human cells and Chinese hamster ovary (CHO) cells [46]. Table 7.3 presents the list of mammalian cell lines and their application in production of respective therapeutic mAbs.

#### *7.5.4.1 Cell Lines*

Chinese hamster ovary (CHO) cells lines account for 70% of industrially produced therapeutic proteins [50]. In recent times, genetically modified CHO cells are capable

**Table 7.3** List of MAbs-producing cell lines along with their respective products.

| S.No | Cell lines | Products | References |
|---|---|---|---|
| 1. | Hybridoma | Muromonab CD3, Satumomab, Capromab | [23] |
| 2. | CHO | Rituximab, Adalimumab, Trastuzumab, Ipilimumab, | [23] |
| 3. | NS0 | Ramucirumab, Sulesomab, Palivizumab, Ofatumumab | [23] |
| 4. | Sp2/0 | Golimumab, Indfliximab, Belimumab, Cetuximab | [23] |
| 5. | PER.C6 | CR57 and CRJB (rabies virus-neutralizing human MAbs) | [47] |
| 6. | HEK 293 | CD-30 specific scFv-Fc-4E3 antibody | [48] |
| 7. | Fibrosarcoma HT-1080 | BCD-F9 (murine IgG-2a) | [49] |

of producing large volumes of humanized mAbs. These cell lines have been reported to secrete up to 100 pg/cell of mAbper day [51]. Second most common mammalian cells used for production of mAbsare murine NS0 cells. NS0 are non-antibody secreting myeloma cells which require the presence of cholesterol in culture medium for growth [52]. Strains independent of cholesterol in the medium have also been developed [53]. Recombinant NS0 cells are selected using with glutamine synthetase (GS) as marker since these cells lack the GS enzyme activity [54]. However, murine cell lines are associated with considerable production of N-glycolylneuraminic acid (NGNA; a sialic acid foreign to humans) and alpha-Gal-alpha(1,3) Gal linkages. Both these molecules were found to cause immunogenic challenges in humans [55, 56]. Though, CHO cells show low levels of expression of both these immunogens. Hence, NS0 cells have limited use in antibody production as compared to CHO cells [8].

Advances in cell culture technology have allowed for production of antibodies in human cell lines. The therapeutic mAbs produced in human embryonic kidney 293 (HEK293) and fibrosarcoma HT-1080 cell lines have gained approval from regulators around the world [47, 57]. Additionally, PER.C6, HKB-11,CAP and HuH-7 human cell lines are also being used for production of therapeutic proteins [58–62]. HEK293 and PER.C6cells offer advantage as compared to the non-human cells because of their human identical glycosylation [8]. Another advantage associated with human cells for the production of mAbs is their considerably low sialylation as compared to CHO cells [63]. However, genetically modified HEK cells have low yields of tens to hundreds milligrams of antibodies per litre of batch during cultivation. The yields can be enhanced considerably by fed-batches process wherein, mAb titres may approach to 0.6 g/L of scFv-Fc antibody fragment [64]. DSM and Crucell N.V. of Holland jointly established PER.C6 cells for the large-scale production of therapeutic proteins under serum-free culture conditions [65]. This Cell line has been immortalized using specific adenovector genes, leading to high viable cell densities and high titres [66]. PER.C6 cells have been reported to yield mAb titres of 0.5 g/L during batch cultivation, 8 g/L during fed-batch culture and even 27 g/L using continuous culture (perfusion)[67].

There are many transfection methods used for stable introduction of vector DNA into mammalian cells which include calcium phosphate, electroporation, cationic lipid-based lipofection, and polymer or dendrimer-based methods [68]. However electroporation and lipofection are the most common choices for transfection [68]. Transfected cells are further selected using metabolic and antibiotic selection markers [69].

### *7.5.4.2 Production Media*

Cultivation of mammalian cells is carried out in highly complex media containing several constituents, *viz.* amino acids, vitamins, trace elements, and nucleosides, etc. [70]. In the past, bovine serum was used for providing the cells with essential nutrients and growth factors [70]. However, recently the use of Bovine serum for the manufacturing of therapeutic mAbs has been limited owing to chances of prion contaminations which can cause bovine spongiform encephalitis (BSE) [71, 72]. In this effort, bovine serum was replaced by complex plant or yeast hydrolysates in cultivation media. Since, hydrolysates are undefined addition to the cultivation media; it can be the source of variation amongst the production batches [73, 74]. In latter developments, protein-free chemically defined media has been developed for most production cell lines like CHO/NS0 which support extremely high cell densities [66, 75]. Development and manufacturing of synthetic media appropriate for high cell densities is an expensive and time-consuming process, requiring significant expertise. Still manufacturers choose to develop their own proprietary media to optimize the cost factor, control the production process and cut down the downstream processing cost to avoid dependence on a single raw material source [76]. Considering the high quality consistency and purity of media required for manufacturing of mAbs, the development of cell cultivation media can be more than INR 1000/L, even for industrial scale, thus significantly contributing to the overall cost of therapeutic mAbs [76].

### *7.5.4.3 Production Conditions*

Growth parameters like temperature, pH, $O_2$, $CO_2$ and agitation are critical from the production point of view. Although initial optimization on these parameters were carried out for hybridoma platforms, now the same strategies have been applied to the modern production cell lines like CHO, NS0 and Sp2/0[76]. Mammalian cells usually grow well at physiological conditions, i.e., 37 °C, pH 7.15 while maintaining dissolved $O_2$ [DO] levels of 30–60% and $pCO_2$ at 31–54 mmHg [76]. The accumulation of by-products such as lactate and ammonium can affect the viability of production culture, typically during long manufacturing processes of 12–14 days [77]. These by-products can lead to a shift in pH and temperature of the production media thus causing apoptosis and reduced productivity [77]. This can be remedied through designing the production media (addition of galactose instead of glucose) or choosing the production strain in such a way that there is minimal lactate or ammonia production [76].

The production process can be maximized using a two-phase approach wherein first phase conditions are set to maximize the growth, and once the threshold cell density is achieved the conditions are altered to maximize the production of mAbs [78]. Unlike bacterial production systems the transition from one phase to another is not caused by a chemical promoter, but an acute shift in temperature or pH. Lowering the temperature to 30–35 °C or pH to 6.7–7.0 can reduce the cell growth and effectively trigger the increase in specific productivity in CHO and hybridoma cell systems [79]. At lower temperatures, the G1 stage of the cell cycle is prolonged, thereby allowing the higher protein production, less nutrient depletion and by-product generation. Lowering the pH to 6.7–7.0 in case of CHO cells can trigger the same effect due to lower glucose consumption. However, these cells are still able to proliferate and produce mAbs. Besides, other cell lines have been found to be sensitive to variation in pH and temperature [79].

CHO cultures can also be cultivated at a wide range of DO levels ranging from 3% to 100% without negative effects, though this may affect the glycosylation patterns [80]. In high cell density cultures $pCO_2$ levels can reach more than 150 mmHg and can have a significant effect on cell growth, production and glycosylation [81]. These effects may set in due to the increased osmolality owing to a change in pH during the production strategies. Bicarbonate-buffering results in acidification of the medium due to high $PCO_2$ because of enhanced respiratory cellular activity [76]. Hence, for neutralization of the acidic medium, suitable base is added to the production medium to bring pH to neutral, leading to a rise in osmolality. To evade this process, manufacturers are now using bicarbonate-free buffers in a perfusion manner. This results in 70% decreases in $pCO_2$ levels and consequently favours cell growth and specific productivity. On the other hand, it has been reported that though high $PCO_2$ does not have significant effect on CHO cell growth it does impact the mAbs specific productivity [82].

Apart from pH, temperature DO and $CO_2$, agitation is another key factor during the production process. Earlier it was presumed that agitation plays a negative role in cell culture production due to sheering of mammalian cells. However, later studies confirmed that the sheering takes place only at the air bubble interface. Thus, for enhanced cell growth and specific productivity a constant agitation is given to the production medium to avoid cell sheering due to bubble bursting. Additionally, protective polymers like Methocel, Pluronic F68, or polyvinyl alcohol (PVA) are added [83].

#### 7.5.4.4 *Production Strategies*

Different fermentation processes, including batch, fed-batch and perfusion methods are used for production of therapeutic mAbs, batch being the most common and simple amongst the three [84]. In batch production, all nutrients are supplied in an initial base medium followed by production at sigmoid curve where initially there is slow cell growth and production later transforming to exponential phase. Throughout the production process there is continuous depletion of nutrients and increased waste metabolite production leading to stationary phase and low cell viability. Though the process is simple, it is not the favourite production strategy owing to unfavourable ecosystem created inside the bioreactor during fermentation process. Generally, the cell density of 1–2 × $10^6$cells/mL is achieved during the process [84]. The process is suitable for fermentations which occur in 4–7 days, after which toxic waste collection can usually have detrimental effects.

In fed batch fermentation process, nutrients are added once they are depleted after the initial cell growth. The supplemented medium can be the normal batch medium, concentrate basal medium development or feeding strategy optimization [85]. Furthermore, several approaches may be applied systematically, such as single-component titration, spent medium analysis, and medium blending [86, 87]. The majority of industrial mammalian cell culture based production involves fed-batch strategy. Initially, simple sugars such as glucose and glutamine were fed as nutrients. However, in recent times sophisticated feed solutions were developed [88]. In another strategy, basal medium and concentrated medium may be fed alternately for enhanced production. In some of the advanced fed batch strategies, spent medium analysis is carried out for determining the composition of medium to be designed for fed batch. Media feed can also be optimized to minimize generation of waste by-products, e.g., substitution of galactose with glucose leads to low lactate production during the fermentation [89].

Continuous fermentation is a modern-day technique for economizing the production of antibodies. Earlier, chemist at was used for mAb production wherein a constant flow rate was maintained, fresh medium was fed to the production culture and spent media was removed though the outlet [90]. However, it was difficult to maintain consistent cell density in such cases. Alternately, in the perfusion method, medium is circulated through a growing culture. This allows synchronised removal of waste from the production tank, recharging of nutrients, and harvesting the desired proteins. Strategies are applied to keep the cell loss at the minimum [91]. In one of the methods, alternating tangential-flow (ATF) filtration device having hollow-fibre filters are used which retains cells in production vessel. Alternatively, in another strategy, cells are cultivated in a packed-bed bioreactor in such a way that they are immobilized on a solid support matrix [92]. Although, the ATF perfusion method provides the higher yield of both viable cells and mAbs, the method poses regulatory issues on a per batch basis and can be an expensive process due to wastage of unspent media [93].

### *7.5.4.5 Production Vessel*

One of the most important and difficult aspects of production technology is the scalability of a process from a laboratory scale to pilot scale to industrial scale bioreactor. However, to meet the never-ending demand of these therapeutic proteins, production has to be executed in industrial fermenters so as to make the process commercially viable. The commercial production of mAbs is carried out using adherent cells or suspension cells. Amongst the two, the suspension method is considered more efficient due to ease of its scalability [84]. Another advantage suspension culture has over the adherent culture is that the cell density is not dependent on the area of bioreactor as in the latter case. The different types of fermenters used for industrial production of mAbs are, i) Stirred Tank Bioreactors, ii) Bubble column bioreactor, iii) Airlift Bioreactors, iv) Packed Bed Bioreactors, v) Disposable wave bioreactors, vi) Roller Bottles.

a. **Stirred Tank Bioreactors**
   A stirred tank bioreactor is the most common fermenter design used for production of biomolecules wherein a cylindrical vessel is fitted with central shaft having one or more impellers [94]. The shaft is attached to the bottom of the bioreactor and is driven by a motor. The numbers of impellers are usually variable depending upon the aspect ratio (height to

diameter) and size of the vessel. However, generally the aspect ratio is two or less in case of submerged mammalian cell fermentation. Air is sparged through a ring with many holes or a tube with a single orifice. Though, Mammalian cells are prone to sheering due to air bubbles in media, the bubbles produced by sparger are broken down to smaller effervescence and dispersed in the medium by optimum agitation, facilitating a uniform environment throughout the vessel. Since, all the parameters are well defined; one of the major advantages of this type of production process is easily addressable regulatory challenges [84].

b. **Bubble Column Bioreactors**

Bubble column bioreactor consists of a cylindrical vessel with aspect ratio of less than one. The gas is sparged into the medium from the bottom of the vessel through a perforated air column sparger which may be disc shaped or a square plate. Apart from this the vessel has no internal structures. Hence, the cells may be prone to sheering and this is the main reason that large-scale cultivation of mammalian cells is not much reported in these kinds of bioreactors [95].

c. **Airlift Bioreactor**

Airlift Bioreactors are similar to bubble column bioreactors where transfer of oxygen, circulation and mixing are achieved by bubbling of air from the bottom of the vessel [96]. However, they contain a channel for gas–liquid up-flow which is termed the riser and a separate channel for the down-flow of the medium. This channelization may be created though: i) plate separation in the vessel (Figure 7.5A); ii) placing the draft tube inside the vessel just over the sparger (Figure 7.5B); iii) placing an external draft tube which has inlet from the top of the vessel and outlet at the bottom of the Wmain vessel (Figure 7.5C). The two channels are linked to each other at the bottom and as well as at the top to form a closed loop

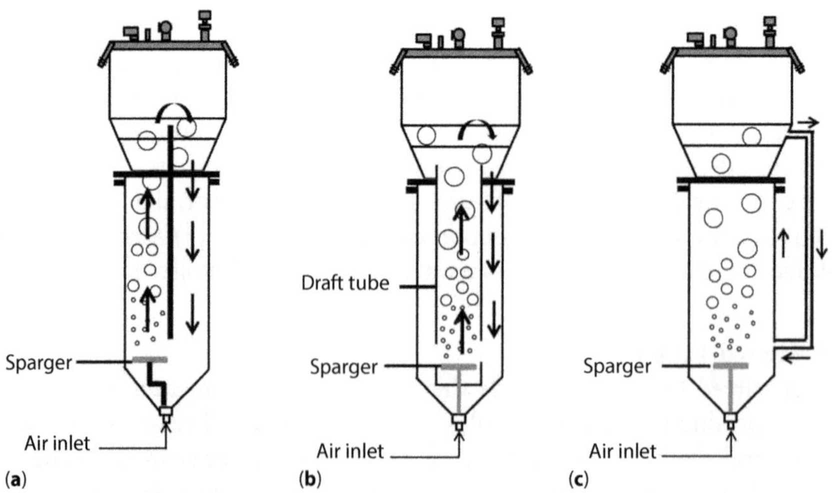


**Figure 7.5** Graphic representation of Airlift Bioreactor.

for uniform mixing of oxygen and nutrients. The gas is injected near the bottom of the riser. Utilization of an airlift system results in cost savings due to elimination of a mechanical agitation and reduction of extra cooling requirements.

d. **Packed-Bed Reactors**

There are number of literature reports on design concepts for packed-bed bioreactors [97–99]. However, only a few have been used in industrial production. A packed-bed bioreactor consists of a packed-bed which supports the cells on or within carriers. The matrix can be a solid glass sphere beads, macroporous glass beads, cross-linked dextran matrix coated with denatured collagen, cellulose porous cubes charged with polyethyleneimine or ceramic pieces. Nutrient broth flows constantly over the immobilised cell matrix. The spent media along with the desired and waste products are released into the fluid on the alternate side and removed (Figure 7.6). Previous studies have revealed that packed-bed bioreactors have achieved cells densities in the range of $1 \times 10^8$ to $5 \times 10^8$ cells per ml [97]. For production of Monoclonal antibodies a reactor system consisting of radial-flow fixed-bed units coupled with a dialysis module have been developed. The dialysis membrane allows the easy transfer of low-molecular-weight nutrients, gaseous exchange and removal of waste metabolites, while high-molecular-weight molecules like monoclonal antibodies are retained and accumulated. However, packed-bed reactors are usually associated with the problem of blockages and poor oxygen transfer [99].

e. **Disposable Wave Bioreactor**

The use of disposable equipment has significant advantages which include the reduction of preparation time and the exclusion of cleaning and sterilization steps because at the end of the process, this is discarded and replaced by a new, clean and sterile bioreactor [84]. Briefly, the wave bioreactor is a sterile disposable plastic bag, half filled with production medium [100]. The rest of the space is filled with gaseous mixture optimum for

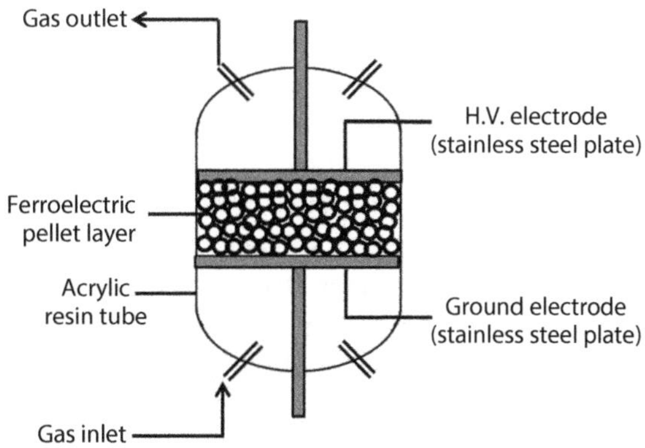


**Figure 7.6** Overview of Packed-Bed Bioreactor.

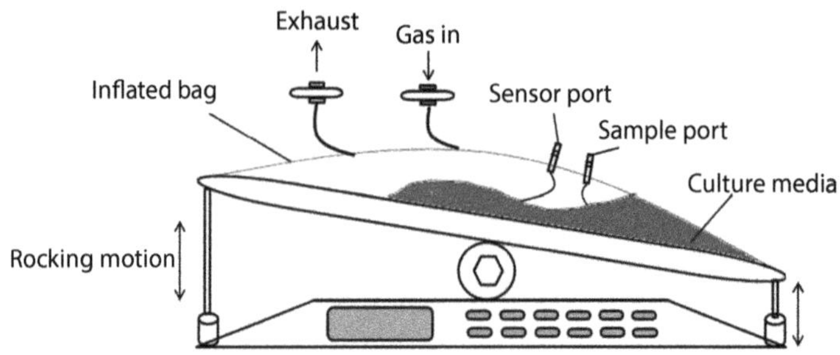


**Figure 7.7** Overview of Disposable Wave Bioreactor.

growth of culture. The bags are positioned on a rocker platform which produced a wave-like motion in medium thus allowing the proper mixing of nutrients and gas in the growing culture (Figure 7.7). Another advantage of this kind of setup is that it allows continuous gassing of medium without formation of bubbles, hence, less sheering of cells. These bioreactors are available in numerous volumes ranging from 2-L to 1,000-L. In recent advancements these sterile plastic vessels are supplied with impellers and ports for typical pH, DO, and temperature probes, e.g., HyClone S.U.B®, Millipore® (CellReady™), or Xcellerex® (XDR™)[76].

f. **Roller Bottles**

Roller bottles are the alternatives to the disposable wave bioreactors where cells are grown in plastic cylindrical disposable bottles kept on the shaking racks thus allowing the rotation of the bottles [84]. The motion of medium inside the bottles allows for the constant mixing of nutrients and gasses without mechanical agitation (Figure 7.8). For commercially viable production, the number of bottles per racks may range from 10 to 100, stacking the racks one above another. Usually, in one production run of one roller bottle with 1L of monoclonal antibody-containing supernatant, 20 to 50 mg (average 20 mg) purified antibody can be obtained [101].

### *7.5.4.6 Microbial Production*

Microorganisms have the ability to produce antibody fragments which are easy to manipulate and cultivate by the recent advancements in the expression systems, strain engineering and production processes. The three main microbes involved in production of antibody fragments are *Saccharomyces cerevisiae*, *Pichia pastoris* and *Escherichia coli* [102]. The reason for using these microbes for antibody production is their enzymatic machinery. Microbes provide the ease of high cell density cultivation and greater productivity in cheaper and defined substrates [102]. These genetically modified microbes have a well-characterized expression system with reproducible production methodology and which makes the validation of the whole process easy [103].

a. ***Saccharomyces cerevisiae***

*Saccharomyces cerevisiae* (yeast) is the most preferred host for recombinant protein production due to its proven history of utilization in

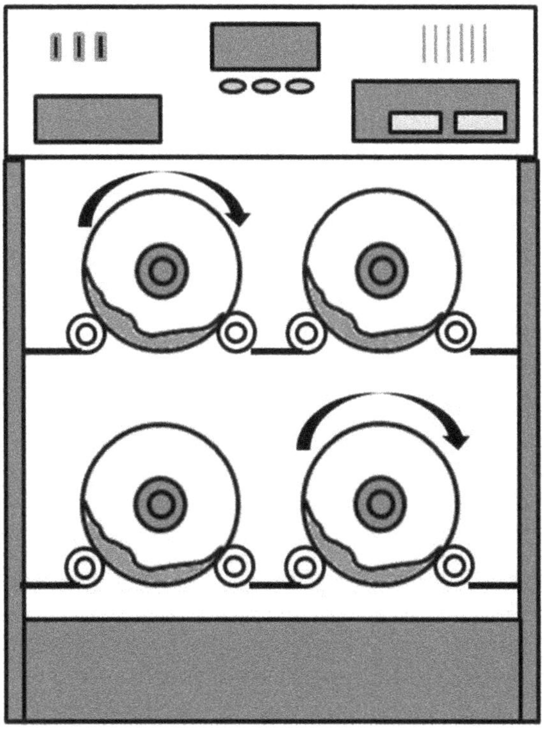

**Figure 7.8** Overview of Roller Bottle Bioreactor.

large-scale conventional industrial processes that lacks measurable endotoxins. *S. cerevisiae* being a eukaryote is potentially capable of post translation modifications which can give rise to biologically active secondary structures [104–106].The genetic modification is carried out in the yeast mainly by using three types of shuttle vectors: a) yeast episomal plasmids; b) yeast centromeric plasmids; and c) yeast integrative plasmids [107]. Different methods are used for the large-scale production of antibodies from yeast such as batch and fed-batch fermentation processes in industrial sectors. These kinds of processes are capable of providing extended growth phase of cells, control over substrate for optimum cell growth, product formation and augmented cell density [108] which enables the use of smaller scale bioreactors like 2000L single-use bioreactors for commercial production. Through fed-batch production titres of more than 5 g/L can be achieved in 14 days [109].

b. ***Pichia pastoris***

*Pichia pastoris* is another modified yeast cell which has great potential for high secretion of heterologous proteins [110]. This yeast cell can be cultivated at high cell densities of around 160 g/L dry weight because it undergoes respiratory growth over fermentative growth without accumulation of ethanol which is a risk factor [111]. The glycol-engineered yeasts are able to produce commercially viable, functionally full length and productive monoclonal antibodies because the N-linked glycosylation can influence the therapeutic functioning of mAbs. When compared with CHO produced marketed therapeutic monoclonal antibody, the results shows similar motilities which are tested by SDS-PAGE, SEC and antigen binding affinities. Monoclonal antibodies produced by glycol-engineered *Pichia pastoris* can be more than 1 g/Lin robust and measurable conditions. The cultivation process condition includes pH (5.5–7.5), temperature (16–24 °C), dissolved oxygen concentration (0.85–3.40 mg $L^{-1}$ and specific methanol feed rate (9–19 mg $g^{-1}$ $h^{-1}$) as well as across different cultivation scales (0.5, 3.0, 15 and 40L) for N-linked glycan uniformity and volumetric productivity [112].

c. ***Escherichia coli***

The commercial production of mAbs with the help of bacteria encompasses several technically complex and lengthy steps. One such organism is a gram-negative bacterium *E. coli* with prokaryotic expression system, well understood genetics and cheap and easy availability. *E.coli* was the first microbe to be used for production of recombinant biopharmaceuticals, therapeutic proteins and antibody fragments e.g., the tumour necrosis factor (TNF)-a Fab.

The limiting glucose fed-batch process is commonly used in mAb production by *E.coli* in bioreactors. There is a need of limiting glucose because it causes the production of acetate in the medium. Different genetically modified methods are also used to control this problem like manipulating the acetate formation pathway [113]. For flexible strategy to optimize early process development of biologics and vaccines, Sartorius Stedim Biotech developed a process using a mini bioreactor system called the ambr250 automated bioreactor chosen for scale-up comparison between microbial and mammalian cell lines. In this process, recombinant *E.coli* was chosen as model organism to represent the microbial cell lines. All vessels were inoculated with 2–5% inoculum at an optical density at 600 nm ($OD^{600}$) between 1 and 5. For 72 hours the cell lines were cultivated at 37 °C in a chemically defined medium with pH 7.0 ± 0.03, 30% dissolved oxygen (DO) and an impeller speed of gassed power per volume between $5.33 \times 10^2$ and $1.43 \times 10^4$ $W/m^3$. The cells were observed at every 12 h using the spectrometer to measure the $OD^{600}$ value of cells. The same clones were cultured in 15-L bench-top fermenters and in 150-L pilot-scale fermenters [114]. However, *E. coli* based therapeutic protein production is associated with the limiting factor of toxicity due to production of endotoxin in the process.

### *7.5.4.7 Plant-Based Production*

Plants offer an advantage over other cell-culture production systems described above. They can be used in both *in vitro* and *in vivo* platforms. Plant species used for production of monoclonal antibodies are tobacco, arabidopsis and alfalfa, etc., since they are easily accessible and have significant leaf biomass [115]. Other plants like maize and soybean can accumulate mAbs in seeds. Tobacco offers significant advantage over other species due to high leaf biomass yield and rapid reproduction [116]. According to several reports, the protein expression levels are found to be consistent throughout the whole tobacco plant biomass eventually resulting in cost efficiency [117]. Furthermore, tobacco expression system has been well characterized even excluding the risk of human pathogen contamination. However, tobacco is associated with nicotine or other alkaloids toxicities and produces heterogeneous N-glycosylated mAbs leading to batch variations in production [118–120].

Though there is an immense potential for therapeutic mAb production in plants, incapability of human N-glycosylation render them unsuitable for production of commercial biopharmaceutical proteins. Besides, there is an associated protease activity in plant expression systems further affecting the final yields of therapeutic proteins [118, 121].

### 7.5.5 Chemical Modifications of Monoclonal Antibodies

In the process of mAb production by recombinant expression systems, final protein products are usually subjected to post-translational modifications (PTMs), such as glycosylation, oxidation, and deamidation [122]. Moreover, degradation of mAbs may occur at various stages of development starting from bioprocessing and purification to formulation, storage, and delivery. The structural or conformational change in antibodies may lead to variation in their binding ability or specificity. All antibodies contain an abundant amount of lysine, which acts as good nucleophiles above pH 8.0 (pKa = 9.18) thus, reacts easily with different reagents to form stable bonds, hence, primarily responsible for these variations [123].

The major chemical pathways responsible for charge heterogeneity in the antibodies are sialylation, deamidation, C-terminal lysine cleavage, adduct formation, amino acid oxidation, asialylation, C-terminal lysine and glycine amidation [124]. Deamidation and sialylation leads to addition of COOH group to the antibodies thus decreasing pIvalues and increasing the net negative charge, leading to formation of the acidic variants. Kaneko *et al.* in their studies proved that antibodies bearing sialylated Fc-glycans have anti-inflammatory properties in a rheumatoid arthritis (RA) murine model [125]. On the contrary, the biological activity of proteins such as IgG1 is adversely affected by deamidation particularly during long-term storage[126]. Cleavage of lysine at C-terminus again results in decreased net positive charge leading to formation of acidic variants [127]. Acidic variants can also be produced through glycation where small sugars like glucose or lactose present in the culture media or formulation covalently binds to the primary amine of a lysine residue [128]. In contrast, basic variants can be produced by C-terminal lysine or glycine amidation, succinimide formation, amino acid oxidation or removal of sialic acid, leading to induction of net positive charge and increase in pI values [129]. These variations in protein composition have a significant impact on pharmaceutical properties (antigen binding) and stability of bio-therapeutics [128].

### 7.5.6 Advances in Antibody Technology

#### *7.5.6.1 Antibody–Drug Conjugates (ADCs)*

In recent times ADCs have been one of the most significant modification of mAbs wherein the powerful cell-killing ability of potent cytotoxic agents are combined with

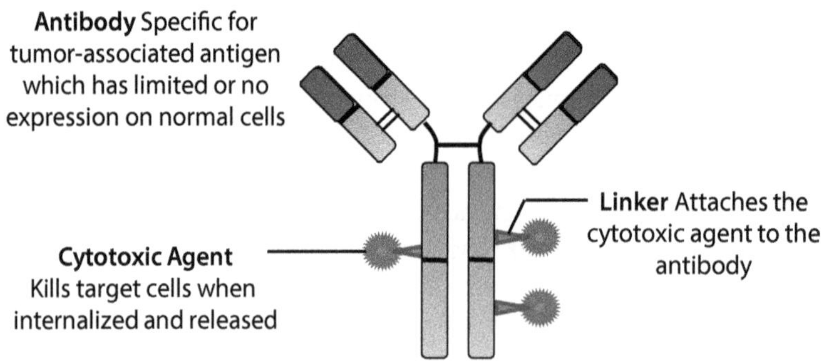


**Figure 7.9** Representation of Antibody-Drug Conjugate.

the target specificity of mAbs [130]. ADCs comprise three components, viz., the antibody, linker and potent drug. All these three components must be fully optimized to realize the goal of a targeted drug therapy (Figure 7.9). The identity and stability of a linker that covalently attaches the immunoglobulin to a potent cytotoxic drug is a crucial component. Conjugation of drug to the antibody can be random or site specific. In case of random conjugation lysine residues present on mAbs are appended to the esters presented by drug-linkers forming stable amide bonds [131]. However, the current focus has been the development of site-specific ADCs. The strategies for their development include insertion of cysteine residues in the antibody sequence by mutation or insertion of an unnatural aminoacid with a bio-orthogonal reactive handle and enzymatic conjugation. Cysteine engineered antibodies present improved pharmacokinetics and a superior therapeutic index as compared to randomly conjugated ADCs. In another study, p-cetylphenylalanine was introduced at one of several positions in the constant region of trastuzu mAb improved pharmacokinetics compared with non-specifically conjugated ADCs [132].

### *7.5.6.2 Antibody Fusion Proteins*

Fc chimeric fusion proteins or simply Fc fusion proteins are the recombinant class of antibodies wherein the protein of interest is directly attached to the Immunoglobulin molecule without any linker [133]. Hence, they are different as compared to ADCs where linkages are results of chemical conjugations whereas Fc fusion proteins are the result of genetic recombination. Sometimes only small effector molecules known as peptibodies are produced recombinantly e.g. Fab, scFv, and single variable VH and VL domains (Figure 7.10).

### *7.5.6.3 Payloads*

In another modification, antibodies are fused to the enzymes or functional proteins which can convert a pro-drug (injected separately) to effector drug inside the tumour cell. These enzymes fused antibodies are known as Payloads (Figure 7.11). Although the pro-drug can also enter the healthy cell, in the absence of payloads (due to specific targeting of mAbs) it is not converted into the effector drug thus evading the side-effects in normal cell population [134].

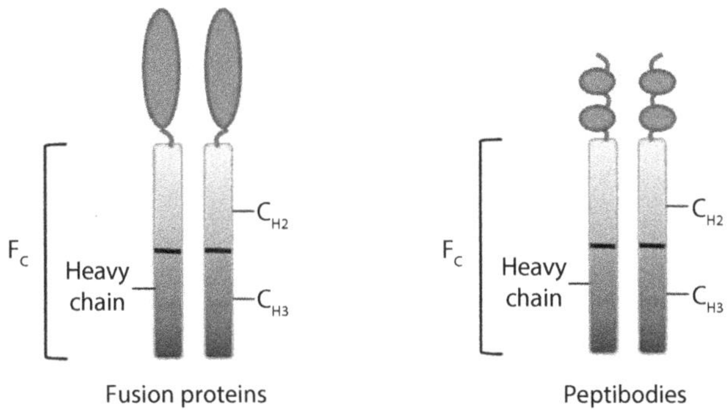


**Figure 7.10** Structure of Fusion Proteins and Peptibodies.

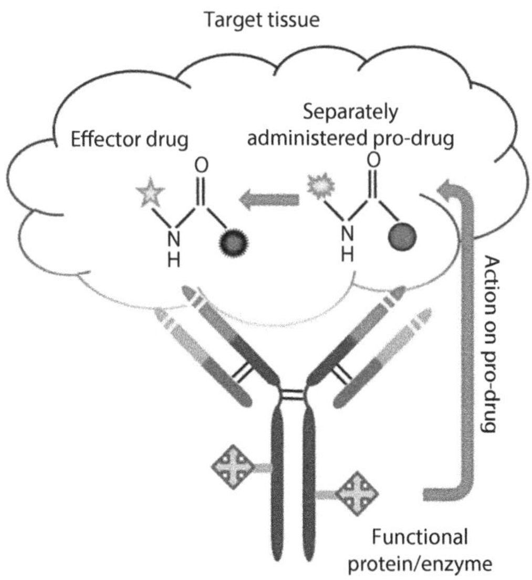


**Figure 7.11** Mechanism of Action of a Payload.

## 7.6 Conclusions

Therapeutic monoclonal antibodies (mAbs) harness the highly evolved specificity of adaptive immunity to fight against disease. mAbs are a highly specific class of antibodies which are produced by identical immune cells that are all clones of a single parent cell. With advances in genetic and bioprocess engineering, many biopharma companies are engaging in production of mAbs. Although the initial MAbs approved for therapeutic use in humans were found to be intolerable, enormous efforts in the field of hybridoma technology have resulted in mAbs that are much more effective and safe. A lot of such next-generation mAbs have successfully gone through clinical trials and are now approved for human therapy. However, even after numerous technical advancements there are inherent biological limitations associated with the production of mAbs. The paramount biological constraint is the antigenicity of the target protein; the human immune system may not recognize certain epitopes and become immunologically blind. Synthetic libraries may check this snag. Moreover, the ability to engineer variable regions that encode multiple specificities into a single molecular entity has been an advantage for optimizing antigen binding capabilities. Moreover, the fresh epitopes on the therapeutic antibody itself may turn out to be immunogenic and result in development of anti-idiotype antibodies having the potential of neutralizing the effects of therapeutic mAbs. Several biotechnological endeavours and breakthroughs like chimerization, humanization and the development of fully human antibodies have resulted in shaping up of therapeutic antibody-based drugs with improved efficacy and safety overcoming the risk of anti-idiotype antibody generation. Another concern is the development of biosimilars where developing economies are required the meet the qualifications of the branded originals. Therefore, all such antibody production shall be done under Good Manufacturing Practices (GMP) conditions, in accordance to the rules and regulations of the Food and Drug Administration (FDA). The use of Disposable Bioreactors has been effective in addressing the regulatory challenges associated with production of mAbs. However, strategies need to be devised to make them cost-effective and accessible to patients from developing countries like India.

There is no doubt that mAbs will dominate the future treatment modalities of conditions like cancers, infectious and autoimmune diseases. With the advances in Biotechnology and antibody engineering, therapeutic mAbs will become commercially affordable and more copious in the near future.

## References

1. D. Focosi, F. Maggi, M. Pistello, U. Boggi, and F. Scatena, *Clinical Microbiology and Infection*, Vol. 17(12), p. 1759, 2011.
2. M. Di Domenico, A. Bisogno, M. Polverino, C. De Rosa, V. Ricci, and A. Capasso, *Inflammation and Allergy Drug Targets*, Vol. 10, p. 2, 2011.
3. H.H. Cai, *MOJImmunology*, Vol. 1(5), p. 28, 2014.
4. J.M. Reichert, *Antibodies to watch in 2016, Microbial Antibodies,* Vol. 8 (2), p. 197, 2016.
5. A.A. Shukla, L.S. Wolfe, S.S. Mostafa, and C. Norman, *Bioengineering and Translational Medicine*, Vol. 2, p. 58, 2017.
6. Y. Yang, G. Wang, T. Song, C.B Lebrilla, and A.J.R Heck, *mAbs*, Vol. 9(4), p. 638, 2017.
7. K. Imai, and A. Takaoka, *Nature Reviews Cancer*, Vol. 6(9), p. 714, 2006.
8. L. Feng, N. Vijayasankaran, A. (Yijuan) Shen, R. Kiss, and A. Amanullah, *mAbs*, Vol. 2(5), p. 466, 2010.
9. O. Spadiut, S. Capone, F. Krainer, A. Glieder, and C. Herwig, *Trends in Biotechnology*, Vol. 32(1), p. 54, 2014.
10. K. Strebhardt, and A. Ullrich, *Nature Reviews Cancer*, Vol. 8(6), p. 54, 2008.
11. G. Marano, S. Vaglio, S. Pupella, G. Facco, L. Catalano, G.M. Liumbruno, and G. Grazzini. *Blood Transfusion*, Vol. 14(2), p. 152, 2016.
12. L.V. Marks, *The Lock and Key of Medicine: Monoclonal Antibodies and the Transformation of Healthcare*, New Haven: Yale University Press, 2015.
13. G. Kohler, and C. Milstein, *Nature*, Vol. 256, p. 495, 1975.
14. O. Leavy, *Nature Reviews Immunology*, Vol. 10(5), p. 297, 2010.

15. G. Horneff, T. Winkler, J.R. Kalden, F. Emmrich, and G.R. Burmester, *Clinical Immunology and Immunopathology*, Vol. 59(1), p. 89, 1991.
16. A.L. Nelson, E. Dhimolea, and J.M. Reichert, *Nature Reviews Drug Discovery*, Vol. 9(10), p. 767, 2010.
17. A. Frenzel, T. Schirrmann, and M. Hust, *mAbs*, Vol. 8(7), p. 1177, 2016.
18. C.A. Janeway, Jr, P. Travers, M. Walport, and M.J. Shlomchik, *Immunobiology: The Immune System in Health and Disease*, New York: Garland Science, 2001.
19. P. Chames, M.V. Regenmortel, E. Weiss, and D. Baty, *British Journal of Pharmacology*, Vol. 157(2), p. 220, 2009.
20. P.T. Jones, P.H. Dear, J. Foote, M.S. Neuberger, and G. Winter, *Nature*, Vol. 321, p. 522, 1986.
21. F.A. Harding, M.M. Stickler, J. Razo, and R.B. DuBridge, *mAbs*, Vol. 2(3), p. 256, 2010.
22. A.J. Murphy, L.E. Macdonald, S. Stevens, M. Karow, A.T. Dore, K. Pobursky, T.T Huang, W.T. Poueymirou, L. Esau, M. Meola, W. Mikulka, P. Krueger, J. Fairhurst, D.M. Valenzuela, N. Papadopoulos, and G.D. Yancopoulos, *Proceedings of the National Academy of Sciences of the United States of America*, Vol. 111(14), p. 5153, 2014.
23. Animal Cell Technology Industrial Platform.http://www.actip.org/products/monoclonal-antibodies-approved-by-the-ema-and-fda-for-therapeutic-use/, 2017.
24. European Medicines Agency.www.ema.europa.eu/ema, 2017.
25. United States Food and Drug Administration. www.fda.gov/drugs/informationondrugs/approveddrugs, 2017.
26. P.W.H.I. Parren, P.J. Carter, and A. Pluckthun, *mAbs*, Vol. 9, p. 898, 2017.
27. Revised monoclonal antibody (mAb) nomenclature scheme, INN Working Doc. 17.416, World Health Organization, Geneva, 2017.
28. S.S. Alkan, *Nature Reviews Immunology*, Vol. 4, p. 153, 2004.
29. B. Alberts, A. Johnson, J. Lewis, M. Raff, K. Roberts, and P. Walters, *Molecular Biology of the Cell*, 4th edition, New York: Garland Science, 2002.
30. M.A. Ritter, in A. J. T. George and C. E. Urch, eds., *Methods in Molecular Medicine*, Humana Press Inc., pp. 23, 2000.
31. J.K.H Liu, *Annals of Medicine and Surgery*, Vol. 3(4), p. 113, 2014
32. M. Tomita, and K. Tsumoto, *Immunotherapy*, Vol. 3(3), p. 371, 2011.
33. L.R. Jackson, L.J Trudel, J.G. Fox, and N.S. Lipman, *Laboratory Animal Science*, Vol. 49, p. 70, 1999a.
34. L.R. Jackson, L.J Trudel, J.G. Fox, and N.S. Lipman, *Laboratory Animal Science*, Vol. 49, p. 81, 1999b.
35. C. Ware, N. Donato, and K. Dorshkind, *Journal of Immunological Methods*, Vol. 85, p. 353, 1985.
36. G. Orive, R.M. Hernández, A.R. Gascón, and J.L. Pedraz, *Methods in Biotechnology™*, Humana Press, p. 345, 2006.
37. P. Muralidhar, E. Bhargav, and K. R. Reddy, *International Journal of Institutional Pharmacy and Life Sciences*, Vol. 7(2), p. 31, 2017.
38. S. Shammah, T.L. Mantovani, R. Dalla-Favera, and P. Casali, *Journal of Immunological Methods*, Vol. 160(1), p. 19, 1993.
39. G.P. Smith, *Science*, Vol. 228, p. 1315, 1985.
40. A. Frenzel, T. Schirrmann, and M. Hust, *mAbs*, Vol. 8(7), p. 1177, 2016.
41. C.M. Hammers, and J.R. Stanley, *The Journal of Investigative Dermatology*, Vol. 134(2), p. 1, 2014.
42. J. Bazan, I. Całkosiński, and A. Gamian, *Human Vaccines & Immunotherapeutics*, Vol. 8(12), p. 1817, 2012.
43. Y. Durocher, and M. Butler, *Current Opinion in Biotechnology*, Vol. 20, p. 700, 2009.

44. D. Ghaderi, M. Zhang, N.H. Ziola, and A. Varki, *Biotechnology and Genetic Engineering Reviews*, Vol. 28, p. 147, 2012.
45. K. Swiech, V.P. Castro, and D.T. Covas, *Protein Expression Purification*, Vol. 84, p. 147, 2012.
46. J. Dumont, D. Euwart, B. Mei, S. Estes, and R. Kshirsagar, *Critical Reviews in Biotechnology*, Vol. 36(6), p. 1110, 2016.
47. W.E. Marissen, R.A. Kramer, A. Rice, W.C. Weldon, M. Niezgoda, M. Faber, J.W. Slootstra, R.H. Meloen, M.C.van der Horst, T.J. Visser, M. Jongeneelen, S. Thijsse, M. Throsby, J. de Kruif, C.E. Rupprecht, B. Dietzschold, J. Goudsmit, and A.B. Bakker, *Journal of Virology*, Vol. 79(8), p. 4672, 2005.
48. C. Heuser, S. Guhlke, A. Matthies, H. Bender, S. Barth, V. Diehl, H. Abken, A, Hombach, *International Journal of Cancer*, Vol. 110(3), p. 386, 2004.
49. M. Popkov, S.S. Ghali, Y. Lusignan, S. Lemieux, and R. Mandeville, *European Journal of Cancer*, Vol. 37(18), p. 2484, 2001.
50. M. Butler, and M. Spearman, *Current Opinion in Biotechnology*, Vol. 30C, p. 107, 2014.
51. M.J.J. Page, and M.A. Sydenham, *Nature Biotechnology*, Vol. 9, p. 64, 1991.
52. M.J. Keen and T.W. Steward, *Cytotechnology*, Vol. 17, p. 203, 1995.
53. S. Gorfien, B. Paul, J. Walowitz, R. Keem, W. Biddle, and D. Jayme, *Biotechnology Progress*, Vol. 16(5), p. 682, 2000.
54. J.R. Birch, and J.R. Racher, *Advanced Drug Delivery Review*, Vol. 58, p. 671, 2006.
55. A. Zhu, and R. Hurst, *Xenotransplantation*, Vol. 9(6), p. 376, 2002.
56. E. Byres, A.W. Paton, J.C. Paton, J.C. Lofling, D.F. Smith, M.C. Wilce, U.M. Talbot, D.C. Chong, H. Yu, S. Huang, X. Chen, N.M. Varki, A. Varki, J. Rossjohn, and T. Beddoe, *Nature*, Vol. 456, p. 648, 2008.
57. E. Casademunt, K. Martinelle, M. Jernberg, S. Winge, M. Tiemeyer, L. Biesert, S. Knaub, O. Walter, and C. Schroder, *European Journal of Haematology*, Vol. 89, p. 165, 2012.
58. N. Enjolras, Y. Dargaud, E. Perot, F. Guillaume, M. Becchi, and C. Negrier, *Thrombosis Research*, Vol. 130(5), p. 266, 2012.
59. S. Estes, and M. Melville, *Advances in Biochemical Engineering/Biotechnology*, Vol. 139, p. 11, 2014.
60. D. Jones, N. Kroos, R. Anema, B. Van Montfort, A. Vooys, S. Van der Kraats, E. Van der Helm, S. Smits, J. Schouten, K. Brouwer, F. Lagerwerf, P. van Berkel, D.J. Opstelten, T. Logtenberg, and A. Bout, *Biotechnology Progress*, Vol. 19, p. 163, 2003.
61. B. Mei, Y. Chen, J. Chen, C.Q. Pan, and J.E. Murphy, *Molecular Biotechnology*, Vol. 34, p. 165, 2006.
62. K. Swiech, A. Kamen, S. Ansorge, Y. Durocher, V.P. Castro, E.M.S. Russo-Carbolante, M.S.A. Neto, and Dimas T Covas, *BMC Biotechnology*, Vol. 11, p. 114, 2011.
63. A. Croset, L. Delafosse, J.P. Gaudry, C. Arod, LoicGlez, C. Losberger, D. Begue, A. Krstanovic, F. Robert, F. Vilbois, L. Chevalet, and B. Antonsson, *Journal of Biotechnology*, Vol. 161(3), p. 336, 2012.
64. V. Jäger, K. Büssow, A. Wagner, S. Weber, M. Hust, A. Frenzel, and T. Schirrmann, *BMC Biotechnology*, Vol. 13, p. 52, 2013.
65. Royal DSM. https://www.dsm.com/content/prcc/cworld/en_US/informationcenter-news/2010/09/41–10-dsm-crucell-announce-expansion-percivia-per-c6-joint-venture.html, 2010
66. Lonza. http://www.lonza.com/products-services/bio-research/cell-culture products / specialty-media/per-c6-media.aspx, 2017.
67. R. Kunert, and D. Reinhart, *Applied Microbiology and Biotechnology*, Vol. 100, p. 3451, 2016.
68. S. Rao, A.A. Morales, and D.D. Pearse, *BioMed Research International*, Vol. 2015, p. 458, 2015.

69. K. Sautter, and B. Enenkel, *Biotechnology and Bioengineering*, Vol. 89(5), p. 530, 2005.
70. R.J. Ertola, A.M. Giulietti, and F.J. Castillo, *Bioprocess Technology*, Vol. 21, p. 89, 1995.
71. D. Barnes, and G. Sato, *Cell*, Vol. 22(3), p. 649, 1980.
72. D.W. Jayme, *Cytotechnology*, Vol. 5(1), p. 15, 1991.
73. C. Burteau, F. Verhoeye, J. Mols, J. Ballez, S. Agathos, and Y.J. Schneider, *In Vitro Cellular and Development Biology-Animal*, Vol. 39(7), p. 291, 2003.
74. B.H. Chun, J.H. Kim, H.J. Lee, and N. Chung, *Bioresource Technology*, Vol. 98(5), p. 1000, 2007.
75. J. Zhang, and D. Robinson, *Cytotechnology*, Vol. 48(1–3), p. 59, 2005.
76. M. Chartrainand, and L. Chu, *Current Pharmaceutical Biotechnology*, Vol. 9, p. 447, 2008.
77. J. Li, C.L. Wong, N. Vijayasankaran, T. Hudson, and A. Amanullah, *Biotechnology and Bioengineering*, Vol. 109(5), p. 1173, 2012.
78. H. Kaufmann, X. Mazur, M. Fussenegger, and J. Bailey, *Biotechnology and Bioengineering*, Vol. 63(5), p. 573, 1999.
79. S. Oguchi, H. Saito, M. Tsukahara, and H. Tsumura, *Animal Cell Technology: Basic & Applied Aspects*, Springer, 2004.
80. V. Restelli, M.D. Wang, N. Huzel, M. Ethier, H. Perreault, and M. Butler, *Biotechnology and Bioengineering*, Vol. 94(3), p. 481, 2006.
81. D. Gray, S. Chen, W. Howarth, D. Inlow, and B. Maiorella, *Cytotechnology*, Vol. 22, p, 65, 1996.
82. M.M. Zhu, A. Goyal, D.L. Rank, S.K. Gupta, T. Vanden Boom, and S.S. Lee, *Biotechnology Progress*, Vol. 21(1), p. 70, 2005.
83. J.D. Michaels, J.E. Nowak, A.K. Mallik, K. Koczo, D.T. Wasan, and E.T. Papoutsakis, *Biotechnology and engineering*, Vol. 47(4), p. 420, 1995.
84. L. S. Carvalho, O.B. da Silva, G. C. de Almeida, J. D. de Oliveira, N. S. Parachin, and T. S. Carmo, in Dr. Angela Jozala, ed., *Fermentation Processes*, InTech, p. 181, 2017.
85. Y.M. Huang, T. Huynh, L. Ly, W. Noe, and D.Y.H. Chang, in F. Gòdia, and M. Fussenegger, eds., *Animal Cell Technology Meets Genomics*, ESACT Proceedings, Springer, 2005.
86. R. Portner, J.O. Schwabe, and B. Frahm, *Biotechnology and Applied Biochemistry*, Vol. 40, p. 47, 2004.
87. L. Xie, G. Nyberg, X. Gu, H. Li, F. Mollborn, and D. Wang, *Biotechnology and Bioengineering*, Vol. 56(5), p. 577, 1997.
88. E. Spens, and L. Haggstrom, *Biotechnology and Bioengineering*, Vol. 98(6), p. 1183, 2007.
89. D. Petch, and M. Butler, *Applied Biochemistry and Biotechnology*, Vol. 59(1), p. 93, 1996.
90. N.H. Simpson, R.P. Sigh, A.N. Emery, and M. Al-Rubeai, *Biotechnology and Bioengineering*, Vol. 64, p. 187, 1999.
91. J. Griffiths, Perfusion systems for cell cultivation, in *Large scale mammalian cell culture technology*, A.S. Lubiniecki, ed., Marcel-Dekker Inc., 1990.
92. R. Spier, J. Griffiths, J. Stephenne, and P. Crooy, *Biotechnology and Bioengineering*, Vol. 82(7), p. 751, 2003.
93. M. Dorceus, S.S. Willard, A. Suttle, K. Han, P.J. Chen, and M. Sha, *Bioprocess International*, 2017.
94. E. Kadic, and T.J. Heindel, Hoboken, *Stirred-Tank Bioreactors, in An Introduction to Bioreactor Hydrodynamics and Gas-Liquid Mass Transfer*, New Jersey, USA: John Wiley & Sons, 2014.
95. J. Varley, and J. Birch, *Cytotechnology*, Vol. 29, p. 177, 1999.
96. Tan-Che Zhou, Wen-Wen Zhou, Weiwei Hu and Jian-Jiang Zhong, in Michael C. Flickinger. eds., John Wiley & Sons, Inc. 2010.
97. F. Meuwly, P.A. Ruffieux, A. Kadouri, and U. von Stockar, *Biotechnology Advances*, Vol. 4(1), p. 45, 2007.

98. C. Fenge, and E. Lullau, in S. Ozturk, and W. Hu, "*Cell Culture Bioreactors, in Cell Culture Technology for Pharmaceutical and Cell-Based Therapies*" in S. Ozturk, and W. Hu eds., CRC Press, p. 155, 2006.
99. R. Portner, and B.O.B.J. Platas, *Methods and Protocols*, Vol. 24, p. 353, 2007.
100. V. Singh, *Cytotechnology*, Vol. 30, p. 149, 1999.
101. F.H.C. Eriksson, I. Erlandsson, N. Norrman, J. Simon, A. Danielsson, A. Milicov, P. Schindler, and J.M. Schlaeppi, *Journal of Chromatography*, Vol. 1418, p. 103, 2015.
102. O. Spadiut, S. Capone, F. Krainer, A. Gliede, and C. Herwig, *Trends in Biotechnology*, Vol. 32(1), p. 54, 2014.
103. T.I. Potgieter, M. Cukan, J.E. Drummond, N.R. Houston-Cummings, Y. Jiang, F. Li, H. Lynaugh, M. Mallem, T.W. McKelvey, T. Mitchell, A. Nylen, A. Rittenhour, T.A. Stadheim, D. Zha, and M. d'Anjou, *Journal of Biotechnology*, Vol. 139, p. 318, 2009.
104. K.S. Kwon, M. Song, M.H. Yu, *Journal of Biotechnology*, Vol. 42, p. 191, 1995.
105. L.D. Arnold, and P. Vaishnav, *Biotechnology Advances*, Vol. 27, p. 297, 2009.
106. A. Graf, M. Dragosits, B. Gasser, and D. Mattanovich, *FEMS Yeast Research*, Vol. 9, p. 335, 2009.
107. M.K. Chee, and S.B. Haase, *G3 (Bethesda),* Vol. 2, p. 515, 2012.
108. S.M. Patrick, M.L. Fazenda, B. McNeil, and L.M Harvey, *Yeast*, Vol. 22, p. 249, 2005.
109. A.A. Shukla, L. S. Wolfe, S. S. Mostafa, and C. Norman, *Bioengineering and Translational Medicine,* Vol. 2, p. 58, 2017.
110. C. Ferndahl, N. Bonander, C. Logez, R. Wagner, L. Gustafsson, C. Larsson, K. Hedfalk, R.A.J. Darby, and R.M. Bill, *Microbial Cell Factory*, Vol. 9, p. 47, 2010.
111. M. Jahic, J.C. Rotticci-Mulder, M. Martinelle, K. Hult, and S.O. Enfors, (2002) *Bioprocess and Biosystems Engineering*, Vol. 24, p. 385, 2002.
112. T.I. Potgieter, M. Cukan, J.E. Drummond, N.R. Houston-Cummings, Y. Jiang, F. Li, H. Lynaugh, M. Mallem, T.W. McKelvey, T. Mitchell, A. Nylen, A. Rittenhour, T.A. Stadheim, D. Zha, and M. d'Anjou, *Journal of Biotechnology*, Vol. 139(4), p. 318, 2009.
113. Y. Tao, Q. Cheng, and A.D. Kopatsis, *Methods in Molecular Biology*, Vol. 834, p. 283, 2012.
114. M. Ngibuini, Sartorius Stedim Biotech, *Bio-Process™ international*, Vol.12(5), 2014.
115. G. Moussavou, K. Ko, J.H. Lee, and Y.K. Choo, *BioMed Research International*, Vol. 2015(Article ID 306164), pages 9, 2015.
116. K. Ko, and H. Koprowski, *Virus Research*, Vol. 111(1), p. 93, 2005.
117. C.Y. Lim, K.J. Lee, D.B. Oh, and K. Ko, *Frontiers in Plant Science*, Vol. 5, p. 778, 2015.
118. K. Ko, M.H. Ahn, M. Song, Y.K. Choo, H.S. Kim, KinaramKo, and H. Joung, *Molecules and Cells*, Vol. 25(4), p. 494, 2008.
119. M.C. Macheteau, A.C. Fitchette-Lain´e, C.L. Bourhis, C. Lange, N.D.Vine, *J.K.* Ma, *P.* Lerouge, and *L.* Faye, *Glycobiology,* Vol. 9, p. 365, 1999.
120. R. Fischer, E. Stoger, S. Schillberg, P. Christou, and R. M. Twyman, *Current Opinion in Plant Biology,* Vol. 7, p. 152, 2004.
121. H.S. Kim, J.H. Jeon, K.J. Lee, and K. Ko, *BioMed Research International*, Vol. 2014, Article ID 249519, 8 pages, 2014.
122. N. Jenkins, L. Murphy, and R. Tyther, *Molecular Biotechnology*, Vol. 39(2), p. 113, 2008.
123. J. Warwicker, S. Charonis, and R.A. Curtis, *Molecular Pharmaceutics*, Vol. 11(1), p. 294, 2014.
124. L.A. Khawli, S. Goswami, R. Hutchinson, Z.W. Kwong, J. Yang, X. Wang, Z. Yao, A. Sreedhara, T. Cano, D. Tesar, I. Nijem, D.E. Allison, P.Y. Wong, Y.H. Kao, C. Quan, A. Joshi, R.J. Harris, and P. Motchnik, *mAbs*, Vol. 2(6), p. 613, 2010.
125. Y. Kaneko, F. Nimmerjahn, and J.V. Ravetch, *Science*, Vol. 313(5787), p. 670, 2006.

126. R.J. Harris, B. Kabakoff, F.D. Macchi, F.J. Shen, M. Kwong, J.D. Andya, S.J. Shire, N. Bjork, K. Totpal, and A.B Chen, *Journal of Chromatography B: Biomedical Science Applications*, Vol. 752, p. 233, 2001.
127. S. Miao, P. Xie, M. Zou, L. Fan, X. Liu, Y. Zhou, L. Zhao, D. Ding, H. Wang and W.S. Tan, *Applied Microbiology and Biotechnology*, Vol. 101, p. 5627, 2017.
128. R.J. Harris, S.J. Shire, and C. Winter, *Drug Development Research*, Vol. 61, p. 137, 2004.
129. B. Hintersteiner, N. Lingg, P. Zhang, S. Woen, K.M. Hoi, S. Stranner, S. Wiederkum, O. Mutschlechner, M. Schuster, H. Loibner, and A. Jungbauer, *mAbs*, Vol. 8(8), p. 1548, 2016.
130. M.S. Sutherland, R.J. Sanderson, K.A. Gordon, J. Andreyka, C.G. Cerveny, C. Yu, T.S. Lewis, D.L. Meyer, R.F. Zabinski, S.O. Doronina, P.D. Senter, C.L. Law, A.F. Wahl, *Journal of Biological Chemistry*, Vol. 281, p. 10540, 2006.
131. S. Goswami, W. Wang, T. Arakawa, and S. Ohtake, *Antibodies*, Vol. 2, p. 452, 2013.
132. J.Y. Axup, K.M. Bajjuri, M. Ritland, B.M. Hutchins, C.H. Kim, S.A. Kazane, R. Halder, J.S. Forsyth, A.F. Santidrian, K. Stafin, Y. Lu, H. Tran, A.J. Seller, S.L. Biroc, A. Szydlik, J.K. Pinkstaff, F. Tian, S.C. Sinha, B.F. Habermann, V.V. Smider, and P.G. Schultz, *Proceedings of the National Academy of Sciences of the United States of America*, Vol. 109(40), p. 16101, 2012.
133. M.P. Hall, C. Gegg, K. Walker, C. Spahr, R. Ortiz, V. Patel, S. Yu, L. Zhang, H. Lu, B. DeSilva, J.W. Lee, *American Association Pharmaceutical Scientists Journal*, Vol. 12, p. 576, 2010.
134. L. Ducry, and B. Stump, *Bioconjugate Chemistry*, Vol. 21, p. 5, 2010.

# 8

# Antimicrobial Peptides from Bacterial Origin: Potential Alternative to Conventional Antibiotics

**Lipsy Chopra, Gurdeep Singh, Ramita Taggar, Akanksha Dwivedi, Jitender Nandal, Pradeep Kumar and Debendra K. Sahoo***

*Biochemical Engineering Research and Process Development Centre*
*CSIR- Institute of Microbial Technology, Chandigarh, (India)*

***Abstract***
Antimicrobial resistance (AMR) is a global public health threat in which microorganisms have generated the capacity to tolerate or withstand the impact of antimicrobials. In recent years, studies of antimicrobial peptides (AMPs), biologically active oligopeptides having bioactivity against broad spectrum of microorganism (from viruses to protozoans), have opened a new field of research and widened the application of AMPs which could also be a potential alternative to conventional antibiotics to combat the multiple drug resistant (MDR) pathogens in future. AMPs produced by bacteria are of two types (i) ribosomally synthesized i.e., bacteriocins, (ii) non-ribosomally synthesized. The non-ribosomally synthesized AMPs kill the bacteria by damaging its membrane, which leads to cell death. The ribosomally synthesised bacteriocins are mostly produced by gram-positive and gram-negative bacteria and have antimicrobial activities against similar or closely related bacterial strain(s). This chapter summarizes the current information on classification, mode of action and potential applications of AMPs including those in food and pharmaceutical industries.

***Keywords:*** Antimicrobials; bacteriocins; bio-preservative; anti-biofilm; therapeutics

## 8.1 Introduction

### Bacteriocins

The study of a few selected bacteriocins in-depth has created a new research field that further broadened the applications of these antimicrobial peptides. The majority of bacteriocins are from gram-positive and gram-negative bacteria which possess the antimicrobial activities [1] e.g., nisin in *Lactococcus lactis,* Colicin in *Escherichia coli.* These are proteinaceous, low molecular weight cationic or amphipathic molecules that restrain the growth of other bacteria; these are phenomenologically equivalent to killing factors produced by paramecium and yeast, defensins of mammals, techyplesins of crabs, cecropins of insects and thionins of plants. These all are structurally, functionally, and ecologically diverse.

**Corresponding author*: debsahoo@imtech.res.in

Saurabh Saran, Vikash Babu, and Asha Chaubey (eds.) High Value Fermentation Products, Volume 1, (193–204) © 2019 Scrivener Publishing LLC

Although bacteriocins could be regarded as antibiotics, unlike antibiotics that have broad-spectrum activity, bacteriocins are observed to have inhibitory activity. This inhibitory activity of bacteriocins is confined to strains of a species closely related to species producing bacteriocins which subsequently kills the competitors so that the producer may flourish. Moreover, bacteriocins are peptides and are synthesized by ribosomes during the primary growth phase, whereas, antibiotics are well-known secondary metabolites [2]. However, bacteriocins do not harm the self-producing bacteria due to specific immunity proteins and thus allowing the producer to survive and compete with other microbes. Most of these are low molecular weight cationic or amphipathic molecules and undergo post-translation modifications (PTMs). Genes encoding these bacteriocins may be located on plasmids, transposons, or bacterial chromosomes. These genes are generally found in the form of an operon responsible for its biosynthesis, post-translation modification, transportation, and imparting immunity to the producer from its own bacteriocins [3].

Since from early 1920s, bacteriocins was identified as an antimicrobial protein from *Escherichia coli* by Gratin in 1925 and named it as colicin [4]., was first identified as an antimicrobial protein from *Escherichia coli* by Garatia in 1925 and named as colicin [4]. Since then, there have been numerous reports on isolation of bacteriocins from strains of eubacteria and archaea sourced food, animals, plants, clinical samples, etc. Some common examples of bacteriocins are nisin (*Lactococcus* sp.), pediocin (*Pediococcus*sp.), lactocin (*Lactobacillus* sp.), enterocin (*Enterococcus* sp.), etc.

## 8.2 Classification of Bacteriocins

### 8.2.1 Bacteriocins from Gram-Negative Bacteria

Bacteriocins derived from gram-negative bacteria can be categorized into the following two groups:

A. **Colicins**: These are considerably large (about 25- 80 kDa) peptides that are bactericidal in nature and their production is stimulated by circumstances prompting the SOS system [5]. e.g., puticidin produced by *Psuedomonas aeruginosa* [6].
B. **Microcins**: Microcins are generally small (<10 kDa) peptides whose production is not induced by SOS system. These are further grouped into two classes:
   Class I microcins: These peptides (<5 kDa) have activity against several intracellular targets and it undergoes post-translational modifications e.g., microcins J25, B17 etc. [7]
   Class II microcins: These are unmodified peptides (about 8–10 kDa) which cause membrane depolarization to kill the bacteria e.g., microcins E492, and colicin V produced by *Enterobacteriaceae* [7].

### 8.2.2 Bacteriocins from Gram-Positive Bacteria

According to Klaenhammer, bacteriocins produced by gram-positive bacteria significantly differ in terms of compositions and structure, which contributes to different thermo-stability patterns and resistance to proteases [8].

Bacteriocins from gram-positive bacteria can be categorized into four different classes based on internal cross-linking, molecular weight, primary structures, thermostability, sensitivity to proteases, the presence of unusual amino acids, and mechanism of action [8].

**Class I:** The members of this group, also known as lantibiotics, are small bacteriocins (19–37 amino acids) containing unusual amino acids such as lanthionine, β-methyllanthionine, dehydroalanine, and dehydrobutyrine, e.g., Nisin [9].

**Class II:** This group consists of a range of small non-lanthionine containing unmodified peptides which acts upon the cell membrane. These have been further grouped in various sub-classes. Based on structural features and shared homologies, such as:

**Class IIa**: Peptides of this group are pediocin-like bacteriocins with conserved N-terminal sequence YGNGVXCXXXXC. The amino acid sequence flanked by the two cysteines is conserved and normally comprises threonine/serine residues and one/two charged residues. All these peptides strongly inhibit the food-borne pathogen *Listeria monocytogenes* and contain a disulfide bridge at N-terminus end e.g., divercin V41, pediocin PA-1, and enterocin A [10].

**Class IIb**: These are pore-forming peptides which can be either active when taken individually or synergistic in the presence of each other (e.g., enterocins L50A and L50B) or in some cases, both may be required for the inhibitory activity (lactococcins Gα/Gβ, lactococcins M/N).

**Class IIc:** These are the sec-dependent bacteriocins (e.g., enterocin P and divergicin A) that are exported via the sec machinery instead of devoted ABC transporters [11].

**Class IId:** The members of this class are leader-less peptides that are synthesized and exported without any additional bio-processing (e.g., enterocin Q and aureocin A70) [12].

**Class III:** These are the large heat-labile bacteriocins that show limited mutual homology. Only a few peptides of this class have been identified (e.g., glycinecin A and enterolysin A).

**Class IV:** These are the complex proteins that require the association with carbohydrate or lipid moieties e.g., lactocin 27.

**Class V:** The bacteriocins belonging to this group are head-to-tail ligated peptides, which form an amide bond between N- and C-termini e.g., gassericin A, enterocin AS-48 [13]. These circular peptides were categorized as Class V, a novel class, instead of originally allocated Class IV.

In 2005, Cotter and co-workers redefined this classification scheme and proposed two classes, lantibiotics [class I] and non-lantibiotics [class II] and these were further sub-categorized. Furthermore, it was suggested that class III should be named as bacteriolysins, instead of bacteriocins. However, this new classification scheme led to a disagreement between Cotter *et al.* [14] and Heng *et al.* [15]. In addition, this proposed classification included only a part of bacteriocins produced by lactic acid bacteria that could be used in food applications [14–16]. Another classification, a combination of the previous classifications by Klaenhammer *et al.* [8] and Cotter *et al.*, in [14], took into account all bacteriocins produced by gram-positive bacteria [15]. This proposed classification is now widely accepted though there are still disagreements on the subdivision of class II and questions on class IV being a separate class [8, 14–18]. However, the fact that class IIa is the pediocin-like bacteriocins and IIb is multi-component (also described as two-component) bacteriocins remained similar in all these suggestions.

## 8.3 Mode of Action

The bactericidal spectrum range is highly dependent on the mode of action of the bacteriocins i.e., the presence of the cellular target in a strain, and specificity of the bacteriocins target binding domain. Based on interaction with membranes, bacteriocins have been divided into two mechanistic classes: (a) Pore-forming bacteriocins (barrel stave, toroidal, carpet, and micellar aggregate models) (b) Non-pore-forming bacteriocins (intracellular targets) (19). The composition of the cell membrane is one of the crucial determinants by which bacteriocins selectively target microbial membrane rather than eukaryotic cell membranes. The presence of lipids with phospholipid head-groups such as cardiolipin (CL), phosphatidylserine (PS), and phosphatidylglycerol (PG) makes the bacterial membranes negatively charged whereas, eukaryotic membranes have neutral zwitterionic phospholipids such as sphingomyelin (SM), phosphatidylcholine (PC), and phosphatidylethanolamine (PE) [20] making it amphipathic in nature. In addition, due to the presence of cholesterol in the mammalian membrane, the activity of bacteriocins is reduced as cholesterol either stabilizes the lipid bilayer or directly interacts and neutralizes the bacteriocins [21].

### 8.3.1 Pore-Forming Bacteriocins

Both gram-positive and gram-negative bacteria produce pore-forming bacteriocins. The molecular mechanism of membrane permeabilization and disruption by bacteriocins is dependent on a number of factors like amino acid sequence, the concentration of bacteriocins and presence of lipids in the membrane. The different size of the pores and bacteriocins both leads to disruption of membrane potential. Antonio group in 2001 has shown that bacteriocins are small cationic peptides having the tendency of pore formation and are produced by Gram-positive bacteria (Figure 8.1).

***Carpet Model:*** This mechanism involves initial binding of the bacteriocins to membrane surface in order to form a high-density cluster that coat the surface. When peptide density reaches a threshold, the membrane undergoes a significant curvature strain resulting in its disruption. Ultimately, a dispersion-like dissolution of the membrane occurs without channel formation, and the bacteriocins do not enter into the hydrophobic core of membrane e.g., Ovispirin, which electrostatically attract the anionic phospholipids head groups at numerous sites [22–23].

***Barrel stave model:*** This mechanism involves the perpendicular insertion and clustering of comparatively less number of bacteriocins (referred to as staves in a barrel-like ring inside the membrane) leads to the formation of cylindrical trans-membrane channel or pore. The hydrophilic surface of the peptides forms the pore lining, whereas the hydrophobic surfaces of α-helical or β-sheet peptides face outward, toward the acyl chains of the membrane [24]. Initially, the binding of monomer bacteriocins occurs at the surface of the membrane, and following a conformational phase transition, the hydrophobic region of the bacteriocins is inserted into the membrane to an extent leading to the hydrophobicity of the outer leaflet of the membrane. Furthermore, on reaching a threshold concentration of bound peptide, bacteriocins monomers could undergo self-aggregation and insert deeper into the hydrophobic core of the membrane. Due to aggregation, the peptides adopt a trans-membrane configuration resulting in negligible exposure of their hydrophilic residues to the hydrophobic membrane interiorly e.g., A lecithin, which form peptide aggregation equilibration by influencing the charge present in lipid composition [23].

***Toroidal pore model:*** This model hypothesizes the formation of well-defined pores by bacteriocins and lipid together. The inward bending of the membrane leads to the hole formation with the head groups facing towards the center of the pore and the bacteriocins lining the pore. However, the critical bacteriocins-lipid ratio determines the formation of the toroidal pore and translocation of the bacteriocins. An excruciating increase in the concentration of bacteriocins would lead to increased electrostatic repulsion between the positively charged side chains and destabilize the pore [23] e.g., Lacticin, binds to negatively charged membrane and form α- helical structures which leads to leakage of the membrane [25].

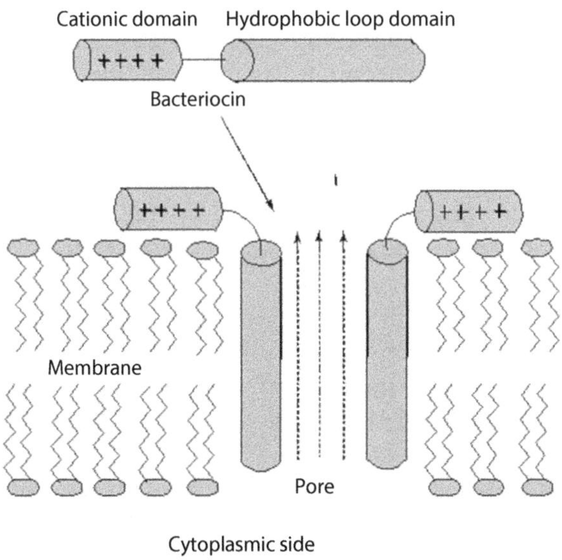


**Figure 8.1** Mechanism of action of bacteriocin.

### 8.3.2 Non-Pore-Forming Bacteriocins: Intracellular Targets

Permeabilization of the bacterial membrane does not always lead to bacterial death. Probably, the action starts when bacteriocins interact with microbial membranes; which ultimately leads to pore formation followed by depolarization and leakage of metabolites and ions. There are many reports which have shown that there is a relationship between permeabilization of the bacterial membrane and cell death, as bacterial killing generally initiated with relatively slight membrane disruption. Further, this suggests that bacteriocins may interact with putative vital intracellular targets. Inhibition of cell wall biosynthesis occurs as the lantibiotics interact with the lipid II cell wall precursor, thus preventing the growth of the peptidoglycan chain. Class IIIa bacteriocins such as lysostaphin, functions as a zinc metalloprotease which causes the hydrolysis of the cell wall in sensitive strains [26]. Alternative modes of action may involve independent or synergistic action resulting in membrane permeabilization [27]. However, the interaction of bacterial cell membrane with bacteriocins seems to be the potential mechanism behind the killing of most bacteria by these bacteriocins [28].

## 8.4 Applications

### 8.4.1 Food Bio Preservative

At present, a lot of concerns about the presence of chemical preservatives in foods causing adverse health effects and affecting food safety and quality leads to a focus on the discovery and development of new methods of food preservation. As a consequence of this, consumer attention has been drawn towards natural foods, which have no chemical additives. This consciousness, together with the popularity of minimally processed foods with extended shelf life, has inspired research interest in discovering natural but effective food preservatives such as bacteriocins. For over 50 years nisin from *Lactococcus lactis* has been in use as a food preservative with proven record high effectiveness against food spoiling/poisoning microbes. Besides Nisin (Nisaplin® and Novasin™) another bacteriocins, Pediocin (Alta®2341) is also approved as a food additive on an international scale. In addition, the application of bacteriocins has been explored in other food industries especially in vegetable, egg, dairy and meat products. Furthermore, bacteriocins could be used to enhance food quality and sensory properties, such as preventing gas blowing defect in cheese or improving the rate of proteolysis. The application of bacteriocins in food preservation commonly follows three different approaches [29]:

1. Inoculation of food items with bacterial strains producing bacteriocins.
2. Addition of bacteriocins as bio preservatives.
3. Use of a product pre-fermented with a bacteriocins producing organism as an ingredient in food processing.

### 8.4.2 Food Packaging (In Packaging Films)

Application of bacteriocins in food packaging (packaging films), to control food spoilage and restrict the food pathogens, is an active research theme from the last decade.

Packaging film with bioactive agents restricts the microbial growth on food surface as there is a direct contact of the active package with the surface of foods like cakes, meats, bread, and cheese. Moreover, it prolongs the lag phase and further reduces the growth rate of spoilage microorganisms, thus maintaining food product quality and enhancing its shelf life [30]. The sustained release of bacteriocins from packaging film to the surface of food may be beneficial as compared to bacteriocins applied in dipping and spraying form. This is because these two methods could result in losing or reducing antimicrobial activity due to food ingredients inactivating bacteriocins or dilution of bacteriocins below active concentration due to the movement of bacteriocins into the foods [31].

There are two widely used methods for preparation of bacteriocins impregnated packaging films. The first incorporates bacteriocins directly into polymers (e.g., nisin into protein films) and the second involves adsorption or coating of bacteriocins to the surface of polymers (e.g., nisin/methyl-cellulose coatings for polyethylene films and nisin coatings for poultry), which suggests that a polymer-based solution coating could be the most attractive method in terms of stability and adhesiveness of appending a bacteriocins to a plastic film [31–33]. In another study, low-density polyethylene (LDPE) films coated with a mixture of methyl cellulose in ethanol and sonorensin solution provided antimicrobial activity against *S. aureus* [34]. Also, bacteriocins can be used under various physicochemical packaging conditions e.g., vacuum packaging [35].

### 8.4.3 Hurdle Technology to Enhance Food Safety

The major functional inadequacy in use of bacteriocins in foods is that bacteriocins alone are not likely to promote comprehensive safety due to their relatively narrow spectrum and moderate antimicrobial effects, especially against gram-negative bacteria. Therefore, the application of bacteriocins is to coalesce with other technologies capable of disrupting cellular membrane making the bacteriocins destroy pathogens more efficiently. Hurdle technology refers to the management of multiple factors (intrinsic and extrinsic) intended to inhibit bacterial contamination or control growth and endurance in food, which has been used to extend shelf life and to improve food safety [36]. It has been observed that gram-negative bacteria become susceptible to bacteriocins if their outer membrane is compromised in terms of permeability. For example, chelating agents such as EDTA, which attach with magnesium ions of the lipopolysaccharide layer and destabilize outer membrane of gram-negative bacteria, thus permitting bacteriocins to access cytoplasmic membrane [34].

A mixture of various bacteriocins has also been observed to improve their bactericidal/bacteriostatic activity. Nisin, in combination with MT104b, is reported to cause a synergistic effect on the elimination of *S. aureus* whereas, the combination of nisin and pediocin, nisin and MT162b, pediocin and MT104b, and pediocin and MT162b are reported to synergistic effect against *Lactobacillus sakei* [37]. There remains a constant interest in the food processing industry on the application of non-thermal food processing techniques for food preservation such as pulsed electric field and high hydrostatic pressure. It has been reported that bacteriocins, in conjugation with these processing technologies, improve the microbial inactivation. For example, a combination of high-pressure processing with enterocin LM2 has enhanced the anti-listerial activity in sliced cooked ham [38].

### 8.4.4 Therapeutic Potential

Bacteriocins have also been reported to be effective against drug-resistant bacteria e.g., lacticin3147 lantibiotics that have been found to be effective against pathogens such as vancomycin-resistant *Enterococcus faecalis* (VRE) and methicillin-resistant *Staphylococcus aureus* (MRSA) [39]. A lantibiotic produced by *Bacillus* sp. strain HIL Y-85, 54728 named as mersacidin, showed growth inhibition of MRSA strains [40] and *Propioni bacteriumacnes* and thus, could be used for treating acne [41]. There are reports on bacteriocins showing in vitro cytotoxicity against cancer cells at very low concentrations. Table 8.1 summarises the therapeutic potential of various bacteriocins.

### 8.4.5 Effect of Bacteriocins on Biofilms

Biofilms, sessile microbial communities that adhere to various surfaces and engage in a self-produced extracellular polymeric substance (EPS), pose a critical problem in clinical therapeutics [55]. Biofilms might form on living or non-living surfaces and can prevail in natural, industrial, and hospital settings. Bacterial cells dwelling in biofilms are physiologically diverse from planktonic ones and EPS can escalate antibiotic resistance by up to 1,000-fold [56]. These adherent microbial communities impose a negative impact on the global economy and are a serious threat to the industry, marine transportation, public health, and medicine due to increased resistance to antibiotics and chemical biocides [57]. Often biofilms account for nosocomial infections and chronic illnesses which, in many cases, are not resolved by antibiotics and persistently re-occur. Furthermore, biofilms may cause biofouling in dairy and other processed foods resulting in food-borne illnesses [58]. Therefore, development of strategies to combat biofilms is a major area of research. The major challenge of biofilms removal lies in addressing their increased resistance to disinfection and being not efficiently killed by common methods in food preservation such as heat, acidification, and exposure to ethanol compared to planktonic cells [59]. Biofilms are also resistant to many antibiotics including ampicillin, tetracycline, penicillin, erythromycin, and chloramphenicol [60].

Bacteriocins are considered among potential next generation of antibiotics for fighting multi-drug resistance and/or biofilms associated with bacterial infections. They have the potential to be used as anti-biofilm agents because of their diverse mechanisms of action including disruption of the membrane, binding with DNA, functional inhibition of proteins, and detoxification of polysaccharides. A number of bacteriocins can be transferred in EPS of biofilms through pores formed in the lipid section of the EPS, while others can scatter the biofilms bacteriocins such as nisin and sonorensin have been shown to be effective against biofilms of MRSA and *S. aureus*, respectively [34, 61].

**Table 8.1** Therapeutic potential of bacteriocins.

| Producing strain | Bacteriocins | Inhibited strain/targets | Therapeutics | Reference |
|---|---|---|---|---|
| *Enterococcus mundtii* ST4SA | Enterocin | *Enterococcus faecalis, Staphylococcus aureus* | Preventing gastric infection | [42] |
| *Escherichia coli* | Colicin E1 | F18-positive *Escherichia coli* | Preventing post weaning diarrhoea | [43] |
| *Bacillus thuringiensis* DPC 6431 | Thuricin CD | *Clostridium difficile* | Preventing gut infection | [44] |
| *Streptococcus salivarius* K12 | Salivaricin A &B | *Streptococcus pyogenes, Streptococcus mutans* | Preventing dental caries infection in human | [45] |
| *Lactobacillus salivarius* strain NRRL B-30514 | OR-7 | *Campylobacter jejuni* | Treating gastroenteritis | [46] |
| *Streptococcus* sp. Thom 1606 | Tomicid | *Streptococcus* sp. | Treating Streptococcal respiratory infections | [47] |
| *Escherichia coli* | Colicins E1, E4, E7 | *Escherichia coli* 0157:H7 | Treating haemorrhagic colitis and haemolytic uremic | [48] |
| *Lactococcus lactis* subsp. *lactis* | Nisin | Spermicidal activity | Vaginal contraceptive agent | [49] |
| *Lactococcus lactis* subsp. *lactis* | Lacticin 3147 | *Streptococcus uberis, Brucella melitensis, Corynebacterium bovis* | Treating mastitis in cattle | [50] |
| *Staphylococcus epidermidis* | Epidermin | *Propionibacterium acnes* | Treating skin infections such as acne | [51] |
| *Escherichia coli* | Microcins J25 and 24 | *Escherichia coli, Salmonella* sp. | Treating *E. coli* and *Salmonella* infections in chickens | [52] |
| *Lactobacillus salivarius* strain CRL 1328 | BLIS | *Enterococcus* sp., *Neisseria gonorrhoeae* | Preventing Female UTI | [53] |
| *Pediococcus acidi lactici* CP2 MTCC5101 | Pediocin CP2 | Cell lines HepG2, HeLa and MCF7 | Anti-cancer | [54] |

## 8.5 Conclusions

The increase in the number of infections from multi-drug resistant (MDR) bacteria in recent years has become a serious problem, and thus, development of alternative antimicrobial agents is becoming increasingly imperative. Bacteriocins, ribosomally synthesized antimicrobial peptides, are one of the best-studied microbial defense systems. These peptides exert their antimicrobial activity differently from conventional antibiotics and chemical preservatives and thus might substitute antibiotics and chemical preservatives. These bacterial compounds have potential in a number of applications in the food and pharmaceutical industries. Although extensive research over the last few years has significantly enhanced the information about bacteriocins, further studies are required to deeply understand the structure-function relationships, molecular mechanisms, and mode of action of bacteriocins. Research in the above-mentioned areas is essential for commercial exploitation of bacteriocins and development of bacteriocins with better activity, solubility, and stability.

## Acknowledgments

The authors are grateful to the Council of Scientific and Industrial Research (CSIR), Department of Biotechnology (DBT), Department of Science and Technology (DST) and University Grants Commission (UGC), India for financial support.

## Abbreviations

AMR: Antimicrobial Resistance
AMP: Antimicrobial Peptide
MDR: Multidrug Resistance
PTM's: Posttranslational Modifications
kDa: Kilo Dalton
LDPE: low density polyethelene
CL: Cardiolipin
PS: Phosphatidylserine
PG: Phosphatidylglycerol
SM: Sphingomyelin
PC: Phosphatidylcholine
PE: Phosphatidylethanolamine
EPS: Extracellular Polymeric Substance

## References

1. M.L. Chikindas, R. Weeks, D. Drider, V.A. Chistyakov, and L.M. Dicks, *Current Opinion in Biotechnology*, Vol. 8, p. 23, 2018.
2. S.S Beasley, and P.E.J. Saris, *Applied and environmental microbiology*, Vol. 70(8), p. 5051, 2004.

3. G.N. Moll, W.N. Konings, and A.J. Driessen, *Antonie van Leeuwenhoek*, Vol. 76, p. 185, 1999.
4. Gratia, *Comptes Rendus Biologie*, Vol. 93, p. 1040, 1925.
5. A.M. Pons, I. Lanneluc, G. Cottenceau, and S. Sable, *Biochimie,* Vol. 84, p. 531, 2002.
6. A.H.A. Parret, G. Schoofs, P. Proost, and R. De Mot, *Journal of bacteriology*, Vol. 185(3), p. 897, 2003.
7. G.G. Sandrine, V. David, M.G.C. Gilles, Z. Nathalie Zorn, and V.D. Alain, *FEMS Microbiology Letters*, p. 95, 2000.
8. T.R. Klaenhammer, *FEMS Microbiology Reviews*, Vol. 12, p. 39, 1993.
9. O. McAuliffe, R.P. Ross, and C. Hill, *FEMS Microbiology Reviews*. Vol. 25, p. 285, 2001.
10. G. Fimland, L. Johnsen, B. Dalhus, and M.J. Nissen, *Journal of Peptide Science*, Vol. 11(11), p. 688, 2005.
11. R. Kemperman, A. Kuipers, H. Karsens, A. Nauta, O. Kuipers O, and J.Kok., *Applied and environmental microbiology*, Vol. 69, p. 1587, 2003.
12. L.M. Cintas, P. Casaus, C. Herranz, L.S. Håvarstein, H. Holo, P.E. Hernández, and I.F. Nes, *Journal of Bacteriology*, Vol. 182(23), p. 6806, 2000.
13. Y. Kawai, R. Kemperman, J. Kok, and T. Saito, *Current Protein & Peptide Science*, Vol. 5(5), p. 393, 2004.
14. P.D. Cotter, C. Hill, and R.P. Ross, *Nature Reviews Microbiology*, p. 1058, Vol. 4, 2006.
15. H.C.K. Heng, and J.R. Tagg, *Nature Review of Microbiology*, p. 621, Vol.4, 2006.
16. P.D. Cotter, C. Hill, and R.P. Ross, *Nature Review Microbiology*, Vol. 3, p. 777, 2005.
17. I.F. Nes, D.B. Diep, and H.Holo, *Journal of Bacteriology*, Vol. 189(4), p. 1189, 2007.
18. M.J. Nissen, P.Rogne, C. Oppegård, H.S. Haugen, and P.E. Kristiansen, *Current pharmaceutical biotechnology*, Vol. 10, p. 19, 2009.
19. J.P.S. Powers, and R.E. Hancock, *Peptides,* Vol. 24, p. 1681, 2003.
20. M.R. Yeaman, and N.Y. Yount, *Pharmacological Reviews,* Vol. 55, p. 27, 2003.
21. K. Matsuzaki, *Biochimica et biophysica acta* Vol. 1462, p. 1, 1999.
22. Y. Shai, and Z.Oren Z. *Peptides*, Vol. 22, p. 1629, 2001.
23. K.A. Brogden, *Nature Reviews Microbiology*, Vol. 3, p. 238, 2005.
24. E.Breukink E, and B. de Kruijff B, *Biochimica et biophysica acta,* Vol. 462, p. 223, 1999.
25. F. Yoneyama, Y.Imura, k. Ohno, T. Zendo, J. Nakayama, and K. Matsuzaki, *Antimicrobial agents and chemotherapy*, Vol. 533, p. 3211, 2009.
26. J.K. Kumar *Applied Microbiology and Biotechnology*, Vol. 80, p. 555, 2008.
27. W.V. T. Hof, E.C.I. Veerman, E.J. Helmerhorst, and A.V.N. Amerongen, *Biological Chemistry*, Vol. 382, p. 597, 2001.
28. O. Toke, *Biopolymers - Peptide Science Section*, Vol. 80, p. 717, 2005.
29. U. Schillinger, R. Geisen, and W.H. Holzapfel, *Trends in Food Science and Technology*, Vol. 7, p. 158, 1996.
30. J.H. Han, *Food Technology*, Vol. 54, p. 56, 2000.
31. P. Appendini, and J.H. Hotchkiss, *Innovative Food Science & Emerging Technologies*, Vol. 3, p. 113, 2002.
32. T. Padgett, I.Y. Han, and P.L. Dawson, *Journal of food protection*, Vol. 61, p. 1330, 1998.
33. D.S. An, Y.M. Kim, S.B. Lee, H.D. Paik, and D.S. Lee, *Food Science and Biotechnology*, Vol. 9, p.14, 2000.
34. L. Chopra, G. Singh, K. Kumar Jena, and D. K. Sahoo, *Scientific Reports*, 13412 Vol. 5, 2015.
35. E. M. Johnson, D. Y.G. Jung, D.Y.Y. Jin, D. R. Jayabalan, D. S. H. Yang, and J. W Suh, *Critical Reviews in Food Science and Nutrition*, p. 1, 2017.
36. L. H. Deegan, P.D. Cotter, C. Hill, and P. Ross, *International Dairy Journal*, Vol. 16, p. 1058, 2006.
37. M. Turgis, K.Vu, M.Jamshidian, B.Maharani, and M. Lacroix, *Microbiology Research International*, Vol. 4, p. 1, 2016.

38. G. Liu, Y. Wang, M. Gui, H. Zheng, R. Dai, and P. Li, *Food Control*, Vol. 24, p. 64, 2012.
39. M. Galvin, C. Hill, and R. P. Ross, *Letters in Applied Microbiology*, Vol. 28, p. 355, 1999.
40. D. Kruszewska, H - G Sahl, G. Bierbaum, U. Pag, S. O. Hynes, and Å. Ljungh, *Journal of Antimicrobial Chemotherapy*, Vol. 54, p. 648, 2004.
41. G. Jung, *Angewandte Chemie International Edition in English*, Vol. 30, p. 1051, 1991.
42. M. Granger, C. A. van Reenen, and L. M. T. Dicks, *International Journal of Food Microbiology*, Vol. 3, p. 277, 2008.
43. S. A. Cutler, S. M. Lonergan, N. Cornick, A. K. Johnson, and C. H. Stahl, *Antimicrobial agents and chemotherapy*, Vol. 51, p. 3830, 2007.
44. M. C. Rea, A. Dobson, O. O'Sullivan, F. Crispie, F. Fouhy, and P. D. Cotter, *Proceedings of the National Academy of Sciences of the United States of America*, Vol. 108, p. 4639, 2011.
45. M. Balakrishnan, R. S. Simmonds, and J. R. Tagg, *Australian Dental Journal*, Vol. 45, p. 235, 2000.
46. N. J. Stern, E. A. Svetoch, B. V. Eruslanov, V. V. Perelygin, E. V. Mitsevich, and I. P. Mitsevich, *Antimicrobial agents and chemotherapy*, Vol.50, p. 3111, 2006.
47. N. I. Briko, M. V. Zhuravlev, *Zhurnal mikrobiologii, epidemiologii, i immunobiologii*, Vol. 4, p. 17, 2004.
48. B. J. A. M. Jordi, K. Boutaga, C. M. E Heeswijk, F. Knapen, and L. J. A. Lipman, *FEMS Microbiology Letter*, Vol. 204, p. 329, 2001.
49. P. M. Sears, B. S. Smith, W. K. Stewart, R. N. Gonzalez, S. D. Rubino, and S. A. Gusik, *Journal of Dairy Science*, Vol. 75, p. 3185, 1992.
50. M. P. Ryan, W. J. Meaney, R. P. Ross, and C. Hill, *Applied and environmental microbiology*, Vol.64, p. 2287, 1998.
51. H. Allgaier, G. Jung, R. G. Werner, U. Schneider and H. Zahner, *European Journal of Biochemistry*, Vol. 160, p. 9, 1986.
52. S. Sable, A.M. Pons, S. Gendron-Gaillard, and G. Cottenceau, *Applied and environmental microbiology*, Vol. 66, p. 4595, 2000.
53. V. S. Ocaña, A. A Pesce De Ruiz Holgado, and M. E Nader-Macías, *Applied and environmental microbiology*, Vol. 65, p. 5631, 1999.
54. Kumar, *Pharmaceut Anal Acta*, Vol. 3 (5), p. 184, 2012.
55. D. Monroe, *PLoS Biology*, Vol. 5, e.307, 2007.
56. P. S. Stewart, and J. William Costerton, *The Lancet*, Vol. 358, p. 135, 2001.
57. E. D. Giaouris, G. J. E. Nychas *Food Microbiology*, Vol. 23, p. 747, 2006.
58. Bagge, M. Hjelm, C. Johansen, I. Huber, and L. Gram, *Applied and environmental microbiology*, Vol. 67, p. 2319, 2001.
59. H. Kubota, S. Senda, N. Nomura, H. Tokuda, and H. Uchiyama, *Journal of Bioscience and Bioengineering*, Vol. 106, p.:381, 2008.
60. T. May, A. Ito, and S. Okabe, *Antimicrobial agents and chemotherapy*, Vol. 53, p. 4628, 2009.
61. E. Mataraci, and S. Dosler, *Antimicrobial agents and chemotherapy*, Vol. 56, p. 6366, 2012.

9

# Non-Ribosomal Peptide Synthetases: Nature's Indispensable Drug Factories

**Richa Sharma, Ravi S. Manhas and Asha Chaubey***

*Fermentation Technology Division, CSIR-Indian Institute of Integrative Medicine, Canal Road, Jammu, (India)*

***Abstract***

The development of methods for rational drug design using non-ribosomal machinery called non-ribosomal peptide synthetases (NRPS) has emerged as an important area of research, as peptide natural products form the largest and diversified therapeutics. This may prove helpful to meet ever-increasing demand for the new antibiotics and therapeutics. Non-ribosomal peptides (NRPs) are a family of unique naturally produced secondary metabolites by microorganisms mainly bacteria i.e., Actinomycetes and *Bacilli*, fungi and marine organisms. NRPs are explicitly targeted in the frontier of the search for therapeutic agents and are naturally a promising pool of candidates. Approximately 5% of 205 grouped families of NRPs have become clinically approved drugs to treat emerging health threats. Recent breakthroughs in the structure, molecular mechanism and chemical biology underlying the A, C, T and TE domains within NRPS and their roles within NRPS assemblies have been extended to performing novel chemistry. This knowledge is crucial with regards to the potential to redesign NRPS-based biosynthetic pathways and tackle new and emerging challenges in human health for overcoming increasing antimicrobial resistance and improved anticancer therapies.

***Keywords:*** Non-ribosomal peptide, non-ribosomal peptide synthetases, A domain, C domain, T domain, TE domain

## 9.1 Introduction

### 9.1.1 Non-Ribosomal Peptides as Natural Products

Microorganisms are important sources of bioactive natural products with enormous potential for the discovery of new molecules for drug discovery, industrial use and agricultural applications [1–3]. In the late 1920s, discovery of penicillin by Alexander Fleming introduced the antibiotic era to the world. According to a report published in *Science* [4], approximately 50% of all new drug approvals in the past 25 years either came from natural products or their variants. Microorganisms produce a huge

**Corresponding author*: achaubey@iiim.ac.in

Saurabh Saran, Vikash Babu, and Asha Chaubey (eds.) High Value Fermentation Products, Volume 1, (205–224) © 2019 Scrivener Publishing LLC

plethora of natural bioactive compounds including antibiotics, immunosuppressants, siderophores, antitumor agents and toxins [5]. A large group of secondary metabolites comprises peptide natural products derived from precursors from the primary amino acid metabolism; there are two fundamentally distinct mechanisms by which they are assembled. One is a ribosomal route and the other is a non-ribosomal route [6]. The initialisation of the analysis of peptide antibiotics that are biosynthesized by mega enzyme complexes called non-ribosomal peptide synthetases (NRPSs) happened unknowingly [7]. Natural products biosynthesized fully or partly by NRPSs are some of the most important drugs currently used clinically for the treatment of a variety of diseases. Within the category of natural products, non-ribosomal peptides (NRPs) are a family of small molecules that are naturally produced as secondary metabolites by microorganisms as well as eukaryotes like fungi [8] and invertebrates [9]. Optimized through natural selection, many non-ribosomal peptides have been evolved to specific functions in certain biological processes. As secondary metabolites, non-ribosomal peptides are not directly necessary in an organism's growth cycle, but often have a vital impact on the organism's continuing existence in adverse situations and play important roles in interspecies defense. As such, non-ribosomal peptides are always targeted in the frontier of the search for therapeutic agents and they are naturally a promising pool of candidates.

### 9.1.2 Non-Ribosomal Peptides as Drugs

Since non-ribosomal peptides are in the forefront of the search for therapeutic agents, a wide range of drugs have been developed from non-ribosomal, as described in Table 9.1. Approximately 5% of 205 grouped families of NRPs have become clinically approved drugs [27–28] to treat emerging health threats, such as daptomycin (to treat life-threatening infections caused by methicillin-resistant *Staphylococcus aureus* (MRSA) and the vancomycin based Oritavancin approved by the FDA in 2014 for treatment of drug resistant skin infections [18, 29–30]. Many candidate NRPs are in various phases of clinical trials like Plitidepsin, a cyclic depsipeptide produced by *Aplidium albicans,* which is undergoing Phase III trials against Cancer; Trabectedin analogue (PM01183), a NRPS alkaloid from *E. turbinata,* which is in Phase I trials against cancer; Vitilevuamide, a NRP produced by *Didemnum cuculiferum* and *Polysyncranton lithostrotum,* which is undergoing preclinical trials against cancer [9] and Ramoplanin, a lipopeptide, which has potentially found a role in the treatment of *Clostridium difficile* infections (undergoing phase III trials), active against multi-drug resistant Gram-positive pathogens produced by *Actinoplanes* sp. ATCC 33076 [31]. There is a surprising rise in the rediscovery of lipopeptides synthesised by NRPS machineries as antibiotics. Even though daptomycin has represented the poster child of the family since the past decade, acyl depsipeptides (ADEPS) illustrated unanticipated mechanisms deviated from working through membrane disruption. Also, the polymyxins have been reentered with prominence, given the multidrug-resistant profiles of Gram-negative pathogens [32]. Certainly abundance of lipopeptide natural products in the biosphere may be starting points for combination therapies. More research efforts are spent in finding novel non-ribosomal originated compounds with bioactivity beyond those already known [33].

**Table 9.1** Non-ribosomal peptides based drugs reported in literature.

| Drug | Biosynthetic class | Source | Disease/molecular target | Reference |
|---|---|---|---|---|
| Polymyxin | Polypeptides | *Bacillus polymyxa* | Antibiotic/Alters bacterial Outer membrane | [10] |
| Pristinamycin | Depsipeptide | *Streptomyces pristinaespiralis* | Antibiotic/protein synthesis inhibitor | [11] |
| Gramicidin | Linearpenta-decapeptide | *Bacillus bovis* | Antibiotic/Alter bacterial Outer membrane | [12] |
| Bacitracin | Cyclicpeptide | *Bacillus subtilis* | Antibiotic/dephosphorylation of C55 isoprenylpyrophosphate | [13] |
| Capreomycin | Cyclicpeptide | *Streptomyces capreolus* | Antibiotic/protein synthesis inhibitor | [14] |
| Teicoplanin | Glycopeptide | *Actinoplanes teichomyceticus* | Antibiotic/inhibit cell wall synthesis | [15] |
| Vancomycin | Glycopeptide | *Amycolatopsis orientalis* | Antibiotic/inhibit cell wall synthesis | [16] |
| Cephalosporin C | β-lactam | *Acremonium* sp. | Antibiotic/Alters bacterial outer membrane | [17] |
| Oritavancin | - | Semi synthetic | Antibiotic/disrupts the cell membrane | [18] |
| Bleomycin | Hybrid peptide | *Streptomyces verticillus* | Antibiotic/inhibition of DNA synthesis | [19] |
| Daptomycin | Lipopeptide | *Streptomyces roseosporus* | Antibiotic/disrupts the cell membrane | [20] |
| Cyclosporine A | Cyclicpeptide | *Tolypocladium inflatum* | Immunosuppressant/lower the activity of T cells | [21] |
| Actinomycin D | Polypeptide | *Streptomyces* sp. | Antitumor/inhibit transcription | [22] |
| Romidepsin | Depsipeptide | *Chromobacterium violaceum* | Antitumor/Histone deacetylase inhibitor | [23] |
| Ceftaroline fosamil | β-lactam | Semi synthetic | Antibiotic/Altersbacterial outer membrane/active against MRSA | [24] |
| Ceftolozane | β-lactam | Semi synthetic | Antibiotic/Alters bacterial outermembrane/ Stability against β-lactamases | [25] |
| Trabectedin | NRPS derived alkaloid | *Ecteinascidia turbinata* | Anticancer/minor groove of DNA | [26] |

## 9.2 NRPS Machinery

Non-ribosomal peptides are synthesized via a multicarrier thiotemplate mechanism on large enzymes called non-ribosomal peptide synthetases (NRPSs), plausibly existing as multi-domain enzyme complexes [34–39]. Briefly, NRPSs are large multifunctional enzymes of 100–1700 k*Da* [31] that have a modular structural organization and are thought to link amino acid residues according to the modular assembly line manner [40–42]. These enzymes are organized into sets of iterative catalytic units called modules [43, 44]. Each module is responsible for specific incorporation of an amino acid or any other building block into the peptide backbone. Different modules can be part of the same or different polypeptides, and the organization and order of modules usually correspond to the amino acid sequence of the peptide product co-linearity rule [45]. The minimal module required for a single monomer addition consists of: (i) an adenylation domain (A, approximately 550 residues); (ii) T, a thiolation or peptidyl carrier domain (PCP, about 80 residues) to which a phosphopantetheine cofactor is covalently attached; (iii) a condensation domain (C, approximately 450 residues) [44, 46]. More recently, multi-domain structures have been determined and are providing clues to the features of these enzyme systems that govern the functional interaction between multiple domains [47].

## 9.3 Catalytic Domains of NRPSs

Non-ribosomal peptide is assembled by NRPS involving a series of repeating steps catalyzed by the coordinated actions of three core catalytic domains: the adenylation, thiolation (or peptidyl carrier protein), and condensation domain (Figure 9.1). These core domains, together, comprise a minimal NRPS module. A fourth domain, thioesterase, is often found at the C terminus of the NRPS and catalyzes the release of the peptide from the NRPS.

### 9.3.1 Adenylation (A) Domains

A domain is involved in the selection of the amino acid (AA) substrate and in its subsequent activation as an aminoacyl adenylated intermediate covalently attached to the

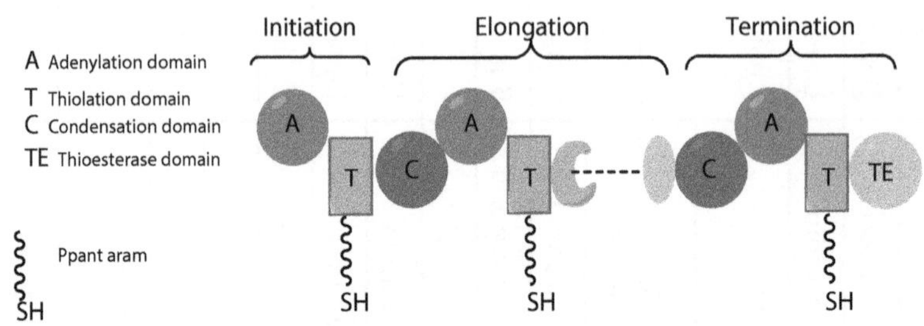


**Figure 9.1** NRPS module set up with core domains for initation, elongation and termination.

enzyme via a thioester bond with the phosphopantetheine arm of the PCP domain, thus considered as gatekeeper domains [42, 48]. The enormous structural diversity of NRPs is contributed by A domain as its substrates are not limited to the standard 20 proteinogenic amino acids. The substrate-binding pocket of A domains is lined by 10 AA residues that determine the nature of the AA that is incorporated into the NRP backbone [42, 49]. Recent breakthroughs in the structure, molecular mechanism, and chemical biology underlying the A domains within NRPS has been demonstrated by Ishikawa and Kakaye [50]. Recent crystal structures of complete NRPS modules in different states of peptide biosynthesis have demonstrated the effects of motion of the carrier protein and support the importance of ATP hydrolysis in orchestrating carrier protein motion during NRPS-catalysed peptide synthesis [38].

### 9.3.2 Thiolation (T) or PCP Domains

T domains must be post-translationally modified with a 4'-phosphopantethiene group [31, 43]. Activation of the T domain to its holo-form is achieved by the addition of the 4'-Ppant on a conserved seryl residue at the center of the T domain by a 4'-Ppant transferase. T domains play a central role in the function of NRPSs as they must interact not only with A domains forming aminoacylthioester but also with other domains involved in peptide bond formation, peptide modification, or peptide release from the megasynthase.

### 9.3.3 Condensation (C) Domains

C domain catalyzes formation of peptide bond between the carboxyl group of the nascent peptide and the amino acid on the flanking module, thus allowing the translocation of the growing peptide chain to the next module [51]. C domain is most commonly located at the start of each NRPS module, positioned precisely for catalyzing peptide bond formation between two aminoacids tethered to the adjacent modules. Two distinct substrate-binding sites are proposed with one site binding the amino acid tethered to the corresponding T domain of the module as aminoacylthioester that will accept the growing peptide from the preceding module, binding at the acceptor site. C-domains role within NRPS assemblies have recently been extended to performing novel chemistry on their substrates, either via direct catalysis or through recruitment of external modifying enzymes to the NRPS-bound substrate, for example, the biosynthesis of the β-lactam antibiotic norcardicin, in which the origin of the β-lactam moiety is demonstrated as the NRPS machinery itself and also the biosynthesis of the glycopeptide antibiotic (GPA) teicoplanin, where a modified C-domain acts as a recruitment domain for external modifying enzymes that target their peptide substrates bound to the NRPS machinery [38].

### 9.3.4 Thioesterase (Te) Domains

The final NRPS module typically contains a Te domain at the C-terminus. This domain catalyzes the release of the NRP either by hydrolysis of thioester bond between the assembled peptide and the phosphopantetheine cofactor, resulting in the release of the final product or by intramolecular cyclization of the peptide from the NRPS [51, 52].

In bacterial NRPS, C-terminal thioesterase (TE) domains catalyze the peptide release [53]. Recent studies have showed further functions of TE domains, which extend the catalytic versatility of these domains. Novel NRPS chemistry from norcardicin biosynthesis showed terminal TE-domain not only to be highly selective for β-lactam containing peptides, but also responsible for the epimerization of the terminal Hpg residue, and also TE-mediated substrate modification during NRPS biosynthesis was recently identified in salinamide biosynthesis, a family of unusual bicyclic depsipeptide antibiotics [38]. Modules may also contain accessory domains for epimerization, cyclization or methylation of the amino acid substrates.

## 9.4 Types of NRPS

Three types of NRPS have been recognized based upon modules composition and usage as discussed below:

### 9.4.1 Type A (Linear NRPS)

In this NRPS, each enzymatic domain, and thus each module, is used once during the synthesis of the non-ribosomal peptide (Figure 9.2). Tyrocidine, a cyclic decapeptide biosynthesized by *B. brevis* [54], ACV synthetase producing β-lactams, a large class of antibiotics which includes the penicillins and the cephalosporins [46], Daptomycin, a cyclic lipopeptide antibiotic, cyclosporins, a series of undecapeptides produced by the fungal species *Tolypocladium inflatum*, vancomycin, produced by the actinomycete *Amycolatopsis orientalis* 128 are examples of type A, linear NRPSs.

### 9.4.2 Type B (Iterative NRPS)

NRPS that re-use its modules to biosynthesize peptide chains from repeats of smaller sequences (Figure 9.2) the quinoxaline antibiotics (echinomycin, thiocoraline)[55], and anti-HIV potential (luzopeptins, quinoxapeptins) [56].

### 9.4.3 Type C (Non-linear NRPS)

Nonlinear NRPSs with certain domains but not all reused during the biosynthesis of a single non-ribosomal peptide that deviate from the C-A-T rule of module formation (Figure 9.2). Capreomycin, a mixture of four structurally related NRP antibiotics produced by the bacterium *Saccharothrix mutabilis* subsp. *capreolus,* used with other antituberculosis drugs to treat multidrug-resistant tuberculosis (MDR-TB) [57]. Bleomycins are hybrid NRPS/PKS, a family of glycosylated peptides that were found to have anticancer activities in the 1960s by Umezawa and colleagues [18, 58].

## 9.5 Working of NRPSs

The non-ribosomal mechanism of peptide biosynthesis was first studied by Fritz Lipmann in the 1970s [59, 60]. NRPSs are megasynthases comprising multiple protein

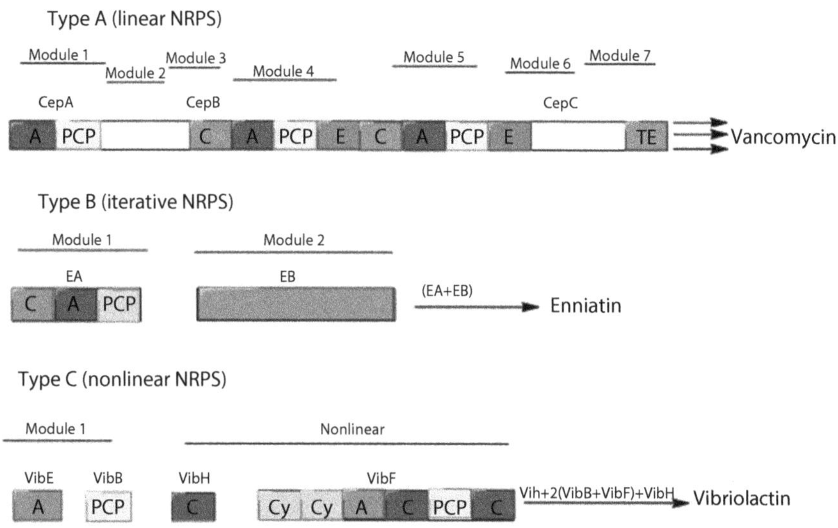


**Figure 9.2** Types of non-ribosomal peptide synthetase (NRPS).

domains with distinct functions forming modules required for the incorporation of a single monomer into the growing peptide (Figure 9.3).

### 9.5.1 Priming Thiolation Domain of NRPS

Prerequisite for NRPS functioning is a posttranslational modification involving transfer of 4'-Ppant prosthetic group from coenzyme A to the hydroxyl group of a conserved serine residue of the apo-thiolation domain by forming ester bond catalyzed by a dedicated PPTase. The thiol group of the Ppant arm serves the site of covalent tethering for amino acids and nascent peptide during NRP assembly [61].

### 9.5.2 Substrate Recognition and Activation

Adenylation (A) domains specifically recognize and activate NRP monomers via adenylation with ATP to yield acyl-AMP which is transferred to the Ppant thiol group of the adjacent thiolation domain forming a carboxythioester [53]. Other carboxylic acids are also recognized and activated as substrates during NRP biosynthesis (Figure 9.3).

### 9.5.3 Peptide Bond Formation between NRP Monomers

Condensation domains catalyze peptide bond formation between covalently tethered amino acids on adjacent T-domains (Figure 9.3) [61]. The deprotonated amine group of the amino acid on the downstream T domain acts as a nucleophile and attacks the acyl group of the upstream bound amino acid or peptide and if the acyl acceptor is a hydroxy acid, acts as nucleophile and an ester bond is formed resulting in a depsipeptide product [62]. The T-domain assist the interaction with either the A or C domains.

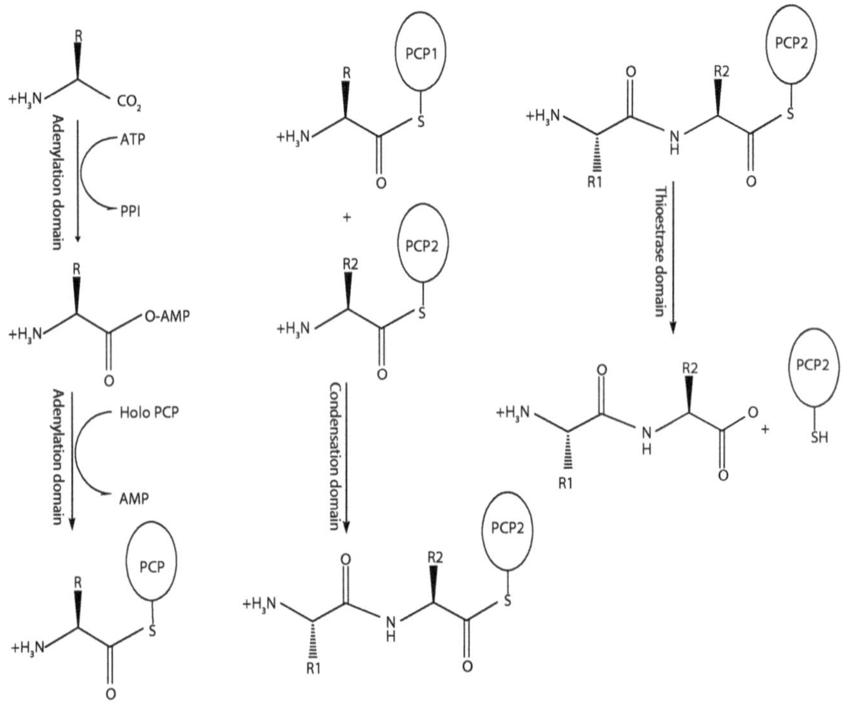


**Figure 9.3** Reactions catalyzed by Adenylation, Condensation, Thiolation (PCP) and Thioesterase domains of NRPS.

### 9.5.4 Chain Termination of NRP Synthesis

In bacterial NRPSs, TE domain is responsible for peptide release commonly found at the C-terminus of termination modules [63]. Peptide is translocated from the thioester linkage on the upstream T domain to a conserved serine residue in the active site of the TE domain via ester linkage. Peptide release occurs subsequently, either by hydrolysis leads to the formation of a linear peptide acid or the reaction with an intramolecular nucleophile results in cyclic structures (Figure 9.3) [64]. In fungal NRPSs, specialized terminal C domains cause macrocyclization or cyclooligomerization which evolved as an alternative peptide termination mechanism [65] or by a thioesterase NADP(H) dependent reductase (R) domain which catalyzes reduction with NADPH to form an aldehyde [52].

### 9.5.5 NRP Tailoring

Additional structural modifications on NRPs require further enzyme activities [53]. Epimerase (E) domains responsible for the incorporation of D-amino acids acts on the amino acyl bound to the upstream T domain and the stereospecific peptide bond formation of the downstream C domain incorporates the D-form [66]. Cyclization (Cy) or heterocyclization domains, catalyze the addition of heterocycles to the peptide backbone by acting on Cys, Ser or Thr as acceptor substrates and catalyze peptide bond formation prior to addition of the side chain thiol (Cys) or hydroxyl (Ser, Thr) and subsequent dehydration to form a thiazoline or oxazoline ring, respectively [67]. An oxidase (Ox) domain was identified in some Cy domain containing modules (e.g. bleomycin A2 synthesis), which catalyzes the flavin mononucleotide (FMN) dependent oxidation of a thiazoline to a thiazole ring. Reduction(R) domain catalyzes NADPH dependent formation of a thiazolidine ring for pyochelin [48].

N-methyltransferase (N-MT) domains transfer the methyl group from S-adenosylmethionine (SAM) usually integrated at the C terminus of the A domains and act on PCP-S-Ppant bound substrates. Formyltransferase(F) domain with N10-formyl-tetrahydrofolate (N10-THF) as the formyl donor formylate the first amino acid. Enzymes further contribute to structural diversification, such as glycosylation, acylation, halogenation, oxidative cross-linking [38].

## 9.6 Sources of NRPs

Bacteria mainly *Actinomycetes, Bacilli*, fungi and marine organisms are found to be major producers of non-ribosomal secondary metabolites and their biosynthetic machineries are well studied (Figure 9.3). Apart from them, non-ribosomal synthesis has been detected in diverse microorganisms, including myxobacteria, cyanobacteria and invertebrates mainly sponges and tunicates [9] and higher eukaryotes [68]. Some of the potential NRPs with sources and bioactivities are presented in Tables 9.2 (a-d)

**Table 9.2 (a):** Examples of bacterial/actinobacterial metabolites along with their uses and producing strains.

| Peptide | Organism | Function | Reference |
|---|---|---|---|
| Daptomycin | *Streptomyces roseosporus* | Antibiotic | [69] |
| Feglymycin | *Streptomyces* sp. DSM 11171 | Atibacterial, antiHIV | [70] |
| Bialaphos | *Streptomyces hygrosco -pic/Streptomyces virid -ochrmogene* | Herbicide/Antibiotic | [71] |
| Viginiamycin S | *Streptomyces pristinaes -piralis/Streptomyces virginia* | Antibiotic | [72] |
| Pristinamycin I | *Streptomyces pristinaespiralis* | Antibiotic | [73] |
| Bleomycin | *Streptomyces verticillus* | Antitumor antibiotic | [74] |
| Streptothricin | *Streptomyces rochei* | Biocide | [75] |
| Ansatrienin | *Streptomyces collinus* | Antibiotic | [76] |
| Rapamycin | *Streptomyces hygroscopicus* | Immunosuppressant | [77] |
| Immunosuppressant FK506 | *Streptomyces* | Immunosuppressant | [78] |
| *Rakicidin D* | *Streptomyces* sp. MWW064 | Antitumor | [79] |
| Coelichelin | *Streptomyces coelicolor* | Siderophore | [80] |

(*Continued*)

**Table 9.2** Cont.

| Peptide | Organism | Function | Reference |
|---|---|---|---|
| Calcium-dependent antibiotic | *Streptomyces coelicolor A3 (2)* | Antibiotic | [81] |
| Vancomycin family glycopeptides | *Streptomyces orientalis* | Antibiotic | [82] |
| Catechoserine | *Streptomyces* sp.TP-A0874 | Antitumor | [83] |
| Prajinamide | *Streptomyces* sp. SPMA113 | | [84] |
| Geldanamycin | *Streptomyces* sp. SPMA113 | Antineoplastic, immunosuppresant | [83] |
| Elaiophylin | *Streptomyces.* SPMA113 | Antibiotic | [84] |
| Bacitracin | *Bacillus licheniformis* | Antibiotic | [41] |
| Fengycin | *Bacillus subtilis* 168 | biosurfactant, Antimicrobial, antiviral | [85] |
| Mycosubtilin | *Bacillus subtilis* | Antifungal antibiotic | [86] |
| Pyochelin | *Pseudomonas aeruginosa* | Siderophore | [87] |
| Dihydroaeruginoic acid | *Pseudomonas aeruginosa* | Antibiotic | [87] |
| Syringomycin | *Psuedomonas syringae* | Phytotoxin | [88] |
| Kalimantacin | *Pseudomonas fluorescens* | Antibiotic | [42] |
| Ornibactin | *Burkholderia cepacia* | Virulence factor | [89] |
| Serrawettin W2 | *Serratia liquefaciens* | Swarming | [90] |
| Swarming peptide | *Proteus mirabilis* | Swarming | [91] |
| Vibriobactin | *Vibrio cholerae* | Siderophore | [92] |
| Yersiniabactin | *Yersinia pestis* | Siderophore, Virulence factor | [93] |
| Mycobactin | *Mycobacterium tuberculosis* | Virulence factor | [94] |
| Exochelin | *Mycobacterium smegmatis* | Siderophore | [95] |
| Glycopeptidolipids | *Mycobacterium smegmatis* | cell wall structure | [96] |
| Microcystin | *Microcystis aeruginosa* | Toxin | [97] |
| Saframycin Mx1 | *Myxococcus xanthus* | Antibiotic | [98] |
| TA | *Myxococcus Xanthus* | Antibacterial, antibiotic | [99] |
| Myxothiazol | *Stigmatella aurantiaca* | Electron transport Inhibitor | [100] |
| Epithilone | *Sorangium cellulosum* | Antitumor | [101] |
| Ergopeptine | *Claviceps purpurea Neotyphodium coenophialum* | Toxin | [102] |
| Balhimycin | *Amycolatopsis mediterranei* | Antibiotic | [103] |
| Chloroeremomycin | *Amycolatopsis orientalis* | Antibiotic | [16] |

**Table 9.2 (b):** Fungal strains and their secondary metabolites (NRPs).

| Fungal Species | NRPs | References |
|---|---|---|
| *Trichoderma atroviride* | Atroviridin, Trichorzin | [105] |
| *Trichoderma aureoviride* | Trichoaureocins | [106] |
| *T. harzianum / T. atroviride P1* | Harzianin,Trichorzianin | [107] |
| *Trichoderma koningii* | Trichokonin,Trichokoningin | [108] |
| *Trichoderma longibrachiatum* | Longhibrachin | [109] |
| *Trichoderma pseudokoningii* | Pseudokonin | [107] |
| *Trichoderma reesei / T. jecorina* | Paracelsin | [110] |
| Trichoderma viride | Alamethicin,Trichovirin, Trichotoxin | [110] |
| Trichoderma polysporum | Trichosporin,Polysporin | [111] |
| *Trichoderma brevicompactum* | Trichothecene | [105] |
| *Trichoderma sp.* | Trichoderins | [112] |
| *Emericellopsis donezkii* | Bergofungin | [113] |
| *Apiocrea chrysosperma* | Chrysospermins | [114] |
| *Acremonium tubakii DSM 12774* | Cephaibols | [115] |
| *Boletus ssp.* | Boletusin | [116] |
| *Sepedonium ampullosporum* | Ampullosporin A | [117] |
| *Stilbella erythrocephala* | Antiamoebin | [118] |
| *Stilbella flavipes* | Stilboflavins | [119] |
| *Acremonium exuviarum* | Acrebol | [120] |
| *Tylopilus neofelleus* | Tylopeptin | [116] |
| *Penicillium lilacinum* | Leucinostatin | [121] |
| *Cephalosporium pimprina* | Antiamoebin | [122] |
| *Trichoderma asperellum* | Asperelines A-F | [123] |
| *Culicinomyces clawisporus* | Culicinin D | [124] |
| *Trichoderma sp.* | Trichoderins A, A1, B | [125] |
| *Trichoderma velutinum ACR-P1* | Two groups of NRPs | [126] |
| *Trichoderma atroviride* | 19-residue peptaibols | [127] |
| *Trichoderma phellinicola* | Hypophellins | [128] |
| *Hypocrea pulvinata* | Hypopulvins | [129] |
| *Hypocrea thelephoricola* | Thelephoricolins | [125] |
| *Hypocrea gelatinosa* | Gelatinosins B | [125] |
| *Hypocrea voglmayrii* | Voglmayrins | [125] |
| *Hypocrea minutispora* | Minutisporins | [125] |
| *Hypocrea citrine* | Hypocitrins | [125] |

**Table 9.2 (c):** Biological activities of NRPs isolated from marine sponges and tunicates [9].

| **NRPs** | **Chemical class** | **Origin** |
|---|---|---|
| MiraziridineA | Linear pentapeptide | *Theonella* aff. *mirabilis* |
| HaligramidesA-B | Cyclichexapeptides | *Haliclonanigra* |
| PrepatellamideA | Cyclicpeptide | *Lissoclinumpatella* |
| TamandarinsA-B | Depsipeptides | *Didemnidascidian* |
| MicroscleroderminsF–I | Cyclicpeptides | *Microscleroderma* sp. |
| Wainunuamide | Cyclichexapeptide | *Stylotellaaurantium* |
| LeucamideA | Cyclichexapeptide | *Leucettamicroraphis* |
| AxinellinC | Cyclicoctapeptide | *S.aurantium* |
| MilnamideD | Linearpeptide | *Cymbastela* sp. |
| KapakahinesE–G | | *Cribrochalinaolemda* |
| DidmolamidesA-B | Cyclichexapeptides | *Didemnummolle* |
| BistratamidesE–J | Cyclichexapeptides | *Lissoclinumbistratum* |
| MilnamideC | | *Auletta* sp. |
| ScleritoderminA | Cyclicpeptide | *Sclerito dermanodosum* |
| MicrocionamidesA-B | | *Clathria abietina* |
| KendarimideA | Linear peptide | *Haliclona* sp. |

**Table 9.2 (d):** Representatives of non-ribosomal peptidic products synthesized non-ribosomally.

| **Non-Ribosomal peptides** | |
|---|---|
| **Modification** | **Examples** |
| Fatty acids substituted | Surfactin, Syringomycin, Mycosubtilin |
| Non-proteinogenic aminoacids | Gramicidin S, ACV Penicillin precursor, CDA |
| Heterocyclization | Bacitracin, Vibriobactin |
| N-Methylation | Cyclosporin, Pristinamycin |
| Carboxy acids & sugars | Vancomycin, Bacillibactin |

## 9.7 Production of Non-Ribosomal Peptides

Biosurfactants are biosurface-active entities largely produced by microorganisms, secreted either extracellularly or attached to cells, predominantly during growth on water-immiscible substrates. The production of Rhamnolipids by *Pseudomonas aeruginosa* mutant strains grown in blackstrap molasses with or without supplementary nitrogen source was of 1.45 g/L [130]. *P. aeruginosa* UCP0992 grown in 100 ml mineral medium (aeration of 80%) supplemented with 3% glycerol and 0.6% $NaNO_3$, as the nitrogen source, at 28 °C after 96 h reached a production of 8.0 g/L [131], an indigenous strain *P. aeruginosa* EM1 originating from an oil-contaminated site located in southern Taiwan grown in inorganic nitrogen ($NaNO_3$) obtained a productivity of 8.63 g/L [132]. Heterologous expression in *Pseudomonas putida* KT42C1, a certified safe strain produced up to 1.5 g/L of rhamnolipid [133], whereas a production of 0.6 g/L in a recombinant *P. putida* strain KT2442 [134]. Engineered *E. coli* TnERAB produced 65–80 mg/L in MS plus glucose media, and 150–185 mg/L in LB plus glucose media, respectively [135]. A wide range of carbon sources and culture conditions have been reported in order to increase the production of lipopeptides like iturins, surfactins and fengycins. A variety of cheap raw materials have been used in their production like rice bran, soybean, potato peels, molasses, etc. In addition, it has been demonstrated that divalent cations, mainly $Mn^{2+}$ and $Fe^{2+}$ have an important influence on LP production. The manganese addition to the medium culture increased yield rates from 0.33 to 2.6 g/L. Also, the presence of $ZnSO_4$, $FeCl_3$ and $MnSO_4$ increased surfactin production in *Bacillus subtilis* [136]. Since the discovery of penicillin in 1928 by Alexander Fleming to 1940, the efforts to produce penicillin have been reported in bioreactors of 100,000 L, achieving 36 g/L at 250 h [137], although upto 50 g/L of penicillin can be produced [138, 139]. Thus, the penicillin fermentation process is a development strategy model to follow into a large scale of non-ribosomal peptide production process. However, the residual concentrations obtained of many NRP are from one to three orders of magnitude below compared to that of penicillin. For example, Cyclosporin A is synthesized by cyclosporine synthetase, one of the most complex and largest modular enzymes described [140]. Cylcosporine A produced by fungus *Tolypocladium inflatum*, *Beauveria nivea*, *Fusarium roseum* and *Tolypocladium niveum* have anti-inflammatory, inmunosupressive, antifungal and antiparasitic properties [141]. Maximum Cyclosporin A production of 1274 mg/L has been reported with *Tolypocladium inflatum* in submerged fermentation [142]. Echinocandins, novel antimycotics produced by ascomycota fungi having cyclic lipo-hexapeptide structure act as β-1, 3-glucan synthase inhibitors [143]. Echinocandin B, the precursor of anidulafungin, produced by *Aspergillus nidulans* had reached 1.5 g/L in potato dextrose broth [144]. The caspofungin precursor, pneumocandin $B_0$ have reached about 2 g/L in the fermentation broth of *Glarea lozoyensis* [145]. The mutant strain of *Coleophoma empetri* and improved medium conditions for production of the antibiotic FR901379, the precursor of micafungin had a 30-fold higher productivity compared to the wild type which produced 1 U/mL, the study demonstrated the production of FR901379 with optimal conditions in a fermenter of 15,000 L reaching 50 U/mL [146].

On the other hand, the *Actinomycetes* are known to produce novel bioactive compounds with more than 10,000 compounds described from genus *Streptomyces* only [147]. Bleomycin, a glycopeptide produced by *Streptomyces verticillus* with antibacterial and antitumor properties [148] produced at 10 mg/L [149]. Daptomycin, a lipopeptide produced by *Streptomyces roseosporus* effective for skin care [150] and a production of 812 mg/L has been achieved through fed-batch fermentation with feedback control of dextrin [151]. In case of the antibiotics bacitracins, produced by *Bacillus licheniformis* and *Bacillus subtilis* [152], bacitracin A is produced approximately 900 mg/L in cultures of *B.licheniformis* NCIMB 8874 with supplementation of oligosaccharides as elicitor [153]. Polymixns produced by *B. subtilis* and *Peanibacillus polymyxa* consist of ten amino acids with a characteristic polycationic heptapeptide ring and a N-terminal fatty acid [154]. Although this antibiotic is commercial, information about its level of concentration reached in production is vague. Ribosome targeting drugs like chloramphenicol at sub-inhibitory levels was most effective at enhancing the production of antibiotics such as actinomycin, calcium dependent antibiotic and piperidamycin resulting in an almost two fold increase in production. Recently, the effect of rare earth elements on bacterial physiology studied and found elicited its capabilities and thus exerts marked effects on secondary metabolism in *Streptomycetes.* Rare earth elements, scandium and/or lanthanum, at low concentrations markedly activate, ranging from 2.5 to 12-fold, the expression of nine genes belonging to nine secondary metabolite–biosynthetic gene clusters of *Streptomyces coelicolor* A3 [155]. A significant increase of antibiotic valinomycin from engineered *E. coli* with biosynthetic constituents from *S. tsusimaensis* was found from 0.3 to 2.4 mg/L which increased to 6.4 mg/L in parallel fed batch mode. Repeated glucose polymer feeding resulted in a valinomycin titer of 10 mg/L [156]. This represents a 33-fold increase in titer and is the highest concentration of NRP produced in *E. coli.* Red pigment pulcherrimin, a biocontrol agent produced by *Bacillus licheniformis* DW2 is optimized by addition of 1.0 g/L Tween 80 and glucose and $(NH_4)_2SO_4$ were served as the optimal carbon and nitrogen sources for its synthesis, respectively. Under optimized conditions, the maximum yield of pulcherrimin was 331.17 mg/L, 5.30 fold higher than initial condition, and is maximum yield of pulcherrimin [157, 158].

## 9.8 Future Scope

The fascinating array of natural products produced by non-ribosomal peptide synthetase assembly lines makes understanding the enzymology behind their action of great importance. From recent reports it has become clear that linear NRPS systems are highly diverse machineries with far greater catalytic potential than has been previously demonstrated. These new insights demonstrate the vital importance of rigorous characterization of NRPS assembly lines in order to build up a complete picture of the catalytic potential of these amazing biological machines. This knowledge is crucial given the potential to redesign NRPS-based biosynthetic pathways from important medicinal compounds and thus reinvent these compounds to tackle new and emerging challenges in human health, such as overcoming increasing antimicrobial resistance and delivering improved anticancer therapies.

With the ever-increasing need for new antibiotics and pharmaceutical products, it has been necessary to develop methods for rational drug design using non-ribosomal peptides as the basis. The modular structure of peptide synthetases gives them the advantage of being suitable candidates for the combinatorial biosynthesis of non-ribosomal peptides by rearranging or substituting the modules such that a new substrate amino acid or modification is placed within the final peptide product. Peptide synthetases have the ability to incorporate more than 500 proteinogenic, non-proteinogenic and hydroxyl amino acids which may be further modified by intrinsic peptide synthetase functions, such as epimerization or N-methylation. Furthermore, the resulting peptide backbone may be structurally altered by engineering of the individual modules or by changing the order of the modules in the synthetase thiotemplate. With increasing understanding of the molecular biology and combinatorial genetics of non-ribosomal peptide biosynthesis, the number of potentially bioactive substances designed using these tools appears to be almost limitless. Thus, it has become imperative to explore microorganisms for NCEs and lead drug molecules to run sustainable programs like drug discovery.

## Acknowledgements

The authors are thankful to the Director of CSIR-IIIM for his interest in this work. We are also thankful to the Fermentation Technology Group for useful discussions and suggestions for the chapter. RS thanks CSIR for Senior Research Fellowship.

## References

1. A.L. Demain, and R.P. Elander, *Antonie Van Leeuwenhoek*, Vol. 75, p. 5, 1999.
2. N.P. Keller, G. Turner, J.W. Bennett, *Nature Reviews Microbiology*, Vol. 3, p.937, 2005.
3. A. Porras-Alfaro, and P. Bayman, *Annual Review of Phytopathology*, Vol. 49, p.291, 2011.
4. Li. Jesse W-H, and John C. Vederas, *Science*, Vol. 325, p. 161, 2009.
5. D.J. Newman and G.M. Cragg, *Journal of Natural Products*, Vol. 79, p. 629, 2016.
6. E.M. Nolan, C.T. Walsh, *ChemBioChem*, Vol. 10, p. 34, 2009.
7. H. Wang, D.P. Fewer, L. Holm, L. Rouhiainen, K. Sivonen, *Proceedings of National Academy of Sciences*, Vol. 111 p. 9259, 2014.
8. R.K. Singh, P. Singh, T.M. Mohapatra, Nonribosomal peptide synthesis in microbes. In *Recent advances in microbiology*, Nova Science Publishers, Inc, New York, 2012.
9. S. Aggrawal, A. Adholeya, and S.K. Deshmukh, *Front Pharmacol*, Vol. 7, p. 333, 2016.
10. H. Paulus, and E. Gray, *Journal of Biological Chemistry*, Vol. 239, p. 865, 1964.
11. V. deCrecy-Lagard, W. Saurin, D. Thibaut, P. Gil, L. Naudin, J. Crouzet, and V. Blanc, *Antimicrobial Agents and Chemotherapy*, Vol. 41, p. 1904, 1997.
12. H. Kleinkauf, and H. von Dohren, *Antonie Van Leeuwenhoek*, Vol. 67, p. 229, 1995.
13. B.A. Johnson, H. Anker, and F.L. Meleney, *Science*, Vol. 102, p. 376, 1945.
14. E.B. Herr, Chemical and biological properties of capreomycin and other peptide antibiotics. In Antimicrobial Agents and Chemotherapy 1962, pp. 201–212. Edited by J.C. Sylvester, Michigan: American Society for Microbiology, 1963.
15. S. Somma, L. Gastaldo, and A. Corti, *Antimicrobial Agents and Chemotherapy*, Vol. 26, p. 917, 1984.
16. A.M.A. vanWageningen, P.N. Kirkpatrick, D.H. Williams, B.R. Harris, J.K.Kershaw, N.J Lennard, M. Jones, S.J.M. Jones, and P.J. Solenberg, *Chemistry & Biology*, Vol. 5, p. 155, 1998.
17. E. Abraham, and G. Newton, *Biochemical Journal*, Vol. 79, p. 377, 1961.
18. A. Markham, *Drugs*, Vol. 74, p. 1823, 2014.
19. H. Umezawa, K. Maeda, T. Takeuchi, Y. Okami, *Journal of Antibiotics*, Vol. 19, p. 200, 1966.
20. V. Miao, M. F. Coeffet-Legal, P. Brian, R. Brost, J. Penn, A. Whiting, *Microbiology*, Vol. 151, p. 1507, 2005.

21. M.R. Murthy, E. Mohan, and A. Sadhukhan, *Process Biochemistry,* Vol. 34, p. 269, 1999.
22. S.A. Waksman, and H.B. Woodruff, Experimental *Biology and Medicine,* Vol. 45, p, 604, 1940.
23. H. Ueda, H. Nakajima, Y. Hori, T. Fujita, M. Nishimura, T. Goto, *Journal of Antibiotics*, 47, p. 301, 1994.
24. G.G. Zhanel, G. Sniezek, F. Schweizer, S. Zelenitsky, P.R. Lagacé-Wiens, E. Rubinstein, A.S. Gin, D.J. Hoban, J.A. Karlowsky, *Drugs*, Vol. 69, p. 809, 2009.
25. G.G. Zhanel, P. Chung, H. Adam, S. Zelenitsky, A. Denisuik, F. Schweizer, P.R. Lagace-Wiens, E. Rubinstein, A.S. Gin, A. Walkty, D.J. Hoban, *Drugs*, Vol. 74, p. 31, 2014.
26. A.M. Mayer, K.B.Glaser, C. Cuevas, R.S. Jacobs, W. Kem, R.D. Little, J.M. McIntosh, D.J. Newman, B.C. Potts, D.E. Shuster, *Trends in pharmacological Sciences*, Vol. 31, p. 255, 2010.
27. S. Caboche, M. Pupin, V. Leclere, A. Fontaine, P. Jacques, G. Kucherov, *Nucleic Acids Research* Vol. 36, p. D326, 2008.
28. M.A. Martinez Nunez, and V.E. Lopez y Lopez, *Sustainable Chemical Process,* Vol. 4, p.13, 2016.
29. V.G. Fowler, N Engl, *Journal of Medicinal Chemistry,* Vol. 355, p. 653, 2006.
30. P.J. Rutledge, G.L. Challis, *Nature Reviews Microbiology,* Vol. 13, p. 509, 2015.
31. C.T. Walsh, T.A. Wencewicz, *Journal of Antibiotics*, Vol. 67, p. 7, 2014.
32. M.J. Winn, K. Fyans, Y. Zhuo, and J. Micklefield, *Natural Products Reports*, Vol. 33, p. 317, 2016.
33. J.B. McAlpine, *Journal of Natural Products*, Vol. 72, p.566, 2009.
34. KMJ de Mattos-Shipley, C Greco, DM Heard, G Hough, NP Mulholland, JL Vincent, J Micklefield, TJ Simpson, CL Willis, Cox RJ, AM Bailey. *Chemical Science.* Vol. 10, p. 4109. 2018.
35. G.H. Hur, C.R.Vickery, and M.D. Burkart, *Natural Reports*, Vol. 29, p.1074, 2012.
36. M. Pupin, Q. Esmaeel, A. Flissi, Y. Dufresne, P. Jacques, V. Leclere, *Synthetic and Systems Biotechnology,* Vol. 1, p. 89, 2016.
37. M.A. Marahiel, *Natural Product Reports*, Vol. 33, p.136, 2016.
38. J.A.E. Payne, M. Schoppet, M.H. Hansen, and M.J. Cryle, *Molecular BioSystems,* Vol. 13, p. 9, 2017.
39. MA Skiba, FP Maloney, Q Dan, AE Fraley, CC Aldrich, JL Smith, WC Brown. *Methods Enzymology.* Vol. 6, p. 45, 2018.
40. J. Grunewald, and M.A. Marahiel, *Microbiology and. Molecular Biology Reviews*, Vol. 70, p.121, 2006.
41. T. Stein, *Molecular Microbiology*, Vol. 56, p. 845, 2005.
42. B. Uytterhoeven, K. Appermans, L. Song, J. Masschelein, T. Lathouwers, C.W. Michiels, R. Lavigne, *Microbiology Open*, Vol. 5, p. 279, 2016.
43. M. Strieker, A. Tanovic, M.A. Marahiel, *Current Opinion in Structural Biology*, Vol. 20, p. 234, 2010.
44. G.D. Amoutzias, A. Chaliotis, D. Mossialos, *Marine drugs*, Vol. 14, p. 80, 2016.
45. F. Tiburzi, P. Visca, F. Imperi, *IUBMB life*, Vol. 59, p. 730, 2007.
46. E.J. Drake, B.R. Miller, C. Shi, J.T. Tarrasch, J.A. Sundlov, C.L. Allen, G. Skiniotis, C.C. Aldrich, A.M. Gulick, *Nature*, Vol. 529, p. 235, 2016.
47. B.R. Miller, A.M. Gulick, "Structural biology of nonribosomal peptide synthetases." *Nonribosomal Peptide and Polyketide Biosynthesis: Methods and Protocols*, p. 3, 2016.
48. R. Finking, M.A. Marahiel, *Annual. Review of Microbiology*, Vol. 58, p. 453, 2004.
49. T. Stachelhaus, H.D. Mootz, M.A. Marahiel, *Chemisty & Biology*, Vol. 6, p. 493, 1999.
50. F. Ishikawa, and H. Kakeya, *Current Organic Chemistry*, Vol. 19, p. 1204, 2016.
51. *A. Tanovic, S.A. Samel, L.O. Essen, M.A. Marahiel, Science,* Vol. 321, *p. 659, 2008.*

52. K.E. Bushley, B.G. Turgeon, *BMC Evolutionary Biology*, Vol. 10, p. 26, 2010.
53. M.A. Fischbach, C.T. Walsh, *Chemical Reviews*, Vol. 106, p.3468, 2006.
54. A. Paladini, L.C. Craig, *Journal of American Chemical Society*, Vol. 76, p. 688, 1954.
55. H.D. Mootz, D. Schwarzer, & M.A. Marahiel, *ChemBioChem*, Vol. 3, p. 490, 2002.
56. F. Lombo, A. Velasco, A. Castro, F. de la Calle, A.F. Brana, J.M. Sanchez-Puelles, C. Mendez, J.A. Salas, *ChemBioChem*, Vol. 7, p. 366, 2006.
57. E.A. Felnagle, M.R. Rondon, A.D. Berti, H.A. Crosby, M.G. Thomas, *Applied and Environmental Microbiol*ogy, Vol. 73, p. 4162, 2007.
58. H. Umezawa, *Antimicrobial Agents Chemotherapy*, Vol. 5, p. 107, 1965.
59. F. Lipmann, *Science*, Vol. 173, p. 875, 1971.
60. F. Lipmann, W. Gevers, H. Kleinkauf, and R. Roskoski, *Advances in Enzymology*, Vol. 35, p.1, 1971.
61. T. Stachelhaus, H.D. Mootz, V. Bergendahl, M.A. Marahiel, *Journal of Biological Chemistry*, Vol. 273, p. 22773, 1998.
62. M.A. Marahiel, T. Stachelhaus, and H.D. Mootz HD, *Chemistry Reviews*, Vol. 97, p. 26511, 1997.
63. T.A. Keating, C.G. Marshall, C.T. Walsh, A.E. Keating, *Nature Structural Biology*, Vol. 9, p. 522, 2002.
64. F. Kopp, M.A. Marahiel, *Natural Product Reports*, Vol. 24, p. 735, 2007.
65. R. Süssmuth, J. Müller, H. Von Dohren, I. Molnár, *Natural product Reports*, Vol. 28, p. 99, 2011.
66. U. Linne, and M.A. Marahiel, *Biochemistry*, Vol. 39, p. 10439, 2000.
67. K. Bloudoff, C.D. Fageb, M.A. Marahiel, and T.M. Schmeinga, *Proceedings of National Academy of Sciences*, 2017.
68. S.A. Sieber, M.A. Marahiel, *Chemical Reviews*, Vol. 105, p. 715, 2005.
69. J.N. Steenbergen, J. Alder, G.M. Thorne, F.P. Tally, *Journal of Antimicrobial Chemotherapy*, Vol. 55, p. 283, 2005.
70. M. Gonsior, A. Muhlenweg, M. Tietzmann, S. Rausch, A. Poch, R.D. Sussmuth, *ChemBioChem* Vol. 16, p. 2610, 2015.
71. D. Schwartz, R. Alijah, B. Nussbaumer, S. Pelzer, and W. Wohlleben, *Applied Environmental Microbiology*, Vol. 61, p. 570, 1996.
72. V. deCrecy-Lagard, W. Saurin, D. Thibaut, P. Gil, L. Naudin, J. Crouzet, and V. Blanc, *Antimicrobial Agents Chemotherapy*, Vol. 41, p. 1904, 1997.
73. Y. Mast, T. Weber, M. Golz, *Microbial Biotechnology*, Vol. 2, p. 192, 2011.
74. S.M. Hecht, *Journal of Natural Products*, Vol. 63, p. 158, 2000.
75. M.A. Fernandezmoreno, C. Vallin, and F. Malpartida, *Journal of Bacteriology*, Vol. 179, p. 6929, 1997.
76. S. Chen, D. Vonbamberg, V. Hale, M. Breuer, B. Hardt, R. Muller, H.G. Floss, K.A. Reynolds, and E. Leistner, *European Journal of Biochem*istry, Vol. 261, p. 98, 1999.
77. A. Konig, T. Schwecker, I. Molnar, G.A. Bohm, P.A.S. Lowden, J. Staunton, and P.F. Leadlay, *European Journal of Biochemistry*, Vol. 247, p. 526, 1997.
78. H. Motamedi, S. Cai, A. Shafee, and K.O. Elliston, *European Journal of Biochemistry*, Vol. 244, p. 74, 1997.
79. H. Komaki, A. Ishikawa, N. Ichikawa, A. Hosoyama, M. Hamada, E. Harunari, T. Nihira, W. Panbangred, and Y. Igarashi, *Standards in Genomic Sciences*, Vol. 11, p.83, 2016.
80. S. Lautru, R.J. Deeth, L.M. Bailey, G.L. Challis, *Nature Chemical Biology*, Vol. 1, p. 265, 2005.
81. P.P. Chong, S.M. Podmore, H.M. Kieser, M. Redenbach, K. Tirgay, M.A. Marahiel, D.A. Hopwood, and C.P. Smith, *Microbiology*, Vol. 144, p. 193, 1998.

82. D.P. Levine, *Clinical Infectious Diseases,* Vol. 42, p. S5, 2006.
83. Y. Igarashi, T. Iida, T. Fukuda, S. Miyanaga, H. Sakurai, I. Saiki, K. Miyanouchi, *Journal of Antibiotics* Vol. 65, p. 207, 2012.
84. Y. Igarashi, Y. Tanaka, M. Ikeda, T. Oikawa, S. Kitani, T. Nihira, P. Mongkol, M. Janhom, W. Panbangred, *Journal of Antibiotics*, Vol. 65, p.157, 2012.
85. V. Tosato, A.M. Albertini, M. Zotti, S. Sonda, and C.V. Bruschi, *Microbiology,* Vol. 143, p. 3443, 1997.
86. E.H. Duitman, L.W. Hamoen, M. Rembold, G. Venema, H. Seitz, W. Saenger, F. Bernhard, R. Reinhardt, M. Schmidt, C. Ullrich, T. Stein, F. Leenders, and J. Vater, *Proceedings of National Academy of Sciences* USA, Vol. 96, p. 13294, 1999.
87. L. Serino, C. Reimann, P. Visca, M. Beyeler, V. Dellachiesa, and D. Haas, *Journal of Bacteriology,* Vol. 179, p. 248, 1997.
88. J.H. Zhang, N.B. Quigley, and D.C. Gross, *Journal of Bacteriology*, Vol. 177, p. 4009, 1995.
89. P.A. Sokol, P. Darling, D.E. Woods, E. Mahenthiralingam, and C. Kooi, *Infection and Immunity*, Vol. 67, p. 4443, 1999.
90. P.W. Lindum, U. Anthoni, C. Christophersen, L. Eberl, S. Molin, and M. Givskov, *Journal of Bacteriology*, Vol. 180, p. 638, 1998.
91. S. Gaisser, and C. Hughes, *Molecular Genetics and Genomics*, Vol. 253, p. 415, 1997.
92. J.R. Butterton, M.H. Choi, P.I. Watnick, P.A. Carroll, and S.B. Calderwood, *Journal of Bacteriology,* Vol. 182, p.1731, 2000.
93. A.M. Gehring, E. Demoll, J.D. Fetherston, I. Mori, G.F. Hayhew, F.R. Blattner, C.T. Walsh, and R.D. Perry, *Chemistry & Biology*, Vol. 5, p. 573, 1998.
94. L.E.N. Quadri, J. Sello, T.A. Keating, P.H. Weinreb, and C.T. Walsh, *Chemistry & Biology*, Vol. 5, p.631, 1998.
95. S.W. Yu, E. Fiss, and W.R. Jacobs, *Journal of Bacteriol*ogy, Vol. 180, p. 4676, 1998.
96. H. Billman-Jacobe, M.J. Mcconville, R.E. Haites, S. Kovacevic, and R.L. Coppel, *Molecular Microbiology,* Vol. 33, p. 1244, 1999.
97. D. Tillett, E. Dittmann, M. Erhard, H. von Döhren, T. Börner, B.A. Neilan, *Chemistry & Biology,* Vol. 7, p. 753, 2000.
98. A. Pospiech, B. Cluzel, J. Bietenhader, and T. Schupp, *Microbiology,* Vol. 141, p. 1793, 1995.
99. Y. Paitan, G. Alon, E. Orr, E.Z. Ron, and E. Rosenberg, *Journal of Molecular Biology*, Vol. 286, p. 465, 1999.
100. B. Silakowski, H.U. Schairer, H. Ehret, B. Kunze, S. Weinig, G. Nordsiek, P. Brandt, H. Blocker, G. Hofe, S. Beyer, and R. Muller, *Journal of Biological Chemistry,* Vol. 274, p. 37391, 1999.
101. I. Molnar, T. Schupp, M. Ono, R.E. Zirkle, M. Milnamow, B. Nowak-Thompson, N. Engel, C. Toupet, A. Stratmann, D.D. Cyr, J. Gorlach, J.M. Mayo, A. Hu, S. Go¡, J. Schmid, and J.M. Ligon, *Chemistry & Biology*, Vol. 7, p. 97, 2000.
102. S.L. Annis, and D.G. Panaccione, *Canadian Journal of Microbiol*ogy, Vol. 44, p. 80, 1998.
103. S. Pelzer, W. Reichert, M. Huppert, D. Heckmann, and W. Wohlleben, *Journal of Biotechnology*, Vol. 56, p.115, 1997.
104. S-U. Oh, S-J. Lee, J-H. Kim, I-D. Yoo, *Tetrahedron Letters,* Vol. 41, p. 61, 2000.
105. H. Bruckner, J. Kirschbaum, A. Jaworski, Sequences of peptaibol antibiotics trichoaureocins from *Trichoderma aureoviride.* In *Peptides* Benedetti E, Pedone C [eds]. Edizione Ziino: Napoli, pp. 362–363, 2002.
106. S. Rebuffat, C. Goulard, S. Hlimi, B. Bodo, B. *Journal of Peptide Science*, Vol. 6, p. 519, 2000.
107. S. Xiao-Yan, S. Qing-Tao, X. Shu-Tao, C. Xiu-Lan, S. Cai-Yun, Z. Yu-Zhong, *FEMS Microbiology Letters,* Vol. 260, p. 119, 2006.
108. C. Krause, J. Kirschbaum, H. Bruckner, *Chemical Biodiversity,* Vol. 4, p.1083, 2007.

109. H. Brückner, H. Graf, *Experientia*, Vol. 39, p. 528, 1983.
110. C.T. Hou, A. Ciegler, C.W. Hesseltine, *Applied Microbiology*, Vol. 23, p. 183, 1972.
111. M. Iwatsuki, Y. Kinoshita, M. Niitsuma, J. Hashida, M. Mori, A. Ishiyama, M. Namatame, A. Nishihara-Tsukashima, K. Nonaka, R. Masuma, K. Otoguro, H. Yamada, K. Shiomi, S. Ōmura, *Journal of Antibiotics*, Vol. 63, p. 331, 2010.
112. P. Pruksakorn, M. Arai, N. Kotoku, C. Vilchèze, A.D. Baughn, P. Moodley, W.R. Jacobs, M. jr. Kobayashi, *Bioorganic and Medicinal Chemistry Letters*, Vol. 20, p. 3658, 2010.
113. A. Berg, M. Ritzau, W. Ihn, W.F. Fleck, S. Heinze, U. Grafe, *Journal of Antibiotics*, Vol. 49, p. 817, 1996.
114. K. Dornberger, W. Ihn, M. Ritzau, U. Gräfe, B. Schlegel, W.F. Fleck, *Journal of Antibiotics*, Vol. 48, p. 977, 1995.
115. M. Schiell, J. Hofmann, M. Kurz, F.R. Schmidt, L. Vériest, M. Vogel, J. Wink, G. Seibert, *Journal of Antibiotics*, Vol. 54, p. 220, 2001.
116. S-J. Lee, B-S. Yun, D-H. Cho, I-D. Yoo, *Journal of Antibiotics*, Vol. 52, p. 998, 1999.
117. I. Berek, A. Becker, H. Schröder, V. Hollt, G. Greksch, *Behavioural Brain Research*, Vol. 203, p. 232, 2009.
118. N.A. Lehr, A. Meffert, L. Antelo, O. Sterner, H. Anke, R.W.S. Weber, *FEMS Microbiology Ecoogyl* Vol 55, p. 105, 2006.
119. Jaworski, H. Brückner, *Journal of Peptide Sciences*, Vol. 7, p. 433, 2001.
120. M.A. Andersson, R. Mikkola, M. Raulio, L. Kredics, P. Maijala, M.S. Salkinoja-Salonen, *Journal of Applied Microbiology*, Vol. 106, p. 909, 2009.
121. T. Arai, Y. Mikami, K. Fukushima, T. Utsumi, K. Yazawa, *Journal of Antibiotics*, Vol. 26, p. 157, 1973.
122. R.C. Pandey, J.C. Cook, K.L. Rinehart, jr. *Journal of Antibiotics*, Vol. 31, p. 241, 1978.
123. J. Ren, C. Xue, L. Tian, M. Xu, J. Chen, Z. Deng, P. Proksch, W. Lin, *Journal of Natural Products*, Vol. 72, p. 1036, 2009.
124. He, J.E. Janso, H.Y. Yang, V.S. Bernan, S.L. Lin, and K. Yu, *Journal of Natural Products*, Vol. 69, p. 736, 2006.
125. G. Bills, Y. Li, L. Chen, Q. Yue, X.M. Niu, Z. An, *Natural Product Reports*, Vol. 31, p.1348, 2014.
126. R. Sharma, V.P. Singh, D. Singh, F. Yusuf, A. Kumar, R.A. Vishwakarma, A. Chaubey, *Applied Microbiology and Biotechnology*, Vol. 100, p. 9091, 2016.
127. A. Carroux, A.I. Van Bohemen, C. Roullier, T. Robiou duPont, M. Vansteelandt, A. Bondon, A. Zalouk-Vergnoux, Y.F. Pouchus, N. Ruiz, *Chemistry & Biodiversity*, Vol. 10, p. 772, 2013.
128. C.R. Röhrich, A. Iversen, W.M. Jaklitsch, H. Voglmayr, A. Vilcinskas, K.F. Nielsen, U. Thrane, H. vonDöhren, H. Brückner, T. Degenkolb, *Chemisty & Biodiversity*, Vol. 10, p. 787, 2013.
129. C.R. Röhrich, A. Iversen, W.M. Jaklitsch, H. Voglmayr, A. Berg, H. Dörfelt, U. Thrane, A. Vilcinskas, K.F. Nielsen, H. Von Doehren, H. Brückner, *Fungal Biology*, Vol. 116, p. 1219, 2012.
130. Z.A. Raza, M.S. Khan, Z.M. Khalid, *Journal of Environment Science and Health*, Vol. 42, p. 73, 2007.
131. S.N. Silva, C.B. Farias, R.D. Rufino, J.M. Luna, L.A. Sarubbo, *Colloids and Surface B: Biointerfaces*, Vol. 79, p. 174, 2010.
132. J.Y. Wu, K.L. Yeh, W.B. Lu, C.L. Lin, J.S. Chang, *Bio resource Technology*, Vol. 99, p. 1157, 2008.
133. A. Wittgens, T. Tiso, T.T. Arndt, P. Wenk, J. Hemmerich, C. Müller, R. Wichmann, B. Küpper, M. Zwick, S. Wilhelm, R. Hausmann, C. Syldatk, F. Rosenau, L.M. Blank, *Microbial Cell Factories*, Vol. 10, p. 80, 2011.

134. U.A. Ochsner, J. Reiser, A. Fiechter, B. Witholt, *Applied Environmental Microbiology*, Vol. 61, p. 3503, 1995.
135. Q. Wang, X. Fang, B. Bai, X. Liang, P.J. Shuler, W.A. Goddard, Y. Tang, *Biotechnology Bioengineering*, Vol. 98, p. 842, 2007.
136. E. Beltran-Gracia, G. Macedo-Raygoza, J. Villafaña-Rojas, A. Martinez-Rodriguez, Y.Y. Chavez-Castrillon, F.M. Espinosa-Escalant, P. Di Mascio, T. Ogura and M.J. Beltran-Garcia, In *Fermentation Processes*. InTech, 2017.
137. S. Goldrick, A. Ştefan, D. Lovett, G. Montague, B. Lennox, *Journal of Biotechnology*, Vol. 193, p. 70, 2015.
138. L. Gidijal, J.A.K.W. Kiel, R.D. Douma, R.M. Seifar, W.M. van Gulik, R.A.L. Bovenberg, M. Veenhuis, I.J. Van der Klei, *PLOS ONE*, Vol. 4, p. e8317, 2009.
139. H. Jørgensen, J. Nielsen, J. Villadsen, H. Mollgaard, *Biotechnology Bioengineering*, Vol. 46, p. 117, 1995.
140. A. Lawen, R. Zocher, *Journal of Biological Chemistry*, Vol. 265, p. 11355, 1990.
141. S.A. Survase, L.D. Kagliwal, U.S. Annapure, R.S. Singhal, *Biotechnology Advances*, Vol. 29, p. 418, 2011.
142. S. Survase, S.B. Bankar, A.S. Annapure, R.S. Singhal, *Indian Journal of Biotechnology*, Vol. 14, p. 401, 2015.
143. T. Emri, L. Majoros, V. Tóth, I. Pócsi, *Applied Microbiology and Biotechnology*, Vol. 97, p. 3267, 2013.
144. S.P. Zou, M. Liu, Q.L. Wang, Y. Xiong, K. Niu, Y.G. Zheng, Y.C. Shen, *Journal of Chromatography*, Vol. 97, p. 111, 2015.
145. J.M. Balkovec, D.L. Hughes, P.S. Masurekar, C.A. Sable, R.E. Shwartz, S.B. Singh, *Natural Product Reports*, Vol. 31, p. 15, 2014.
146. M. Kanda, M. Tsuboi, K. Sakamoto, S. Shimizu, M. Yamashita, H. Honda, *Journal of Bioscience and Bioengineering*, Vol. 107, p. 530, 2009.
147. J. Bérdy, *Journal of Antibiotics*, Vol. 65, p. 395, 2012.
148. J. Gu, R. Codd, *Journal of Inorganic Biochemistry*, Vol. 115, p. 198, 2012.
149. J. Gu, R. Codd, *RSC Advances*, Vol. 5, p. 3443, 2015.
150. L.B. Arthur, C.F. Peter, D.B. Steven, *Antimicrobial Agents Chemotherapy*, Vol. 45, p.1919, 2001.
151. I.S. Ng, Ye. Chimimg, Z. Zhang, Y. Lu, K. Jing, *Bioprocess and Biosystem Engineering*, Vol. 37, p. 415, 2014.
152. D. Konz, A. Klens, K. Schörgendorfer, M.A. Marahiel, *Chemical Biology*, Vol. 4, p. 927, 1997.
153. T. Murphy, I. Roy, A. Harrop, K. Dixon, T. Keshavarz, *Journal of Biotechnology*, Vol. 131, p. 397, 2007.
154. M. Shaheen, J. Li, A.C. Ross, J.C. Vederas, S.E. Jensen, *Chemical Biology*, Vol. 18, p. 1640, 2011.
155. Y. Tanaka, T. Hosaka & K. Ochi, *Journal of Antibiotics*, Vol. 63, p. 477, 2010.
156. Li Jian, J. Jaitzig, F. Hillig, R. Süssmuth, and P. Neubauer, *Applied Microbiology and Biotechnology*, Vol. 98, p. 591, 2014.
157. X. Li, D. Wang, D. Cai, Y. Zhan, Q. Wang and S. Chen, *Applied Biochemistry and Biotechnology*, p. 1, 2017.
158. R Chen, Q Zhang, B Tan, L Zheng, H Li, Y Zhu, C Zhang. *Organic Letters*. Vol 20, p. 5697, 2017

# 10

# Enzymes as Therapeutic Agents in Human Disease Management

**Babbal, Adivitiya, Shilpa Mohanty and Yogender Pal Khasa***

*Department of Microbiology, University of Delhi South Campus, New Delhi, (India)*

***Abstract***

Enzymes are biological macromolecules that can be used as biotherapeutic agents due to high specificity, affinity and efficacy. They are commonly used for the treatment of diseases like Alzheimer's, lymhphoblastic leukemia, refractory gout, lysosomal storage disorders, cardiovascular diseases and pancreatic disorders. Moreover, the introduction of the Orphan Drug Act allowed the development of enzyme-based drugs for the treatment of rare genetic disorders which resulted in the improvement of patients' quality of life with significant reduction in morbidity. Further, the techniques like PEGylation, release of therapeutic molecules at the target site, polymer-enzyme conjugates, nano-immobilization and masking of antigenic region have been employed to design drugs with a high potency and stability. Several enzyme based biotherapeutics have been approved by FDA and many are in clinical trials. The advancement in recombinant DNA technology has further helped in the development of efficient disease-specific enzyme-based therapies cost-effectively.

***Keywords:*** Therapeutic enzymes, orphan drugs, immunogenicity, enzyme replacement therapy, Food and Drug Administration (FDA)

## 10.1 Introduction

The use of enzymes as therapeutic molecules has marked the beginning of a new era in the bio-pharmaceutical industry. The application of enzymes for disease management began around 40 years back when the crude extract from *Bacillus pyocyaneus* was demonstrated as an effective agent for the treatment of anthrax due to the action of enzymes present in the crude extract. Currently, enzymes are being used as an excellent mucolytic, thrombolytic, oncolytic, debriding and anti-inflammatory agents as well as effective digestive aids (Table 10.1). Alteplase (Activase®), a recombinant tissue plasminogen activator was the first genetically engineered drug that was FDA (US Food & Drug Administration) approved in 1989 for use as a thrombolytic agent [1]. In 1983, the United States passed the Orphan Drug Act to support the development

*Corresponding author: yogi110@gmail.com

Saurabh Saran, Vikash Babu, and Asha Chaubey (eds.) High Value Fermentation Products, Volume 1, (225–264) © 2019 Scrivener Publishing LLC

**Table 10.1** List of FDA approved therapeutic enzymes.

| Enzyme | Trade name | Company | Therapeutic use |
|---|---|---|---|
| **Pancreatic enzymes** | | | |
| Lipase | TheraCLEC-Total™ | Altus Biologics Inc., Cambridge, USA | Steatorrhea, exocrine pancreatic insufficiency (EPI) |
| Protease | Xenaderm®<br>Granulex® | Healthpoint Biotherapeutics Inc., USA<br>Mylan Institutional Inc., USA | Debriding agent, anti-inflammatory properties, anti-cancer properties |
| | TheraCLEC-Total™ | Altus Biologics Inc., Cambridge, USA | Exocrine pancreatic insufficiency (EPI) |
| Amylase | TheraCLEC-Total™ | Altus Biologics Inc., Cambridge, USA | Exocrine pancreatic insufficiency (EPI) |
| **Oncolytic enzymes** | | | |
| L-asparaginase | Oncaspar® | Enzon Pharmaceuticals Inc., USA | Acute lymphoblastic leukaemia |
| | Erwinaze® | Jazz Pharmaceuticals Inc., USA | Acute lymphoblastic leukaemia |
| Arginine deiminase | Hepacid | Phoenix Pharmacologics, Inc., Lexington, USA | Hepatocellular carcinoma |
| | Melanocid | Phoenix Pharmacologics, Inc., Lexington, USA | Melanoma |
| **Kidney disorders** | | | |
| Uricases | Pegloticase (Krystexxa®) | Savient, East Brunswick, New Jersey, USA | Refractory gout, hyperuricemia |
| | Rasburicase (Elitek®/ Fasturtec®) | Sanofi-Synthelabo research, New York, USA | Tumor lysis syndrome |
| **DNA enzymes** | | | |
| Dornase alpha | Pulmozyme® | Genentech, South San Francisco, USA | Cystic fibrosis (CF) |
| Adenosine deaminase | Adagen®<br>(pegademase bovine) | Enzon Pharmaceuticals Inc., USA | Severe combined immunodeficiency (SCID) |

| | | | |
|---|---|---|---|
| **Enzymes for treatment of clotting diseases** | | | |
| Recombinant von Willebrand Factor | Vonvendi® | Baxalta US Inc., CA, USA | von Willebrand Disease |
| von Willebrand Factor + FVIII concentrate | Wilate®<br>Humate-P®<br>Alphanate® | Octapharma USA, Inc., USA<br>CSL Behring, GmbH, Germany<br>Grifols Biologicals Inc., CA, USA | von Willebrand Disease<br>Hemophilia A, von Willebrand Disease<br>Hemophilia A, von Willebrand Disease |
| Recombinant FVIII | Recombinate™<br>Advate®<br>Kogenate® FS<br>BAY81–8973 (Kovaltry®)<br>Eloctate®<br>Adynovate (BAX855) | Baxter Healthcare Corporation, USA<br>Baxter Healthcare Corporation, USA<br>Bayer HealthCare, USA<br>Bayer HealthCare, USA<br>Biogen Inc., USA<br>Baxalta US, Inc., CA, USA | Hemophilia A |
| Recombinant FIX | rIX-FP (Idelvion®)<br>Alprolix®<br>N9-GP | CSL Behring, GmbH, Germany<br>Biogen Inc., USA<br>Novo Nordisk Inc., Denmark | Hemophilia B |
| Fibrinogen | RiaSTAP™ | CSL Behring, GmbH, Germany | Afibrinogenemia, hypofibrinogenemia |
| Recombinant FVII | NovoSeven® | Novo Nordisk Inc., Denmark | Hemophilia A, haemophilia B, congenital FVII deficiency |
| FX concentrate | Coagadex® | Bio Products Laboratory Limited, UK | FX deficiency |
| FXI concentrate | Hemoleven® | Laboratoire Francais du Fractionnement et des Biotechnologies (LFB), France | Hemophilia C |
| FXIII | Corifact™ (FXIII concentrate) | CSL Behring, GmbH, Germany | Congenital FXIII deficiency |
| | Tretten® (FXIII A-subunit) | Novo Nordisk Inc., Denmark | FXIII A-subunit deficiency |

(*Continued*)

Table 10.1 Cont.

| Enzyme | Trade name | Company | Therapeutic use |
|---|---|---|---|
| **Thrombolytic Enzymes** | | | |
| Streptokinase (SK) | Streptase®<br>Kabikinase | Hoechst Marion Roussel, Germany<br>Pharmacia Upjohn, Sweden | AMI, DVT, Pulmonary Embolism, Arterial Thrombosis |
| Urokinase (uPA) | Abbokinase®<br>Kinlytic™ | Abbott Laboratories, USA<br>ImaRx Therapeutics, Inc., Tucson Arizona, USA | Pulmonary Embolism |
| Anistreplase (APSAC) | Eminase® | Smithkline Beecham, UK | AMI |
| Alteplase | Activase® | Genentech, Inc., CA, USA | AMI, Ischemic stroke, pulmonary embolism |
| **Enzymes for treatment of Lysosomal Storage Disorders** | | | |
| α-galactosidase A | Agalsidase beta (Fabrazyme®) | Genzyme Corporation, USA | Fabry Disease |
| Glucocerebrosidase (acid β-glucosidase) | Imiglucerase (Cerezyme®)<br>Velaglucerase alfa (VPRIV®)<br>Taliglucerase alfa (Uplyso®/ Elelyso®) | Genzyme Corporation, USA<br>Shire Pharmaceuticals Inc., Dublin<br>Pfizer, USA/Protalix Biotherapeutics, Israel | Gaucher Disease |
| Acid alpha-glucosidase (GAA) | Myozyme®/ Lumizyme® | Genzyme Corporation, USA | Pompe Disease |
| α-L-iduronidase | Laronidase (Aldurazyme®) | Biomarin Pharmaceutical, USA and Genzyme Corporation, USA | MPS I |
| **Miscellaneous enzymes** | | | |
| Collagenase | Xiaflex® | Auxilium Pharmaceuticals, USA | Dupuytern's contracture and Peyronie's disease |
| | Santyl® | Advance Biofactures Corporation, USA | Debridement agent |
| Hyaluronidase | Hylenex® | Halozyme Therapeutics, USA | Anti-cancer properties, Substitute or adjunct to conventional vitrectomy |

of therapeutics for the treatment of rare diseases affecting a small population. Since then various therapeutic enzymes have been developed for the treatment of rare genetic disorders. Adenosine deaminase (Adagen®) was the first enzyme to get approval as bio-therapeutic molecule for the treatment of severe combined immunodeficiency (SCID) [2]. Thus, the development of enzyme-based therapies has indeed helped in improving the quality of life and reducing global morbidity and mortality rates.

Enzyme replacement therapy (ERT) involves the intravenous infusion of the deficient enzyme to the patients. The ERT technique has been successfully used in diseases like lysosomal storage disorders, exocrine pancreatic insufficiency, severe combined immunodeficiency (SCID) and hemophilia. Since then several enzymes have been developed for the treatment of genetic disorders such as agalsidase-alpha and agalsidase-beta for the treatment of Fabry disease, β-glucosidase for treating Gaucher disease, acid alpha-glucosidase for Pompe disease and dornase alpha (DNase) for treating cystic fibrosis.

The pancreatic enzymes like trypsin, lipases and amylases have been used successfully for the treatment of pancreatic enzyme insufficiency and multiple food intolerance. They also possess anti-inflammatory as well as debriding properties [1]. The oncolytic properties of enzymes have proven to be a major advancement in treating cancers. The anti-neoplastic enzymes like L-asparaginase and L-glutaminase are the most efficacious treatment strategy for patients having acute lymphoblastic lymphoma. Another enzyme, arginine deiminase, also exhibits anti-cancer properties against hepatocellular carcinomas and melanomas. Enzymelike glucokinase can also act as an anti-diabetic molecule. Superoxide dismutase, alkaline phosphatase and ribozymes display potential to treat cellular disorders, rheumatoid arthritis and osteoarthritis. Kidney disorders like refractory gout and chronic kidney disease can be treated using uricase, whereas urease can effectively prevent the occurrence of dental caries. Moreover, DNA-based enzymes like dornase alpha and adenosine deaminase have been used to treat the genetic disorders, cystic fibrosis and SCID, respectively. RNA-based enzymes like onconase (an RNase) is under clinical trial as a drug, for the treatment of malignant mesothelmia [1, 3, 4].

Cardiovascular disorders are, globally, the leading cause of mortality according to the World Health Organization (WHO). A major breakthrough in this field was achieved with the approval of Activase® for use in inischemic stroke, myocardial infarction and pulmonary embolism. Streptokinase, a thrombolytic agent of bacterial origin, also changed the face of thrombolytic therapy especially in third world countries due to its cost-effectiveness [5]. Similarly, several enzyme-based drugs have been developed to control bleeding and haemorrhage occurring during severely afflicting blood clotting disorders, the most predominant of which is haemophilia. Drugs like Recombinate™, Advate®, Alprolix® are efficient therapeutic agents for treatment of haemophilia A and B [6, 7]. Vonvendi®, a drug containing recombinant Von Willebrand Factor as the active agent is the first and only recombinant vWF for combating von Willebrand disease [8]. Several orphan drugs have also been generated for the rare clotting diseases occurring due to deficiencies in factors V, VII, XIII, II, XI, X, XII and fibrinogen. Coagadex®, NovoSeven™, Corifact™, RiaSTAP™ are a few drugs that have been approved for the management of such rare clotting disorders [9–12]. To circumvent the problem of immunogenicity, the covalent modification of enzymes with molecules such as polyethylene glycol (PEG) is usually carried out. Techniques like microencapsulation and artificial liposomal entrapmentare used to increase the half-life and stability of enzymes. Moreover, strategies such as Polymer Masked–Unmasked Protein Therapy (PUMPT) have proven to be effective in masking the antigenic domain and also allowed the release of enzyme-based drugs at specific site of action [13]. These strategies greatly improved the therapeutic outcomes of various enzymes which in turn helped in the better management of several clinical conditions. Thus, this chapter provides a detailed description of a variety of enzymes that are used for the therapeutic management of human diseases with their market potential and current FDA approved status.

## 10.2 Pancreatic Enzymes

Amylases, lipases, and proteases belong to the class of pancreatic enzymes. Trypsin, a protease, digests the proteins present in food to peptides. Amylase is responsible for converting carbohydrates (except cellulose) into oligosaccharides and disaccharides while lipase carries out hydrolysis of fats into fatty acids and monoglycerides. These pancreatic enzymes have been used for the treatment of several digestive disorders. A mixture of lipase, amylase and protease is commercially available under the trade name Thera CLEC- Total™ (Altus Biologics Inc., USA). It has received the orphan drug status in 2002 [14]. The problem of fat malabsorption in HIV patients is treated using these enzyme combinations. A mixture of these enzymes is also used in pancreatic enzyme replacement therapy, i.e., PERT for treating exocrine pancreatic insufficiency (EPI), a digestive disorder of breakdown and absorbance of carbohydrates, fats and proteins. Besides this, the pancreatic enzymes can also be used in the treatment of constipation, indigestion and bloating. Most of the pancreatic enzyme preparations are from porcine sources and all are acid resistant enteric-coated microspheres except Viokaze® which is not enteric coated. These enteric-coated microspheres allow the release of pancreatic lipase at a pH of 5.5–6 [1]. The FDA approved porcine origin pancrelipase formulations, which is a mixture of amylase, protease and lipase, include Creon®, Pancreaze®, Zenpep®, Ultresa®, Pertyze®and Viokaze®. Each Creon® capsule contains 5,000 USP units of lipase, 16,000 USP units of amylase and 18,750 USP units of protease. The use of these formulations has resulted in significant relief in disease symptoms in patients suffering from EPI and pancreatic disorders and also prevented morbidity with improvement in the quality of life [3]. The other therapeutic properties of these pancreatic enzymes are described in the following section.

### 10.2.1 Trypsin (EC 3.4.21.4)

Trypsin, an endopeptidase, belongs to the class of serine proteases. This enzyme is secreted by the pancreas and converts proteins into oligopeptides by carrying out proteolytic cleavage at lysine or arginine residues. It has many other therapeutic applications apart from its role in digestion. Trypsin acts as a mild debriding agent and aids in the treatment of chronic and conventional wounds. During wound healing, the monocytes migrate from the blood circulation to enter at the injury site where they get differentiated into fibroblast-like cells known as fibrocytes. The differentiation of monocytes into fibrocytes requires the presence of tryptic fragments of albumin, which are formed by the proteolytic activity of trypsin [15]. Two FDA-approved topical preparations Xenaderm® (Healthpoint Biotherapeutics Inc., US) and Granulex® (Mylan Institutional Inc., US), have trypsin as their main constituent [1]. Trypsin also has an anti-inflammatory property since it causes the degradation of immune-response signalling molecules and has been used to treat bronchial asthma.

It has been reported that trypsin causes tumor necrosis and its degeneration [16]. It is also used to remove the fibrous membrane from the throat of patients suffering from diphtheria [1]. Besides this, trypsin has also been shown to produce effective anti-metastatic and anti-tumor effects in animal and human systems [17]. The strategy of polymer-enzyme conjugation has been implemented to improve the enzyme stability in the circulation. Unmasking of the enzyme at the target site has been carried out in accordance to the Polymer Masked–Unmasked Protein Therapy (PUMPT) where increased enzyme stability with masking of immunogenic domain was achieved. The dextrin-trypsin and PEG-trypsin conjugates have been developed for diminishing the adverse effects of chemotherapy in cancer patients [13].

### 10.2.2 Pancreatic Lipase (EC 3.1.1.3)

Lipases, a subclass of the esterase family, are enzymes involved in the hydrolysis of fat molecules. They are commonly employed as digestive aids and also have various therapeutic roles in the treatment of gastrointestinal disturbances, dyspepsia, and cutaneous manifestations of digestive allergies, familial combined hyperlipidemia (FCHL), steatorrhea, pancreatic insufficiency, skin inflammation, malignant tumours and coronary atherosclerosis [18].

*Candida rugosa* lipase has been utilized to produce a drug known as lovastatin, which has the ability to reduce the cholesterol level in the serum. Lovastatin inhibits 3-hydroxy-3-methylglutaryl-coenzymeA reductase (HMG-CoA reductase) which catalyzes the formation of mevalonate, a building block for cholesterol biosynthesis. The lipase produced by *Serratia marcescens* catalyses the hydrolysis of asymmetric 3-phenylglycidic acid ester which is a crucial intermediate for producing diltiazem hydrochloride, a coronary vasodilator. Lipases produced by *Galleria mellonella* (wax moth) have a killing effect on *Mycobacterium tuberculosis* (MBT) H37Rv [19].

Similase and Vitaline® are plant-based enzyme supplements (containing lipase, amylase, protease and other enzymes) having a high potential to support the digestion of carbohydrates, proteins and fats. However, microbial origin lipases exhibit high proteolytic stability and resistance against bile salts thus resulting in their high lipolytic activity. Human gastric lipases are also very stable enzymes and are considered as good tools for enzyme replacement therapy [3].

### 10.2.3 Amylases (EC 3.2.1.1)

Amylases are the enzymes secreted by salivary glands and pancreas that act on polysaccharides like starch and glycogen via the cleavage of α-(1,4)-glycosidic linkages to produce dextrin. They are commonly used as digestive aids in order to have a better digestion of dietary carbohydrate. The concentration of amylase in body fluids is of clinical relevance in the diagnosis of peptic ulcers, inflammation of pancreas and macroamylasemia [19].

Amylases produced by microbial species are generally employed to treat enzyme deficiencies. Digest GOLD Atpro and Digest contain an effective concentration of amylase along with other proteolytic enzymes like lipase, cellulase and protease which help in the breakdown of carbohydrates, proteins, fats and fibers. Another product, Digest Spectrum, is used for the treatment of multiple food intolerance. It contains amylase, cellulase, protease Thera-blend formulas (having more than 3 protease producing strains), lipase and dipeptidyl peptidase-4(DPP-IV) for gluten and casein digestion, as well as invertase and maltase for starch and sugar digestion [20].

## 10.3 Oncolytic Enzymes

The oncolytic enzymes majorly fall into two classes: (i) anti-neoplastic enzymes, which degrade small molecules like amino acids (e.g., asparaginase, glutaminase and arginine deiminase) which are required by neoplastic tissues and (ii) enzymes that degrade polysaccharides, nucleic acids, structural and functional proteins. Currently, the specific action of enzymes on tumor cells has been observed only in the former category [4]. Cancer therapy aims at targeting cancerous cells without affecting the normal cells in the body. This majorly includes the amino acid deprivation method where the tumor cells are starved for a particular amino acid for which they are auxotrophic. This affects the proliferation of cancerous cells leading to their death [1]. The function and clinical applications of asparaginase, glutaminase and arginine deiminase have been discussed in the following section.

### 10.3.1 L-Asparaginase (EC 3.5.1.1)

Asparaginase, an aminohydrolase, is an extensively used antineoplastic enzyme which carries out the asparagine hydrolysis to ammonia and aspartic acid. It has been extensively employed for treating acute lymphoblastic leukemia (ALL) for the past 30 years. Asparaginase has been shown to be the most successful therapeutic for ALL resulting in 60% incidence of complete remission of leukemia in about 6,000 cases. It was first isolated from guinea pig serum and was clinically tested in an 8-year-old boy suffering from multiple relapsed ALL where L-asparaginase activity resulted in a significant reduction of lymphomas. Currently, there are 3 forms of asparaginase viz., the native enzyme, PEGylated asparaginase from *E. coli* (PEGasparaginase), and an *Erwinia chrysanthemi* derived form (crisantaspase). Actinomycetes such as *Streptomyces canus*, *S. phaeochromogenes*, *S. exfoliates* and *S. cyaneus* have shown to be potential sources for the production of glutaminase free L-asparaginase as it showed better therapeutic properties for the treatment of ALL [3]. The bacterial derived asparaginase has been shown to possess the least toxicity compared to other enzymes having antitumor activity. *Bacillus aryabhattai* ITBHU02 is a potential source for L-asparaginase production. Other bacterial sources include *Serratia marcescens, Vibrio succinogenes*, and *Proteus vulgaris*. It has also been isolated from plant sources like *Withania somnifera* and *Pisum sativum* [1].

Asparagine-dependent leukemic cells need a large quantity of asparagine for their proliferation. The leukemic cells cannot synthesize asparagine as they lack the enzyme L-asparagine synthetase. On the other hand, the healthy cells can synthesize asparagine intracellularly with the aid of L-asparagine synthetase. Thus, administration of enzyme L-asparaginase starves the leukemic cells by depleting the circulating serum asparagine, making asparaginase an effective therapeutic enzyme for targeting asparagine-dependent cancers. [4].

The *E.coli* derived L-asparaginase has been tested in phase I and phase II clinical trials. The major problem associated with the use of this enzyme as a drug was the development of a hypersensitive response in certain patients. Thus, Enzon pharmaceuticals Inc., USA, made the PEGylated version of this enzyme in 1981 in order to prevent the immunogenic response and to enhance the stability and efficacy of this enzyme. The two commercially available forms of L-asparaginase are Elspar®(native *E. coli* asparaginase) and Oncaspar®(PEGasparagase). In 2006 FDA approved the use of Oncaspar®(Enzon Inc., USA) for treatment of ALL [1] whereas, in 2011, Erwinaze™(Jazz pharmaceuticals Inc., USA) from *Erwinia chrysanthemi* got FDA approval for the treatment of ALL patients.

### 10.3.2 L-Glutaminase (EC 3.5.1.2)

L-glutaminase, an amidohydrolase, carries out the L-glutamine hydrolysis into ammonia and glutamate. Cancer cells particularly the lymphatic tumor cells lack the ability to produce L-glutamine and need an ample quantity of L-glutamine for cell proliferation and growth. Therefore, L-glutaminase holds the potential to be used as a therapeutic enzyme for the treatment of cancer by depriving the tumor cells of glutamine. L-glutaminase has been shown to reduce the L-glutamine level in serum and tissues. The reduction in the serum level of L-glutamine was shown to decrease the serum reverse transcriptase activity substantially. Thus, L-glutaminase can also be used as an effective anti-retroviral agent against the human immunodeficiency virus (HIV) [21]. L-glutaminase and L-asparaginase are also used in combination to treat cancers, particularly ALL. A glutaminase-asparaginase complex (for combinatorial therapy) from *Achromobacter glutaminasificans* was chemically modified via succinylation, and the modified enzyme showed broader antitumor activity as compared to *E. coli* asparaginase [22].

### 10.3.3 Arginine Deiminase (ADI) (EC 3.5.3.6)

Arginine deiminase is an antineoplastic enzyme with therapeutic potential. It converts arginine, a non-essential amino acid, into citrulline. Arginine is produced in the cells via enzymes argininosuccinate lyase (ASL) and argininosuccinate synthetase (ASS). However, cancer cells like renal cell carcinomas, melanomas and hepatocellular carcinomas do not have the enzyme ASS and depend on the circulating arginine for cell growth and proliferation. Thus, use of arginine deiminase, deprives tumor cells from arginine thereby inhibiting tumor progression [23].

The PEGylated form of this enzyme, ADI-PEG 20 (Phoenix Pharmacologics Inc., Lexington, USA) has been generated to reduce its antigenicity and to increase its efficacy. This commercial form of ADI was generated by cloning the ADI gene of *Mycoplasma hominis* in *E. coli* [2]. ADI-PEG 20 was approved by European Agency for the Evaluation of Medicinal Products (EMEA) and FDA under the Orphan Drug Act for treating conditions like hepatocellular carcinoma and melanoma. Currently, ADI-PEG 20 is available with the trade name Hepacid (for HCC treatment) and Melanocid (for treatment of invasive melanoma). In an *in-vitro* experiment the $IC_{50}$ value for ADI-PEG-20 was determined, where its value ranged from 0.001 to 0.3 ng/ml in melanoma cell lines and 0.01 to 0.1 ng/ml in human hepatoma cell lines. *In-vivo* experiment was done in xenograft mice having human hepatocellular carcinoma. The mice were injected intramuscularly (for two weeks) with 5U/mouse of ADI-PEG-20, non-PEGylated ADI or vehicle. A significant reduction in tumor size was observed in treated animals compared to control mice. The survival rates were found to be more than 50% after 24 weeks in treated animals, whereas the control mice treated with either saline or non-PEGylated ADI did not survive after 7 weeks [24]. The phase I and II trials were also carried out where a single intramuscular dose of ADI-PEG-20 was given to 19 patients suffering from advanced or inoperable hepatocellular carcinoma. Different doses, namely 20, 40, 80 and 160 $U/m^2$ of ADI-PEG-20 were tested. The 160 $U/m^2$ dose resulted in complete depletion of plasma arginine from day 2 to 15. Doses 20 and 40 $U/m^2$ resulted in the elimination of 50 and 70% arginine, respectively. Normal levels of arginine were detected at day 15 [25, 26].

## 10.4 Antidiabetic Enzymes

### 10.4.1 Glucokinase (EC2.7.1.1)

The hexokinase family member, glucokinase (GK or GLK/ hexokinase IV or hexokinase D/ D-glucose-6-phosphotransferase), catalyses the first step in glucose metabolism i.e., phosphorylation of glucose to glucose-6-phosphate. It is expressed in liver parenchymal cells, pancreatic β-cells, brain and gastrointestinal tract. This enzyme is majorly responsible for maintaining the glucose homeostasis inside the body. In liver, the phosphorylation activity of glucokinase enhances the synthesis of glycogen whereas in pancreatic β-cells it causes insulin release. The importance of GK in maintaining glucose homeostasis has been demonstrated using mice as a model system. Mice which had less of pancreatic β-cells and thus less expression of GK die because of hyperglycemia, whereas mice which didn't have hepatic GK had reduced secretion of insulin. Moreover, an improvement in the tolerance of high glucose levels was reported when the GK was overexpressed in diabetic and non-diabetic mice. Another property of GK is that the half maximal activity of GK is achieved at the glucose concentration of 8.0 mM whereas the other three hexokinases get saturated at very low concentration of glucose (<1.0 mM). GK exhibits unique kinetic properties as it does not follow the normal Michaelis–Menten kinetics. It has a low affinity for its substrate glucose with a $K_{0.5}$ of ~7 mmol/L and is not saturated by its product glucose-6-phosphate. Though GK is a monomeric enzyme, it follows the sigmoidal saturation curve pattern for glucose with a Hill slope of ~1.7. These kinetic properties make glucokinase a very sensitive enzyme which can rapidly sense the fluctuation of glucose levels in blood and thus helps in maintaining the metabolic flux [27].

The problem of diabetes is increasing day by day and currently there are 371 million adults suffering from diabetes worldwide. Thus, small molecules which can activate glucokinase (glucokinase activators) can potentially be used as glucose-lowering agents. In 2003, the therapeutic use of the glucokinase activator (GKA), RO-28-1675, was first tested for diabetes treatment by activating glucokinase and increasing its affinity for glucose [28].

Since 2003, many GKA have been established and were tested in animal models suffering from type 2 diabetes. They have proven to be effective in maintaining the glucose levels in diabetic animals. It was also found that GK promotes the production of β-cells. However, the phase-II clinical trials of the GKAMK-0941 have shown that the administration of GKA resulted in hypoglycemic conditions in certain cases. Further, it lost its efficacy within several months of and resulted in development of dyslipidemia. To resolve these issues associated with clinical trials, studies are going on to develop a better knowledge over the role of GK in metabolic pathways [27].

## 10.5 Liver Enzymes

### 10.5.1 Superoxide Dismutase (SOD) (EC 1.15.1.1)

Superoxide dismutase (SOD) is a metal-containing enzyme which converts the reactive oxygen species (ROS) like the superoxide anion produced from oxygen into hydrogen peroxide and molecular oxygen. These ROS species can otherwise cause oxidative stress and cellular damage. In mammalian cells, the SOD exists in three different molecular forms. Two of them are intracellular forms: cytosolic CuZn-SOD and Mn-SOD, present in mitochondrial matrix. The third form is extracellular SOD i.e., EC-SOD, present between the interstitial spaces of tissues, plasma, lymph and synovial fluid [29].

During normal conditions, the levels of superoxide anions are tightly regulated via SOD. However, under inflammatory conditions, rapid production of superoxide anions takes place, resulting in peroxidation of lipids, inhibition of respiratory enzymes present in mitochondria, inhibition of membrane $Na^+/K^+$ ATPase activity, inactivation of membrane sodium channels and of enzyme glyceraldehyde-3-phosphate dehydrogenase and oxidative modifications in proteins.

Oxidative stress conditions are prevalent in certain pathological conditions like atherosclerosis, vascular remodelling and hypertension. SOD has been proven to be effective for controlling inflammatory diseases like osteoarthritis and rheumatoid arthritis. The EC-SOD has been specifically targeted to endothelial cells for the treatment of ischemic injury and atherosclerosis. The motor neuron diseases have been shown to be associated with loss of SOD [30].

The use of SOD for treating cancer patients was first suggested in 1998, as it can significantly eliminate the side effects associated with radiation and chemotherapy by acting as a scavenger of ROS. Thus, the antioxidant property of SOD makes it an enzyme of great therapeutic importance [1]. It can also minimize the organ injury associated with hemorrhagic shock.

Commercially, SOD was available as Orgotein (bovine CuZn-SOD). However, the administration of Orgotein resulted in the production of immune response in patients resulting in its withdrawal from the market [1]. In 1991, Oxsodrol® (Biotechnology General Corp., Israel), a recombinant human SOD, got registered for FDA approval under the orphan drug category for preventing the occurrence of bronchopulmonary dysplasia in premature neonates. It underwent the phase-II clinical trials but the drug failed to get FDA approval [14]. The synthesis of small synthetic molecule of SOD which can mimic SOD's function is a solution to combat its immunogenicity problems. Moreover, these small molecules possess advantages of tissue permeability, *in-vivo* stability and low cost.

For improving the pharmacokinetic properties of SOD, it has been conjugated to different polymers like PEG and amphiphilic polymers, like dextran, pluronic block copolymers, poly(2-oxazoline) block copolymers and poly(N-vinylpyrrolidone) [13].

### 10.5.2 Alkaline Phosphatase (ALP) (EC 3.1.3.1)

Alkaline phosphatase is an enzyme which carries out the hydrolysis of pyrophosphate (PPi) thus releasing inorganic monophosphate (Pi) as the product. The pyrophosphate moiety acts as an inhibitor of hydroxyapatite synthesis whereas monophosphate (Pi) acts as an activator. The hydrolytic activity of ALP is crucial for the synthesis of hydroxyapatite and thus in the regulation of bone development [31].

For medical purposes, the immobilization of ALP has been carried out on different matrices. ALP immobilized on poly (2-hydroxyethyl methacrylate), a hydrophilic chemical polymer, showed promising results in the treatment of hypophosphatasia, that is a rare metabolic bone illness which leads to intense skeletal hypomineralization [32].

Recently, the ALP has been immobilized on nanomatrix using covalent immobilization method. The nanomatrix containing microporous fibrin scaffolds (FSs) was used in developing nanodevice ALP which was found to be effective in bone repair [31]. Other nanodevices used to immobilize ALP include poly(styrene-acrylic-acid) nanospheres, magnetic nanoparticles and nanoporous nickel-titanium films for biosensing applications [32].

Chandrupatla *et al.* (2017) had demonstrated the novel role of ALP in treating arthritis. For this, arthritis was locally induced in the right knee of wistar rats which were then given 200 μg of placental recombinant ALP, subcutaneously. The treated rats showed reduction in knee swelling and a 3.5 fold decline in synovial infiltration of both ED1 and ED2 macrophages in the arthritis knee, as compared to the untreated rats. This study warrants further evaluation at preclinical and clinical levels [33].

Alkaline phosphatase also has a role in treating sepsis-associated acute kidney injury (AKI). Clinical trials were carried out in ICU patients where the effect of dephosphorylating enzyme ALP was demonstrated in improving the kidney function. Initially, ALP was used for the treatment of patients suffering from sepsis. The ALP works by dephosphorylating the endotoxin lipopolysaccharide (LPS) component lipid A which binds to the toll-like receptor-4 and plays a major role in sepsis. Another role of ALP is the dephosphorylation of ATP resulting in the formation of adenosine, which has been shown to protect the body from the tissue injury caused by inflammatory responses [34].

Recently, a chimeric alkaline phosphatase (ChimAP) has been generated for the treatment of acute kidney injury, inflammatory bowel disease and gut dysbiosis. ChimAP was produced by replacing the crown domain of intestinal alkaline phosphatase (IAP) with the crown domain of placental alkaline phosphatase. The molecule thus generated retained the structural folding of IAP with increased stability and transphosphorylation, high turnover number, $Zn^{2+}$ binding in active site, narrower substrate specificity and selectivity for bacterial LPS. In humans, the ALPs belong to the multi-gene family and are encoded by 4 loci. The *ALPL* loci encodes for alkaline phosphatase which is expressed in liver, bone and kidney and is referred to as tissue non-specific AP (TNAP). The mutation in the *ALPL* gene resulted in hypophosphatasia (HPP). The major cause for HPP is the inadequate hydrolysis of pyrophosphate (PPi) and pyridoxal 5'-phosphate (PLP, an isomer of vitamin B6) both of which are substrates for TNAP. Kiffer-Moriera *et al.* in collaboration with Alexion pharmaceuticals have developed the first therapeutic recombinant TNAP (Asfotase alpha) that has shown effective results in patients suffering from life-threatening HPP and has prevented seizures and skeletal manifestations associated with this disease. Asfotase alpha is now under clinical trial for use in adolescents and adults suffering from HPP [35].

## 10.6 Kidney Disorder

### 10.6.1 Uricase (EC 1.7.3.3)

Uricase or urate oxidase (urate oxygen oxidoreductase) is an important enzyme involved in nitrogen metabolism. It is a therapeutically relevant enzyme that can be used for the analysis of uric acid levels in human serum and urine. It also acts as a drug and prevents the accumulation of urate to toxic levels which can otherwise lead to kidney and cardiovascular disorders. It is produced by bacteria such as *Bacillus subtilis, Bacillus fastidious, Microbacterium sp., Pseudomonas aeruginosa, Arthrobacter globiformis,* and *Nocardia farcinica* [36].

Uricase has an immense therapeutic potential, as it is the only saviour drug for patients suffering from refractory gout. It is a distinctive enzyme that specifically acts on urate which is the end product of purine catabolism formed by the oxidation of xanthine by xanthine oxidase. Uricase catalyzes the oxidative hydrolysis of urate into 5-hydroxyisourate, and hydrogen peroxide. 5-hydroxyisourate further decomposes into carbon dioxide and allantoin, which has high plasma solubility and thus can be easily excreted out from the body [37].

Normally, urate is excreted via kidney. However, certain pathophysiological conditions leading to increased purine catabolism can cause decreased excretion of urate thus, resulting in its high level in plasma which ultimately causes hyperuricemia. Hyperuricemia results in the precipitation of urate to form crystals that are associated with diseases like chronic kidney disease (CKD), tumor lysis syndrome (TLS), gout, and various cardiovascular diseases. The drugs used for the treatment of hyperuricemia include allopurinol and febuxostat which inhibit xanthine oxidase activity. However, they are ineffective in the treatment of refractory gout. Probenecid is another anti-hyperuricemia drug but it is found to be toxic to kidney and liver [38].

Pegloticase is a commercially available recombinant mammalian uricase and is sold under the trade name Krystexxa® (Savient, East Brunswick, New Jersey, USA). It was modified using monomethoxyl-poly(ethylene glycol) (mPEG) moiety and was approved in 2010 by FDA for the treatment of refractory gout and hyperuricemia. This commercial formulation is a chimera of baboon and porcine liver uricase and includes 10 kDa PEG chains which are covalently attached via p-nitrophenol. It was found to be effective in the treatment of refractory gout, at a dose of 0.14 mg/kg body weight for three months. However, mPEG elicited an immune response in almost 20% of the patients. Rasburicase is another uricase formulation approved for the treatment of hyperuricemia and treatment of TLS. It is a recombinant fungal uricase from *Aspergillus flavus* expressed in *Saccharomyces cerevisiae.* It was approved in 2009 by FDA and is commercially available as Elitek/Fasturtec® (Sanofi-Synthelabo research, New York, USA). The currently available enzymatic formulations cannot be used for the treatment of chronic refractory gout due to their antigenicity. However, the use of pegloticase at a dose below 4μg/kg body weight showed promising effects. Thus, the future aspects of uricase therapy involve the usage of site-specific PEGylation and molecular engineering so that its immunogenicity can be reduced to minimal levels [38–41].

### 10.6.2 Urease (EC 3.5.1.5)

Urease is an enzyme which catalyses the hydrolytic conversion of urea to ammonia and carbamate that is further decomposed to yield ammonia and carbonic acid. The product of urea hydrolysis i.e., two molecules of ammonia and carbonic acid, remain in equilibrium with their protonated and deprotonated states resulting in an increase of the surrounding pH. The urease activity results in the modulation of pH both intracellularly and extracellularly as the ammonia rapidly diffuses out of the cells. In *Streptococcus thermophilus* the urease activity resulted in its higher catabolic efficiency by increasing the activity of lactate dehydrogenase, β-galactosidase, and glycolytic enzymes. Beside this, urease activity is mainly associated with nitrogen assimilation and recycling of nitrogenous waste in most of the microorganisms [42].

The gene encoding for urease is absent in humans. This enzyme is produced by the microorganisms colonizing the mouth and gut. The ammonia released from the hydrolysis of urea is a source of nitrogen for the microbes colonizing the human body. It has been shown that urea found in human milk can be hydrolyzed by the bacterium *Bifidobacterium longum* subsp. *infantis* present in the gastrointestinal tract (GIT) of infants. This results in the production of amino acids in the infant plasma and provides the essential nitrogen source required for the growth and metabolism of the infant and the bacterial population colonizing the GIT [43].

The urease positive bacterium *Streptococcus thermophilus* has been found in yogurt which is associated with several health benefits. Urea is also present in the oral cavity at a concentration ranging from 3–10 mM, and is a nitrogen source for oral microbiota like *Actinomyces naeslundii, S. vestibularis* and *S. salivarius.* The urease produced by these microbes generates alkaline conditions in dental biofilms and thus prevents the occurrence of dental caries [42, 44].

## 10.7 DNA- and RNA-Based Enzymes

The DNA- and RNA-based enzymes can be used to treat various diseases by blocking the function of a specific gene product. The DNAse enzyme isolated from *Bacillus* sp. and *Nocardia* sp. can act as a mucolytic agent and thus can be used for the treatment of chronic bronchitis. In patients with pleura empyema the breakdown of bacterial cells, phagocytes and intrapleural cells results in the production of enormous amount of DNA causing the production of viscous pus [3]. A DNAse obtained from *Streptococci* known as Streptodornase is used to decrease the viscosity of pus but it tends to elicit an allergic reaction. Streptodornase is one of the constituents of the commercial preparation Varidase® (Wyeth) which also contains streptokinase. It has been recommended for the treatment of wounds and ulcers. The combinatorial effect of both these enzymes has resulted in liquefaction of fibrous and infecting matters and dissolution of blood clots [45]. Schmidts *et al.* (2012) demonstrated that DNA enzymes can be effectively used for the treatment of inflammatory diseases like atopic dermatitis. A nano-sized drug carrier system was also developed for the topical application of DNA enzymes [46].

The emerging RNA-based drugs include ribozymes, aptamers and siRNAs. Hammerhead ribozymes can cause specific cleavage of phosphodiester bonds within a RNA molecule. Hence, they hold therapeutic potential to reduce the intracellular levels of mRNAs encoding for proteins, which plays a role in disease development. The specific inactivation of RNAs by ribozymes makes it an important tool for the treatment of various diseases. Ribozymes have been used to target various genes viz., oncogenes and drug resistance genes, viral diseases like hepatitis, mumps, AIDS and cellular diseases like rheumatoid arthritis, diabetes, and Alzheimer's [3]. However, it should be kept in mind that nucleic acids are antigenic molecules and thus require delivery tools which help prevent immune response and increase the efficacy of these molecules. The therapeutic potential of DNA- and RNA-based enzymes have been discussed in the following section.

### 10.7.1 Dornase

Cystic fibrosis (CF), a genetically inherited autosomal recessive disorder, occurs due to the dysfunctioning of ion channel cystic fibrosis transmembrane regulator (CFTR) protein which is responsible for the regulation of salt and water transport. The defects in this ion channel result in the production of a thick mucus along with the impairment of mucocilliary function which further leads to progressive pulmonary dysfunction ultimately causing respiratory failure. In CF patients, the breakdown of neutrophils results in the congestion of lungs due to the release of vast quantities of DNA resulting in an increased sputum viscosity and decreased ciliary transport [18].

Dornase alpha is an anti-inflammatory recombinant deoxyribonuclease enzyme which carries out the cleavage of extracellular DNA present in the airways of CF patients thereby, resulting in the clearance of mucus in the lungs. It is commercially accessible as Pulmozyme® (Genentech, South San Francisco, USA) and got approval by FDA as an orphan drug [47]. In a study, where dornase alpha was tested in patients suffering from CF, it was found that there was a reduction in the infection of the lower respiratory tract. In another study, the more direct action of dornase alpha was elucidated where it caused the release of antimicrobial peptides which got bound to the DNA or the actin bundles in the sputum. It also causes the degradation of the bacterial extracellular matrix DNA, a constituent of the pro-inflammatory response and is also responsible for the activation of neutrophils. Moreover, it resulted in reduced pulmonary tissue destruction [48].

### 10.7.2 Adenosine Deaminase

Adenosine deaminase carries out the irreversible deamination of adenosine into inosine and 2'-deoxyadenosine into deoxyinosine during purine metabolism. The deficiency of this enzyme can lead to the accumulation of purine nucleotides and other metabolites at toxic levels which can result in blockage of lymphocyte maturation and function. Severe combined immunodeficiency (SCID) is one such disease which occurs due to the deficiency of enzyme adenosine deaminase. It is an autosomal recessive trait characterized by dysfunctioning of T and B cells thus making SCID patients vulnerable to various infections due to the lack of defence mechanisms [1].

ADA deficiency results in increased levels of dATP, which causes the inhibition of ribonucleotide reductase involved in deoxyribonucleotide synthesis and thus obstructing DNA synthesis. Its deficiency also leads to higher levels of S-adinosylhomocysteine, which is toxic to immature lymphocytes. Thus, the enzyme replacement therapy employs the use of adenosine deaminase to combat SCID. ADA carries out the cleavage of excess adenosine levels, present in patients suffering from SCID. It is commercially available as Adagen® (pegademase bovine, Enzon pharmaceuticals Inc., USA) which contains the PEGylated ADA to increase its half-life and to reduce the immunogenicity of this enzyme. It was the first drug under the orphan drug act to receive the approval by FDA in 1990. The modified ADA can be safely used for up to 10 years with a weekly dose of about 20μg/Kg [2].

### 10.7.3 Ribonuclease

Ribonucleases (RNases) are a group of hydrolytic enzymes which carry out the cleavage of phosphodiester bonds in RNA. They have a substantial role in maintaining the cellular metabolism of RNAs, such as rRNA, mRNA and tRNA. The cytotoxicity of RNases is employed as a medical intervention for cancer treatment. There are two classes of ribonucleases i.e., endoribonuclease which acts on RNA endo-ribonucleolytically and can cleave the phosphodiester bond of single stranded or double stranded RNA and the exoribonuclease which acts on RNA exo-ribonucleolytically by cleaving the phosphodiester bonds at the end of the polynucleotide chain in a 3'-5' direction [49].

The use of RNases with other drugs has also been postulated as an effective combinatorial therapy for the treatment of cancer. The anticancer property of RNases has been widely studied using ribonucleolytic enzymes from animals which include bovine pancreatic RNase A, bovine seminal RNase (BS-RNase), onconase and angiogenin.

The BS-RNase from the seminal vesicle and testicles of *Bos taurus* exerts an apoptotic and anti-proliferative effect on cancer cells via the beclin-1 mediated autophagy. The BS-RNase is used in the treatment of thyroid cancer as it can cause the death of thyroid carcinoma cells. This secretory enzyme functions by destabilizing the bilayer membrane and causing the degradation of cellular RNA [50].

Onconase is a RNase from *Rana pipiens* (northern leopard frog) which has the potential to treat malignant mesothelioma. It catalyses the formation of siRNA and miRNA i.e., interfering RNAs, causing the degradation of tRNAs, thereby inhibiting protein synthesis and ultimately leading to apoptosis. It has also displayed anticancer activity against a number of cancers such as those of breast, colon, ovaries, pancreas and prostate. It has been tested at the clinical level where it caused problems such as muscle stiffness, suppression of lymphocyte proliferation, bone marrow toxicity and renal failure. Onconase has an apoptotic effect on lymphocytic leukemia by reducing the expression of NF-κB. In malignant mesothelioma models, it was shown that onconase affects the level of microRNA expression and reduces the oncogenic microRNAs [51].

Binase is a cationic RNase isolated from *Bacillus intermedius* and is the first bacterial RNase that has cytotoxic activity against malignant cells. It cleaves the RNA at purine residues without the aid of any co-factor. It also has antiviral activity against plant viruses, rabies virus and influenza strains. It exhibits antiproliferative and apoptotic activity against ovarian cancer cells. It also suppresses the activity of oncogenes like AML1-ETO, FLT3-ITD and KIT [52, 53].

Angiogenin is an RNase which exhibits anti-bacterial activity via the degradation of tRNA. It is found in *Homo sapiens*, *Mus musculus*, and *Ratus norvegicus*. RNase 7 is another RNase from *Homo sapiens* which has antimicrobial activity against *Pseudomonas aeruginosa*, *Enterococcus farcium* and *Pichia pastoris*. The RNases produced by certain mushrooms like *Ramaria formosa*, *Hygrophorus rusulla*and *Thelephora gabajun* have inhibitory activity against the reverse transcriptase of HIV-1. Granules of eosinophils also secrete two types of ribonuclease, Eosinophil Cationic Protein (ECP) and Eosinophil Derived Neurotoxin (EDN). These RNases can elicit the host defence mechanisms via the activation and maturation of leucocytes. The recombinant version of EDN, EDNsFV has cytotoxic effect against leukemic cells expressing transferrin receptor. Another recombinant version rhEDN causes significant reduction in the infectivity of respiratory syncytial virus [54].

## 10.8 Enzymes for the Treatment of Cardiovascular Disorders

The term hemostasis has a Greek origin which means "stoppage of bleeding". The hemostatic system of our body is responsible for preventing extensive blood loss in the event of any injury via the formation of a blood clot and suppresses its development during normal circulation [5]. It is composed of 3 elements: primary hemostasis or platelet aggregation, secondary hemostasis or blood coagulation and tertiary hemostasis or fibrinolysis [55]. A physiological balance exists in our body between blood coagulation and fibrinolysis. Any disruption in this balance may result in thrombosis or hemorrhage resulting from hyper-coagulation or hypo-coagulation of blood. Indeed, cardiovascular diseases have the highest annual global mortality rate of 17.7 million which may increase to 23.6 million deaths by 2030 [56]. The following section will discuss the different types of hemostasis and the enzymes involved as well as the disorders caused due to deficiencies in these enzymes and how they are currently being combated.

### 10.8.1 The Hemostatic System

As mentioned above, the hemostatic system of our body is composed of primary, secondary and tertiary hemostasis, which have been described in detail as follows:

#### *10.8.1.1 Primary Hemostasis*

It involves platelet aggregation in response to injuries to small blood vessels with the subsequent formation of the platelet plug. The first step of primary hemostasis is vasoconstriction or contraction of the local vascular smooth muscle. This is followed by interaction of the von Willebrand factor (vWF) and the exposed collagen in the subendothelial layer. Recruitment of platelets to the injury site is mediated via the binding of vWF with GP1bα subunit of the GP1b-V-IX complex located on the platelet surface [57]. Platelets attach to the subendothelium via the collagen receptors GPVI and α2β1 integrin and GPIIbIIIa [58]. Platelets are activated by binding to collagen or thrombin causing their aggregation. The GPIIbIIIa then alters its conformation and binds to fibrinogen which cross-links and stabilizes the platelet plug via interactions with GPIIbIIIa receptors present on the neighbouring platelets. The platelets release their contents such as adenosine diphosphate (ADP), α granules, dense granules and thromboxane A2 (TXA2). Both ADP and TXA2 promote platelet aggregation. The platelets then change shape and expose the negatively charged phospholipids to attract coagulation factors to the injury site to begin the secondary hemostasis or blood coagulation [59].

#### *10.8.1.2 Secondary Hemostasis*

In a continuation of the primary hemostatic mechanism, the secondary hemostasis involves blood coagulation to restrict bleeding from large wounds and results in the formation of blood clots (or thrombi) to strengthen the platelet plug. There are two pathways for blood coagulation: the extrinsic pathway and the intrinsic pathway. Several factors are involved in the blood clotting cascades that have been designated by Roman numerals according to their order of discovery. The extrinsic pathway requires the involvement of an external component called the tissue factor. Tissue Factor (TF, thromboplastin or factor III) is a 263 amino acid transmembrane glycoprotein composed of an extracellular N-terminus containing the FVII/FVIIa binding domain (219 residues), a transmembrane domain (23 residues) and an intracellular C-terminus (21 residues) [60]. TF initiates the extrinsic blood coagulation cascade by binding and activating FVII (stable factor) to FVIIa (a serine protease) resulting in the sequential activation of several zymogens. TF is constitutively produced by cells surrounding the blood vessels and comes in contact with the other factors of the clotting cascade only after disruption of the endothelial barrier [61]. The complex of FVIIa-TF activates FX (Stuart Prower factor) to FXa (a serine protease) which binds and activates FV to FVa (Labile factor). This prothrombinase complex (FXa-FVa) activates prothrombin (FII) to thrombin (FIIa, serine protease) which cleaves fibrinogen (FI) into fibrin (FIa) monomers that cross-link to form the blood clot [62].

In the intrinsic pathway, FXII (Hageman factor) undergoes contact activation via negatively charged surfaces like collagen to form the serine protease, FXIIa. FXIIa activates prekallikrein to kallikrein (Fletcher factor, a serine protease) which in turn activates more FXII resulting in a positive feedback loop [63]. Subsequently, a cascade of reactions continues to activate enzyme precursors FXI (plasma thromboplastin antecedent, a serine protease), FIX (Christmas factor, a serine protease), FVIII (antihemophilic factor) and FX.

The two pathways, i.e., intrinsic and extrinsic pathways, converge at FX activation and continue to form the prothrombinase complex for prothrombin activation to thrombin resulting in fibrin formation [64]. Thrombin activates platelets by binding a protease-activated receptor (PAR) and glycoproteins IIb/IIIa and GPIb [65]. Further, thrombin is also responsible for activating FXIII (fibrin stabilizing factor, a transglutaminase) which stabilizes the clot by cross-linking fibrin. Regulation of the clotting cascade occurs by the thrombin activatable fibrinolysis inhibitor (TAFI) that reduces fibrinolysis thereby favouring fibrin formation [66].

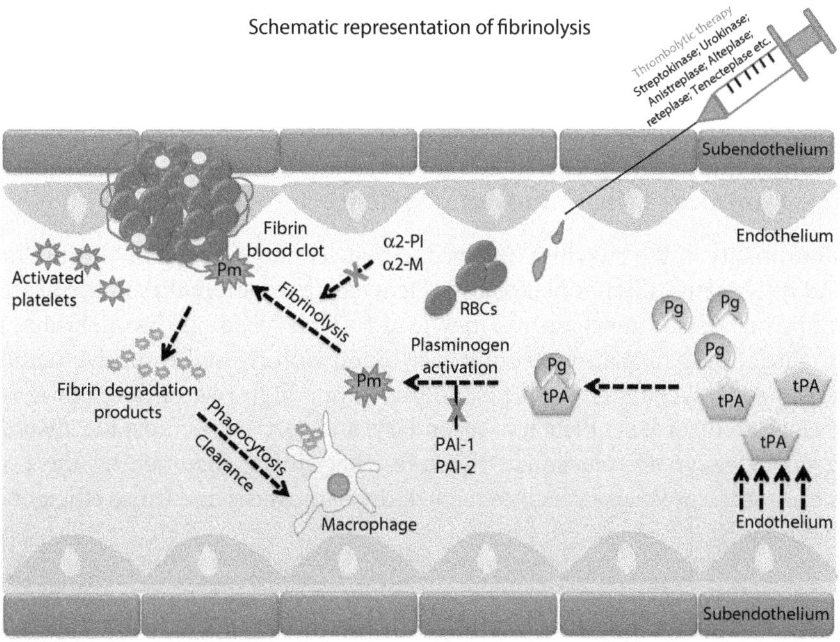


**Figure 10.1** Response to a vascular injury occurs via the formation of a platelet plug as part of primary hemostasis which involves the activation of platelets and their migration to the site of injury. Bleeding is further controlled via secondary hemostasis or blood clotting mechanisms leading to the formation of a fibrin mesh into which the RBCs, platelets and other blood cells get entrapped, forming a blood clot. The fibrin clot may be naturally dissolved by the action of the serine protease plasmin as part of tertiary hemostasis or fibrinolysis. In the absence of the clot, plasmin (Pm) exists as a zymogen known as plasminogen (Pg). Formation of a fibrin clot results in the activation of plasminogen to plasmin via the activity of plasminogen activators such as the naturally occurring tPA and uPA. Under pathological conditions, medical intervention may involve the administration of several other thrombolytic enzymes such as streptokinase, anistreplase, recombinant tPA (alteplase), reteplase, tenecteplase, etc., that activate plasminogen to plasmin which binds to the fibrin blood clot and dissolves it into fibrin degradation products that are removed from the circulation by phagocytosis. This relieves blockage from the vessel allowing the normal flow of blood and oxygen through the affected regions of the body. Regulation is achieved at the level of plasminogen activators via the activity of plasminogen activator inhibitors-1 and 2 (PAI-1, PAI-2) and at the level of plasmin via α2-PI (α2-antiplasmin) and α2-M (α2-macroglobulin).

#### 10.8.1.3 Tertiary Hemostasis

Fibrinolysis or thrombolysis (Figure 10.1) refers to the enzymatic digestion of the fibrin blood clot (or thrombus) so as to remove blockage and restore blood flow and oxygen supply through the blood vessels to the different organs of the body [64]. The enzyme plasmin, a serine protease, is responsible for the dissolution of the blood clot into fibrin degradation products which are eventually removed from circulation by phagocytosis. This enzyme remains in its inactive zymogen form, plasminogen, under normal conditions [67]. However, formation of a blood clot results in its activation to plasmin. Activation is brought about by any one of two naturally occurring plasminogen activators in the human body: (i) tissue plasminogen activator (tPA) produced by the vascular endothelium; (ii) urokinase (uPA) secreted by the kidney cells [5]. Plasminogen activators have now been discovered from other sources such as bacteria, snakes, earthworms etc., which shall be discussed subsequently. Several inhibitors are involved in the regulation of the fibrinolytic pathway such as plasminogen activator inhibitors-1 and 2 (PAI-1 and PAI-2) that act at the level of plasminogen activation inhibition and α2-antiplasmin (α2-PI) and α2-macroglobulin (α2-M) which are involved in plasmin inhibition [64].

### 10.8.2 Enzymes of the Hemostatic System

Any abnormality in the enzymes involved in platelet plug formation, thrombus formation and its lysis may disrupt hemostasis. Deficiencies or anomalies in the primary and secondary hemostatic mechanisms may lead to fatal bleeding disorders due to inefficient platelet plug formation or improper blood clotting and extensive hemorrhage. Similarly, people lacking efficient fibrinolysis have a high risk of developing ischemic cardiovascular disorders. Primary, secondary and tertiary hemostatic disorders are addressed by enzyme replacement therapies that aim at maintaining the respective enzyme at sufficient levels to maintain a healthy hemostasis and cause efficient clotting or thrombolysis.

#### 10.8.2.1 Clotting Enzymes

Coagulation disorders are anomalies in the body's ability to control blood clotting. The major clotting deficiencies lead to hemophilia and von Willebrand disease. However, 3–5% of all congenital clotting deficiencies are autosomal disorders and come under the category of rare bleeding disorders [12]. These include deficiencies of FXIII, FV, FVII, FII, fibrinogen, FX and FXI.

a. **von Willebrand Factor (vWF)**: It is a multimeric protein (500–20,000 kDa) produced by megakaryocytes and endothelial cells as a pre-pro-vWF. In the plasma, vWF is also responsible for binding and stabilizing the antihemophilic factor, FVIII [68]. von Willebrand disease (VWD) occurs due to a deficiency or a defect in vWF responsible for promoting platelet aggregation. Due to the high polymorphism in the vWF gene, the disease can be classified into 6 types which can result from qualitative or quantitative deficiency of pro-vWF [68, 69]. Type 1 results from a quantitative vWF deficiency and is an autosomal dominant trait. The disease presents with symptoms of bruising, wound bleeding, epistaxis as well as menorrhagia in females. Types 2A, 2B, 2M and 2N VWD are caused by qualitative defects of vWF and maybe autosomal dominant or recessive with type 2N presenting symptoms of mild hemophilia A. The most severe form of VWD is type 3 that stems from the complete deficiency of vWF combined with an associated FVIII deficiency and is an autosomal dominant trait. Symptoms include mucocutaneous hemorrhage and prolonged bleeds post-surgery [59]. The current clinical intervention aims at correcting the vWF deficiency and the concomitant FVIII deficit [70]. In 2015, the FDA approved the first recombinant vWF produced in CHO cells (Vonvendi®, Baxalta US Inc., CA, USA) to control bleeding in adults diagnosed with VWD [8]. vWF/FVIII concentrates may be administered under severe cases or under conditions where desmopressin, the first line of treatment is ineffective or contraindicated. Some of such commercially available complexes are Wilate® (Octapharma USA, Inc., USA), Humate-P® (CSL Behring, GmbH, Germany) and Alphanate® (Grifols Biologicals Inc., CA, USA) [71–73].

b. **Factor VIII (antihemophilic factor)**: The FVIII protein (2332 residues) has a multi-domain structure composed of A1-A2-B-A3-C1-C2. A part of the B domain is removed via proteolytic processing producing a heterodimer that remains complexed with vWF in circulation promoting its stability. FVIIIa, a heterotrimer possessing the structure A1/A2/A3-C1-C2 is produced via further proteolytic processing [74]. A congenital X-linked deficiency in FVIII is termed classical hemophilia or hemophilia A affecting 1 in 5,000 males [75]. The common symptoms include spontaneous and prolonged bleeding, bleeding into the joints, hematomas and bleeding of the oral cavity, as well as menorrhagia, heavy periods and post-partum haemorrhage in females [76]. The mainstay of hemophilia A therapy aims at maintaining FVIII at a sufficient level to prevent bleeding. This replacement therapy involves the administration of FVIII concentrate. Prophylactic treatment is the preferred mode in comparison with on-demand replacement therapy since it reduces the frequency of joint bleeding and chronic arthropathy [77–79]. The FDA-approved drug, Recombinate™ (recombinant antihemophilic factor) is manufactured using a Chinese Hamster Ovary (CHO) cell line for patients with hemophilia A (Baxter Healthcare Corporation, USA) [6]. In 2003, Advate® (recombinant antihemophilic FVIII, Plasma/Albumin-Free Method, Baxter Healthcare Corporation, USA) produced in the CHO cell line was FDA approved for use in patients with classical hemophilia. It provides the benefits of Recombinate™ with additional safety by surpassing the inclusion of human and animal plasma proteins and albumin to remove any risk of viral infections present in these proteins. Kogenate® FS is a sucrose formulated version of recombinant FVIII (rFVIII) [80]. Another FVIII product, BAY 81–8973 (Octocog Alfa;

Kovaltry®, Bayer HealthCare, USA) that has been approved for hemophilia A involves the introduction of the heat shock protein 70 into the Kogenate® FS producing cell line that allows efficient protein folding [81].

The relatively short half-life of the antihemophilic factor limits its efficient therapy owing to its recurrent administration to at least three times per week. Improvement in treatment strategy is based on the emergence of longer lasting replacements so as to increase the gap between such transfusions from days to weeks [82]. In 2014, the FDA approved Eloctate®(Biogen Inc., USA), which is a recombinant FVIII-Fc fusion protein, for use in children and adults suffering from hemophilia A. The active ingredient of this drug contains the human FVIII with its B-domain deleted that is fused to the IgG1 Fc domain sequence expressed in a human embryonic kidney (HEK) cell line. Fusion to the Fc domain enhanced its circulating half-life by 1.5 fold in comparison with Advate® [6]. Adynovate (BAX 855, Antihemophilic factor, recombinant, PEGylated, Baxalta US, Inc., CA, USA) is indicated for patients with congenital hemophilia A. It is a PEGylated form of the full length recombinant FVIII produced in CHO cells with an extended half-life of 14.3 h. PEGylation results in reduced binding to its clearance receptor, low density lipoprotein receptor related protein 1. Other PEGylated forms of the molecule such as BAY94–9027 (Bayer Pharma, Germany) and N8-GP (turoctocog alfa pegol, a glycoPEGylated FVIII, Novo Nordisk Inc., Denmark) are under clinical trials [6, 83].

c. **Factor IX (EC 3.4.21.22)**: This serine protease is composed of a γ-carboxyglutamic acid domain at the N-terminus, two epidermal growth factor-like domains with a serine protease catalytic region at the C-terminus [84]. Congenital X-linked deficiency in FIX is known as hemophilia B (Christmas disease). It occurs in 1 in 40,000 males [75]. The mild form of the disease manifests as bleeding after trauma or surgery, moderate hemophilia results in muscle and joint bleeding and excessive bleeding post-surgery while spontaneous bleeding is observed in severe cases. Enzyme replacement therapy using FIX concentrate has helped in improving the quality of patients' life. A long-acting FIX fusion protein, rIX-FP (Idelvion®, CSL Behring, GmbH, Germany), has been developed where recombinant albumin has been fused with recombinant FIX to provide a long-acting treatment option. An Fc fusion of FIX expressed in HEK cell line, Alprolix® (Biogen Inc., USA), was approved in 2014 for adults and children with hemophilia B [7]. A glycoPEGylated form of FIX, i.e., N9-GP (Novo Nordisk Inc., Denmark), expressed in CHO cells was developed to increase its half-life [85].
d. **Fibrinogen**: This protein is synthesized mainly by the hepatocytes and is encoded by 3 genes FGA, FGB and FGG forming the three fibrinogen chains Aα, Bβ and γ. Fibrinogen (FI) disorders may be classified as: (i) afibrinogenemia: absence of fibrinogen (autosomal recessive); (ii) hypofibrinogenemia: partial fibrinogen deficiency (autosomal dominant or recessive) or (iii) dysfibrinogenemia: abnormal functioning of fibrinogen

resulting in improper clotting (autosomal dominant). While the first two are quantitative disorders, dysfibrinogenemia is a qualitative defect [86]. Congenital fibrinogen disorders are rare and occur in 1 in 1,000,000 individuals due to mutations in any of the 3 fibrinogen genes FGA, FGB or FGG [87]. The symptoms commonly include umbilical stump bleeding, nose and mouth bleeds, musculoskeletal bleeds, menorrhagia, recurrent abortion with infrequent episodes of gastrointestinal, urinary tract and intracranial bleeding. Platelet adhesion and aggregation is also impaired along with delayed wound healing [88]. There is an increased risk of blood clot formation in the veins in dysfibrinogenemia. In 2009, the FDA approved RiaSTAP™, a human fibrinogen concentrate made from pooled human plasma (CSL Behring, GmbH, Germany), for patients with afibrinogenemia or hypofibrinogenemia but not for dysfibrinogenemia due to the increased risk of developing blood clots [89].

e. **Thrombin (EC 3.4.21.5)**: Congenital factor II deficiency may be of 2 types: (i) Type I: hypoprothrombinemia resulting from hypo-production of prothrombin or (ii) Type II: dysprothrombinemia caused due to an abnormality in prothrombin structure. Symptoms include joint bleeding, umbilical cord bleeding, bruising and nose-bleeds [90, 91]. Replacement therapy involves the use of fresh frozen plasma or prothrombin complex concentrates (PCCs).
f. **Factor V (proaccelerin)**: Congenital deficiency of FV is a rare disorder occurring in 1 per million individuals and is termed parahemophilia which is an autosomal recessive trait [92]. Patients present symptoms of prolonged epistaxis, easy bruising, prolonged bleeding after trauma, and menorrhagia. The disease can result from mutation in the FV gene. Deficiency of FV may also result from presence of FV inhibitors [93]. Acquired FV+FVIII deficiency results due to mutations in genes encoding for the proteins LMAN1 or MCFD2 involved in processing and transport of the two factors [94, 95]. Since factor V concentrate is not available, fresh frozen plasma (FFP) is the only replacement therapy available for FV deficiency. The goal is to reach FV levels of above 20% by giving an infusion of FFP at 15–20 ml/kg daily [94, 96].
g. **Factor VII (EC 3.4.21.21)**: Congenital FVII deficiency affects 1 in 300,000 to 500,000 individuals. It is an autosomal recessive trait presenting with symptoms of epistaxis and oral cavity bleeds, menorrhagia, hematomas and even central nervous system and gastrointestinal tract bleeding [97–99]. NovoSeven® (Novo Nordisk Inc., Denmark), a recombinant human FVIIa produced using a baby hamster kidney (BHK) cell line was FDA approved for treatment of congenital FVII deficiency and hemophilia A or B with inhibitors [10, 100].
h. **Factor X (EC 3.4.21.6)**: FX deficiency occurs in 1 in 1,000,000 individuals [101]. The mild form of the disease presents as nose or mouth bleeds, easy bruising and bleeding post-surgery. Severe FX deficiency is associated with umbilical cord bleeding, joint and muscle bleeds, gastrointestinal bleeding and a high risk of intracranial hemorrhage. Women may

experience menorrhagia, miscarriage and post-partum hemorrhage. Coagadex® (Bio Products Laboratory Limited, UK) was the first FX concentrate that has been approved by the FDA in 2015 for human use [9].

i. **Factor XI (EC 3.4.21.27)**: Hemophilia C (FXI deficiency) is an autosomal recessive or autosomal dominant trait also known as Rosenthal syndrome. It usually presents as a mild form of hemophilia showing minor clotting defects [90, 102, 103]. However, there is a poor correlation between FXI level and clinical bleeding since even a mild deficiency has been seen to cause extensive bleeding while a severe deficiency may not show extreme symptoms [104]. Under severe conditions, therapy is achieved by the use of the FXI concentrate, Hemoleven®(LFB, France) [105–109].

j. **Factor XII(EC 3.4.21.38)**: FXII deficiency (Hageman Factor Deficiency) is a rare genetic trait that is autosomal recessive. Humans deficient in FXII are asymptomatic and no excessive or abnormal bleeding is seen but they may experience poor wound healing due to a prolonged clotting period [110].

k. **Factor XIII(EC 2.3.2.13)**: Fibrin stabilizing factor (FXIII) is a tetrameric pro-transglutaminase composed of 2 active A-subunits (~83 kDa) produced in the bone marrow and 2 inhibitor/carrier B subunits (~80 kDa) synthesized by the hepatocytes. In the absence of clot formation, FXIII A-subunit remains complexed in the plasma while FXIII B-subunit freely circulates and protects FXIII A-subunit from activation and clearance [111, 112]. The enzyme plays a role in the final step of the clotting pathway and mechanically stabilizes the fibrin clot. Its deficiency is very rare, occurring in 1 per 4 million individuals in an autosomal recessive manner [113]. Patients can commonly experience umbilical cord bleeding, intracranial hemorrhage, nose and mouth bleeds, muscle bleeds and prolonged bleeding post-surgery. Women display symptoms of menorrhagia, heavy menstrual periods and repeated miscarriages [112, 114]. Replacement therapy using factor XIII is recommended for preventing hemorrhage under severe conditions. In 2011, the FDA approved Corifact™, a FXIII concentrate (CSL Behring, GmbH, Germany) for preventing bleeding episodes in adults and children with congenital FXIII deficiency [11]. In 2013, the FDA approved Tretten® (Novo Nordisk Inc., Denmark) came into use for routine prophylaxis in people with hereditary FXIII A-subunit deficiency [115].

#### *10.8.2.2 Thrombolytic Enzymes*

Anomalies in efficient fibrinolysis increase the possibility of developing ischemic cardiovascular events. Depending upon the organ affected they may manifest in the form of various thromboembolic disorders like pulmonary embolism, acute myocardial infarction (AMI) and deep vein thrombosis (DVT), etc. [5]. Thrombolytic therapy to remove the blood clots and restore blood and oxygen flow to the different organs constitutes the use of enzymatic drugs and has helped bypass the need for surgical intervention. Thrombolytic enzymes may belong to one of the two categories: plasminogen activators or plasmin-like enzymes [64]. Plasminogen activators act via limited proteolysis and activation of the zymogen plasminogen into plasmin. The plasmin-like thrombolytic enzymes, on the other hand, directly degrade fibrin without involving the intermediate step of plasminogen activation.

Plasminogen activators are divided into: (a) first generation plasminogen activators that do not show fibrin clot specificity and cause systemic plasminogen activation, e.g., Streptokinase (SK) and Urokinase (uPA); (b) second-generation plasminogen activators developed for targeted thrombolysis e.g. Anistreplase, Pro-urokinase, tissue plasminogen activator (tPA) and (c) third generation plasminogen activators possessing improved safety, efficacy, half-life, resistance to inhibitors and fibrin specificity e.g. Reteplase, Tenecteplase, Desmoteplase, Staphylokinase [116].

Several plasmin derivatives have also been created as direct-acting thrombolytic enzymes such as mini-plasmin, micro-plasmin and delta-plasmin. These enzymes carry modifications in the plasmin molecule and have been recombinantly produced in *E. coli* and *Pichia pastoris*[64]. A food-derived direct thrombolytic is Nattokinase from the Japanese dish natto (soybeans fermented with *Bacillus subtilis* var. *natto*) [117]. It has been shown to possess the highest fibrinolytic activity among all food-derived fibrinolytic agents and has been observed to fight the development of blood clots. Thrombinase from *Bacillus sphaericus* sero type H5a and 5b and subtilisin from *B. amyloliquefaciens* are other potential direct-acting fibrinolytic enzymes from microbial sources [118].

Animals and plants also produce fibrinolytic proteases. Snake venoms such as those from the Chinese green tea viper snake (*Trimeresurus stejnejeri*), Siberian pit viper (*Agkistrodon halys*) and bushmaster snake (*Lachesis muta muta*) are also excellent sources of proteases exhibiting fibrinolytic activity [119–121]. Fibrolase is a direct-acting metalloprotease thrombolytic agent derived from the copperhead snake (*Agkistrodon contortrix contortrix*) venom. Its derivative, Alfimeprase (Nuvelo), has been recombinantly produced for treatment of acute peripheral arterial occlusion [122]. Lumbrokinase is a fibrinolytic agent from the earthworm *Lumbricus rubellus* [123]. Several plant-based thrombolytics such as those from *Euphorbia hirta* (hirtin, a serine protease) and *Euphorbia milii var. hislopii* (Eumilin) have also been described in literature [64]. The current section will describe a few of these thrombolytic enzymes.

a. **Streptokinase (SK, EC 3.4.99.22)**: Streptokinase is a 47 kDa protein that is naturally secreted by several strains of β-hemolytic streptococci and exhibits an indirect fibrinolytic activity. Contrary to its name, streptokinase has no enzymatic activity by itself. Rather a 1:1 stoichiometric (activator) complex of streptokinase and plasmin (ogen) possesses serine protease activity [124]. It is a single chain polypeptide with a pI of 4.7 lacking any cystine, cysteine, phosphorus, conjugated carbohydrates and lipids. It has a longer half-life (18 min) in comparison with tPA. However, being of bacterial origin it can elicit immunogenic reactions in individuals limiting its administration in multiple doses. Despite its limitations, the protein remains the prime choice for thrombolytic therapy in developing countries due to its cost-effectiveness [5]. Naturally occurring streptokinase from β-hemolytic streptococci is available under the trade names of Streptase® (Hoechst

Marion Roussel, Frankfurt, Germany) and Kabikinase (Pharmacia Upjohn, Stockholm, Sweden). In India, biosimilars produced in *E. coli* are available under several trade names, e.g., Lupiflo™ (Lupin), Lifokinase™ (Shasun), Glanikinase™ (Gland Pharma), Icikinase™ (Abbott), Thrombosolv™ (VHB), etc. The drug is FDA approved for use in AMI, DVT, pulmonary embolism and arterial thrombosis [64].

A second-generation plasminogen activator, called anistreplase (anisoylated plasminogen-streptokinase activator complex, APSAC), has been prepared via the modification of streptokinase to increase its plasma half-life and fibrin binding. The active ingredient contains an equimolar complex of streptokinase and human lys-plasminogen (plasminogen with N-terminal lysine residue has a more open conformation thereby making it the preferred substrate for activation) with its active site acylated using p-amidinophenyl p-anisate hydrochloride (APAN) [125, 126]. It is marketed under the trade name Eminase® (Smithkline Beecham, UK) for AMI treatment. The acylation process aids in protecting the enzyme's active site against neutralization by plasmin inhibitors. Upon administration, the p-anisoyl group is lost via hydrolysis freeing the activator complex to activate other plasminogen molecules and degrade fibrin. This strategy helped in increasing its half-life to 90–105 minutes with improved fibrin binding and shorter duration of administration from 60 min to 2–5 minutes [125]. However, being a streptokinase derivative, the drug can cause systemic fibrinogenolysis [64].

b. **Urokinase (uPA, EC 3.4.21.73)**: It is a first-generation plasminogen activator that was originally isolated from the urine where it is synthesized by the kidney cells as a 53 kDa single chain pro-enzyme (pro-uPA, scu-PA). It is converted to the active two chain serine protease (tcu-PA) form via cleavage at the $Lys^{158}$-$Ile^{159}$ bond by plasmin or kallikrein [127]. There are 3 domains in the molecule: (a) a serine protease catalytic domain; (b) a kringle domain and (c) an epidermal growth factor-like domain. The kringle domain is responsible for its interaction with PAI-1 while the EGF-like domain allows specific interaction with the uPAR/CD87 receptor [128, 129]. A low molecular weight form (33 kDa) of uPA is produced by consecutive cleavage at the $Lys^{158}$-$Ile^{159}$ and $Lys^{135}$-$Lys^{136}$ bonds [127, 129]. This is the commercially available form sold under the trade names Abbokinase®(Abbott Laboratories, USA) and Kinlytic™ (ImaRx Therapeutics, Inc., Tucson Arizona, USA) produced using human neonatal kidney cells for treatment of pulmonary embolism [64]. The unglycosylated recombinant scu-PA, termed Saruplase (r-scuPA, pro-uPA), is being produced in *E. coli* as a second-generation thrombolytic since the molecule mediates clot specific lysis in the presence of fibrin unlike tcu-PA that does not exhibit fibrin specificity. It can be cleaved *in vivo* to the two chain form and has been found to be safe and effective when administered to patients with AMI [130]. The drug is being produced under the trade name Prolyse® (Abbot Laboratories) but has not yet been approved by the FDA [131].

c. **Tissue plasminogen activator (EC 3.4.21.68)**: tPA is produced from the vascular endothelial cells as a 70 kDa single chain polypeptide that undergoes plasmin-mediated cleavage across the $Arg^{275}$-$Ile^{276}$ bond to form the active two-chain molecule [129, 132, 133]. It is composed of 4 domains: (i) F-domain (fibronectin-like or finger domain), (ii) E-domain (epidermal growth factor-like domain), (iii) kringle domains (K1 and K2) and (iv) a serine protease catalytic domain [133, 134]. The K2 and F domains are responsible for its high fibrin affinity. The E and F domains as well as the carbohydrate side chains are responsible for its half-life of 6 minutes and clearance from the hepatocytes [132]. Further, the $Asn^{117}$ in the K1 domain is recognized by mannose receptors in hepatocytes leading to its rapid removal from circulation. tPA is more fibrin specific than the other plasminogen activators contributing to its popularity. Recombinant t-PA (alteplase) is produced using the CHO cell line and has been approved by the FDA for use during episodes of AMI, acute ischemic stroke and pulmonary embolism. In Europe, the drug is sold under the trade name of Actilyse® while in the US it is being marketed as Activase® (Genentech, Inc., CA, USA). Several t-PA variants such as reteplase, tenecteplase, duteplase, monteplase, lanoteplase, palmiteplase and amediplase have been generated as third-generation plasminogen activators that possess properties of enhanced plasma half-life and fibrin specificity [64]. Such modifications allow easy administration of these drugs owing to their slower clearance rates.

d. **Desmoteplase (Bat-PA, DSPAα1, v-PA)**: A third-generation plasminogen activator of animal origin, bat-PA has been produced from the vampire bat (*Desmodus rotundus*) saliva [135]. DSPAα1 (50 kDa) possesses a half-life of 2.8 h and has been recombinantly produced in insect and CHO cells [118]. It is composed of a finger, EGF and kringle domains and shows negligible activity in the absence of fibrin which is enhanced 12,900 fold in the presence of fibrin(ogen) [135, 136].

e. **Staphylokinase (SAK)(EC 3.4.99.22)**: Secreted from *Staphylococcus aureus*, SAK is an indirect third-generation plasminogen activator (15.5 kDa) that works by binding plasminogen in a 1:1 stoichiometry like streptokinase [117, 137]. The presence of a fibrin clot results in 4-fold enhancement of plasminogen activation by SAK and does not cause systemic fibrinogenolysis [138]. Its therapeutic potential has been shown in several clinical trials like STAR trial as well as the CAPTORS I and II trials [139–141].

## 10.9 Lysosomal Storage Disorders

### 10.9.1 α-Galactosidase A (EC 3.2.1.22)

Mutations within the GLA gene, coding for α-galactosidase A, result in a rare X-linked lysosomal storage disorder called Fabry disease. The enzyme is responsible for cleaving glycolipids by removing their terminal α-D galactosyl residues. A deficiency of this enzyme promotes the build-up of globotriaosylceramide (GL-3/ $Gb_3$) and glycosphingolipids in the body [142–144]. Symptoms of the disease include abdominal pain, gastrointestinal disorders, angiokeratoma, hypohidrosis and proteinuria. In adults, the disease progresses to renal dysfunction and other complications like stroke, cardiac fibrosis, arrhythmias and could also be fatal [145]. There are two recombinant drugs available for combating Fabry disease: agalsidase alfa (Replagal®, Transkaryotic Therapies, Cambridge) which is produced in human fibroblasts [143] and agalsidase beta (Fabrazyme®, Genzyme Corporation, Cambridge, MA, USA) which is produced in CHO cells [146] and has been approved for its use in Europe, United States and Japan [143]. The recombinant enzyme preparations have shown to reduce the levels of $Gb_3$ in tissues, decrease hypertrophy and also help in improving renal function [144]. These enzymatic therapies for Fabry disease have greatly improved the diseased condition and have also helped in reducing Fabry neuropathy in many patients, especially in younger ones [147].

### 10.9.2 Glucocerebrosidase (EC 3.2.1.45)

Gaucher disease, a rare lysosomal storage disorder, occurs due to dysfunctional metabolism of glycosphingolipids. The disease is a consequence of mutations in the GBA1 gene located on chromosome 1q21 which encodes for acid β-glucosidase enzyme (glucocerebrosidase). The consequences of deficiency of this lysosomal enzyme are glycosphingolipid and glycosylceramide accumulation within the cells of monocyte-macrophage lineage, thereby leading to a number of pathological conditions such as hepatosplenomegaly, i.e., enlargement of liver and spleen, anaemia, thrombocytopenia and other bone associated diseases [148, 149]. Currently, three types of Gaucher disease have been identified, the most common being the GD type 1 (GD1), which is prevalent in the western world. The current enzyme therapy available for treating GD1 involves the intravenous administration of the recombinant active form of β-glucosidase, which helps in relieving the visceral symptoms of the disease [149].

In early years, β-glucosidase enzyme was derived from the human placenta and administered as an unmodified form. However, this approach was found to be futile as the enzyme did not enter the target cells. A better uptake of β-glucosidase and consequent degradation of glycosphingolipids was achieved by targeting the mannose receptors located on the macrophages that enabled the efficient entry of exoglycosidases treated β-glucosidase enzyme having exposed mannose residues into the target cells [149–152]. In 1991, a clinical trial was conducted involving 12 patients suffering from type 1 Gaucher disease who were subjected to placenta-derived β-glucocerebrosidase, that had been modified using glycosidases to expose its mannose residues. It was observed that within a year of treatment, there was reduction in hepatosplenomegaly and a significant rise in haemoglobin and platelet count, which resulted in its approval for clinical use [153, 154]. The original extraction of β-glucosidase from the human placenta commenced at National Institutes of Health (NIH) laboratories by Brady and co-workers, after which the large-scale enzyme production was undertaken by Genzyme Corporation, USA, and was available under the trade name of Ceredase® (alglucerase). However, the usage of Ceredase® ceased due to the reduced availability of human placentae and the remote possibility of spread of infectious agents [150]. Therefore, the mass production of recombinant β-glucosidase was done in different expression systems and is available commercially.

Imiglucerase sold under the trade name of Cerezyme® by Genzyme Corporation, USA, is produced in CHO cells that has been licensed both in the United States (1994) and Europe (1998). Another mammalian-cell based recombinant β-glucosidase has been developed by Shire Pharmaceuticals Inc., Dublin, known as velaglucerase alfa (VPRIV®, Shire HGT), which is produced in human fibrosarcoma cell line. Recently, a novel approach for the recombinant β-glucosidase enzyme production was devised involving a plant-based system, in which taliglucerase alfa (Uplyso®/Elelyso®, Pfizer, USA/Protalix Biotherapeutics, Israel) is produced in carrot cells [149, 155, 156].

### 10.9.3 Acid Alpha-Glucosidase (GAA) (EC 3.2.1.20)

Acid maltase or acid alpha-glucosidase (GAA) is essential for the cleavage of α-1,4 and α-1,6 linkages of glycogen present within the cells of the human body to glucose [157]. Absence of GAA or its inefficient activity causes glycogen accumulation in cells thereby leading to glycogen storage disease type 2 or Pompe disease. It is a rare, autosomal recessive muscle disorder, in which glycogen accumulates within the lysosomes and results in irreparable damage to the muscle cells [158, 159].

In 2004, the therapy, which involved the use of recombinant human GAA from rabbit milk, was proposed for patients suffering from Pompe disease [159]. Genzyme Corporation, USA, has been credited for the production of the human recombinant GAA (rhGAA) in CHO cells that is marketed under the trade name of Myozyme®/Lumizyme®. The recombinant GAA enzyme therapy in patients suffering from Pompe disease showed a marked improvement in the functioning of several organs such as liver and heart. In these organs, glycogen was found to be cleared efficiently, which aided in the improvement of motor skills and reduced cardiac hypertrophy. This enzymatic therapy has been approved for clinical use in many countries including Canada and the United States as well as Europe in 2006 [160].

### 10.9.4 α-L-iduronidase (Laronidase) (EC 3.2.1.76)

Mucopolysaccharidosis I (MPS I) are a set of inherited diseases that belong to the group of lysosomal storage disorders [161]. This autosomal recessive disorder is a consequence of mutations in the α-L-iduronidase gene located on chromosome 4p16.3. The lysosomal enzyme, α-L-iduronidase plays a critical role in the breakdown of glycosaminoglycans (GAGs), which otherwise may accumulate in the cells and tissues thereby leading to a number of pathological conditions such as joint problems, cardiovascular diseases, pulmonary disorders and ophthalmic diseases such as retina damage. Other manifestations of the disease include mental retardation, hepatosplenomegaly, cognitive impairment, cervical myelopathy and naval herniation [162, 163].

GAGs are mucopolysaccharides which are required for structural support and cellular communication. They are the critical component of the connective tissues which are present in the form of large complexes known as proteoglycans. They basically comprise of sugars and sulphated residues of uronic acid and hexosamine associated with specific proteins. Thus, for the degradation of such large macromolecules, the lysosomal enzyme α-L-iduronidase is required, which cleaves dermatan sulphate and heparan sulphate and removes their terminal α-L-iduronic acid residues [164, 165]. This mode of treatment has been established with a view to reduce the accumulation of GAGs in tissues and thereby prevent the progression of the disease. The CHO cells derived recombinant form of human α-L-iduronidase has been developed by BioMarin Pharmaceutical Inc., USA, and commercialized by Genzyme Corporation, USA, and is available under the trade name of Aldurazyme® (laronidase) [166, 167]. The approval for the use of laronidase for therapy in Europe and the United States was given in 2003. It has been shown that laronidase supplementation has helped in improving the respiratory conditions, organomegaly, walking efficiency and a number of other pathological conditions associated with MPS 1 [167, 168].

## 10.10 Miscellaneous Enzymes

### 10.10.1 Phenylalanine Hydroxylase (EC 1.14.16.1)

Phenylalanine hydroxylase (PAH) is chiefly a hepatic enzyme that catalyses the hydroxylation of L-phenylalanine to L-tyrosine, an irreversible reaction that occurs in the presence of the co-factor tetrahydrobiopterin (BH4) in the liver [169, 170]. This is a critical step as it helps in the removal of the extra phenylalanine obtained from diet [171]. Around 500 mutations have been detected in the PAH gene which render the enzyme ineffective. Hence, deficiency of phenylalanine hydroxylase leads to phenylalanine metabolic disorder known as phenylketonuria (PKU) and its milder version collectively referred to as hyperphenylalaninemias (HPA) [171, 172]. The high-level accumulation of L-phenylalanine in blood and tissue leads to a number of neurological conditions such as seizures, intellectual impairment, microcephaly and mental retardation [170, 172]. Apart from a controlled diet, enzyme replacement therapy and liver transplantation are some of the promising approaches which can aid in the treatment of patients suffering from PKU. However, liver transplantation itself is plagued with a number of issues such as risks associated with surgery and the use of immunosuppressant drugs [172].

Enzyme therapy can be done using phenylalanine hydroxylase (PAH) enzyme, which acts as a substituent for the defective PAH enzyme. However, the large-scale production and subsequent purification of PAH is a challenging task as the enzyme is unstable. In addition, its sensitivity to protease, co-factor requirement and immunogenic nature further contributes to its limited use. A number of modifications of PAH have been attempted in order to make it therapeutically viable [171]. These include the development of truncated versions of PAH with the aim of improving its stability and catalytic hydroxylating activity [173]. Gamez *et al.* [174] attempted PEGylation in order to improve its stability and immunogenicity. However, for this system to be efficient, it is necessary to supplement BH4, the co-factor required for the activity of PAH.

Recently, Ramirez *et al.* proposed the use of probiotic Lactic Acid Bacteria (LAB) as a vehicle for the expression of genetically engineered human PAH into the gut (*in situ*). This can confer a number of advantages as now the intravenous administration of

PAH will not be required and the enzyme would also be protected from degradation [170].

Phenylalanine lyase (PAL, EC 4.3.1.24) is a viable substitute for PAH enzyme that catalyses the non-oxidative deamination of L-phenylalanine, thereby forming the products *trans*-cinnamic acid and ammonia [171]. It is an autocatalytic enzyme and does not require any co-factor for the degradation and subsequent removal of L-phenylalanine [171, 175]. The degradation product, *trans*-cinnamic acid, possessing a very low level of toxicity is further degraded to benzoic acid in the liver and finally excreted from the body through urine as hippurate [175]. However, it was observed that intravenous administration of PAL elicits an unfavourable immune response and it was more prone to proteolytic degradation when given orally. The concept of PEGylation was applied for the development of *Rhodotorula glutinis* PAL based therapy in order to minimize the immune reactions. When the half-lives of native PAL and linear PEGylated PAL after the first injection were compared, it was found that the half-life of PEGylated PAL was approximately three times more as compared to that of its native form. Nevertheless, these PEGylated forms of the enzyme also require additional modifications such as enteric coating or silk fibroin entrapment before their use as oral supplements [175, 176].

### 10.10.2 Collagenase (EC 3.4.24.3)

Collagenase is a metalloproteinase enzyme that catalyses the breakdown of collagen which may help in debridement, a process that involves the removal of debris accumulated in wounds, ulcers and burns. Collagenase also finds application in enzymatic fasciotomy and in the treatment of partial-layer burn wounds. Ointments whose major component is collagenase are found to be more efficient as compared to those which are based on petrolatum. One of the major advantage of collagenase is that it specifically targets collagen while the other enzymes are non-specific [3, 177].

Collagenase from *Clostridium histolyticum* has been approved by the FDA as a medical intervention for Dupuytren's contracture, a condition which results from an abnormal accumulation of collagen in the fingers and palm, thereby hindering their normal functioning. Several *in vitro* studies highlight the ability of the *Clostridium* collagenase to catalyse the breakdown of the extracellular matrix of the cords found in Dupuytren's disease, without causing cell toxicity or damaging other tissue components. The drug is available under the trade name of Xiaflex® (Auxilium Pharmaceuticals, USA) and is also recommended for the treatment of Peyronie's disease in males, which is a type of connective tissue disorder [3, 177]. Santyl®(Advance Biofactures Corporation, USA) is another commercially available preparation of collagenase and other non-specific proteases that is derived from *Clostridium.* It has been approved for clinical use by the FDA and is being used as a debridement agent [178]. The abnormal deposition of collagen in eye channels through which fluid drains causes their closure, thereby leading to glaucoma, one of the leading causes of permanent blindness. Collagenase-based treatment after glaucoma filtration surgery prevents fibrosis and intraocular scarring. The deposits of collagen tend to minimize the vertebral spine distances and hence induce lower back pain. Collagenase injections may help in reducing such aggregations and thereby ameliorate the conditions of patients suffering from disc herniation. In addition, it has also been implicated in the treatment of cellulite and uterine fibroid [177].

### 10.10.3 Hyaluronidase

Hyaluronic acid (HA) is a polymer of N-acetyl glucosamine and D-glucuronic acid found in the extracellular matrix, joints and loose connective tissues [179, 180]. The excessive accumulation of HA causes the formation of a gel-like extracellular matrix which does not allow the penetrance of fluids [180]. Overexpression of HA is also linked with certain solid tumours such as breast cancer where the periphery of the tumour is surrounded by stroma deposited with HA [179]. The degradation of HA is facilitated by a glycosidase, known as hyaluronidase. It possesses hydrolytic as well as transglycosidase activities and hence is responsible for the breakdown of HA, thereby leading to an increase in the permeability of the tissues [3]. As a substitute or an adjunct to the conventional vitrectomy, the hyaluronidase enzyme can be used to treat vitreous haemorrhage. It is used as a drug adjunct for efficient delivery and absorption of intramuscular or subcutaneous injections with reduced discomfort [3]. The recombinant human hyaluronidase (PH20) developed by Halozyme Therapeutics, USA gained FDA approval in 2005 and is available under the trade name of Hylenex® [180].

### 10.10.4 Bromelain

One of the members of the Bromeliaceae family, *Ananas comosus,* commonly known as the pineapple plant, contains several endopeptidases along with an incompletely characterized fraction containing carbohydrates, glycoproteins, phosphatases and cellulases. The complex extract, i.e., bromelain also contains a number of protease inhibitors and can be divided into two groups: stem and fruit derived bromelain with Enzyme Commission Nos. EC 3.4.22.32 and EC 3.4.22.33 respectively [181]. Bromelain has anti-inflammatory properties and may facilitate the regulation of the immune system. Furthermore, it can be used as a skin debridement agent and may also act as a prophylactic drug for prevention of metastasis. Bromelain is responsible for enhancing the permeability of the tissues by fibrinolysis and aids in reabsorption of the accumulated fluid in cases of edema [181]. A majority of the cardiovascular diseases can be treated efficiently using bromelain as the therapeutic agent since it inhibits the aggregation of blood platelets and thereby reduces the chances of embolism and arterial thrombosis [182]. It has also been reported that the process of debridement may be enhanced by the application of bromelain based creams. Bromelain is sold as a food supplement in Europe and the United States under the trade name of Bromelain-POS® developed by Ursapharm GmbH, Saarbrucken, Germany [182, 183].

## 10.11 Conclusions

Enzymes have an enormous potential to be used as therapeutic agents. Since the introduction of the first genetically engineered enzyme, alteplase, tremendous advancements have been made in the field of biotherapeutic enzymes. They now find applications in the fields of cancer therapy, thrombolytic therapy, clotting disorders and other genetic disorders like SCID and lysosomal storage diseases. Their use as debriding agents, mucolytics, anti-inflammatory agents and digestive aids has helped in improving the living standards of several patients. Extensive research has been conducted to develop technologies such as polymer-enzyme conjugates (eg. PEGylation), microencapsulation and antigenic region masking to improve the safety and potency of enzyme-based drugs. These advanced methods allow the pharmaceutical companies to manufacture safe and better enzyme-based therapies of future generation.

## References

1. S. S. Kumar and S. Abdulhameed, in *Bioresources and Bioprocess in Biotechnology*, Springer, Vol. 2, p. 45, 2017.
2. F. W. Holtsberg, C. M. Ensor, M. R. Steiner, J. S. Bomalaski, and M. A. Clark, *Journal of controlled release*, Vol. 80, p. 259, 2002.
3. T. UmaMaheswari, T. Hemalatha, P. Sankaranarayanan, and R. Puvanakrishnan, *Indian Journal of Experimental Biology*, Vol. 54, p. 7, 2016.
4. A. Sabu, *Indian Journal of Biotechnology*, Vol. 2, p. 334, 2003.
5. A. Banerjee, Y. Chisti, and U. C. Banerjee, *Biotechnology Advances*, Vol. 22, p. 287, 2004.
6. K. Lieuw, *Journal of Blood Medicine*, Vol. 8, p. 67, 2017.
7. J. M. Ducore, M. G. Miguelino, and J. S. Powell, *Expert Review of Hematology*, Vol. 7, p. 559, 2014.
8. M. Franchini and P. M. Mannucci, *Expert Review of Hematology*, Vol. 9, p. 825, 2016.
9. A. Shapiro, *Expert Opinion on Drug Metabolism and Toxicology*, Vol. 13, p. 97, 2017.
10. U. Hedner and C. A. Lee, *Haemophilia*, Vol. 17, p. e172, 2011.
11. D. Nugent, *Thrombosis Research*, Vol. 130 Suppl 2, p. S12, 2012.
12. F. Peyvandi, P. H. Bolton-Maggs, A. Batorova, and P. De Moerloose, *Haemophilia*, Vol. 18 Suppl 4, p. 148, 2012.
13. A. Scomparin, H. F. Florindo, G. Tiram, E. L. Ferguson, and R. Satchi-Fainaro, *Advanced Drug Delivery Reviews*, Vol. 118, p. 52, 2017.
14. M. Vellard, *Current Opinion in Biotechnology*, Vol. 14, p. 444, 2003.
15. M. J. White, M. Glenn, and R. H. Gomer, *PLoS One*, Vol. 8, p. e70795, 2013.
16. J. Beard, *British Medical Journal*, Vol. 1, p. 140, 1906.
17. J. F. NOVAK and F. TRNKA, *Anticancer Research*, Vol. 25, p. 1157, 2005.
18. S. A. Pawar, V. B. Pokharkar, and N. R. Solanki, *Pharma Science Monitor*, Vol. 4, p. 2013.
19. N. Gurung, S. Ray, S. Bose, and V. Rai, *BioMed Research International*, Vol. 2013, p. 2013.
20. Enzymedica, https://enzymedica.com/products/digest-gold-enzymes-digestive- enzyme.
21. T. A. Elzainy and T. H. Ali, *Journal of Applied Sciences*, Vol. 6, p. 1389, 2006.
22. K. Sarada, *Asia Pacific Journal of Research*, Vol. 1, p. 1, 2013.
23. R. H. Kim, J. M. Coates, T. L. Bowles, G. P. McNerney, J. Sutcliffe, J. U. Jung, R. Gandour-Edwards, F. Y. Chuang, R. J. Bold, and H.-J. Kung, *Cancer Research*, Vol. 69, p. 700, 2009.
24. C. M. Ensor, F. W. Holtsberg, J. S. Bomalaski, and M. A. Clark, *Cancer Research*, Vol. 62, p. 5443, 2002.
25. P. A. Ascierto, S. Scala, G. Castello, A. Daponte, E. Simeone, A. Ottaiano, G. Beneduce, V. De Rosa, F. Izzo, and M. T. Melucci, *Journal of Clinical Oncology*, Vol. 23, p. 7660, 2005.
26. F. Izzo, P. Marra, G. Beneduce, G. Castello, P. Vallone, V. De Rosa, F. Cremona, C. M. Ensor, F. W. Holtsberg, and J. S. Bomalaski, *Journal of Clinical Oncology*, Vol. 22, p. 1815, 2004.
27. A. Nakamura and Y. Terauchi, *Journal of diabetes investigation*, Vol. 6, p. 124, 2015.

28. J. Grimsby, R. Sarabu, W. L. Corbett, N.-E. Haynes, F. T. Bizzarro, J. W. Coffey, K. R. Guertin, D. W. Hilliard, R. F. Kester, and P. E. Mahaney, *Science,* Vol. 301, p. 370, 2003.
29. I. N. Zelko, T. J. Mariani, and R. J. Folz, *Free Radical Biology and Medicine,* Vol. 33, p. 337, 2002.
30. S. Cuzzocrea, C. Thiemermann, and D. Salvemini, *Current Medicinal Chemistry,* Vol. 11, p. 1147, 2004.
31. T. Osathanon, C. M. Giachelli, and M. J. Somerman, *Biomaterials,* Vol. 30, p. 4513, 2009.
32. V. E. Bosio, G. A. Islan, Y. N. Martínez, N. Durán, and G. R. Castro, *Critical Reviews in Biotechnology,* Vol. 36, p. 447, 2016.
33. D. Chandrupatla, C. Molthoff, E. Elshof, W. Ritsema, M. Verlaan, R. Vos, A. Hammond, A. Lammertsma, C. van der Laken, and R. Brands, *BMJ Publishing Group Ltd,* Vol. 76, p. 1075, 2017.
34. E. Peters, A. van Elsas, S. Heemskerk, L. Jonk, J. van der Hoeven, J. Arend, R. Masereeuw, and P. Pickkers, *Journal of Pharmacology and Experimental Therapeutics,* Vol. 344, p. 2, 2013.
35. T. Kiffer-Moreira, C. R. Sheen, K. C. da Silva Gasque, M. Bolean, P. Ciancaglini, A. van Elsas, M. F. Hoylaerts, and J. L. Millán, *PLoS One,* Vol. 9, p. e89374, 2014.
36. T. Poovizh, P. Gajalakshmi, and S. Jayalakshmi, *International Journal of Advanced Research,* Vol. 2, p. 34, 2014.
37. I. Ramazzina, C. Folli, A. Secchi, R. Berni, and R. Percudani, *Nature Chemical Biology,* Vol. 2, p. 144, 2006.
38. X. Yang, Y. Yuan, C. G. Zhan, and F. Liao, *Drug Development Research,* Vol. 73, p. 66, 2012.
39. N. Schlesinger, U. Yasothan, and P. Kirkpatrick, *Nature Reviews Drug Discovery,* Vol. 10, p. 17, 2011.
40. C. S. Yue, W. Huang, M. Alton, A. N. Maroli, R. W. Waltrip, D. Wright, and M. Marco, *The Journal of Clinical Pharmacology,* Vol. 48, p. 708, 2008.
41. J. S. Sundy, N. J. Ganson, S. J. Kelly, E. L. Scarlett, C. D. Rehrig, W. Huang, and M. S. Hershfield, *Arthritis & Rheumatology,* Vol. 56, p. 1021, 2007.
42. D. Mora and S. Arioli, *PLoS pathogens,* Vol. 10, p. e1004472, 2014.
43. R. G. LoCascio, P. Desai, D. A. Sela, B. Weimer, and D. A. Mills, *Applied and Environmental Microbiology,* Vol. 76, p. 7373, 2010.
44. Y. Zhu, H. Wang, J. H. Hollis, and P. F. Jacques, *European Journal of Nutrition,* Vol. 54, p. 543, 2015.
45. G. Simpson, D. Roomes, and M. Heron, *CHEST Journal,* Vol. 117, p. 1728, 2000.
46. T. Schmidts, K. Marquardt, P. Schlupp, D. Dobler, F. Heinz, U. Mäder, H. Garn, H. Renz, J. Zeitvogel, and T. Werfel, *International Journal of Pharmaceutics,* Vol. 431, p. 61, 2012.
47. A. P. Jones and C. Wallis, *Cochrane Database of Systematic Reviews,* Vol. 3, p. 2010.
48. M. W. Konstan and F. Ratjen, *Journal of Cystic Fibrosis,* Vol. 11, p. 78, 2012.
49. G. Shruti, S. Sukhdev, and K. S. Singh, *Insight,* Vol. 1, p. 1, 2016.
50. D. Spalletti-Cernia, R. Sorrentino, S. Di Gaetano, A. Arciello, C. Garbi, R. Piccoli, G. D'alessio, G. Vecchio, P. Laccetti, and M. Santoro, *The Journal of Clinical Endocrinology & Metabolism,* Vol. 88, p. 2900, 2003.
51. M. Zwolińska and P. Smolewski, *Postepy Hig Med Dosw (Online),* Vol. 64, p. 58, 2010.
52. A. R. Garipov, A. A. Nesmelov, H. A. Cabrera-Fuentes, and O. N. Ilinskaya, *Toxicon,* Vol. 92, p. 54, 2014.
53. R. Shah Mahmud and O. Ilinskaya, *Acta Naturae,* Vol. 5, p. 49, 2013.
54. S. Li and G.-f. Hu, *International Journal of Biochemistry and Molecular Biology,* Vol. 1, p. 26, 2010.
55. J. M. Stassen, J. Arnout, and H. Deckmyn, *Current Medical Chemistry,* Vol. 11, p. 2245, 2004.

56. World Health Organization, *Global atlas on cardiovascular disease prevention and control,* 2014.
57. M. R. de Queiroz, B. B. de Sousa, D. F. da Cunha Pereira, C. C. N. Mamede, M. S. Matias, N. C. G. de Morais, J. de Oliveira Costa, and F. de Oliveira, *Toxicon,* Vol. 133, p. 33, 2017.
58. M. Holinstat, *Cancer and Metastasis Reviews,* Vol. 36(2), p. 195, 2017.
59. C. H. van Ommen and M. Peters, *Europian Journal of Pediatrics,* Vol. 171, p. 1, 2012.
60. A. J. Chu, *Archieves of Biochemistry and Biophysics,* Vol. 440, p. 123, 2005.
61. N. Mackman, *Anesthesia and Analgesia,* Vol. 108, p. 1447, 2009.
62. S. A. Smith, R. J. Travers, and J. H. Morrissey, *Critical Reviews in Biochemistry and Molecular Biology,* Vol. 50, p. 326, 2015.
63. F. Muller, D. Gailani, and T. Renne, *Current Opinion in Hematology,* Vol. 18, p. 349, 2011.
64. Adivitiya and Y. P. Khasa, *Bioengineered,* Vol. 8, p. 331, 2017.
65. A. J. Chu, *International Journal of Inflammation,* Vol. 2011, p. 367284, 2011.
66. B. N. Bouma and J. C. Meijers, *Seminars in Hematology,* Vol. 41, p. 13, 2004.
67. F. J. Castellino and V. A. Ploplis, *Journal of Thrombosis and Haemostasis,* Vol. 93, p. 647, 2005.
68. A. Sabih and H. M. Babiker, *von Willebrand Disease*, StatPearls Publishing, 2017.
69. M. Bowman, W. M. Hopman, D. Rapson, D. Lillicrap, M. Silva, and P. James, *Pediatric Blood and Cancer,* Vol. 55, p. 171, 2010.
70. G. Castaman, A. Goodeve, and J. Eikenboom, *Haematologica,* Vol. 98, p. 667, 2013.
71. P. Batty, Y. H. Chen, L. Bowles, D. P. Hart, S. Platton, and K. J. Pasi, *Haemophilia,* Vol. 20, p. 846, 2014.
72. P. Kouides, K. Wawra-Hehenberger, A. Sajan, H. Mead, and T. Simon, *Transfusion,* Vol. 57, p. 2390, 2017.
73. V. Jimenez-Yuste, J. Oldenburg, S. Rangarajan, M. H. Kurth, J. Bozzo, and E. Santagostino, *Haemophilia,* Vol. 22, p. e71, 2016.
74. B. W. Shen, P. C. Spiegel, C. H. Chang, J. W. Huh, J. S. Lee, J. Kim, Y. H. Kim, and B. L. Stoddard, *Blood,* Vol. 111, p. 1240, 2008.
75. P. M. Mannucci and M. Franchini, *Blood Transfusion,* Vol. 11 Suppl 4, p. s77, 2013.
76. E. Santagostino and M. R. Fasulo, *Seminars in Thrombosis and Hemostasis,* Vol. 39, p. 697, 2013.
77. M. J. Manco-Johnson, T. C. Abshire, A. D. Shapiro, B. Riske, M. R. Hacker, R. Kilcoyne, J. D. Ingram, M. L. Manco-Johnson, S. Funk, L. Jacobson, L. A. Valentino, W. K. Hoots, G. R. Buchanan, D. DiMichele, M. Recht, D. Brown, C. Leissinger, S. Bleak, A. Cohen, P. Mathew, A. Matsunaga, D. Medeiros, D. Nugent, G. A. Thomas, A. A. Thompson, K. McRedmond, J. M. Soucie, H. Austin, and B. L. Evatt, *New England Journal of Medicine,* Vol. 357, p. 535, 2007.
78. L. M. Aledort, R. H. Haschmeyer, and H. Pettersson, *Journal of Internal Medicine,* Vol. 236, p. 391, 1994.
79. A. Gringeri, B. Lundin, S. von Mackensen, L. Mantovani, and P. M. Mannucci, *Journal of Thrombosis and Haemostasis,* Vol. 9, p. 700, 2011.
80. T. C. Abshire, H. H. Brackmann, I. Scharrer, K. Hoots, C. Gazengel, J. S. Powell, E. Gorina, E. Kellermann, and E. Vosburgh, *Journal of Thrombosis and Haemostasis,* Vol. 83, p. 811, 2000.
81. M. Maas Enriquez, J. Thrift, S. Garger, and Y. Katterle, *Protein Expression and Purification,* Vol. 127, p. 111, 2016.
82. N. Savage, *Nature,* Vol. 515, p. S162, 2014.
83. S. Meunier, J. Alamelu, S. Ehrenforth, H. Hanabusa, F. Abdul Karim, K. Kavakli, M. Khodaie, J. Staber, O. Stasyshyn, D. L. Yee, and L. Rageliene, *Journal of Thrombosisand Haemostasis,* Vol. 117, p. 1705, 2017.

84. A. E. Schmidt and S. P. Bajaj, *Trends in Cardiovascular Medicine,* Vol. 13, p. 39, 2003.
85. G. Young, P. W. Collins, T. Colberg, A. Chuansumrit, H. Hanabusa, S. R. Lentz, J. Mahlangu, E. P. Mauser-Bunschoten, C. Negrier, J. Oldenburg, T. Patiroglu, E. Santagostino, R. Tehranchi, M. Zak, and F. A. Karim, *Thrombosis Research,* Vol. 141, p. 69, 2016.
86. S. S. Acharya and D. M. Dimichele, *Haemophilia,* Vol. 14, p. 1151, 2008.
87. D. H. Bevan, *Thrombosis Research,* Vol. 124 Suppl 2, p. S12, 2009.
88. K. Tziomalos, S. Vakalopoulou, V. Perifanis, and V. Garipidou, *Vascular Health and Risk Management,* Vol. 5, p. 843, 2009.
89. F. Peyvandi, *Thrombosis Research,* Vol. 130 Suppl 2, p. S7, 2012.
90. B. A. Konkle, H. Huston, and S. Nakaya Fletcher, *GeneReviews,* 1993.
91. T. Livnat, A. A. Barg, S. Levy-Mendelovich, and G. Kenet, *Blood Cells, Molecules, and Diseases,* Vol. 67, p. 63, 2017.
92. S. Boujrad, B. El Hasbaoui, H. Echahdi, M. Malih, and A. Agadr, *Pan African Medical Journal,* Vol. 27, p. 182, 2017.
93. M. B. Streiff and P. M. Ness, *Transfusion,* Vol. 42, p. 18, 2002.
94. J. N. Huang and M. A. Koerper, *Haemophilia,* Vol. 14, p. 1164, 2008.
95. G. Lippi, E. J. Favaloro, M. Montagnana, F. Manzato, G. C. Guidi, and M. Franchini, *Blood Coagulation and Fibrinolysis,* Vol. 22, p. 160, 2011.
96. R. Asselta and F. Peyvandi, *Seminars in Thrombosis and Hemostasis,* Vol. 35, p. 382, 2009.
97. D. J. Perry, *British Journal of Haematology,* Vol. 118, p. 689, 2002.
98. G. Mariani and F. Bernardi, *Seminars in Thrombosis and Hemostasis,* Vol. 35, p. 400, 2009.
99. Z. Salcioglu, A. Akcay, H. S. Sen, G. Aydogan, F. Akici, D. Tugcu, N. A. Ayaz, and Z. Baslar, *Clinical and Applied Thrombosis/ Hemostais,* Vol. 18, p. 588, 2012.
100. U. Hedner, *Seminars in Hematology,* Vol. 38, p. 43, 2001.
101. D. L. Brown and P. A. Kouides, *Haemophilia,* Vol. 14, p. 1176, 2008.
102. K. Gomez and P. Bolton-Maggs, *Haemophilia,* Vol. 14, p. 1183, 2008.
103. S. Duga and O. Salomon, *Seminar in Thrombosis and Hemostasis,* Vol. 39, p. 621, 2013.
104. A. P. Wheeler and D. Gailani, *Expert Review of Hematology,* Vol. 9, p. 629, 2016.
105. M. O'Connell N, *Seminars in Hematology,* Vol. 41, p. 76, 2004.
106. P. H. Bolton-Maggs, R. T. Wensley, P. B. Kernoff, C. K. Kasper, L. Winkelman, R. S. Lane, and J. K. Smith, *Thrombosis and Haemostasis,* Vol. 67, p. 314, 1992.
107. P. M. Mannucci, K. A. Bauer, E. Santagostino, E. Faioni, S. Barzegar, R. Coppola, and R. D. Rosenberg, *Blood,* Vol. 84, p. 1314, 1994.
108. F. Bauduer, E. de Raucourt, C. Boyer-Neumann, M. Trossaert, P. Beurrier, A. Faradji, J. Peynet, J. Y. Borg, P. Chamouni, C. Chatelanaz, C. Henriet, F. Bridey, and J. Goudemand, *Haemophilia,* Vol. 21, p. 481, 2015.
109. G. Ling, H. Kagdi, B. Subel, P. Chowdary, and K. Gomez, *Haemophilia,* Vol. 22, p. 411, 2016.
110. A. Remkova and M. Remko, *Vnitrni Lekarstvi,* Vol. 61, p. 5S63, 2015.
111. M. Karimi, Z. Bereczky, N. Cohan, and L. Muszbek, *Seminar in Thrombosis and Hemostasis,* Vol. 35, p. 426, 2009.
112. A. Tahlan and J. Ahluwalia, *Archives of Pathology and Laboratory Medicine,* Vol. 138, p. 278, 2014.
113. A. Biswas, V. Ivaskevicius, A. Thomas, and J. Oldenburg, *Hamostaseologie,* Vol. 34, p. 160, 2014.
114. V. Schroeder and H. P. Kohler, *Seminar in Thrombosis and Hemostasis,* Vol. 39, p. 632, 2013.
115. M. M. Goldenberg, *P T,* Vol. 39, p. 112, 2014.
116. C. Longstaff, S. Williams, and C. Thelwell, *Cardiovascular and Hematological Agents in Medical Chemistry,* Vol. 6, p. 212, 2008.
117. E. Kotb, *Biotechnology Progress,* Vol. 30, p. 656, 2014.

118. M. Flemmig and M. F. Melzig, *Journal of Pharmacy and Pharmacology,* Vol. 64, p. 1025, 2012.
119. R. Gao, Y. Zhang, Q. X. Meng, W. H. Lee, D. S. Li, Y. L. Xiong, and W. Y. Wang, *Toxicon,* Vol. 36, p. 457, 1998.
120. D. Park, H. Kim, K. Chung, D. S. Kim, and Y. Yun, *Toxicon,* Vol. 36, p. 1807, 1998.
121. E. F. Sanchez, C. I. Santos, A. Magalhaes, C. R. Diniz, S. Figueiredo, J. Gilroy, and M. Richardson, *Archieves of Biochemistry and Biophysics,* Vol. 378, p. 131, 2000.
122. S. R. Deitcher, W. D. Funk, J. Buchanan, S. Liu, M. D. Levy, and C. F. Toombs, *Expert Opinion on Biological Therapy,* Vol. 6, p. 1361, 2006.
123. H. Mihara, H. Sumi, T. Yoneta, H. Mizumoto, R. Ikeda, M. Seiki, and M. Maruyama, *Japanese Journal of Physiology,* Vol. 41, p. 461, 1991.
124. A. Kunamneni, T. T. Abdelghani, and P. Ellaiah, *Journal of Thrombosis and Thrombolysis,* Vol. 23, p. 9, 2007.
125. S. Sherry, *Clin Cardiol,* Suppl 5, p. V3, 1990.
126. M. A. Munger and E. A. Forrence, *Clinical Pharmacology,* Vol. 9, p. 530, 1990.
127. J. P. Irigoyen, P. Munoz-Canoves, L. Montero, M. Koziczak, and Y. Nagamine, *Celullar and Molecular Life Sciences,* Vol. 56, p. 104, 1999.
128. V. Bansal and P. K. Roychoudhury, *Protein Expression and Purification,* Vol. 45, p. 1, 2006.
129. D. Collen and H. R. Lijnen, *Blood,* Vol. 78, p. 3114, 1991.
130. M. Moser and C. Bode, *Expert Opin. Investig. Drugs,* Vol. 8, p. 329, 1999.
131. P. K. Roychoudhury, S. S. Khaparde, B. Mattiasson, and A. Kumar, *Biotechnology Advances,* Vol. 24, p. 514, 2006.
132. D. Collen and H. R. Lijnen, *Arteriosclerosis, Thrombosis and Vascular Biology,* Vol. 29, p. 1151, 2009.
133. D. Pennica, W. E. Holmes, W. J. Kohr, R. N. Harkins, G. A. Vehar, C. A. Ward, W. F. Bennett, E. Yelverton, P. H. Seeburg, H. L. Heyneker, D. V. Goeddel, and D. Collen, *Nature,* Vol. 301, p. 214, 1983.
134. D. Collen and H. R. Lijnen, *Journal of Thrombosis and Haemostasis,* Vol. 2, p. 541, 2004.
135. J. Kratzschmar, B. Haendler, G. Langer, W. Boidol, P. Bringmann, A. Alagon, P. Donner, and W. D. Schleuning, *Gene,* Vol. 105, p. 229, 1991.
136. P. Bringmann, D. Gruber, A. Liese, L. Toschi, J. Kratzchmar, W. D. Schleuning, and P. Donner, *Jounal of Biological Chemistry,* Vol. 270, p. 25596, 1995.
137. E. Kotb, *Applied Microbiology and Biotechnology,* Vol. 97, p. 6647, 2013.
138. H. R. Lijnen, B. Van Hoef, F. De Cock, K. Okada, S. Ueshima, O. Matsuo, and D. Collen, *Journal of Biological Chemistry,* Vol. 266, p. 11826, 1991.
139. S. Vanderschueren, L. Barrios, P. Kerdsinchai, P. Van den Heuvel, L. Hermans, M. Vrolix, F. De Man, E. Benit, L. Muyldermans, D. Collen, and *et al., Circulation,* Vol. 92, p. 2044, 1995.
140. P. W. Armstrong, J. R. Burton, D. Palisaitis, C. R. Thompson, F. Van de Werf, B. Rose, D. Collen, and K. K. Teo, *American Heart Journal,* Vol. 139, p. 820, 2000.
141. P. W. Armstrong, J. Burton, S. Pakola, P. G. Molhoek, A. Betriu, M. Tendera, C. Bode, A. A. Adgey, F. Bar, A. Vahanian, and F. Van de Werf, *American Heart Journal,* Vol. 146, p. 484, 2003.
142. R. J. Hopkin, G. Cabrera, J. Charrow, R. Lemay, A. M. Martins, M. Mauer, A. Ortiz, M. R. Patel, K. Sims, S. Waldek, D. G. Warnock, and W. R. Wilcox, *Molecular Genetics and Metabolism,* Vol. 119, p. 151, 2016.
143. H. Sakuraba, M. Murata-Ohsawa, I. Kawashima, Y. Tajima, M. Kotani, T. Ohshima, Y. Chiba, M. Takashiba, Y. Jigami, T. Fukushige, T. Kanzaki, and K. Itoh, *Journal of Human Genetics,* Vol. 51, p. 180, 2006.
144. B. E. Smid, S. M. Rombach, J. M. Aerts, S. Kuiper, M. Mirzaian, H. S. Overkleeft, B. J. Poorthuis, C. E. Hollak, J. E. Groener, and G. E. Linthorst, *Orphanet Journal of Rare Diseases,* Vol. 6, p. 69, 2011.

145. D. P. Germain, J. Charrow, R. J. Desnick, N. Guffon, J. Kempf, R. H. Lachmann, R. Lemay, G. E. Linthorst, S. Packman, C. R. Scott, S. Waldek, D. G. Warnock, N. J. Weinreb, and W. R. Wilcox, *Journal of Medical Genetics,* Vol. 52, p. 353, 2015.
146. K. Lee, X. Jin, K. Zhang, L. Copertino, L. Andrews, J. Baker-Malcolm, L. Geagan, H. Qiu, K. Seiger, D. Barngrover, J. M. McPherson, and T. Edmunds, *Glycobiology,* Vol. 13, p. 305, 2003.
147. R. Skrunes, E. Svarstad, K. Kampevold Larsen, S. Leh, and C. Tondel, *Nephrology Dialysis Transplant,* Vol. 32, p. 807, 2017.
148. G. A. Grabowski, N. W. Barton, G. Pastores, J. M. Dambrosia, T. K. Banerjee, M. A. McKee, C. Parker, R. Schiffmann, S. C. Hill, and R. O. Brady, *Annals of Internal Medicine,* Vol. 122, p. 33, 1995.
149. G. A. Grabowski, M. Golembo, and Y. Shaaltiel, *Molecular Genetics and Metabolism,* Vol. 112, p. 1, 2014.
150. M. Connock, A. Burls, E. Frew, A. Fry-Smith, A. Juarez-Garcia, C. McCabe, A. Wailoo, K. Abrams, N. Cooper, A. Sutton, A. O'Hagan, and D. Moore, *Health Technology Assessment,* Vol. 10, p. 3, 2006.
151. G. A. Grabowski, *American Society Hematology Education Program,* Vol. 2012, p. 13, 2012.
152. G. A. Grabowski, N. Leslie, and R. Wenstrup, *Blood Review,* Vol. 12, p. 115, 1998.
153. M. Beck, *IUBMB Life,* Vol. 62, p. 33, 2010.
154. N. W. Barton, R. O. Brady, J. M. Dambrosia, A. M. Di Bisceglie, S. H. Doppelt, S. C. Hill, H. J. Mankin, G. J. Murray, R. I. Parker, C. E. Argoff, *et al., New England Journal of Medicine,* Vol. 324, p. 1464, 1991.
155. R. J. Desnick and E. H. Schuchman, *Annual Review of Genomics and Human Genetics,* Vol. 13, p. 307, 2012.
156. E. Shemesh, L. Deroma, B. Bembi, P. Deegan, C. Hollak, N. J. Weinreb, and T. M. Cox, *Cochrane Database of Systematic Reviews,* p.CD010324, 2015.
157. C. Angelini, C. Semplicini, S. Ravaglia, B. Bembi, S. Servidei, E. Pegoraro, M. Moggio, M. Filosto, E. Sette, G. Crescimanno, P. Tonin, R. Parini, L. Morandi, G. Marrosu, G. Greco, O. Musumeci, G. Di Iorio, G. Siciliano, M. A. Donati, F. Carubbi, M. Ermani, T. Mongini, and A. Toscano, *Journal of Neurology,* Vol. 259, p. 952, 2012.
158. J. L. Schneider and S. V. Balu-Iyer, *Journal of Pharmaceutical Sciences,* Vol. 105, p. 3097, 2016.
159. A. Toscano and B. Schoser, *Journal of Neurology,* Vol. 260, p. 951, 2013.
160. E. Richard, G. Douillard-Guilloux, and C. Caillaud, *IUBMB Life,* Vol. 63, p. 979, 2011.
161. A. Tylki-Szymanska, J. Marucha, A. Jurecka, M. Syczewska, and B. Czartoryska, *Journal of Inherited Metabolic Disease,* Vol. 33, p. 151, 2010.
162. H. Harada, H. Uchiwa, M. Nakamura, S. Ohno, H. Morita, A. Katoh, M. Yoshino, and H. Ikeda, *Molecular Genetics and Metabolism,* Vol. 103, p. 215, 2011.
163. S. Laraway, J. Mercer, E. Jameson, J. Ashworth, P. Hensman, and S. A. Jones, *Journal of Pediatrics,* Vol. 178, p. 219, 2016.
164. E. Jameson, S. Jones, and J. E. Wraith, *Cochrane Database Syst Rev,* p. CD009354, 2013.
165. E. D. Kakkis, J. Muenzer, G. E. Tiller, L. Waber, J. Belmont, M. Passage, B. Izykowski, J. Phillips, R. Doroshow, I. Walot, R. Hoft, and E. F. Neufeld, *New England Journal of Medicine,* Vol. 344, p. 182, 2001.
166. G. Walsh, ed., *Directory of Therapeutic Enzymes,* CRC Press, 2005.
167. R. Giugliani, V. M. Rojas, A. M. Martins, E. R. Valadares, J. T. Clarke, J. E. Goes, E. D. Kakkis, M. A. Worden, M. Sidman, and G. F. Cox, *Molecular Genetics and Metabolism,* Vol. 96, p. 13, 2009.

168. V. Valayannopoulos, N. Boddaert, V. Barbier, M. Le Merrer, C. Caillaud, and P. de Lonlay, *Molecular Genetics and Metabolism,* Vol. 100, p. 20, 2010.
169. R. Eavri and H. Lorberboum-Galski, *Journal of Biological Chemistry,* Vol. 282, p. 23402, 2007.
170. A. M. Ramirez, A. Rodriguez-Lopez, A. Ardila, L. Beltran, C. A. Patarroyo, A. D. P. Melendez, O. F. Sanchez, and C. J. Almeciga-Diaz, *European Journal of Pharmaceutical Sciences,* Vol. 109, p. 48, 2017.
171. C. N. Sarkissian and A. Gamez, *Molecular Genetics and Metabolism,* Vol. 86 Suppl 1, p. S22, 2005.
172. R. A. Williams, C. D. Mamotte, and J. R. Burnett, *Clinical Biochemical Reviews,* Vol. 29, p. 31, 2008.
173. H. Erlandsen, M. G. Patch, A. Gamez, M. Straub, and R. C. Stevens, *Pediatrics,* Vol. 112, p. 1557, 2003.
174. A. Gamez, L. Wang, M. Straub, M. G. Patch, and R. C. Stevens, *Molecular Therapy,* Vol. 9, p. 124, 2004.
175. W. Kim, H. Erlandsen, S. Surendran, R. C. Stevens, A. Gamez, K. Michols-Matalon, S. K. Tyring, and R. Matalon, *Molecular Therapy,* Vol. 10, p. 220, 2004.
176. K. J. Wieder, N. C. Palczuk, T. van Es, and F. F. Davis, *Journal of Biological Chemistry,* Vol. 254, p. 12579, 1979.
177. A. R. Hamzeh Alipour, Sedigheh Zakeri, Navid Dinparast Djadid, *Asian pecific Journal Tropical Biomedicine,* Vol. 6, p. 975, 2016.
178. A. R. Sheets, T. N. Demidova-Rice, L. Shi, V. Ronfard, K. V. Grover, and I. M. Herman, *PLoS One,* Vol. 11, p. e0159598, 2016.
179. E. J. Menzel and C. Farr, *Cancer Letters,* Vol. 131, p. 3, 1998.
180. P. Scodeller, *Carcinogenesis and Mutagenesis,* Vol. 5, p. 2014.
181. H. R. Maurer, *Cell Mol Life Sci,* Vol. 58, p. 1234, 2001.
182. R. Pavan, S. Jain, Shraddha, and A. Kumar, *Biotechnology Research International,* Vol. 2012, p. 976203, 2012.
183. J. M. Braun, B. Schneider, and H. J. Beuth, *In Vivo,* Vol. 19, p. 417, 2005.

# 11

# Erythritol: A Sugar Substitute

**Kanti N. Mihooliya, Jitender Nandal, Himanshu Verma, and Debendra K. Sahoo***

*Biochemical Engineering Research and Process Development Centre*
*CSIR-Institute of Microbial Technology, Chandigarh, (India)*

***Abstract***
Erythritol belongs to the category of sugar alcohols/polyols and consists of four carbon sugar alcohols. Erythritol occurs naturally in many food items such as grapes, wine, beer, sake, soy sauce, peers, mushrooms, and watermelon, etc. Among different sugar alcohols, erythritol is non-caloric and non-fermentable polyol, which makes it as an attractive sugar substitute. It has a small molecular size, and because of this reason, it is mostly absorbed in the small intestine. As it is not metabolized in the body, it does not show any glycaemic and insulinaemic effect. These properties make erythritol an essential sweetener for those people who want to reduce their blood sugar level. Presently, erythritol is used in baked products, confectionery, beverages, manufacturing of chewing gums, etc., and it has potential to be used as a sugar additive in larger global demand. Herein, this chapter focuses on potential applications of erythritol as a sugar substrate in the current global scenario.

***Keywords:*** Erythritol; sweetener; sugar alcohol; calorie; sugar substitute

## 11.1 Introduction

### 11.1.1 Background of Erythritol

Sweetness is a perception. One of the five primary sensations, sweetness is probably the most popular sensation of all due to its intensity. Sweeteners provide sweet flavor when added to food products, maintain freshness and product quality and also act as a preservative. Sugar substitutes mimic the effect of sugar but usually have lower food energy [1] and are classified into four categories, i.e., natural sweeteners, artificial sweeteners, sugar alcohols (or polyols), and novel sweeteners [2] (Table 11.1).

Some of these sugar substitutes having therapeutics/non-therapeutics applications are available in the market (Figure 11.1).

Sugar alcohols, also known as polyols, have increased in prominence as they have been exploited as sugar replacers and sweeteners in many foods, with particular importance in diet and health-oriented food items (Table 11.2).

**Corresponding author*: debsahoo@imtech.res.in

Saurabh Saran, Vikash Babu, and Asha Chaubey (eds.) High Value Fermentation Products, Volume 1, (265–284) © 2019 Scrivener Publishing LLC

**Table 11.1** Types of sugar substitutes [2].

| Sugar substitutes | Definition | Examples |
|---|---|---|
| Natural sweeteners | These are found naturally and healthier options than processed table sugar | Molasses, honey, monatin, curculin, etc. |
| Artificial sweeteners | These are synthetic sweeteners that are derived from natural substances | Aspartame, sucralase, saccharin, etc. |
| Sugar alcohols (or polyols) | These are carbohydrates which occur naturally in some fruits and vegetables | Erythritol, maltitol, xylitol, isomalt, sorbitol, etc. |
| Novel sweeteners | These are combinations of various types of sweeteners | Tagatose, stevia extracts, trehalose, etc. |

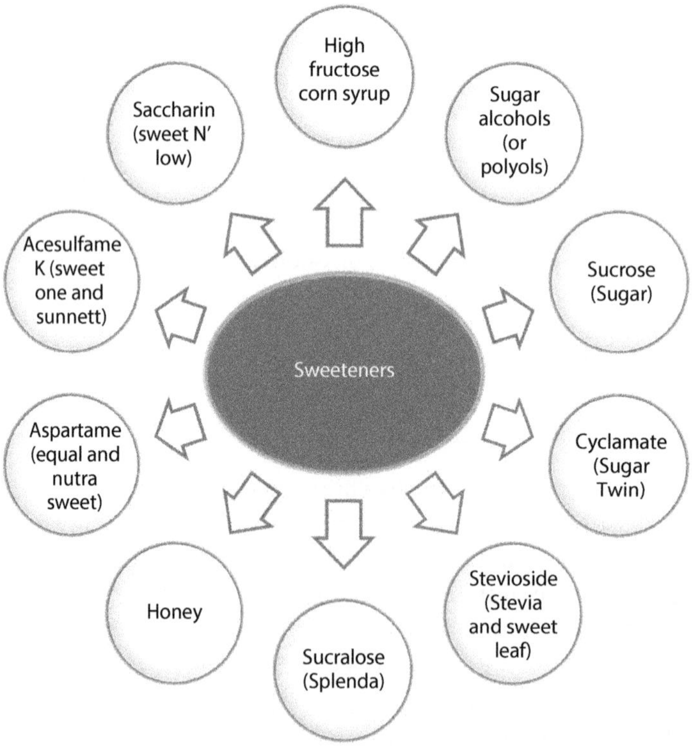


**Figure 11.1** List of sugar substitutes available in market.

Due to technological advancements, a range of sugar alcohols are now used as food additives, and commercially available sugar alcohols include erythritol, xylitol, sorbitol, mannitol, lactitol, and isomalt (Table 11.3). These polyols have various properties including:

**Table 11.2** Comparison of sugar alcohols and sugar.

| Sugar alcohols or polyols | Sugar |
|---|---|
| No or low calories suitable for diabetic patients | High calories not ideal for diabetic patients |
| No effect on blood sugar level | Raises or increases blood sugar level |
| Non-cariogenic, therefore, prevents tooth decay | Cariogenic, thus, causes tooth decay |
| Free radicals scavenger or good antioxidants | Promotes free radicals |
| Protects blood vessels | Damages blood vessels |
| No effect on weight | Promotes weight gain |
| No effect on bone density | Decreases bone density |

**Table 11.3** Types of sugar alcohols and their applications.

| Polyols | Sweetness (Relative to 100% sucrose) | Calories/gram | Applications |
|---|---|---|---|
| Erythritol ($C_4H_{10}O_4$) | 60–70% | 0.2 | Used in pharmaceuticals, confectioneries, bakery, beverages etc. |
| Sorbitol ($C_6H_{14}O_6$) | 60% | 2.6 | Used in confectioneries, cough drops, etc. |
| Xylitol ($C_5H_{12}O_5$) | 100% | 2.4 | Used in toothpaste, confectioneries, fruit spreads, beverages, etc. |
| Lactitol ($C_{12}H_{24}O_{11}$) | 40% | 2.0 | Used in ice creams, confectioneries etc. |
| Mannitol ($C_6H_{14}O_6$) | 50% | 1.6 | Used in confectioneries items etc. |
| Maltitol ($C_{12}H_{24}O_{11}$) | 90% | 2.1 | Used in confectioneries items, sugar-free syrups, etc. |
| Isomalt ($C_{12}H_{24}O_{11}$) | 55% | 2.0 | Used in confectioneries items etc. |

- Polyols mostly have a clean, sweet, neutral, and pleasant taste.
- Most polyols have less sweetness than that of sucrose.
- Different polyols have different solubility depending on the range of temperature.
- Different polyols have different hygroscopicity depending on the relative humidity.
- Polyols have a negative heat of solution which results in a cooling effect in the mouth.

The present chapter focused on the erythritol, a 4-carbon sugar polyol having a relatively low molecular weight that provides it a unique characteristic.

### 11.1.2 History of Erythritol

Erythritol ($C_4H_{10}O_4$) or 1,2,3,4-butanetetrol is a four carbon and zero-calorie bulk sweetener which occurs naturally in many fruits and vegetables in small quantity. It was first discovered and extracted by Scottish chemist John Stenhouse in 1848 from lichen [3]. In 1852, Lamy isolated the erythritol from algae *Protococcus vulgaris* (now named *Apatococcus lobatus*) [4]. Later in the 1900s, *Trentepohlia jolithus* was also used for the isolation of erythritol [5]. In 1960, Pfeifer *et al.* used the chemical process for erythritol production [6]. They converted periodate-oxidized starch or dialdehyde starch into erythritol by using high-temperature chemical reaction in the presence of nickel catalyst. In the early 1980s, fermentation process for the production of erythritol was developed [7]. In this process, natural raw materials were used for increasing the fermentation efficiency and improving yield. It has been first approved to be used at commercial level as a food additive in Japan and then spread throughout the world. In Japan, erythritol has been a part of many sugar-based commercial products like candies, chewing gum, chocolates, soft drinks, etc., since 1990 and due to the immense potential of erythritol in the food industry, its demand has been exponentially increased, and it has been approved for use in more than 55 countries [8].

### 11.1.3 Occurrence of Erythritol

Erythritol is widely distributed in nature among many fruits including melons, grapes, pears, etc., and many vegetables. It is present in many fermented food products like wine, soy sauce, etc. Erythritol also presents in tissue and body fluids of human and animals. A summary of the amount of erythritol naturally presents in various foods shown in Table 11.4.

### 11.1.4 General Characteristics

Erythritol is a tetracarbon sugar alcohol ($C_4H_{10}O_4$) having a molecular weight of 122.12 (Figure 11.2). It is an asymmetric molecule and hence, exists only in meso-form. 1,2,3,4-butanetetrol is its IUPAC name. It is a white to off-white, anhydrous, non-hygroscopic substance available in powdered or crystals forms with a similar appearance to sucrose and has a moderately sweet taste of 60–70% as compared to that of sucrose [8, 9] (Figure 11.3).

Among sugar alcohols (or polyols), erythritol is a zero or low-calorie and non-fermentable polyol which makes it more exciting sugar substitute in the recent time. Due to small molecular size, erythritol is mostly absorbed in the small intestine, does not metabolize in the body, and approximately 90% is excreted unchanged in the urine [10, 11] (Figure 11.4). Thus it does not have any glycemic and insulinemic effect [9] (Figure 11.5). As erythritol is well absorbed, its consumption becomes very high without raising side effects. Some clinical studies have also shown that out of all polyols erythritol is the best polyol in terms of causing least side effects. Erythritol is also more digestible than many fibers (like inulin) and sugars (like lactose and tagatose) [8]. These properties make erythritol a vital sweetener for those people who want to reduce their blood sugar level.

**Table 11.4** Amount of erythritol in various foods [8].

| Food items | Content of erythritol |
|---|---|
| Watermelons | 22–25 mg/kg |
| Melons | 47–50 mg/kg |
| Wine | 130–300 mg/L |
| Soy sauce | 910 mg/L |
| Pears | 0–40 mg/kg |
| Grapes | 0–42 mg/L |
| Sake | 1550 mg/L |
| Sherry Wine | 70 mg/L |
| Miso bean paste | 1310 mg/kg |

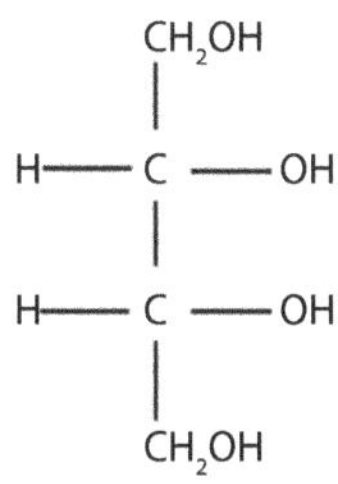

**Figure 11.2** Chemical structure of erythritol.

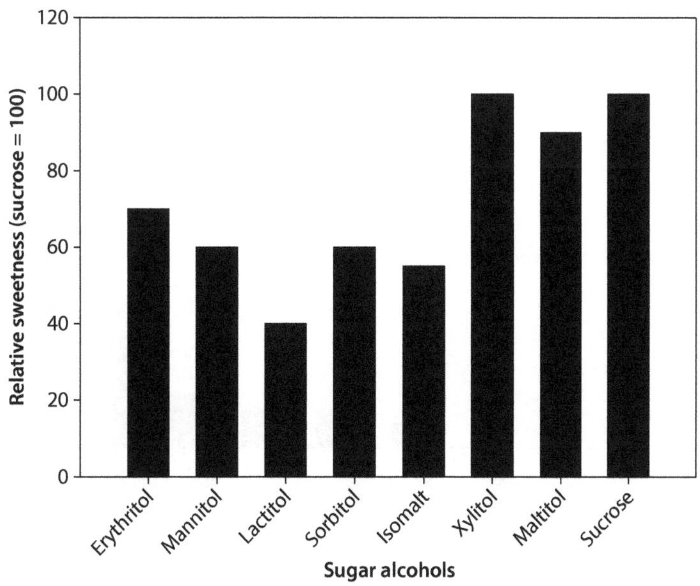


**Figure 11.3** Relative sweetness of different sugar alcohols (Sucrose = 100) [8, 9].

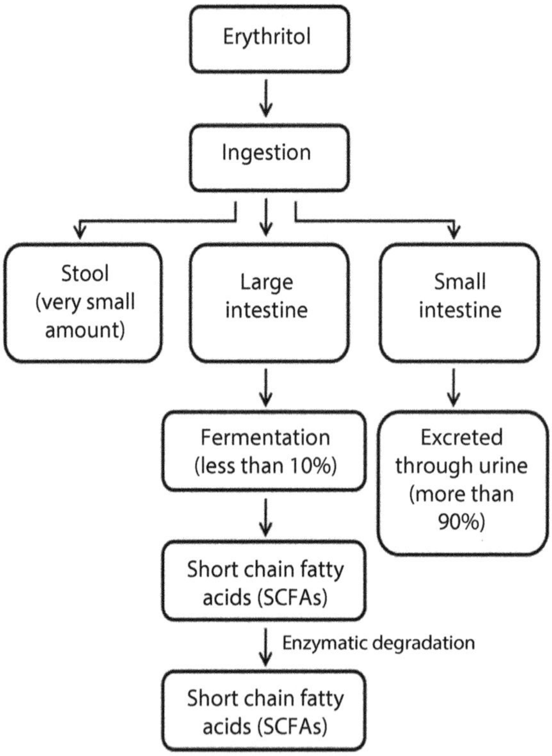


**Figure 11.4** Schematic view of Erythritol metabolism [11].

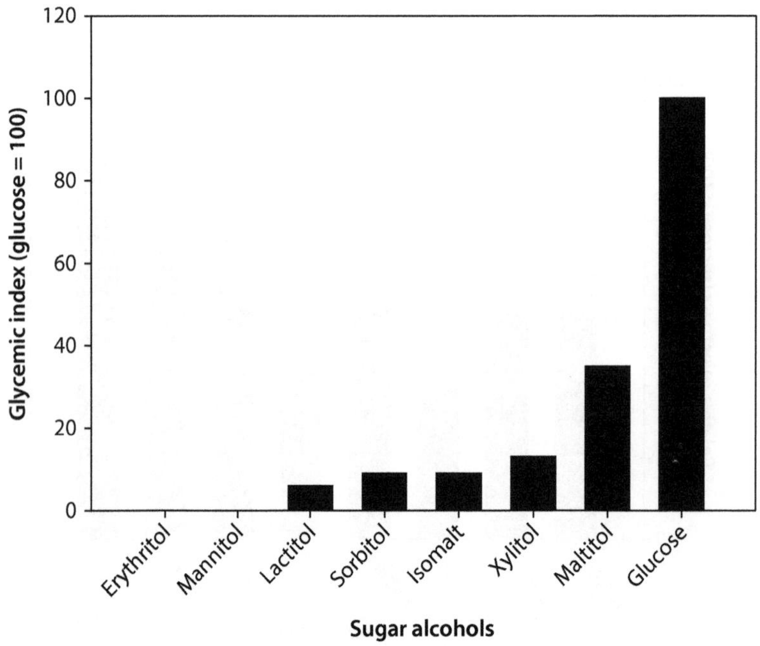


**Figure 11.5** Glycemic index of different sugar alcohols (Glucose = 100) [9].

Like other polyols, erythritol prevents tooth decay and shows antioxidant properties. Erythritol usage in food prevents dental caries – like the xylitol because the microorganism such as *Streptococcus mutans* which causes dental caries, cannot use erythritol as a substrate [12]. As mentioned before, it is well absorbed in the small intestine and not fermented. Therefore, it shows potential free radical scavenger activity while circulating throughout the body.

## 11.2 Chemical and Physical Properties of Erythritol

Erythritol is highly soluble in water (saturated solution contains about 66% w/w) [13] (Figure 11.6). However, it has low solubility in pyridine (saturated solution contains 2.5% w/w) and alcohol [14] and insoluble in ethyl ether and benzene [15]. The melting and boiling point of erythritol is 117–123 °C and 330.5 °C, respectively and has a specific gravity of 1.451 g/cu cm at 20 °C [15]. Because of its small molecular size, erythritol is less viscous than sucrose or other sugar alcohols at similar concentrations [8], and it is non-hygroscopic as well.

Erythritol is exceptionally stable in heat and acid because it has no reducing groups and therefore it also does not take part in Maillard or browning reactions like other polyols. As it is stable at about 180 °C, it can handle an extended period of heat without browning (Table 11.5).

## 11.3 Estimation of Erythritol

In the fermentation process, erythritol production may accompany with the production of several other components, and hence, efficient estimation methods are required

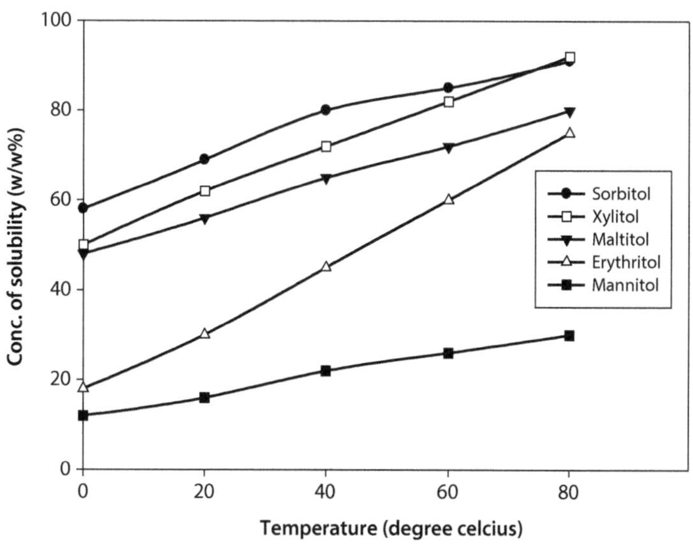


**Figure 11.6** Comparison of water solubility of erythritol with other sugar alcohols [13].

**Table 11.5** Chemical and physical properties of different sugar alcohols [8].

| | Sucrose | Erythritol | Xylitol | Mannitol | Sorbitol | Isomalt | Lactitol | Maltitol |
|---|---|---|---|---|---|---|---|---|
| **Carbon (*n*°)** | 12 | 4 | 5 | 6 | 6 | 12 | 12 | 12 |
| **Molecular weight** | 342 | 122.10 | 152 | 182 | 182 | 344 | 344 | 344 |
| **Melting point (°C)** | 190 | 116–121 | 94 | 165 | 97 | 145–150 | 122 | 150 |
| **Heat stability (°C)** | 160 | 60 | 60 | 60 | 60 | 60 | 60 | 60 |
| **Solubility (% w/w at 25 °C)** | 67 | 37 | 64 | 20 | 70 | 25 | 57 | 60 |
| **Heat of solution (cal/g)** | -4.3 | -43 | -36.5 | -28.5 | -26 | -9.4 | -13.9 | -18.9 |
| **pH stability** | Hydrolyses | 2–10 | 2–10 | 2–10 | 2–10 | 2–10 | 3 | 2–10 |
| **Viscosity** | Low | Very low | Very low | Low | Medium | High | Very low | Medium |
| **Hygroscopicity** | Medium | Very low | High | Low | Medium | Low | Medium | Medium |

for quantitative estimation of erythritol production. As polyols used in several low-sugar products, their determination becomes very vital for providing nutritional information and also helps in controlling the quality. There are several methods reported in the literature for the estimation of erythritol, and some of them are summarized below.

### 11.3.1 Thin Layer Chromatography (TLC)

The estimation/detection of polyols is carried out by TLC. In this method, a mixture containing erythritol or other polyols is spotted onto silica plates, and development of plates perform with a solvent mixture of ethyl acetate: 2-butanol: water in the ratio of 6:3:1. After development, the plates are allowed to dry and spray with the 1% sodium periodate ($NaIO_4$) followed by 1% potassium permanganate ($KMnO_4$) [16]. Erythritol/other polyols produce yellow color on TLC plates. The principle behind the detection of erythritol/other polyols by TLC is that polyols such as erythritol, mannitol, etc., contain the vicinal OH groups that are oxidatively broken down into formaldehyde and formic acid by spraying of the periodic acid which gets reduced to iodic acid [17].

### 11.3.2 Colorimetric Assay for Detection of Polyols

The colorimetric assay is another traditional method for the detection of erythritol. This technique involves the detection of erythritol by the oxidation of alditols to formaldehyde by acidic periodate and then reduction of the periodate with l-rhamnose. The formaldehyde is then determined by using the Nash reagent [18].

### 11.3.3 High-Performance Liquid Chromatography (HPLC)

Another method which is used for the detection of erythritol/other polyols is high-performance liquid chromatography. It is the most common method used for the detection or estimation of erythritol and many reports in literature also suggest that HPLC is the preferable method for evaluation of sugar alcohols. Many different types of columns are reported for the detection of sugar alcohols including HPX-87H organic acid column [19–23], amino based carbohydrate column [24, 25], and ion-exclusion column [26, 27]. The absence of chromophoric and fluorophoric moieties in erythritol makes it impossible for UV or fluorescence-based methods for detection of erythritol. Therefore, detection of erythritol by HPLC-based methods involves the use of less sensitive refractive index detector (RID). Though HPLC-based methods are efficient, the main drawback associated with it is the limited life of the column that makes it expensive [28].

### 11.3.4 Capillary Electrophoresis (CE)

Recently, Coelho and De Jesus [29] reports an analytical method, i.e., Capillary Electrophoresis (CE) for the determination of erythritol and other polyols. In this technique, the capacitively coupled contactless conductivity ($C_4D$) is used for detection of erythritol/other polyols. This method involves the interaction of polyols with the borate ions in the background electrolyte resulting in the formation of negatively charged borate esters. The polyols from the sample were extracted by using the ultra-pure water and ultrasonic energy. The efficiency of this method ranges from 70% to 116%.

## 11.4 Production Methods for Erythritol

Erythritol is produced by the chemical as well as fermentative methods. It is the first sugar alcohol to be produced and commercialized by using natural fermentation process [8].

### 11.4.1 Chemical Methods for Erythritol Production

There are several chemical processes for erythritol production. In the early 1940s, the erythritol was synthesized from 2-butene-1,4-diol, and this method was first patented in 1943 [30]. In this process, the substrate dialkyl ester or tartaric acid gets hydrogenated in methanol and ethanol solution under high pressure (i.e., 20 MPa) and temperature (i.e., 125 to 200 °C) which subsequently results in the formation of an erythritol and threitol mixture [31]. Another chemical method for the production of erythritol makes use of dialdehyde starch and high temperature in the presence of a catalyst such as nickel [6]. All these chemical methods have not been used commercially for the production of erythritol because of the high cost and very low efficiency or specificity.

### 11.4.2 Fermentative Methods for Erythritol Production

A number of yeast-like fungi and yeasts have been reported for erythritol production by fermentation processes. *Moniliella pollinis*, *Trichosporonoides megachiliensis*, and *Yarrowia lipolytica* yeasts are commonly used for the production of erythritol at industrial level and the primary substrate for erythritol production is glucose [32].

#### *11.4.2.1 Erythritol Production in Bacteria*

Erythritol is formed in some bacteria such as lactic acid bacteria through phosphoketolase pathway [33]. In this pathway, common sugars like fructose and glucose are converted into xylulose 5-phosphate after the steps of glucose 6-phosphate, 6-phosphogluconate, and ribulose 5-phosphate and accompanied with the transformation of two NAD(P) into NAD(P)H. This is followed by the breakdown of the xylulose 5-phosphate into acetyl phosphate and glyceraldehyde 3-phosphate by phosphoketolase enzyme. In this pathway, ATP is produced along with lactate and acetate production and ethanol is produced by the transfer of NAD(P)H to acetyl-CoA. At the end of the pathway, the excessive NAD(P)H is reoxidized with the help of electron acceptors which subsequently results in the production of polyols or sugar alcohol [34] (Figure 11.7).

#### *11.4.2.2 Erythritol Production in Yeasts*

The production of erythritol by yeasts was first suggested by Binkley and Wolfrom in 1950 [35]. Spencer and Sallans (1956) demonstrated the production of polyols such as

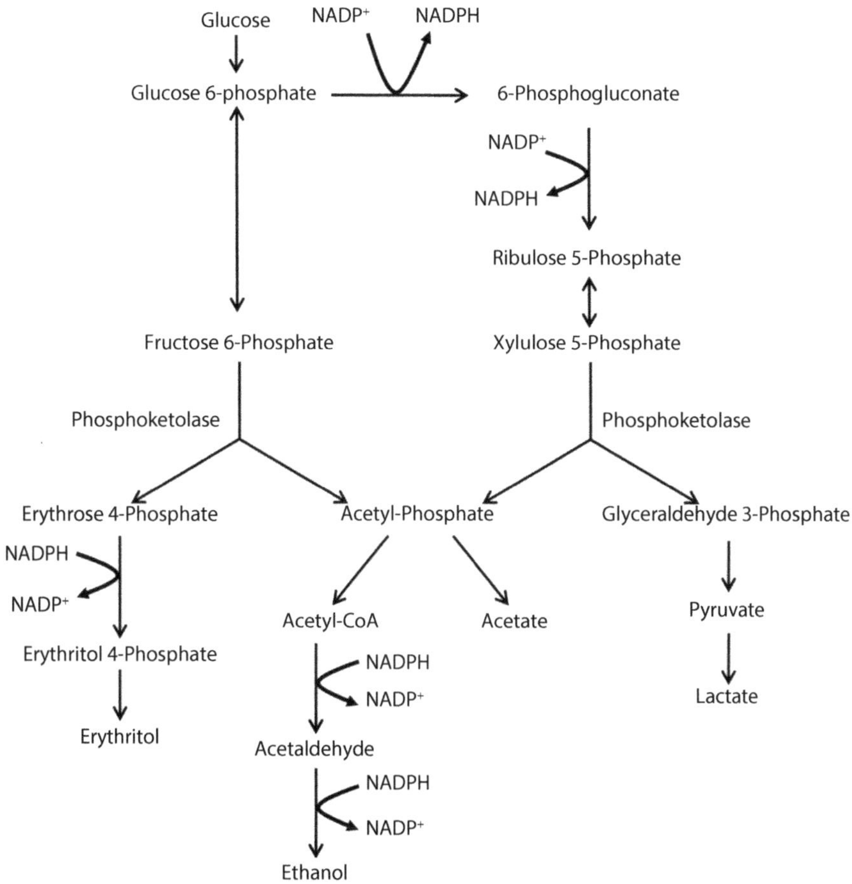


**Figure 11.7** Erythritol biosynthesis in bacteria from glucose [31].

glycerol, erythritol, D-arabitol, and mannitol under aerobic conditions by using osmophilic yeasts [35]. (Table 11.6).

In yeasts and fungi, erythritol is produced via the pentose phosphate pathway where D-erythrose 4-phosphate is converted to erythritol by using dephosphorylation and reduction reactions [31] (Figure 11.8). Erythritol production from yeasts involves the conversion of erythrose to erythritol by erythrose reductase using NAD(P)H as cofactor resulting in the hydrogenation of the erythrose.

## 11.5 Optimization of Erythritol Production

Optimization is the critical step to enhance the production of erythritol using fermentation process. Factors such as nutritional and physiological have shown major impact on the production of biomolecules [45, 46]. Optimization studies involve multiple steps to enhance the production of a desirable biomolecule such as one factor at a time method and statistical methods.

**Table 11.6** List of various erythritol producing microorganisms [36].

| Microorganism | Erythritol ($gL^{-1}$) | Yield (%) | Reference |
|---|---|---|---|
| Ustilaginomycetes 618A-01 | 100 | 39 | 37 |
| *Moniliella tomentosa* var. *pollinis* | 133 | 37 | 35 |
| *Moniliella* sp 440 | 116 | 39 | 38 |
| *Candida magnoliae* | 200 | 43 | 39 |
| *Yarrowia lipolytica* | 170 | 56 | 40 |
| *Pseudozyma tsukubaensis* | 241–243 | 60–61 | 41 |
| *Trichosporon* sp. | 138–150 | 45–46 | 42 |
| *Torula* sp. | 192 | 48 | 43 |
| *Aureobasidium* sp. | 175 | 44 | 44 |

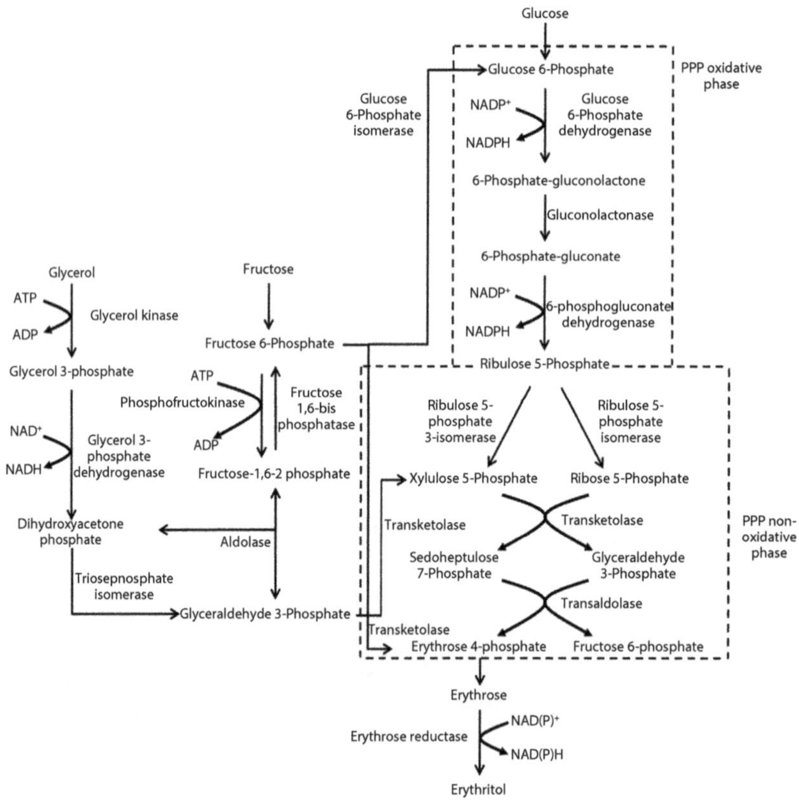


**Figure 11.8** Erythritol biosynthesis in yeasts [31].

## 11.5.1 One Factor at a Time

In this approach, the level or concentration of one factor is varied keeping other factors constant to study the effect of this factor on production profile. This process proceeds till the optimization of remaining factors and obtaining final optimized condition. Initially, different physiological and nutritional parameters are optimized by using this method to achieve maximum production of any biomolecule. In case of erythritol, the temperature is shown to have a significant impact on growth and metabolism of yeasts [47], and erythritol production was found to be maximum at a temperature of 30 °C as at this temperature sugar utilization is very fast [35]. The pH is also crucial for metabolic activities, and pH of 5.6 was reported as optimum for the production of erythritol from *Aureobasidium* [44]. Other factors like aeration and age of inoculums also play vital roles in enhancing its production.

Nutritional factors like glucose have significant effects on erythritol production and, therefore, it is important to determine the optimum concentration of glucose. Mutant of *Aureobasidium* has been used for commercial production of erythritol in a medium containing 40% glucose [44]. 59 $gL^{-1}$ of erythritol is produced from *Candida magnoliae* mutant R23 by using glucose (250 $gL^{-1}$) as a carbon source [48]. Nitrogen source such as yeast extract also plays a significant role in the production of erythritol as nitrogen limitation or exhaustion triggers the erythritol production [49]. Therefore, optimization of nitrogen concentration is also important for increasing the erythritol production.

### 11.5.2 Statistical Design Approaches

Statistical design approaches (such as response surface methodology) are efficient and essential tools for significantly enhancing the production of any biomolecule by studying the interaction of various parameters involved in production. These statistical experimental designs have frequently been used for the optimization of media composition and process parameters for enhancing the product formation [50–53]. Erythritol production by *Candida magnoliae* mutant R23 has been optimized by using the response surface methodology approach by studying the combined effect of four media components (such as yeast extract, glucose, magnesium sulfate, and di-hydrogen phosphate) [48].

## 11.6 Toxicology of Erythritol

Erythritol is endogenously present in body fluids of human and animals such as urine, human plasma [54]. As the erythritol is well absorbed in small intestine, it causes very limited gastrointestinal side effects. Some of the reports in the literature suggest that erythritol only causes diuresis and does not involve any other effects [55, 56]. Studies on effects of erythritol on reproductive, mutagenesis, and teratology did not show any adverse side effects on rats and mice [57–58]. In the diabetic patient, erythritol does not have any effect on the level of glucose, free fatty acid, 3-hydroxybutyric acid, and insulin [59]. Erythritol is known to be very safe and beneficial for use at a commercial level.

## 11.7 Applications of Erythritol

Erythritol has a wide range of applications in different fields including bakery, confectioneries, beverages, pharmaceuticals, etc. Due to the physical and chemical properties of erythritol, it has been widely used as flavor enhancer, nutritive sweetener, formulation aid, texturizer, and stabilizer and thickener at uttermost amounts of 100% in sugar substitutes, this including 60% in fat-based cream for use in pastries, cakes, and cookies, 40% in soft candies, 50% in hard candies, and 1.5% in low and reduced-calorie beverages [60]. Because of the non-cariogenic property of erythritol, it has been used as a non-cariogenic sweetener [61]. Erythritol as a zero-calorie sweetener has a positive enthalpy of solubilization (23.3 kJ $mol^{-1}$) and hence gives an excellent cooling effect on ingestion [62] (Figure 11.9). Some various applications of erythritol in different area are summarized below.

### 11.7.1 Confectioneries

Erythritol is used in several kinds of confectioneries items such as candies, chewing gums, chocolate, etc. In chocolate, erythritol reduces calorie level up to 34% in comparison of sucrose. In candies, usage of erythritol is more beneficial and safe than the xylitol and sorbitol [63]. Erythritol is also used in fresh mints as it provides the substantial cooling effect. Due to the low-calorie value, it is used for coating of chewing gums for delivering sweet flavor and also used in the center of chewing gums for providing flexibility.

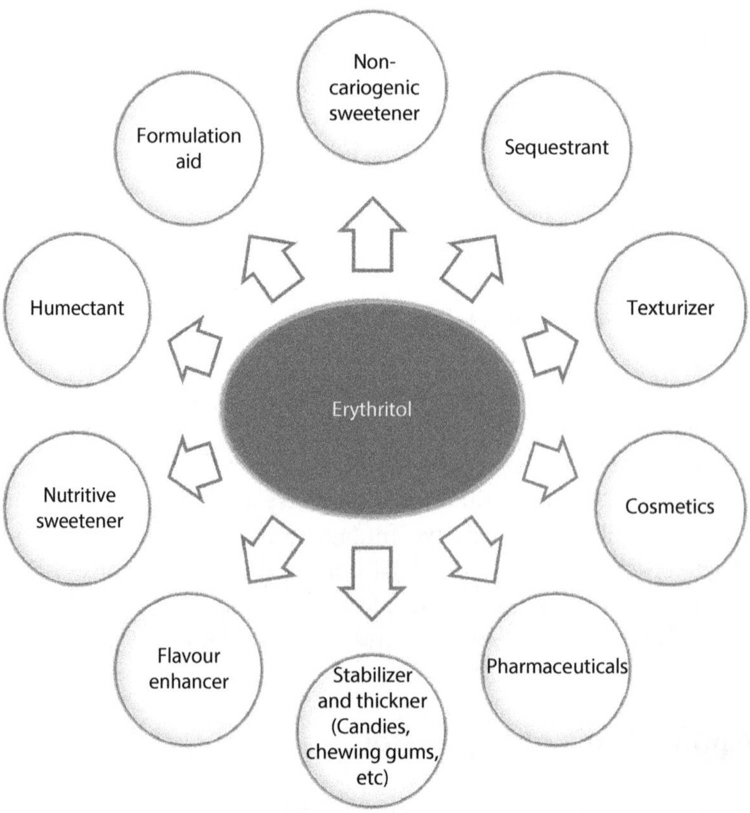


**Figure 11.9** Applications of erythritol.

### 11.7.2 Bakery

In bakery industries, erythritol is used in the manufacturing of pastries and biscuits. In pastry, erythritol is used in the cream, where it provides a sweet flavor with a refreshing sensation with a cooling effect, and at the level of 60%, it offers the best texture of the fat cream. Also, it gives more thick dough, less color formation, better moisture control, etc., than the sucrose. In some cases, for providing higher sweetness erythritol is often used in combination with maltitol.

### 11.7.3 Pharmaceuticals

Erythritol is different from other polyols as it is completely absorbed in small intestine and does not metabolize in the human body [10]. Because of its bioavailability profile, erythritol has additional health benefits for diabetic patients, since it reduces arterial stiffness and causes improvement in small vessel endothelial functionality [64]. Erythritol also provides better stability and flowability to the powder formulations so that the drug is not metabolized before its mechanism of action.

### 11.7.4 Cosmetics

Erythritol prevents the formation of plaque; therefore, it reduces the risk of dental caries. Thus, it is often used in oral care applications like toothpaste and mouthwash. It is well documented that erythritol inhibits activity and growth of cariogenic bacteria like *Streptococcus mutans*; therefore, it helps in the prevention of dental caries. It decreases the production of lactic acid and adherence of common streptococcal bacteria to tooth surfaces and therefore results in decreased dental plaque weight and biofilm formation [63]. Erythritol prevents the caries with higher efficacy than xylitol and sorbitol [63]. Erythritol also helps in masking the irritant, astringent, and bitter tastes associated with some mouthwash.

### 11.7.5 Beverages

In 2015, erythritol was finally approved for use in beverages by the European Food Safety Authority (EFSA) with a maximum limit of 1.6% [65]. Its use in low or zero-calorie diet beverages provides sweet flavor and also helps in masking some off-flavors in drinks like coffee, tea, and grapefruit juice. Commercially, erythritol is used in many beverages like soft drinks, energy drinks, yogurt, non-alcoholic drinks, coffee, tea, grapefruits, etc.

## 11.8 Precautions for Erythritol Usage

The intake quantity for erythritol per unit weight is different in male and female adults. The intake quantity reported as 0.80 g/kg body weight of female adults and 0.66 g/kg body weight of male adults [66]. At these concentrations no laxation (diarrhea) would occur after the complete ingestion of erythritol. But the consumption of erythritol at higher concentrations may cause diarrhea. Among different sugar alcohols, erythritol is least likely to cause diarrhea. The acceptable daily intake (ADI) for erythritol for use as sweetening agent was established as "NOT SPECIFIED" by the 53rd report of the joint FAO/WHO expert committee on food additives (JECFA). It suggested that erythritol is very safe to be used as a sweetener in manufacturing different food products at commercial level.

## 11.9 Global Market for Erythritol

Sugar cane is used for the production of 80% of the world's sugar, and sugar beet meets remaining demand. Brazil (around 39%) is the largest sugar producer followed by India (about 26%). The market for sugar substitutes is dominated by North America, which accounts for roughly 49% of the total market.

Among all sugar alcohols, erythritol is attractive as a sugar substitute for industries because of its zero or very low-calorie value. It was first commercially produced in 1993 in Japan, and since then, it has been used in the manufacturing of bakery products, candies, chewing gums, cold drinks, etc. Erythritol was approved as GRAS (generally regarded as safe) by the US Food and Drug Administration (FDA) on September 11, 2001. As per WHO, over 1.9 billion people were overweight in 2014 because of the more calorie intake; out of these 600 million were obese, and this situation is more severe in children. Problems like this increase the demand for production of low-calorie sugar substitutes, and among all sugar alternatives, erythritol is getting popular because of its unique properties.

Several companies commercially produce erythritol including Cargill Food & Pharma Specialties (Minneapolis, MN), Mitsubishi Chemical Corporation (Tokyo, Japan), Baolingbao Biotechnology (Dezhou Shi, China), O'Laughlin Corporation (Tianjin, China), Foodchem International Corporation (China), Zhongshi Gerui Biotech (China), Jungbunzlauer (Switzerland).

## 11.10 Conclusions

Erythritol falls into the group of sugar alcohols and has characteristics of sweetness, non-reducing, very high temperature and acid stability, low molecular size, zero-calorie, and tooth-friendly. It shares many characteristics with other sugar alcohols, but it is the only sugar alcohol which is produced commercially by natural fermentation processes. It has a very significant metabolic profile that provides no calories to the body and is also highly digestive tolerant. After commercial production of erythritol in 1990, its usage and demand significantly increased in many applications such as bakery products, candies, chewing gums, cosmetics, therapeutics, etc. In 2015, it was also approved for use in beverages like cold drinks, coffee, etc. Commercially, it is produced in about 55 countries, and these numbers are still increasing. Due to its zero calorie characteristics, it is highly beneficial for a diabetic patient without compromising on best sensation "the sweetness," which also makes it an attractive candidate for industries.

## References


1. K.R. Tandel, Sugar substitutes: Health controversy over perceived benefits, *Journal of Pharmacology & Pharmacotherapeutics*, Vol. 2, p. 236, 2011.
2. M. Kroger, K. Meister, and R. Kava, Low-calorie Sweeteners and Other Sugar Substitutes: A Review of the Safety Issues, *Comprehensive Reviews in Food Science and Food Safety*, Vol. 5, p. 35, 2006.
3. J. Stenhouse, Examination of the Proximate Principles of Some of the Lichens, *Philosophical Transactions of the Royal Society of London*, Vol. 138, p. 63, 1848.
4. A. Lamy, Ueber einige Bestandtheile desProtococcus vulgaris, *Journal Fur Praktische Chemie*, Vol. 57, p. 21, 1852.
5. M. Bamberger, and A. Landsiedl, Erythritol in Trentepohlia jolithus, *Monatshefte Fuer Chemie*, Vol. 21, p. 571, 1900.
6. V.F. Pfeifer, V.E. Sohns, H.F. Conway, E.B. Lancaster, S. Dabic, and E.L. Griffin, Two-stage process for dialdehyde starch using electrolytic regeneration of periodic acid, *Industrial and Engineering Chemistry*, Vol. 52, p. 201, 1960.
7. H. Roeper, and J. Goossens, Erythritol: a new raw material for food and non-food applications, *Starch-Staerke*, Vol. 45, p. 400, 1993.
8. P.D. Cock, "Erythritol" in K. O'Donnell and M.W. Kearsley, eds. 2, *Sweeteners and Sugar Alternatives in Food Technology*, John Wiley & Sons, pp. 215, 2012.
9. M. Grembecka, Sugar alcohols-their role in the modern world of sweeteners: a review, *European Food Research and Technology*, Vol. 241, p. 1, 2015.
10. F.R. Bornet, A. Blayo, F. Dauchy, and G. Slama, Plasma and urine kinetics of erythritol after oral ingestion by healthy humans, *Regulatory Toxicology and Pharmacology*, Vol. 24, p. S280, 1996.
11. T. Oku, and A.S. Nakamura, Digestion, absorption, fermentation, and metabolism of functional sugar substitutes and their available energy, *Pure and Applied Chemistry*, Vol. 74, p. 1253, 2002.
12. K.K. Mäkinen, M. Saag, K.P. Isotupa, J. Olak, R. Nõmmela, E. Söderling, and P.L. Mäkinen, Similarity of the effects of erythritol and xylitol on some risk factors of dental caries, *Caries Research*, Vol. 39, p. 207, 2005.
13. R.G.M. van der Sman, Predicting the solubility of mixtures of sugars and their replacers using the Flory-Huggins theory, *Food & Function*, Vol. 8, p. 360, 2017.
14. M.J. O'Neil, "an encyclopedia of chemicals, drugs, and biologicals" in The Merck index, eds. 14, Merck, 2006.
15. D.R. Lide, *CRC Handbook of Chemistry and Physics*, Florida, Taylor & Francis, 2007.
16. K.J. Lee, J.Y. Lim, Optimized conditions for high erythritol production by *Penicillium* sp. KJ-UV29, mutant of *Penicillium* sp. KJ81, *Biotechnology and Bioprocess Engineering*, Vol. 8, p. 173, 2003.
17. F. Feigl, and V. Anger, *Spot tests in organic analysis*, Amsterdam, Elsevier, 2012.
18. S.H. Bok, and A.L. Demain, An improved colorimetric assay for polyols, *Analytical Biochemistry*, Vol. 81, p. 18, 1977.
19. J. Lu, L.B. Tsai, C.S. Gong, and G.T. Tsao, Effect of nitrogen sources on xylitol production from D-xylose by *Candida* sp. L-102, *Biotechnology Letters*, Vol. 17, p. 167, 1995.
20. A. Converti, and J.M. Domínguez, Influence of temperature and pH on xylitol production from xylose by *Debaryomyces hansenii*, *Biotechnology and Bioengineering*, Vol. 75, p. 39, 2001.
21. F.C. Sampaio, H.C. Mantovani, F.J.V. Passos, C.A. de Moraes, A. Converti, and F.M.L. Passos, Bioconversion of d-xylose to xylitol by *Debaryomyces hansenii* UFV-170: Product formation versus growth, *Process Biochemistry*, Vol. 40, p. 3600, 2005.

22. S.M. Park, B.I. Sang, D.W. Park, and D.H. Park, Electrochemical reduction of xylose to xylitol by whole cells or crude enzyme of *Candida peltata*, *Journal of Microbiology*, Vol. 43, p. 451, 2005.
23. S.I. Mussatto, J.C. Santos, W.C. Ricardo Filho, and S.S. Silva, A study on the recovery of xylitol by batch adsorption and crystallization from fermented sugarcane bagasse hydrolysate, *Journal of Chemical Technology and Biotechnology*, Vol. 81, p. 1840, 2006.
24. E.C.H. Wan, and J.Z. Yu, Determination of sugar compounds in atmospheric aerosols by liquid chromatography combined with positive electrospray ionization mass spectrometry, *Journal of Chromatography*, Vol. 1107, p. 175, 2006.
25. P. Bhandari, N. Kumar, B. Singh, and V.K. Kaul, Simultaneous determination of sugars and picrosides in *Picrorhiza* species using ultrasonic extraction and high-performance liquid chromatography with evaporative light scattering detection, *Journal of Chromatography*, Vol. 1194, p. 257, 2008.
26. K. Ohsawa, Y. Yoshimura, S. Watanabe, H. Tanaka, K. Imaeda, A. Yokota, and K. Tamura, Determination of xylitol in the human serum and saliva by ion chromatography with pulsed amperometric detection, *Analytical Sciences*, Vol. 2, p. 165, 1986.
27. C. Cheng, C.S. Chen, and P.H. Hsieh, On-line desalting and carbohydrate analysis for immobilized enzyme hydrolysis of waste cellulosic biomass by column-switching high-performance liquid chromatography, *Journal of Chromatography*, Vol. 1217, p. 2104, 2010.
28. T.H. Wang, Z.H. Wu, S.L. Liu, and Y.B. Qu, Construction of heterogeneous genes expression system of filamentous fungus *Trichoderma reesei*, *Chinese Journal of Biochemistry and Molecular Biology*, Vol. 19, p. 736, 2003.
29. A.G. Coelho, and D.P. de Jesus, A simple method for determination of erythritol, maltitol, xylitol, and sorbitol in sugar-free chocolates by capillary electrophoresis with capacitively coupled contactless conductivity detection, *Electrophoresis*, Vol. 37, p. 2986, 2016.
30. L. Nabors, *Alternative sweeteners*, New York, Marcel Dekker, Inc., 2004.
31. N.R. Bacher, and F.A., Trenner, Hydrogenation of tartaric acid esters to erythritol, US Patent 2571967A, 1951.
32. D.A. Rzechonek, A. Dobrowolski, W. Rymowicz, and A.M. Mirończuk, Recent advances in biological production of erythritol, *Critical Reviews in Biotechnology*, p. 1, 2017.
33. M. Veiga-da-Cunha, H. Santos, and E.V. Schaftingen, Pathway and regulation of erythritol formation in *Leuconostoc oenos*, *Journal of Bacteriology*, Vol. 175, p. 3941, 1993.
34. T. Zaunmüller, M. Eichert, H. Richter, and G. Unden, Variations in the energy metabolism of biotechnologically relevant heterofermentative lactic acid bacteria during growth on sugars and organic acids, *Applied Microbiology and Biotechnology*, Vol. 72, p. 421, 2006.
35. G.J. Hajny, J.H. Smith, and J.C. Garver, Erythritol production by a yeast like fungus, Applied Microbiology, Vol. 12, p. 240, 1964.
36. M. Ishikawa, M. Miyashita, Y. Kawashima, T. Nakamura, N. Saitou, and J. Modderman, Effects of oral administration of erythritol on patients with diabetes, Regulatory Toxicology and Pharmacology, Vol. 24, p. S303, 1996.
37. Y. Hirata, K. Igarashi, S. Ezaki, H. Atomi, and T. Imanaka, High-level production of erythritol by strain 618A-01 isolated from pollen, *Journal of Bioscience and Bioengineering*, Vol. 87 (5), p. 630, 1999.
38. S.J. Lin, C.Y. Wen, C.C. Huang, and W.S. Chu, Erythritol producing *Moniliella* strains, US Patent 0008378 A1, 2003.
39. E.S. Kohl, T.H. Leet, Y.D. Kim, H.J. Ryu, Y.W. Seo, and J. Ho, Scale-up of erythritol production by an osmophilic mutant of *Candida magnolia*, *Biotechnology Letters*, Vol. 25 (24), p. 2103, 2003.

40. W. Rymowicz, R. Anita, and M. Marta, High-yield production of erythritol from raw glycerol in fed-batch cultures of *Yarrowia lipolutica*, *Biotechnology Letters*, Vol. 31 (3), p. 377, 2009.
41. M. Jeya, K.M. Lee, M.K. Tiwari, J.S. Kim, P. Gunasekaran, S.Y. Kim, and J.K. Lee, Isolation of a novel high erythritol-producing *Pseudozyma tsukubaensis* and scale-up of erythritol fermentation to industrial level, *Applied Microbiology and Biotechnology*, Vol. 83 (2), p. 225, 2009.
42. J. Park, B. Seo, J. Kim, and Y. Park, Production of erythritol in fed-batch cultures of *Trichosporon* sp., *Journal of Fermentation and Bioengineering*, Vol. 86 (6), p. 577, 1998.
43. D.K. Oh, C.H. Cho, J.K. Lee, and S.Y. Kim, Increased erythritol production in fed-batch cultures of *Torula* sp. by controlling glucose concentration, *Journal of Industrial Microbiology and Biotechnology*, Vol. 26 (4), p. 248, 2001.
44. H. Ishizuka, K. Wako, T. Kasumi, and T. Sasaki, Breeding of a mutant of *Aureobasidium* sp. with high erythritol production, *Journal of Fermentation and Bioengineering*, Vol. 68 (5), p. 310, 1989.
45. L. Agarwal, J. Isar, G.K. Meghwanshi, and R.K. Saxena, Influence of environmental and nutritional factors on succinic acid production and enzymes of reverse tricarboxylic acid cycle from *Enterococcus flavescens*, *Enzyme and Microbial Technology*, Vol. 40, p. 629, 2007.
46. R.K. Saxena, S. Misra, I. Rawat, P. Gupta, K. Dutt, and V.S. Parmar, Production of 1,3 regiospecific lipase from *Bacillus* sp. RK-3: Its potential to synthesize Cocoa Butter Substitute, *Malaysian Journal of Microbiology*, Vol. 7, p. 41, 2011.
47. C.S. Gong, T.A. Claypool, L.D. McCracken, C.M. Maun, P.P. Ueng, and G.T. Tsao, Conversion of pentoses by yeasts, *Biotechnology and Bioengineering*, Vol. 25, p. 85, 1983.
48. L.S. Savergave, R.V. Gadre, B.K. Vaidya, and K. Narayanan, Strain improvement and statistical media optimization for enhanced erythritol production with minimal by-products from *Candida magnoliae* mutant R23, *Biochemical Engineering Journal*, Vol. 55, p. 92, 2011.
49. Y.W. Ryu, C.Y. Park, J.B. Park, S.Y. Kim, and J.H. Seo, Optimization of erythritol production by *Candida magnoliae* in fed-batch culture, *Journal of Industrial Microbiology and Biotechnology*, Vol. 25, p. 100, 2000.
50. L.P. Ooijkaas, E.C. Wilkinson, J. Tramper, and R.M. Buitelaar, Medium optimization for spore production of *coniothyrium minitans* using statistically-based experimental designs, *Biotechnology and Bioengineering*, Vol. 64, p. 92, 1999.
51. K.J. Rao, C.H. Kim, S.K. Rhee, Statistical optimization of medium for the production of recombinant hirudin from *Saccharomyces cerevisiae* using response surface methodology, *Process Biochemistry*, Vol. 35, p. 639, 2000.
52. S. Saxena, and R.K. Saxena, Statistical optimization of tannase production from *Penicillium* variable using fruits (chebulic myrobalan) of Terminalia chebula, *Biotechnology and Applied Biochemistry*, Vol. 39, p. 99, 2004.
53. R. Kaushik, R.G. Marwah, P. Gupta, S. Saran, L. Saso, V.S. Parmar, and R.K. Saxena, Optimization of lipase production from *Aspergillus terreus* by response surface methodology and its potential for synthesis of partial glycerides under solvent free conditions, *Indian Journal of Microbiology*, Vol. 50, p. 456, 2010.
54. K. Noda, K. Nakayama, and T. Oku, Serum glucose and insulin levels and erythritol balance after oral administration of erythritol in healthy subjects, *European Journal Clinical Nutrition*, Vol. 48, p. 286, 1994.
55. F.F. Beck, C.J. Carr, and J.C. Krantz, Acute toxicity of certain sugar alcohols and their anhydrides, *Proceedings of the Society for Experimental Biology and Medicine*, Vol. 35, p. 98, 1936.
56. F.F. Beck, C.J. Carr, and J.C. Krantz, Sugar alcohols XVI. The fate of erythritol and erythritan in the animal body, *Quarterly Journal of Pharmacy and Pharmacology*, Vol. 11, p. 234, 1938.

57. M. Shimizu, M. Katoh, M. Imamura, and J. Modderman, Teratology study of erythritol in rabbits, *Regulatory Toxicology and Pharmacology*, Vol. 24, p. S247, 1996.
58. D.H. Waalkens-Berendsen, A.E. Smits-van Prooije, M. V Wijnands, and A. Bär, Two-generation reproduction study of erythritol in rats, *Regulatory Toxicology and Pharmacology*, Vol. 24, p. S237, 1996.
59. M. Ishikawa, M. Miyashita, Y. Kawashima, T. Nakamura, N. Saitou, and J. Modderman, Effects of oral administration of erythritol on patients with diabetes, *Regulatory Toxicology and Pharmacology*, Vol. 24, p. S303, 1996.
60. H.J. Moon, M. Jeya, I.W. Kim, and J.K. Lee, Biotechnological production of erythritol and its applications, *Applied Microbiology and Biotechnology*, Vol. 86, p. 1017, 2010.
61. K.K. Mäkinen, K.P. Isotupa, T. Kivilompolo, P.L. Mäkinen, J. Toivanen, and E. Söderling, Comparison of erythritol and xylitol saliva stimulants in the control of dental plaque and mutans streptococci, *Caries Research*, Vol. 35, p. 129, 2001.
62. J. Goossens, and H. Roper, Erythritol: A new bulk sweetener, *International Food Ingredients*, Vol. 1/2, p. 27, 1994.
63. P. de Cock, K. Mäkinen, E. Honkala, M. Saag, E. Kennepohl, and A. Eapen, Erythritol is more effective than xylitol and sorbitol in managing oral health endpoints, *International Journal of Dentistry*, Vol. 2016, p. 15, 2016.
64. D.M.P.H.J. Boesten, G.J.M. den Hartog, P. de Cock, D. Bosscher, A. Bonnema, and A. Bast, Health effects of erythritol, *Nutrafoods*, Vol. 14, p. 3, 2015.
65. Scientific Opinion on the safety of the proposed extension of use of erythritol (E 968) as a food additive, *European Food Safety Authority Journal*, Vol. 13, p. 23, 2015.
66. T. Oku, and M. Okazaki, Laxative threshold of sugar alcohol erythritol in human subjects, *Nutrition Research*, Vol. 16, p. 577, 1996.

# 12

# Sugar and Sugar Alcohols: Xylitol

**Bhumica Agarwal[1] and Lalit Kumar Singh[2]***

[1]*Department of Biotechnology, Meerut Institute of Engineering and Technology, Meerut, (India)*
[2]*Department of Biochemical Engineering, School of Chemical Technology, Harcourt Butler Technical University Kanpur, (India)*

***Abstract***
Xylitol, a five carbon sugar alcohol, also known as wood or birch sugar, is obtained by reduction of xylose. It is a rare sugar that occurs naturally, albeit in small amounts, in fruits, vegetables, lichens, mushrooms, etc. Thus extracting xylitol from these sources is difficult and not economical. Xylitol is a sweetener with outstanding anticariogenic properties. It has attracted global interest due to its sweetening power similar to that of sucrose and besides this there is no insulin requirement for its digestion. Xylitol has recently become attractive as an alternative sweetener for the treatment of diabetics and glucose-6-phosphate dehyrogenase deficient diseases. The extraction of xylose from the lignocellulosic biomass and bioconversion of xylose to xylitol via fermentation is encapsulated in the present chapter.

***Keywords:*** Xylose, fermentation, xylitol, biomass, sweetener

## 12.1 Introduction

Use of sugar in households seems to be irreplaceable, especially in Indian systems. Although a new generation is becoming familiar with associated health disadvantages, table sugar remains a main component of our diet. It has been well established now that diseases like epidemic obesity, diabetes, and arthritis are due to high consumption of sugar in the diet and sedentary lifestyle. Although non-caloric sweeteners have been available commercially since the late 1800s [1], the increase in their unprecedented use has been witnessed only recently, with the advent of commercial biotechnological production of these sweeteners. Most of the alternative sweeteners were prepared on a large scale by chemical routes leading to ethical and toxicity issues in the minds of layman. It is only with the advent of advances in fermentative production that their use has been widely accepted in common households. Another factor initiating the same is routing

**Corresponding author*: lalitdbtiitr@gmail.com

Saurabh Saran, Vikash Babu, and Asha Chaubey (eds.) High Value Fermentation Products, Volume 1, (285–308) © 2019 Scrivener Publishing LLC

of lignocellulosic sugar into commercial production of non-caloric sweeteners making their production cheap and thus within reach of masses. The role of aggressive marketing strategies in promoting their capacity to induce weight loss and maintenance could not be undermined [2].

Sugar alcohols, also known as alditols, are a class of polyols in which sugar's carbonyl group (aldehyde or ketone) is reduced to the corresponding primary or secondary hydroxyl group. Characteristically, they are similar to sugar and are therefore increasingly being used to accentuate the nutritional profile of food products on account of their other health-associated benefits [3]. Sugar alcohols, such as xylitol, mannitol, sorbitol, and erythritol are emerging sugar-alternatives owing to their lower calorie content than sucrose with similar/ better sensory properties. Such polyols are non-cariogenic, possess low glycemic index and insulin response. Due to these properties, sugar alcohols are widely and increasingly used in food, beverage, confectionery, cosmetic, animal nutrition and pharmaceutical industries [4]. In nature they are found in minimal quantities in fruits, vegetables or mushrooms as well as in human organism. They are produced by microorganisms, where they serve as carbohydrate and reducing power reserves, translocatory compounds, and osmoprotectants. Chemical synthesis of sugar alcohols pertains to catalytic hydrogenation of crude carbohydrates whereas biotechnological production route offers non-toxic and sustainable supply from renewable biomass.

### 12.1.1 Lignocellulosic Biomass

Lignocellulosic biomass has served as a substrate in microbial processes for the production of alcoholic beverages for thousands of years. It is only recently that broader applications of this material have been envisaged. Lignocellulosics are an abundant, inexpensive and renewable source of organic material. They are composed of cellulose, hemicellulose and lignin. On hydrolysis, they serve as a rich source of sugar monomers like glucose, galactose, mannose, xylose and arabinose. Biotechnologists are now developing efficient systems for the production of liquid fuels, pharmaceuticals, foods and chemical feedstocks from such waste organic materials. Since hydrolysis of xylan (polymer of xylose) component of hemicellulosic fraction of lignocellulosics is comparatively easier than that of cellulose yielding a xylose rich hydrolysate, more research is now focused on development of novel products being derived from xylose. Products which can be derived from this rich source of xylose include ethanol [5], lactic acid [6], butanediol [7], microbial protein [8], acetic acid [9] and xylitol [10]. In addition to production of some high-value products, bioconversion of lignocellulosics helps combating environmental threats posed by large amounts of lignocellulosic wastes generated through forestry and agricultural practices, paper-pulp industries, timber industries and many agro-industries. Generally they are incinerated or disposed off in the open. This poses a threat to the environment along with wastage of valuable lignocellulosic materials which can otherwise be converted into useful products like biofuels, chemicals, and cheap energy sources for fermentation, improved animal feeds and human nutrients [11]. The composition of various extensively studied lignocellulosic materials in terms of cellulose, hemicellulose and lignin is presented in Table 12.1 [12].

**Table 12.1** Composition of various lignocellulosic materials [12].

| Lignocellulosic biomass | Cellulose (%) | Hemicellulose (%) | Lignin (%) |
|---|---|---|---|
| Hardwoods stems | 40–55 | 24–40 | 18–25 |
| Softwood stems | 45–50 | 25–35 | 25–35 |
| Nut shells | 25–30 | 25–30 | 30–40 |
| Corn cobs | 45 | 35 | 15 |
| Grasses | 25–40 | 35–50 | 10–30 |
| Wheat straw | 30 | 50 | 20 |
| Sorted refuse | 60 | 20 | 20 |
| Leaves | 15–20 | 80–85 | 0 |
| Cotton seed hairs | 85–95 | 5–20 | 0 |
| Newspaper | 40–55 | 25–40 | 18–30 |
| Waste papers from chemical pulps | 60–70 | 10–20 | 5–10 |
| Solid cattle manure | 1.6–4.7 | 1.4–3.3 | 2.7–5.7 |
| Coastal Bermuda grass | 25 | 35.7 | 6.4 |
| Switch grass | 45 | 31.4 | 12.0 |
| Water-hyacinth | 18.4 | 49.2 | 3.55 |

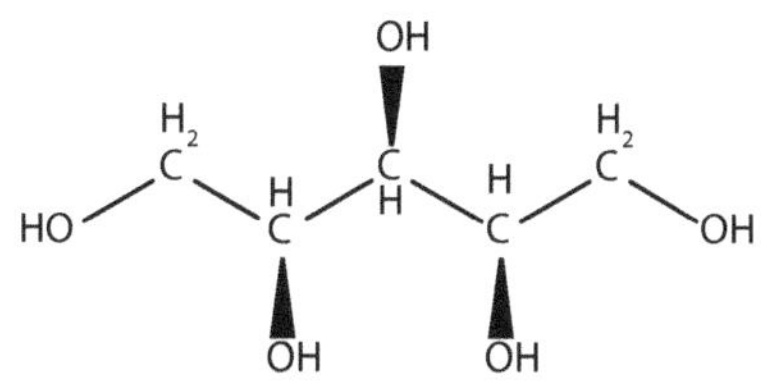

**Figure 12.1** Chemical Structure of Xylitol [14].

## 12.1.2 Properties of Xylitol

Xylitol ($C_5H_{12}O_5$) is a pentose sugar alcohol used as an alternate sweetener to sucrose. Emil Fisher and Stahel first synthesized xylitol in 1891 [13]. The chemical structure of xylitol is presented in Figure 12.1 [14]. However, interest in xylitol production and its use increased after almost five decades of its discovery, during the Second World War when sugar shortages forced scientists to locate alternate sweeteners. Gradually, interest in xylitol increased, when its insulin-independent characteristics and involvement in essential pentosuria (harmless, genetic disorder) was revealed. Since 1970 xylitol has been studied intensively due to its benefits in preventing dental caries [15]. It is also a normal metabolic intermediate in mammalian carbohydrate metabolism and is endogenously produced and further utilized (about 5–15 g) daily in adult human beings. Recently use of xylitol has increased manyfold in mainly three industries: food (for its sweetening power and insulin independent metabolism), odontological (for its anti-carcinogenicity, tooth re-hardening and demineralization properties) and pharmaceutical (for its capability of preventing otitis and use in syrups, tonics and vitamins formulations as an excipient) [16, 17]. The physical properties of xylitol is given in Table 12.2 [18, 19].

Primary interest in xylitol, therefore, centers on its properties and potential uses as an alternative noncaloric sweetener. In contrast to other noncaloric sweeteners, such as

**Table 12.2** Physical properties of xylitol [18, 19].

| Property | Xylitol characteristics |
|---|---|
| Formula | $C_5H_{12}O_5$ |
| Molecular Weight | 152.15 |
| Appearance | White, crystalline powder |
| Odor | None |
| Solubility at 20 °C | 169 g/100 g $H_2O$ |
| pH in water (1 g/ 10 ml) | 5–7 |
| Melting Point(°C) | 93–94.5 |
| Boiling Point(at 760 mm Hg) | 216 °C |
| Density at 15 °C | 1.50 g/ L |
| Caloric value | 4.06 cal/ g (16.88 J/g) |
| Moisture absorption (%) (4 days, 20–22 °C)<br>at 60% relative humidity<br>at 92% relative humidity | <br>0.05<br>90 |
| Density of aqueous solution (20 °C)<br>10%<br>60% | <br>1.03<br>1.23 |
| Heat of solution | Endothermic, 36.61 cal/g (153.76 J/g) |
| Viscosity (cP) (20 °C) | |
| 10% | 1.23 |
| 40% | 4.18 |
| 50% | 8.04 |
| 60% | 20.63 |
| Relative sweetness | Equal to sucrose; greater than sorbitol and mannitol |
| Specific rotation | Optically inactive |

saccharine, xylitol has many properties similar to those of sucrose. It dissolves readily in water, it is as sweet as sucrose and approximately twice as sweet as sorbitol and nearly three times as sweet as mannitol. Its caloric content of 17 kJ/kg, is the same as that of sucrose. In addition, it gives a pleasant cool and fresh sensation due to its high negative heat of solution. Perhaps the most significant characteristic of xylitol, however, is the fact that it is not utilized by the acid producing, cariogenic bacteria of the human oral cavity and therefore inhibits demineralization of tooth enamel. It has been demonstrated by various researchers that consumption of even small amounts of xylitol can significantly reduce the formation of new dental caries.

Xylitol finds a broad application either as the sole sweetener or in conjunction with other sweeteners in the preparation of a wide variety of full-and reduced-energy sugarless confectionary products suitable for infants and diabetics. Additionally, xylitol extends the shelf life of food by inhibiting saprophytes such as *Clostridium butyrican* and *Salmonella typhii*. Therefore, xylitol is incorporated as an ingredient in various consumer products, such as chewing gum, soft drink, ice cream, and toothpaste.

### 12.1.3 Occurrence and Production of Xylitol

In nature, xylitol occurs widely in many fruits and vegetables in low concentration (<1% on dry basis) rendering xylitol extraction from them economically unattractive. The availability of xylitol in various fruits and vegetables is given in Table 12.3. Currently large-scale production of xylitol has been carried out by catalytic reduction of pure xylose obtained from hardwood hydrolysates, involving acid hydrolysis of xylan rich lignocellulosic material and subsequent conversion of D-xylose to xylitol in the presence of a Raney nickel catalyst [20].

Because the hemicellulose hydrolysate fraction also contains other sugars, the chemical process includes expansive separation and purification steps to remove these byproducts from xylose or xylitol as only pure xylose can be used for chemical reduction [21]. Also, overall xylitol yield is relatively low (50–60%) from the total xylan content of the wood hemicellulose [22]. These two factors increase the commercialization cost thereby decreasing xylitol's consumer market.

Alternatively, xylitol can be produced by biotechnological routes using similar feedstocks. One advantage of using a microbiological xylitol production is the possibility to use industrial side-streams as a raw material, which contains hemicellulose fractions without any multiple purification steps. This enables better use of renewable natural resources as well as ensures overall cost efficiency of the process.

## 12.2 Biomass Conversion Process

### 12.2.1 Pretreatment Methodologies

Complete substrate utilization is one of the prerequisites to render lignocellulosic xylitol process economically competitive. But the closely knitted structure of cellulose, hemicellulose and lignin in lignocelluloses offers resistance to microorganisms from acting on fermentable sugars. The biochemical production of xylitol from hemicellulosic

**Table 12.3** Xylitol presence in fruits and vegetables [19].

| Product | Xylitol (mg/100 g dry substance) |
|---|---|
| Brewer's Yeast | 4.5 |
| Carrot Juice | 12 |
| Chestnut | 14 |
| Banana | 21 |
| Leek | 53 |
| Carrot | 86.5 |
| Onion | 89 |
| Fennel | 92 |
| Kohlrabi | 94 |
| Lettuce | 96.5 |
| Pumpkin | 96.5 |
| Spinach | 107 |
| White mushroom | 128 |
| Eggplant | 180 |
| Endive | 258 |
| Raspberry | 268 |
| Lamb's Lettuce | 273 |
| Cauliflower | 300 |
| Strawberry | 362 |
| Yellow plum | 935 |

portion of lignocellulosic biomass involves conditioning the residues by preliminary treatment, hydrolysis of hemicellulosic components to sugars and fermentation of these to xylitol.

#### *12.2.1.1 Physical Pretreatment*

Physical pretreatment of lignocellulosic biomass excludes the use of any chemical agent. It includes pyrolysis, liquid hot water treatment, mechanical comminution [23, 24]. Pretreatment with gamma-ray, electron-beam and microwave irradiation are also under investigation. Feedstocks can be comminuted by an arrangement of chipping, grinding and milling to reduce cellulose crystallinity. The size of the materials reduced to 10–30 mm after chipping and 0.2–2 mm after milling or grinding. Vibratory ball milling is also an effective method to reduce cellulose crystallinity of spruce and aspen chips. The power requirement of mechanical comminution of lignocellulosic materials depends on the final particle size. Mechanical disruption of lignocellulosic material also depolymerizes lignin [25]. Pyrolysis of lignocellulosic materials, at temperatures greater than 300 °C rapidly break down cellulose and produce gaseous products and char [26, 27]. The break down may be much slower and less volatile products are formed at lower temperatures. Zinc chloride or sodium carbonate is used to break down pure cellulose at lower temperature. Irradiations are also investigated and found to be an effective means of break down the complex structure of lignocellulosic biomass. Yang *et al.* [28] used gamma irradiation at 500 kGy and glucose yield of 13.40% was obtained from wheat straw. Microwave irradiation of 700 W caused significant weight loss in lignocellulosic biomass due to break down of its constituents [29].

### *12.2.1.2 Physico-Chemical and Chemical Pretreatment*

The physico-chemical pretreatment process includes mainly steam explosion (autohydrolysis), ammonia fiber explosion (AFEX), $CO_2$ explosion, $SO_2$ explosion [30–32]. The steam explosion is one of the most extensively used pretreatment methods [23]. Steam at 0.69–4.83MPa used for few second to a few min then material is exposed to the atmospheric pressure. Chip size, moisture content steam pressure and residence time are the key factors to affect steam explosion [33]. Major drawback of steam explosion includes the destruction of xylan and incomplete lignin transformation, and generation of degradation compounds that are inhibitory to microbial growth, enzymatic hydrolysis, and fermentation needs to be washed by water thus reducing the overall yield of saccharification. Similar to steam explosion ammonia fiber explosion is used for pretreatment of biomass. In an AFEX process, 1–2 kg ammonia per kg dry biomass is used at 90 °C for 30 minutes of residence time, and then the pressure is suddenly reduced. This process is effective only for the biomasses of low lignin content [34, 35]. Further ammonia must be recovered and recycled after pretreatment to protect the environment. $CO_2$ explosion is also used for pretreatment of biomass in similar way to the above mentioned two processes but the sugar yield is low as compared to steam or ammonia explosion pretreatment after enzymatic hydrolysis. $CO_2$ explosion was found cost effective in comparison to steam explosion and AFEX processes [36].

The chemical pretreatment process includes mainly ozonolysis, dilute acid hydrolysis, alkaline hydrolysis, organosolv [37–39]. The ozonolysis pretreatment process effectively removes lignin and does not produce toxic compounds. Although reaction is carried out at room temperature and pressure [40] but the large quantity of ozone requirement makes this process very expensive [41]. The dilute acid hydrolysis is one of the most extensively studied methods of pretreatment [42–45]. The dilute acid process for hydrolysis of lignocellulosic biomass consist essentially two stages; whereas the first stage is conducted under mild process conditions (e.g., 0.7% sulfuric acid, 190 °C) to produce 5-carbon sugars from hemicellulosic fraction, while in the second stage remaining solid i.e., more resistant cellulose is hydrolyzed under harsher conditions (215 °C, 0.4% acid) to produce 6- carbon sugars. The sugar yield of this process is low as compared to the other processes because of the severe conditions used the produced sugars gets converted to further degradation products like xylose converted to furfural and glucose converted to hydroxymethyl furfural which are also inhibitory to the microorganisms used in the subsequent fermentation process [46]. Therefore, an additional process step known as detoxification of hydrolysate is used to remove or reduce the concentration of these inhibitory compounds prior to fermentation leading to enhance the overall cost of the process. Alkaline pretreatment uses bases like sodium hydroxide or calcium hydroxide. This process is mainly used for delignification but it also solubilizes hemicellulose to some extent, thus reducing the overall fermentable sugar yield. Furthermore, this process is effective only for the low lignin content biomasses like straws, herbaceous biomass but not for high lignin content biomasses like softwood [47]. Iyer *et al.* [48] used ammonia recycled percolation process at 170 °C for 1 h for treatment of corn stover and switchgrass. The concentration of ammonia was ranged from 2.5 to 20%. The delignification of both the substrate was found 60–85%. The organosolv pretreatment process typically use an organic or aqueous solvent mixture like methanol, ethanol, acetone, ethylene glycol, etc., in combination with dilute mineral acids like sulfuric acid, hydrochloric acid as catalyst [41, 49]. Some organic acids can also be used as catalysts in like oxalic acid, salicylic acid, etc. It is used for fractionation of biomass to hemicellulose, cellulose and lignin leaving a highly digestible substrate. The recovery and recycling of solvents from the process vessel is essentially required as the solvents may be inhibitory to the microorganisms and enzymes which make this process too expensive [41].

#### *12.2.1.3 Microbial Pretreatment*

In the biological pretreatment processes, mainly fungi are used to degrade lignin known as biodelignification. Fan *et al.* [50] and Akin *et al.* [51] reported that white-rot fungi like *Phanerochaete chrysosporium* are the most effective *Basidiomycetes* for biological pretreatment of lignocellulosic biomasses. Although biological pretreatment has advantages of low energy use at mild reaction conditions, the very low hydrolysis rate (few days to weeks) impedes its implementation. Other well-studied microorganisms are *Aspergillus terreus, Trichoderma spp., Pleurotus spp., Lentinus edodes, Streptomyces griseus* and genetically engineered *Escherichia coli* and *Klebsiella oxytoca* [52–56].

### 12.2.2 Enzymatic Hydrolysis

The abovementioned pretreatment processes are intended to disturb the crystalline structure of cellulose, remove lignin and in some cases hemicellulose too. All these phenomena lead to increase the porosity of the lignocellulosic biomass. Thus hydrolytic enzymes are allowed to act into the fibers for efficient hydrolysis of the loosen polymer chains of cellulose and hemicellulose [57]. Three main groups of cellulases are required to hydrolyze cellulose viz. endoglucanase (act on low crystalline region to release free chain ends), exoglucanase (to produce cellobiose units from the free chain ends), and β-glucosidase (to produce glucose from cellobiose units). For hydrolysis of hemicellulose a number of enzymes are required viz. xylanase, β-xylosidase, galactomannanase, glucomannanase and others [58, 59]. The pretreatment processes in which lignin is not removed ligninolytic enzymes are used to lignin degradation. Lignin peroxidase, manganese peroxidase and laccase are the main lignin hydrolyzing enzymes [60].

### 12.2.3 Detoxification Techniques

Regardless of the method employed for hydrolysate preparation from lignocellulosic materials, a number of inhibitory compounds are generated that are toxic to microorganisms used for xylose fermentation. This may result in inefficient utilization of sugars with increased fermentation time and decreased product formation. Comparison of fermentation yields obtained with commercial sugars, detoxified hydrolysate and untreated hydrolysate reveals that fermentation of untreated hydrolysates are prone to slow microbial kinetics along with decreased productivities. In order to overcome the inhibitory effects of such compounds a number of detoxification methods— chemical, physical and biological—have been suggested. Detoxification is achieved by either conversion of inhibitors into nontoxic compounds or by reducing their concentrations in the hydrolysates. The suitability of any detoxification method depends upon the composition of hemicellulosic hydrolysate which in turn depends upon the raw material and the hydrolysis conditions employed [61]. Detoxification methods are specific to particular type of inhibitors present in the hydrolysate. Therefore it is necessary to study the composition of hydrolysate before submitting it to any detoxification process. It has also been demonstrated that a combination of various physical, chemical and biological methods is more effective than detoxification by a single treatment process [62].

#### *12.2.3.1 Chemical Methods*

***Change of pH***

Precipitation of toxic compounds by pH adjustments using acids and bases either alone or in combinations is a low-cost treatment giving good fermentation results [63]. Detoxification by pH adjustments serves two functions. Primarily, at pH 6.5 acetate and acidic molecules exist in their unprotonated, less toxic form and secondarily change in pH leads to precipitation of toxic compounds including heavy metals, furfurals, acetate, tannins, terpenes and phenolics [61]. A number of alkalis can be used for precipitation of toxic compounds. Comparing the effect of three bases namely $Ca(OH)_2$, CaO and KOH in detoxifying the hemicellulose hydrolysate it has been established that over titrating with $Ca(OH)_2$ is the best approach leading to 95% consumption of xylose as compared to only 56% consumption when KOH is used. Although overtitrating with CaO lead to 98% xylose consumption, xylitol yield obtained with CaO was only 0.36 g xylitol produced/ g xylose consumed as compared to 0.48 g xylitol produced/ g xylose obtained with $Ca(OH)_2$ [63].

Overliming with $Ca(OH)_2$ in combination with sulphite addition has been shown to be the most efficient method to detoxify willow hemicellulose hydrolysate prior to fermentation by recombinant *Escherichia coli* [64]. When the hydrolysate was supplemented with sodium sulphite (0.1%) and heated (90 °C, 30 min), in addition to the overliming treatment, the fermentation time was reduced by a factor three along with complete utilization of xylose in 40 h, whereas only 24% of xylose was consumed in the untreated hydrolysate. On the other hand, adjustment to pH 10.0 with KOH, readjustment to pH 6.5 with HCl and addition of 1% sodium sulphite at room temperature has been found to be the most efficient method of improving fermentation of a hemicellulose hydrolysate of sugar cane bagasse with *Pichia stipitis* [5]. In another experiment, an appreciable reduction of 61.15% in lignin-derived compounds have been achieved by treating the hydrolysate firstly with $Ca(OH)_2$ upto pH 10.0 followed by filtration and addition of $H_2SO_4$ upto pH 5.5. Finally, addition of sodium sulfite (0.1% w/w) at room temperature leads to precipitation of the phenolic inhibitors [65].

***Adsorption on ion exchange resin***

Anionic exchange resin is considered to be one of the best detoxification methods. Using weak anion exchange resins equilibrated with 0.1M NaOH, at pH 5.5 and hydrolysate: resin ratio of 5:1, 83% reduction in phenolic compounds could be achieved after a period of 1 h [66]. Wood hydrolysate, when submitted to anion exchange resins, corresponded to removal of high percentages of toxic compounds. The treated hydrolysate contained reduced quantities of inhibitors: acetic acid (96% removal), phenolic compounds (91%), furfural (73%) and hydroxymethylfurfural (70%) [67].

***Adsorption on activated charcoal or carbon***

The effect of adsorption by activated charcoal and carbon pellets on inhibitors and xylose concentrations have been compared. It has been established that after 6 h of contact time, activated carbon adsorbed considerable amount of xylose along with inhibitors decreasing its concentration from 22.7 g/l to 8.4 g/l. However, activated charcoal pellets selectively removed inhibitors and water thus increasing xylose concentration slightly in the hydrolyzate. Fermentation of charcoal treated hydrolysate resulted in a yield of 0.74 g xylitol/ g xylose within 36 h [68]. By treatment of 3% (w/v) activated charcoal for 1 h at 30 °C, significant reduction in acetic acid, furfural and hydroxymethylfurfural concentrations could be achieved, final concentrations being 1.23, 0.05 and 0.1 g/kg respectively. Fermentation of above treated hydrolysate after been concentrated by four folds, yielded 0.80 g xylitol/g xylose with *Candida guillermondii* after 72 h [69].

Comparative studies have been carried out to determine the optimum concentration of activated charcoal required for effective detoxification of acid treated and three-fold concentrated wheat straw hydrolysate by its adsorption on 2.5% and 10% activated charcoal for 1 h at 30 °C and 200 rpm. With 2.5% activated charcoal, slightly higher xylitol production was obtained (28.6 g xylitol /h) with a productivity of 0.40 g/l suggesting that only 2.5% activated charcoal is sufficient for detoxification at prevailing conditions [70].

***Solvent extraction***

Salient features of solvent extraction method for detoxification of hemicellulose hydrolysate are: good fermentation performance, easy recovery of solvents which leads to improved process economy and easy recovery of solubilized lignin fractions which might be used as a cheap, renewable source of food anti-oxidants. Extraction efficiency depends upon the nature of solvent employed for extraction, operating pH and hydrolysate to solvent volume ratio. It has been found that ethyl acetate is more effective in removal of phenolics (88.1% removal) than diethyl ether (79.9% removal). With ethyl acetate as extraction solvent best results were obtained at pH 3 with hydrolysate: solvent ratio of 1:3 as it resulted in highest polyphenolics removal and a maximum xylitol yield of 0.511 g/ g xylose consumed after 47 h of fermentation with hemicellulose hydrolysate obtained by single stage extraction with ethyl acetate [71].

Use of alkaline hydrogen peroxide (AHP) has also been suggested for extraction of hemicellulose fractions from destarched corn fiber. Yields of the water-soluble hemicellulose B ranged from 35% (24 h extraction at 25 °C) to 42% (2 h extraction at 60 °C). AHP treatment caused delignification and facilitated the alkaline extraction of hemicellulose from the lignocellulosic fiber matrix. In the absence of $H_2O_2$, yields were reduced by more than one-third when using otherwise identical extraction conditions of time, temperature and pH. In the standard protocol, corn fiber, NaOH solution, and $H_2O_2$ were mixed in a 1:25:0.25 (w/v/w) ratio. Extractions were conducted at pH 11.5 at 25 °C or 60 °C. The pH was adjusted to 11.5 by addition of NaOH at ambient and elevated temperatures. The optimum hemicellulose yield (51.3%; dry, starch-free basis) was obtained when the pH was increased to 12.5 for the final one-half of the extraction period. Under most conditions, the yields of hemicellulose B, potentially the most useful form for food and industrial applications exceeded those of hemicellulose A by more than 10-fold [72].

### 12.2.3.2 *Physical Method*

Evaporation of the hydrolysate is an important step as it leads to an increase in sugar concentration as well as acts as a detoxifying agent by reducing the concentration of volatile compounds such as acetic acid, furfural and vanillin. Though not suitable for removal of phenolics, complete removal of furfurals and partial removal of aliphatic acids have been reported by evaporation of raw and detoxified hydrolysate [67, 73, 74]. However, this method also moderately increases the concentration of non-volatile toxic compounds (extractives and lignin derivatives) and consequently the degree of fermentation inhibition. As reported, though acetic acid concentration was slightly higher than the critical acetic acid concentration (4.0 g/l), evaporation resulted in improved fermentation characteristics. In another study total removal of furfural from wood hemicellulosic hydrolysate has been attained by reducing its volume by 90% through vacuum evaporation [67].

### 12.2.3.3 *Biological Method*

***Use of enzymes or microorganisms***

Detoxification of wood hydrolysate can be achieved with laccase and peroxidase enzymes of white-rod fungus *Trametes versicolor* as they are capable of oxidative polymerization of low molecular weight phenolic compounds rendering them inactive and non-toxic [75]. Treatment of willow hydrolysates with the filamentous soft-rot fungus *Trichoderma reesei* has been reported to degrade acetic acid, furfural and benzoic acid derivatives along with complete removal of phenolics [76]. A mutant of *Saccharomyces cerevisiae* has been used for selective detoxification of acetic acid decreasing its concentration from 6.8 g/l to less than 0.4 g/l in wood hydrolysate [77].

***Adaptation of microorganism***

Adaptation of microorganism to hydrolysate by carrying out several successive batch cultures is an efficient and cost-effective technique to abase the inhibitory effect of toxic compounds on microbial growth [64, 73]. The adaptation of *Candida guilliermondii* to rice straw hemicellulose hydrolysate for xylitol production has been found to be an

**Table 12.4** Effect of various detoxification methods on xylitol productivity and yield obtained by fermentation of acid hydrolyzed hemicellulose fraction.

| Raw material | Detoxification method | Xylitol productivity (g/l.h) | Xylitol yield (g/g) | Reference |
|---|---|---|---|---|
| Rice straw | None | 0.18 | 0.69 | [79] |
| Rice straw | Charcoal + 4 fold conc. | - | 0.78 | [69] |
| *Eucalyptus globulus* | Ethyl acetate + vacuum conc. | 0.66 | 0.51 | [71] |
| *Eucalyptus globulus* | $CaCO_3$ (pH 6.5) + Charcoal + vacuum conc. | 0.50 | 0.73 | [80] |
| Brewer's spent grain | Activated charcoal | 0.33 | 0.50 | [66] |
| Corn cob | Neutralization by NaOH (pH 10) + $H_2SO_4$ (pH 5.3) | 0.60 | 0.71 | [81] |
| Sugarcane bagasse | Neutralization by CaO (pH 7) + $H_3PO_4$ (pH 5.5) + charcoal | 0.68 | 0.54 | [82] |
| Sugarcane bagasse | Overliming | - | 0.49 | [63] |
| Sugarcane bagasse | CaO + $CaCO_3$ (pH 4.5–6) + cation exchange | 0.195 | 0.26 | [83] |

effective and inexpensive method to alleviate the inhibitory effect of toxic compounds on the xylose-to-xylitol bioconversion [78]. The effect of various detoxification methods of acid hydrolyzed hemicellulose fraction of lignocellulosic materials is given in Table 12.4 along with the xylitol yield and productivity.

## 12.3 Utilization of Xylose

### 12.3.1 Microorganisms Utilizing Xylose

Microorganisms preferentially utilize glucose over xylose; a small number of bacteria, yeasts and fungi are capable of assimilating and fermenting xylose to xylitol [84]. A few bacteria such as *Corneybacterium* sp. [85], *Enterobacter liquefaciens* [86, 87] and *Mycobacterium smegmatis* [88] have been reported to produce xylitol although in relatively small quantities, therefore bacteria do not attract researchers' attention for economical production of xylitol. Regarding the fungi, there is only one significant report

**Table 12.5** Xylitol producing yeast strains.

| Yeast | Xylose (g/l) | Xylitol (g/l) | Reference |
|---|---|---|---|
| *C. guilliermondii* FTI 20037 | 54.4 | 37 | [91] |
| *C. boidinii* NRRL-Y17213 | 150 | 58.9 | [92] |
| *C. tropicalis* IFO-0618 | 170 | 101.6 | [93] |
| *C. parapsilosis* KFCC 10875 | 300 | 210 | [94] |
| *Debaryomyces hansenii* NRRL-Y 7426 | 279 | 221 | [95] |
| *Pachysolen tannophilus* ATCC 32691 | 15–60 | 9.4–15 | [96] |
| *C. shehatae* ATCC 22984 | 92 | 28 | [97] |
| *C. peltata* NRRL-Y 6888 | 50 | 23.6 | [98] |

regarding *Petromyces albertensis* [89] which accumulated 39.8 g/l of xylitol after 10 days of fermentation with initial xylose concentration of 100 g/l. Since then a number of fungi has been screened, like *Penicillium chrysogenum, P. citriinum, P. expansum, P. griseoroseum, P. italicum, P. purpurogenum, Aspergillus niger, P. bravicompactum, P. roqueforti, P. crustosum*, etc. for optimum xylitol production. *Penicillium crustosum* gave the highest xylitol yield of 0.52 g/l from 11.5 g D-xylose which was much less than that obtained with *P. albertensis*. A possible explanation for the lower xylitol production in filamentous fungi than in yeast is the preeminence of aerobic metabolism in the former. Even under low oxygenation rates, aerobic respiration is more efficient than fermentation leading to utilization of D-xylose as carbon and energy source, with considerable xylitol dehydrogenase activity due to efficient recovery of NAD(P)+ and subsequent production of D-xylulose-5 phosphate via D-xylulose [90].

In general, the yeasts are considered to be the best xylitol producers. *Candida* strains have been extensively studied for xylitol production. Some of the best xylitol producers have been summarized in Table 12.5.

## 12.3.2 Metabolism of Xylose

### *12.3.2.1 Transport of Xylose*

The first step in the metabolism of D-xylose is the transport of the sugar across the cell membrane. A number of researchers have studied the factors affecting xylose transport and found availability of oxygen to be the most important parameter deciding the kind of transport that takes place. However, the manner in which a microorganism reacts to its environment is specific for that microorganism. A number of investigations have shown that under aerobic and oxygen limited conditions the rate of transport can limit the utilization of D-xylose in *Pichia stipitis* CBS 7126 and *C. shehatae* ATCC 22984 [99–101], whereas D-xylose metabolism is transport independent in both under anaerobic conditions. Instead, the limitation occurs in the two initial steps of D-xylose metabolism, reduction of D-xylose and subsequent oxidation of xylitol. On the other hand, presence of oxygen induces or activates a transport system in *P. stipitis* CBS 6054 [102]. C. *shehatae* CBS 2779 follows both proton symport and a facilitated xylose uptake diffusion system when starved of oxygen [103] and a facilitated diffusion system otherwise. Simultaneous operation of a low-affinity and a high-affinity xylose proton symport in both starved and non-starved cells of *P. stipitis* IGC 4374 has been reported [100]. In contrast to other microorganisms, the D-xylose uptake rate in *C. mogii* ATCC 18364 follows Michaelis-Menten kinetics which suggests a carrier mediated facilitated diffusion transport system [104].

### *12.3.2.2 Metabolic Pathways*

The major metabolic pathways for xylose fermentation are believed to be similar in bacteria, yeast and fungi (Figure 12.2), despite important differences in transport, regulation, cofactor requirements and pyruvate fermentation pathways [105, 106]. Xylose metabolism employs pathways distinctly different from those involved in the utilization of glucose. In 1960, it was found that the filamentous fungus *Penicillium chrysogenum* carried out the conversion of D-xylose to D-xylulose through a two-step reduction and oxidation [107]. It possessed enzymes that differed from xylose isomerase in bacteria. This finding, as well as some further investigations [108] led to the conclusion that the two-step conversion of D-xylose to D-xylulose is specific for yeasts and fungi, whereas in bacteria the same conversion is catalyzed by xylose isomerase in a single step. Yeasts *Rhodotorula* [109] and C. *boidinii* no. 2201 [110] are the only exceptions to this generalization in that they carry the enzyme xylose isomerase.

In bacteria, D-xylose is first transported across the cell membrane where it is isomerised to D-xylulose [10]. D-xylulose is then phosphorylated to D-xylulose 5- phosphate connecting the initial pentose metabolism to the central metabolism through the pentose phosphate pathway (PPP). Once inside the yeast cell, D-xylose is reduced to xylitol by either NADH- or NADPH-dependent xylose reductase (aldose reductase EC 1.1.1.21). Xylitol is either secreted from the cell or oxidized to xylulose by NAD- or

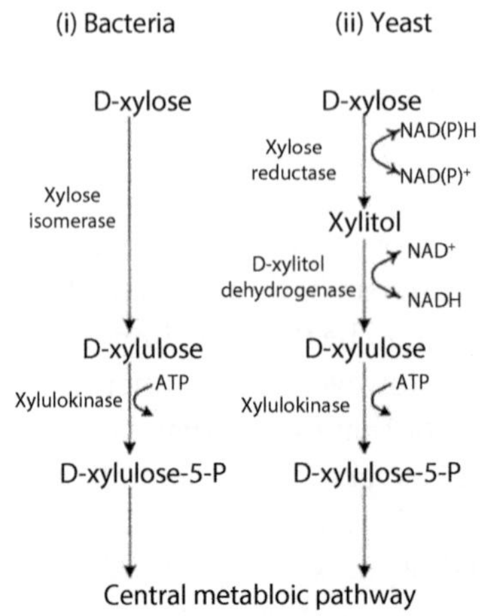


**Figure 12.2** The metabolic pathway to utilize xylose in (i) bacteria and (ii) fungi.

NADP-dependent xylitol dehydrogenase (EC 1.1.1.9). The first two reactions are considered to be limiting in D-xylose fermentation. The phosphorylation of xylulose to xylulose 5-phosphate is catalyzed by xylulokinase (EC 2.7.1.17) [111, 112]. The conversion of pentoses to xylulose-5-phosphate is a prerequisite for its utilization by the central catabolic pathways [113]. Xylulose-5-phosphate can subsequently enter the pentose phosphate pathway. This pathway consists of an oxidative phase that converts hexose phosphates to pentose phosphates providing NADPH needed in biosynthetic pathways and a non-oxidative phase in which the pentose phosphates are converted into hexose and triose phosphates like glyceraldehydes-3-phosphate and fructose-6-phosphate [105]. Both of them can be converted to pyruvate in the Embden-Meyerhof-Parnas pathway. Pyruvate can either be decarboxylated and reduced to ethanol or can enter the tricarboxylic acid cycle. The conversion of xylulose-5-phosphate into glyceraldehyde-3-phosphate and acetyl phosphate by xylulose-5-phosphate phosphoketolase presents an alternative route for the utilization of xylulose-5-phosphate [114]. The pentose phosphate pathway also yields ribose-5-phosphate used for the synthesis of nucleic acids and histidine and of erythrose-4-phosphate necessary for the synthesis of aromatic amino acids. Xylose metabolism in yeasts yields a variety of carbon- containing products which include carbon dioxide, ethanol, acetic acid and polysaccharides. Product yields are dependent upon the regulation of carbon flow through available metabolic routes [113]. D-Xylose conversion to xylitol in yeasts cannot be separated from the conversion of D-xylose to these products. The process of xylitol formation cannot be stopped after the first step, when D-xylose is converted to xylitol. Cell growth depends on some of the above metabolic products and it is also necessary that the cofactors be regenerated through different steps in the metabolic pathway. Therefore, for obtaining good yields of xylitol, the amount of xylose being converted to xylitol and the amount of xylitol which is available for further metabolism, have to be well balanced.

## 12.4 Process Variables

### 12.4.1 Temperature and pH

The most suitable temperature for xylitol production in yeast was shown to be 30 °C. Small variations in cultivation temperature do not significantly affect xylitol production in most of the yeast species. The xylitol yield of *Candida tropicalis* DSM 7524 was almost temperature independent when the yeast was cultivated in a temperature range between 30 °C and 37 °C but above 37 °C xylitol yield decreased sharply [115]. Similar effect was observed in *C. guilliermondii* FTI 20037 where constant xylitol yield was obtained at temperatures between 30 °C and 35 °C but yield decreased when temperature was increased to 40 °C [116]. A loss in activities of both NADPH and NADH-dependent xylose reductase as temperature increases can be a possible cause of decrease in xylitol yields with increase in temperature [113].

Change in temperature also affects relative production of xylitol and ethanol as xylitol production is favored over ethanol at higher temperatures. Xylitol production of *C. shehatae* CSIR-Y492 increased 8-fold as the temperature increased from 22 °C to 36 °C. *P. stipitis* CSIR-Y633 produced xylitol at 36 °C but no detectable amounts at lower temperatures. The yeasts are generally cultivated at pH values between 4 and 6 [117]. *C. parapsilosis* ATCC 28474 and *C. guilliermondii* NRC 5578 [118, 119] were grown at pH 6, *C. mogii* ATCC [104] and *P. stipitis* NRRL Y-7124 at pH 5 and 5.5, respectively, while pH 4 was optimum for *C. tropicalis* IF0 0618 [120]. In contrast, it was reported that *C. tropicalis* DSM 7524 was not very sensitive to pH and attained a maximum xylitol yield at pH 2.5. Increasing the pH from 2.5 to 4 led to an increase in xylitol productivity but a decrease in xylitol yield [115]. During the fermentation pH drops sharply, if uncontrolled, therefore the initial pH values have to be higher than under controlled conditions. The optimum initial pH value for best xylitol yield in *C. boidinii* was 7 [121], whereas under controlled conditions, a pH of 5.5 was better [92]. Batch culture of *C. parapsilosis* ATCC 28474 [122] was performed at pH 6, while for continuous culture, a pH of 4.5 was used [123].

### 12.4.2 Substrate Concentration

Yeast growth as well as xylitol production is critically dependent on D-xylose concentration. Together with aeration, D-xylose concentration induces xylitol formation in yeasts. Increased xylose concentration favors xylitol production at the expense of ethanol production, resulting in an increase in the xylitol/ethanol ratio, a decrease in the ethanol yield and without exception, an increase in the xylitol yield [92, 102, 119, 124]. As the initial D-xylose concentration increases, the specific growth rate decreases, demonstrating substrate inhibition, whereas the effect on overall xylitol productivity depends on the yeast type. Therefore, determination of optimum initial xylose concentration is essential for growth and xylitol production. Initial xylose concentrations in the range of 20–50 g/l produced the highest specific growth in *C. guilliermondii* [119] while maximum specific growth in *C. mogii* was observed at an initial xylose concentration of 5–10 g/1 [104]. The optimum initial xylose concentration reported for xylitol production was 60 g/l for *P. tannophilus*, 200 g/l [125] and 100 g/l [115] for *C. tropicalis* and 200 g/1 [119] for *C. guilliermondii*. It can be concluded that for most yeasts, the initial D-xylose concentrations resulting in the highest yields are between 100 and 200 g/l, with C. *guilliermondii* NRC 5578 being an exception for which 300 g/l is the most suitable concentration. However, all yeasts need a relatively long time for the conversion of D-xylose to xylitol. The effect of initial substrate concentration on fermentation parameters is given in Table 12.6.

Presence of hexoses also effect D-xylose utilization by yeasts. It has been stated that D-galactose, D-cellobiose and L-arabinose are not inhibitory to D-xylose assimilation, while D-mannose and particularly D-glucose considerably slow down the utilization of D-xylose [126, 127, 128, 115]. In mixtures of glucose and D-xylose, the yeasts first consume the glucose resulting in faster growth. After the depletion of glucose, the assimilation of xylose starts leading to the production of xylitol; however, yields decrease considerably [92, 129]. It is assumed that in the presence of glucose, there is a partial inhibition of xylose reductase and consequently a smaller amount of xylitol is produced (Silva, 1996). Higher amounts of glucose relative to D-xylose reduced the yield further [130].

When a mixture of glucose and xylose was used as a substrate for the fed-batch culture of *C. boidinii* NRRL Y-17213 with the former being l/10 of the concentration of the

**Table 12.6** Influence of initial substrate concentration on the fermentation parameters for some xylitol producing yeasts.

| Microorganism | D-xylose(g/l) | $C_P$ (g/L) | $Y_{P/S}$ (g/g) | Time (h) | Reference |
|---|---|---|---|---|---|
| *C. boidinii* 2201 | 50 | 17.1 | 0.53 | 48 | [110] |
| | 100 | 36.0 | 0.44 | 144 | |
| | 150 | 17.0 | 0.22 | 144 | |
| *C. boidinii* NRRL Y-17213 | 50 | 4.8 | 0.12 | 96 | [92] |
| | 100 | 25.2 | 0.29 | 132 | |
| | 150 | 53.1 | 0.47 | 336 | |
| *C. guilliermondii* NRC 5578 | 50 | 30.9 | 0.63 | 165 | [119] |
| | 150 | 110.3 | 0.74 | 238 | |
| | 300 | 221.0 | 0.74 | 406 | |
| *C. parapsilosis* ATCC 28474 | 50 | 29.5 | 0.59 | 266 | [118] |
| | 100 | 61.0 | 0.61 | 477 | |
| | 200 | 116.0 | 0.58 | 1009 | |
| | 300 | 93.0 | 0.31 | 1860 | |

latter, glucose was consumed first, resulting in faster growth with a maximum specific growth rate of 0.067 l/h, compared to 0.023 l/h in the process using xylose alone. In contrast, xylitol accumulation, and consequently xylitol yield were lower; 39.4 g/l and 0.57 g/g, compared to 46.5 g/l and 0.64 g/g, respectively [131].

### 12.4.3 Aeration

Aeration is another parameter of paramount importance for determining the extent to which xylitol accumulates in culture. Under fully aerobic conditions, xylitol production does not take place. Under anaerobic conditions, yeasts failed to grow on D-xylose to any appreciable extent, reaching about one doubling at best [132]. It was found that fermentation and growth occur simultaneously only under oxygen limitation [133]. Supply of oxygen to the culture is required for yeast growth even under fermentation for the synthesis of unsaturated fatty acids and ergosterol required for sugar transport through the membrane [134]. The dependence of the fermentative behavior of the xylose-fermenting yeasts on oxygen availability can be presented in three stages: (i) under anaerobic conditions or at very low oxygen transfer rates, the electron transport system of the yeasts is unable to oxidize NADH completely. As a consequence, the intracellular NADH concentration increases and this imbalance between NADH and NAD concentrations lead to xylitol secretion. (ii) increasing the oxygen transfer rate permits the enhancement of xylose fermentation. This is believed to be due to the role of oxygen as a terminal electron acceptor, thus relieving the imbalance of the two initial steps of anaerobic xylose metabolism. This hypothesis is supported by the inverse relationship between the degree of aeration and xylitol production observed for some yeasts. Under these circumstances, the main metabolic product is ethanol. (iii) When oxygen is supplied in excess, a deviation in the pyruvate flow from the fermentative pathway to the tricarboxylic acid cycle is observed resulting in increase in cell mass. [135].

A survey of xylitol production by various yeasts under different oxygenation levels is presented in Table 12.7. It is evident that yeasts producing xylitol require very small amounts of oxygen, which appear to be specific for each yeast strain. As can be seen, *D. hansenii* DTIA-77 has the highest demand for oxygen compared to the other yeasts. Under aerobic conditions, as D-xylose is depleted, some yeasts, such as *D. hansenii* DTIA-77 [136], C. *tropicalis* ATCC 32113 [137] and C. *boidinii* NRRL Y- 17213 [138] can reassimilate both xylitol and ethanol. It has further been found that the interaction between D-xylose concentration and aeration rate was related to cell concentration. If the cell concentration was low, the dissolved oxygen concentration was maintained at a high level, resulting in low xylitol production. [120]. This finding indicates that the most relevant parameter defining the oxygenation of a culture is the specific oxygen uptake rate. Therefore it can be concluded that oxygen supply rate is a key parameter which determines whether D-xylose will be fermented or respired under given oxygen supply condition. It is very important, therefore, for an effective process, to determine the oxygen flux that will enable balanced utilization of carbon both for growth and fermentation. Since xylitol formation in yeasts is most sensitive to substrate concentration and aeration rate, it would be best if both parameters were simultaneously taken into consideration when optimizing xylitol production. The effect of aeration on xylitol production in yeast strains is given in Table 12.7 where D-xylose was taken as the sole carbon source.

**Table 12.7** Influence of aeration on xylitol production in yeasts with D-xylose as substrate.

| Microorganism | Substrate concentration $C_s$ (g/L) | Biomass concentration (g/L) | Aeration (vvm) | $K_La$ (l/h) | Reference |
|---|---|---|---|---|---|
| *Pachysolen tannophilus* ATCC 32691 | 50 | 13.5 | 0.08 | 4.8 | [122] |
| | 50 | 11 | 0.9 | 35.4 | |
| *C. parapsilosis* ATCC 28474 | 50 | 22 | 0.08 | 4.8 | [122] |
| | 50 | 32.5 | 0.3 | 16.8 | |
| | 50 | 30.4 | 0.9 | 35.4 | |
| *D. hansenii* DTIA-77 | 90 | 36.0 | - | 112.8 | [136] |
| | 50 | 7.5 | 1 | - | [139] |
| *C. guilliermondii* FTI 20037 | 65 | 4.7 | 0.46 | 5.3 | [140] |
| | 65 | 21.4 | 0.46 | 10.6 | |
| | 65 | 6.4 | 0.46 | 41 | |

## References


1. K.J. Parker, *Nature*, Vol. 271, p 493, 1978.
2. Sugar substitutes, http://www.pepsiproductfacts.com/sugarsub.php
3. T.B. Granstrom, K. Izumori and M. Leisola, *Applied Microbiology and Biotechnology*, Vol. 74, p 277, 2007.
4. M.M. Silveira and R. Jonas, *Applied Microbiology and Biotechnology*, Vol. 59, p 400, 2002.
5. C. van Zyl, B.A. Prior and J.C. du Preez, *Applied Biochemistry and Biotechnology*, Vol. 17, pp 357–369, 1988.
6. S.S. Silva, I.M. Manchilla, M.C.D. Vanetii, and M.F.S. Barbosa, "Use of sugar cane bagasse hemicellulosic hydrolysate for lactic acid production", Proceedings of 5th European Conference on *Biomass for Energy and Industry*, Vol. 2, Ed, Elsevier, London, pp 1005–9, 1990.
7. E.R. Frazer and T.A. McCaskey, *Biomas and Bioenergy*, Vol. 18, pp 31–42, 1989.
8. P.S. Meyer, J.C. du Preez, and S. Kilian, *Applied Microbiology*, Vol. 15, pp 161–165, 1992.
9. J.E. Brownell and J.P. Nakas, *Journal of Industrial Microbiology*, Vol. 7, pp 1–6, 1991.
10. L.F. Chen and C.S. Gong, *Journal of Food Science*, Vol. 50, pp 226–228, 1985.
11. R.L., Howard, E. Abotsi, E.L.J. van Rensburg and S. Howard, *African Journal of Biotechnology*, Vol. 2(12), pp 602–619, 2003.
12. L.K. Singh, G. Chaudhary, C.B. Majumder and S. Ghosh, *Advances in Applied Science Research*, Vol. 2 (5), 508–521, 2011.
13. P.M. Olinger and T. Pepper, Xylitol, in *Alternative Sweeteners* Ed. Lyn O'Brien Nabors Marcel Dekker, Inc. New York, p 335, 2011.
14. H.O. Beutler, "Xylitol", in H.U. Bergmeyer, ed., *Methods of Enzymatic Analysis*, Weinheim, Verlag, Chemie, Vol. 3(6), pp 484–490, 1984.
15. K.K. Makinen, *Medical Hypotheses*, Vol. 54, pp 603–613, 2000(a).
16. A.A. Bakr, *Nahrung Food*, Vol. 41, pp 170–175, 1997.
17. M. Uhari, T. Kontiokari and M. Niemela, *Pediatric*, Vol. 102(4), pp 879–884, 1998.
18. J.N. Counsell, *Dental Caries Prevention*, Applied Science Publishers, Vol. 14, p. 191, 1978.
19. G.M. Jaffe, *Sugary Azucar*, Vol. 93, pp 36–42, 1978.
20. C. Aminoff, E. Vanninen and T.E. Doty, "The occurrence, manufacture and properties of xylitol", in J.N. Counsell, ed., *Xylitol*, Applied Sciences Publishers, London, pp 1–9, 1978.
21. L. Hyvonen and R. Torma, *Journal of Food Science*, Vol. 48, pp 183–185, 1983.
22. P. Nigam and D Singh, *Process of Biochemistry*, Vol. 30, pp 117–124, 1995.
23. J.D. McMillan, *ACS Symposium Series*, Vol. 566(15), pp 292–324, 1994.
24. C.E. Wyman, *Annual Review of Energy and the Environment*, Vol. 24, pp 189–226, 1999.
25. M.A. Millet, A.J. Baker and L.D. Scatter, *Biotechnology and Bioengineering Symposium*, Vol. 6, pp 125–153, 1976.
26. F.J. Kilzer and A. Broido, *Pyrodynamics*, Vol. 2, pp 151–163, 1965.
27. A. Demirbas, *Energy Sources*, Vol. 27(4), pp 327–337, 2005.
28. C. Yang, Z. Shen, G. Yu and J. Wang, *Bioresource Technology*, Vol. 99(14), pp 6240–6245, 2008.
29. S. Zhu, Z. Yu, Y. Wu, X. Zhang, H. Li and M. Gao, *Chemical Engineering Communications*, Vol. 192(12), pp 1559–1566, 2005.
30. H. Alizadeh, F. Teymouri, T. I. Gilbert and B.E. Dale, *Applied Biochemistry and Biotechnology*, Vol. 121, pp 1133–1141, 2005.
31. I. Ballesteros, M.J. Negro, J.M. Oliva, A. Cabanas, P. Manzanares and M. Ballesteros, *Applied Biochemistry and Biotechnology*, Vol. 130(1–3), pp 496–508, 2006.
32. R. Eklund and G. Zacchi, *Enzyme and Microbial Technology*, Vol. 17, pp 255–259, 1995.
33. S.J.B. Duff and W.D. Murray, *Bioresource Technology*, Vol. 55, pp 1–33, 1996.

34. S. Reshamwala, B.T. Shawky and B.E. Dale, *Applied Biochemistry and Biotechnology*, Vol. 51–52 (1), pp 43–55, 1995.
35. M.T. Holtzapple, J. Jun, G. Ashok, S.L. Patibandla and B.E. Dale, *Applied Biochemistry and Biotechnology*, Vol. 28–29(1), pp 59–74, 1991.
36. Y.Z. Zheng, H.M. Lin and G.T. Tsao, *Biotechnology Progress*, Vol. 14, pp 890–896, 1998.
37. K. Karimi, G. Emtiazi and M.J. Taherzadeh, *Enzyme and Microbial Technology*, Vol. 40, pp 138–44, 2006.
38. D. Sidiras and E. Koukios, *Bioresource Technology*, Vol. 94(1), pp 91–98, 2004.
39. G. Chaudhary, L.K. Singh and S. Ghosh, *Bioresource Technology*, Vol. 124, pp 111–118, 2012.
40. P.F. Vidal and J. Molinier, *Biomass*, Vol. 16, pp 1–17, 1988.
41. Y. Sun and J. Cheng, *Bioresource Technology*, Vol. 83, pp 1–11, 2002.
42. R. Millati, L. Edebo and M.J. Taherzadeh, *Enzyme and Microbial Technology*, Vol. 36(2–3), pp 294–300, 2005.
43. A. Sues, R. Millati, L. Edebo and M.J. Taherzadeh, *FEMS Yeast Res.*, Vol. 5, pp 669–676, 2005.
44. G. Sanchez, L. Pilcher, C. Roslander, T. Modig, M. Galbe and G. Liden, *Bioresource Technology*, Vol. 93(3), pp 249–256, 2004.
45. M.J. Taherzadeh, R. Eklund, L. Gustafsson, C. Niklasson and G. Liden, *Industrial & Engineering Chemistry Research*, Vol. 36(11), pp 4659–4665, (1997a).
46. S. Larsson, A. Quintana-Sainz, A. Reimann, N.O. Nilvebrant, and L.J. Jonsson, *Applied Biochemistry and Biotechnology*, Vol. 84–6, pp 617–632, 2000.
47. A.B. Bjerre, A.B. Olesen, T. Fernqvist, A. Ploger and A.S. Schmidt, *Biotechnology and Bioengineering*, Vol. 49, pp 568–577, 1996.
48. P.V. Iyer, Z.W. Wu, S.B. Kim and Y.Y. Lee, *Applied Biochemistry and Biotechnology*, Vol. 57/58, pp 121–132, 1996.
49. H. L. Chum, D. K. Johnson, S. Black, J. Baker, K. Grohman, K.V. Sarkanen, K. Wallace, and H.A. Schroeder, *Biotechnology and Bioengineering*, Vol. 31, pp 643–649, 1988.
50. L.T. Fan, M.M. Gharpuray and Y.H. Lee, *Cellulose Hydrolysis Biotechnology*, Monographs, Springer, Berlin, Vol. 3, 1987.
51. D.E. Akin, L.L. Rigsby, A. Sethuraman, W.H. Morrison 3rd, G.R. Gamble and K.E. Eriksson, *Applied and Environmental Microbiology*, Vol. 61, pp 1591–1598, 1995.
52. G. Emtiazi, N. Naghavi and A. Bordbar, *Biodegradation*, Vol. 12, pp 259–263, 2001.
53. R. Mukherjee and B. Nandi, *International Biodeterioration and Biodegradation*, Vol. 53(1), pp 7–12, 2004.
54. M. Brienzo, V. Arantes, A.M.F. Milagres, *Fungal Biology Reviews*, Vol. 22(3–4), pp 120–130, 2008.
55. M. Mukhopadhyay, A. Kuila, D.K. Tuli and R. Banerjee, *Biomass and Bioenergy*, Vol. 35, pp 3584–3591, 2011.
56. J.D. Peterson and L.O. Ingram, *Annals of the New York Academy of Sciences*, Vol. 1125(1), pp 363–372, 2008.
57. M. Galbe and G. Zacchi, "Pretreatment of lignocellulosic materials for efficient bioethanol production" in *Biofuels*, Springer Berlin Heidelberg, pp 41–65, 2007.
58. H.J. Roman, J.E. Burgess and B.I. Pletschke, *African Journal of Biotechnology*, Vol. 5(10), pp 963–967, 2006.
59. T. Georgieva, M. Mikkelsen and B. Ahring, *Applied Biochemistry and Biotechnology*, Vol. 145(1–3), pp 99–110, 2008.
60. G. Mtui and R. Masalu, *Scientific Research and Essay*, Vol. 3, pp 154–61, 2008.
61. S.I. Mussato, and I.C. Roberto, *Process Biochemistry*, Vol. 39, pp 1433–1439, 2004.
62. R.S. Rao, C.P. Jyothi, R.S. Prakasham, C.S. Rao, P.N. Sharma and L.V. Rao, *Bioresource Technology*, Vol. 97, pp 1974–1978, 2006.

63. I.C. Roberto, L.S. Lacis, M.F.S. Barbosa and I.M. Manchilla, *Process Biochemistry*, Vol. 26, pp 15–21, 1991.
64. L. Olesson and B. Hahn–Hagerdal, *Enzyme and Microbial Technology*, Vol. 18, pp 312–331, 1996.
65. J.M. Dominguez, C.S. Gong, and G.T. Tsao, *Bioresource Technology*, Vol. 61(1), pp 85–90, 1997 (a).
66. F. Carvalhiero, L.C. Durate, S. Lopes, J.C. Parajo, H. Pereira, and F.M. Girio, Process *Biochemistry*, Vol. 40, pp 1215–1223, 2005.
67. S. Larsson, E. Palmqvist, B. Hahn–Hagerdal, C. Tengborg, K. Stenberg, G. Zacchi, and N-O. Nilvebrant, *Enzyme and microbial Technology*, Vol. 24, pp 151–159, 1999.
68. K. Tada, J.J. Horiuchi, T. Kanno and M. Kobayshi, *Journal of Bioscience and Bioengineering*, Vol. 98(3), pp 228–230, 2004.
69. A.A.A. Zeid, M.Z. El-Fouly, Y.A. El-Zawahry, T.M. El-Mongy and A.B.A. El-Aziz, *Journal of Applied Research*, Vol. 4(8), pp 975–986, 2008.
70. L. Canilha, W. Carvalho, M.G.A. Felipe and J.B.A. Silva, *Brazilian Journal of Microbiology*, Vol. 39, pp 333–336, 2008.
71. J.M. Cruz, J.M. Domiguez, H. Domiguez and J.C. Parajo, *Food Chemistry*, Vol. 67, pp 147–153, 1999.
72. L.W. Doner and K.B. Hicks, *Cereal Chemistry*, Vol. 74(2), pp 176–181, 1997.
73. J.C. Parajo, H. Dominguez and J.M. Dominguez, *Bioresource Technology*, Vol. 66(1), pp 25–40, 1998b.
74. P. Perego, A. Convery, E. Palazzi, M.D. Borghi and G. Ferraiole, *Journal of Industrial Microbiology*, Vol. 6, pp 157–164, 1990.
75. L.J. Jonsson, E. Palmiqvist, N.O. Nilvebrant and B. Hahn-Hagerdal, *Applied Microbiology and Biotechnology*, Vol. 49, pp 691–697, 1998.
76. E. Palmqvist, B. Hahn-Hagerdal, Z. Szengyel, G. Zacchi and K. Reczey, *Enzyme and Microbial Technology*, Vol. 20, pp 286–293, 1997.
77. H. Schneider, *Enzyme and Microbial Technology*, Vol. 19, pp 94–98, 1996.
78. C.J.S.M. Silva and I.C. Roberto, *Letters in Applied Microbiology*, Vol. 32, pp 248–252, 2001.
79. S.I. Mussato and I.C. Roberto, *Journal of Applied Microbiology*, Vol. 95, pp 331–337, 2003.
80. J.C. Parajo, H. Dominguez and J.M. Dominguez, *Enzyme and Microbial Technology*, Vol. 21, pp 18–24, 1997.
81. J.H. Kim, Y.W. Ryu and J.H. Seo, *Journal of Industrial Microbiology and Biotechnology*, Vol. 22, pp 181–186, 1999.
82. A. Martinez, M.E. Rodriguez, S.W. York, J.F. Preston and L.O. Ingram, *Biotechnology and Bioengineering Symposium*, Vol. 69, pp 526–536, 2000.
83. J.M. Dominguez, C.S. Gong and G.T. Tsao, *Applied Biochemistry and Biotechnology*, Vol. 57–58, pp 49–56, 1996.
84. J.A. Barnett, R.W. Payne and D. Yarrow, *Yeasts: Characteristics and Identification*, 2nd Ed. Cambridge University Press, New York, 1990.
85. J. Yoshitake, H. Obiwa and M. Shimamura, *Agricultural Biology and Chemistry*, Vol. 35, pp 905–911, 1971.
86. J. Yoshitake, H. Ishizaki and Y. Irie, *Agriculture Biology and Chemistry*, Vol. 37, pp 2261–2267, 1973.
87. J. Yoshitake, M. Shimamura, H. Ishizaki and Y. Irie, *Agriculture Biology and Chemistry*, Vol. 40, pp 1493–1503, 1976.
88. K. Izumori and K. Tuzaki, *Journal of Fermentation Technology*, Vol. 66, pp 33–36, 1988.
89. J.S. Dahiya, *Canadian Journal of Microbiology*, Vol. 37, pp 14–18, 1991.
90. F.C. Sampaio, W.B. Silveira, V.M. Chaves-Alves, F.M.L. Passos and J.L.C. Coelho, *Brazilian Journal of Microbiology*, Vol. 34, pp 325–328, 2003.

91. M.G.A. Felipe, I.M. Mencila, M. Vitolo, I.C. Roberto and S. Rosa, *Arquivos de Biologica E Technologia*, Vol. 36, pp 103–114, 1993.
92. E. Vandeska, S. Amaetey, S. Kuzmanova and T.W. Jeffries, *World Journal of Microbiology and Biotechnology*, Vol. 11, pp 213–218, 1995(a).
93. Y. Yahashi, M. Hatsu, H. Horitsu, K.K. Kawai, T. Suzuki and K. Takamizawa, *Biotechnology Letters*, Vol. 18(12), pp 1395–1400, 1996.
94. S.Y. Kim, J.H. Kim and D.K. Oh, *Journal of Fermentation and Bioengineering*, Vol. 83(3), pp 267–270, 1997.
95. J.M. Dominguez, C.S. Gong and G.T. Tsao, *Applied Biochemistry and Biotechnology*, Vol. 63(1), pp 117–127, 1997(b).
96. P. Thonart, X. Gomez, J. Dideelez and M. Paquot, "Biological conversion of xylose to xylitol by *Pachysolen tannophilus*", in *Anaerobic Digestion and Carbohydrate Hydrolysis of Waste*, G.L. Ferrero, M.P. Ferranti and H. Naveau, (eds.), Commission of the European Communities, Luxembourg, pp 429–431, 1984.
97. H.K. Sreenath and T.W. Jeffries, *Biotechnology Letters*, Vol. 9(4), pp 293–298, 1987.
98. B.C. Saha and R.J. Bothast, *Journal of Industrial Microbiology and Biotechnology*, Vol. 22, pp 633–636, 1999.
99. M.A. Alexander, V.W. Yang and T.W. Jeffries, *Applied Microbiology and Biotechnology*, Vol. 29, pp 282–288, 1988.
100. S.G. Kilian and N.V. Uden, *Applied Microbiology and Biotechnology*, Vol. 27, pp 545–548, 1988.
101. M.E. Ligthelm, B.A. Prior, J.C. du Preez and V. Brandt, *Applied Microbiology and Biotechnology*, Vol. 28, pp 293–296, 1988.
102. K. Skoog and B. Hahn-Hagerdal, *Enzyme and Microbial Technology*, Vol. 10, pp 66–80, 1988.
103. C. Lucas and N.V. Uden, *Applied Microbiology Biotechnology*, Vol. 23, pp 491–495, 1986.
104. S. Sirisansaneeyakul, M. Staniszewski and M. Rizzi, *Journal of Fermentation and Bioengineering*, Vol. 80(6), pp 565–570, 1995.
105. T.W. Jeffries, *Advance Biochemical Engineering and Biotechnology*, Vol. 27, pp 1–32, 1983.
106. R.J. Magee and N. Kosariac, *Advance Biochemical Engineering and Biotechnology*, Vol. 32, pp 61–93, 1985.
107. C. Chaing and S.G. Knight, *Nature*, Vol. 188, pp 79–81, 1960.
108. M. Chakravorty, L.A. Veiga, M. Bacila and B.L. Horecker, *The Journal of Biological Chemistry*, Vol. 237(4), pp 1014–1020, 1962.
109. M, Hofer, A. Betz and A. Kotyk, *Biochimica et Biophysica Acta*, Vol. 252(1), pp 1–12, 1971.
110. V. Vongsuvanlert and Y. Tani, *Agricultural and Biological Chemistry*, Vol. 52, pp 1817–1824, 1988.
111. K.L. Smiley and P.L. Bolen, *Biotechnology Letters*, Vol. 4, pp 607–610, 1982.
112. A.H. Lache and T.W. Jeffries, *Enzyme and Microbial Technology*, Vol. 8, pp 353–359, 1986.
113. P.J. Slininger, P.L. Bolen and C.P. Kurtzman, *Enzyme and Microbial Technology*, Vol. 9, pp 5–15, 1987.
114. B.A. Prior, S.G. Killian and J.C. du Preeze, *Process Biochemistry*, Vol. 24, pp 21–32, 1989.
115. S.S. Silva and A.S. Afschar, *Bioprocess Engineering*, Vol. 11, pp 129–134, 1994.
116. M.F.S. Barbosa, M.B. de Medeiros, I.M. Mancilha, H. Schneider and H. Lee, *Journal of Industrial Microbiology*, Vol. 3, pp 241–251, 1988.
117. J.C. du Preez, M. Bosch and B.A. Prior, *Enzyme and Microbial Technology*, Vol. 8, pp 360–364, 1986.
118. V. Nolleau, L. Preziosi-Belloy, J. P. Delgenes and J.M. Navarro, *Current Microbiology*, Vol. 27, pp 191–197, 1993.

119. V. Meyrial, J.P. Delgenes, R. Moletta and J.M. Navarro, *Biotechnology Letters*, Vol. 13(4), pp 281–286, 1991.
120. H. Horitsu, Y. Yahashi, K. Takamizawa, K. Kawai, T. Suzuki and N. Watanabe, *Biotechnology and Bioengineering*, Vol. 40(9), pp 1085–1091, 1992.
121. V. Vongsuvalert and Y. Tani, *Journal of Fermentation and Bioengineering*, Vol. 67, pp 35–39, 1989.
122. S.A. Furlan, P. Boutlloud and H.F. de Castro, *Process Biochemistry*, Vol. 29, pp 657–662, 1994.
123. S.A. Furlan, P. Boutlloud, P. Strehaiano and J.P. Riba, *Biotechnology Letters*, Vol. 13, pp 203–206, 1991.
124. J.C. Roserio, M.A. Peito, F.M. Girio and M.T. Amaral-Collaco, *Archives of Microbiology*, Vol. 156, pp 484–490, 1991.
125. C.H. Gong, L.F. Chen and G.T. Tsao, *Biotechnology Letters*, Vol. 3, pp 130:135, 1981.
126. H. Lee, C.R. Sopher and K.Y.F. Yau, *Journal of Chemical Technology and Biotechnology*, Vol. 66, pp 375–379, 1996.
127. J.K. Sugai and J.P. Delgenes, *Journal of Industrial Microbiology*, Vol. 14, pp 46–51, 1995.
128. P.A. Bichio, P.L. Runnals, J.D. Cunningham and H. Lee, *Applied and Environmental Microbiology*, Vol. 54, pp 50–54, 1988.
129. J.P. Delegenes, R. Moletta and J.M. Navarro, *Applied Microbiology and Biotechnology*, Vol. 29, pp 155–161, 1988.
130. S.S. Silva, A. Quesada-Chanto and M. Vitolo, *Zeitschrift fur Naturforschung*, Vol. 52c, pp 359–363, 1997.
131. E. Vandeska, S. Amartey, S. Kuzmanova and T.W. Jeffries, *Process Biochemistry*, Vol. 31, pp 265–270, 1996.
132. H. Schneider, *Critical Reviews in Biotechnology*, Vol. 9, pp 1–40, 1989.
133. M.A. Alexander, T.W. Chapman and T.W. Jeffries, *Applied Microbiology and Biotechnology*, Vol. 28: 478–486, 1988.
134. H. Kuriyma and H. Kobayashi, *Journal of Fermentation and Bioengineering*, Vol. 75(5), pp 364–367, 1993.
135. J.M. Laplace, J.P. Delgenes, R. Moletta and J.M. Navarro, *Applied Microbiology and Biotechnology*, Vol. 36(2), pp 158–162, 1991.
136. J.C. Roserio, M.A. Peito, F.M. Girio and M.T. Amaral-Collaca, *Archives of Microbiology*, Vol. 156, pp 484–490, 1991.
137. E.M. Lohmeier-Vogel, B. Hahn-Hagerdal and H.J. Vogel, *Applied and Environmental Microbiology*, Vol. 61(4), pp 1420–1425, 1995.
138. E. Vandeska, S. Kuzmanova and T.W. Jeffries, *Journal of Fermentation and Bioengineering*, Vol. 80(5), pp 513–516, 1995(a).
139. F.M. Girio, J.C. Roserio, P. SA-Machaodo, A.R. Durate-Reis and M.T. Amaral-Collacco, *Enzyme and Microbial Technology*, Vol. 16(12), pp 1074–1078, 1994.
140. S.S. Silva, I.C. Roberto, M.G.A. Felipe and I.M. Manchilla, *Process Biochemistry*, Vol. 31(6), pp 549–553, 1996.

# 13

# Trehalose: An Anonymity Turns Into Necessity

**Manali Datta*** **and Dignya Desai**

*Amity Institute of Biotechnology, Amity University Rajasthan, Jaipur, (India)*

***Abstract***
Trehalose, a non-toxic disaccharide of glucose, also known as the "sugar of life," is a ubiquitous molecule that occurs in all higher and lower life-forms except in mammals. In cryptobionts, it has been found to play a major role in dessication alleviation. This and other unique properties have enabled this ideal compound to be utilized in robust and versatile formulations in the food, beverage, pharmaceutical and cosmetic industries. In this chapter, we discuss the pathways and production parameters at laboratory and industrial scale. Both the enzymatic and microbe mediated production of trehalose has been studied in the design of the production process.

***Keywords:*** Trehala manna, kosmotrope, recombinnt TreS, anhydrobiosis, trehalose, sweetener

## 13.1 Introduction

A rare non-reducing disaccharide composed of glucose linked together in a α-1,1-glycosidic linkage was first discovered in rye ergot in 1832. Later, in 1858, a similar compound was isolated from mushrooms and was named mycose. Simultaneously in the Middle East, a novel sugar present in "trehala-manna" was identified as "trehalique glucose".

Trehalose, as it now commonly named, is found to be conserved in prokaryotes (bacteria and archaea), lower eukaryotes, plants, and invertebrates, but is notably non-existent in vertebrates and—more specifically—in mammals. Different derivatives of this saccharide tend to occur naturally like trehalose-6 phosphate, an intermediate product of its own biosynthetic pathway from uridine diphosphate glucose and glucose- 6 phosphate. Initially named as cord factor, trehalose 6,6'- dimycolate forms an integral part of glycolipids formed by mycolates. Virulent mycobacteria additionally contains the sulphated variant of trehalose in sulphoglycolipids. Each of the species have developed pivotal roles for trehalose in vivo [1, 2].

In 1925, Koch and Koch successfully managed to isolate trehalose from *Saccharomyces cerevisae*. Subsequently, in the 1990s, Hayashibara, a Japanese company, divulged a methodology enabling mass production from starch using maltooligosyl–trehalose

**Corresponding author*: mdatta@jpr.amity.edu

Saurabh Saran, Vikash Babu, and Asha Chaubey (eds.) High Value Fermentation Products, Volume 1, (309–322) © 2019 Scrivener Publishing LLC

synthase (MTSase) and maltooligosyl–trehalose trehalohydrolase (MTHase) [3]. Trehalose may be segregated and isolated from biological mixtures by ethanolic extraction followed by heating. Trehalose dehydrate tends to crystallize at melting point of around 97 °C.

Both US Food and Drug Administration (USFDA) and European regulation committees have recognized trehalose as a safe food ingredient. Trehalose is widely accepted as an ingredient in major markets, as its aroma, texture and color has satisfied consumer expectations. Its favourable sensory attributes have made it publicly acceptable and thus, trehalose has been extensively used in research applications and forms an important constituent of therapeutics like Herceptin®, Avastin®, Lucentis®, and Advate®. With the advent in technology, augmented utilization of trehalose is expected [4].

## 13.2 Trehalose Metabolism Pathways

Five pathways (Table 13.1) are involved in synthesis of trehalose in various organisms, namely TPS-TPP (OTSA-OTSB) pathway, TREY-TREZ pathway, TRES pathway, TRET pathway, and TREP pathway. Omics-based technologies have provided a glimpse of trehalose biology and discovery of trehalose metabolism has been one of the most exciting developments. In bacteria, extensive studies have been done for trehalose metabolism in significant species like *Mycobacterium, Escherichia coli, Corynebacterium* and *Propionibacterium freudenreichii.* [5]

In plants, a bidirectional network has been observed for trehalose and sucrose in which the key mediator is trehalose 6 phosphate (T6P). Research in the late 1990s led to identification of trehalose synthesizing enzymes in angiosperms. Surprisingly, meta analysis studies have led to discovery of multiple copies of T6P synthase gene (11 in Arabidopsis and 14 in maize), probably indicating the level of necessity of the molecule

**Table 13.1** Prevalent pathways for trehalose synthesis in living systems.

| Pathway | Prevalence | First substrate | Type |
|---|---|---|---|
| TPS-TPP (OTSA-OTSB) | Archaea, bacteria, fungi, plants and arthropods | Glucose 6 Phosphate+ Uridine 6 Phosphate/ glucose-6-phosphate and UDP-glucose | Irreversible |
| TREY-TREZ | *Arthrobacter sp., Rhizobium, Bradyrhizobium japonicum, Corynebacterium, Sulfolobus* | Maltopentose | Irreversible |
| TRES | Bacteria | Maltose | Reversible |
| TRET | Archaea and bacteria | ADP-glucose and glucose | Reversible |
| TREP | Prokaryotes and eukaryotes | Glucose-1-phosphate + glucose | Reversible |

in plant survival [6, 7]. T6P has been associated with numerous metabolic and developmental processes starch turnover, stomatal opening, abiotic stress tolerance and increased biomass. Trehalose has also been incriminated as the molecule governing the dessication tolerance in "resurrection plants" [8].

In humans, the enzyme is found in the brush border cells of the epithelial membrane of the small intestine and in the proximal tubules of the kidneys. Alternatively, some clinical data point to trehalose as a toxic molecule when stored inside the human body [2].

## 13.3 Physicochemical Properties and its Biological Significance

Trehalose tends to act as a positive modulator of many physiological activities, thus increasing its potential application in biotechnology. The success of trehalose is attributed to its unique physical properties like polymorphic variants having varied temperature stability and dessicant properties. Surprisingly, anhydrous crystalline state (Ta) upon absorbing moisture tends to undergo a reversible transition to the crystalline dihydrate form. As the three-dimensional structure remains unaltered, it confers a significant protective action against water stress.

With proteins, it has been observed to increase the surface tension of the medium, hence impacting decrease in surface tension of the soluble protein and maintenance at hydration state Figure 13.1a [9, 5]. Molecular dynamics based study of ternary complex of protein-water- trehalose indicated consistency in hydrogen bond formation with protein-water complex, but the protein tends to preferably form hydrogen bonds with trehalose. Thus, the replacement hypothesis indicated a prominent exclusion of water from surface of protein with simultaneous enrichment of trehalose [30].

Spectroscopic studies have identified a distinct assembly of trehalose on the protein shell enabling its cryopreservation due to its kosmotropic properties [10] (Figure 13.1b). It is non-reactive to amino groups and hence does not contribute to Maillard's reaction but enables binding of lipids to its shell via hydrogen bonding [11].

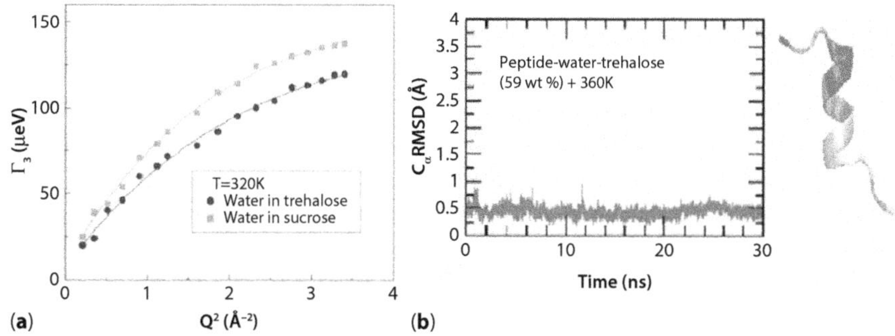


**Figure 13.1** (a) Diffusion coefficient and residence time as a function of temperature for trehalose in comparison with water [26] (b) Molecular dynamics of thermal denaturation process in presence of trehalose of a 15-residue S-peptide analogue at 360 K temperature.

Another property associated with trehalose is glass transition ($T_g$) capability. Amongst the naturally occurring carbohydrates, trehalose has the highest $T_g$ and does not change on addition of water. According to preferential exclusion theory, trehalose tends to entrap water in crystalline pockets thus creating reserves of water in case of scarcity. Positron annihilation lifetime spectroscopy (PALS) has confirmed morphing of trehalose from crystalline to amorphous form on addition of water maintains the crystalline pockets as there is no change in the structure and bonding pattern of the molecule.

Each of the property ascribed are not mutually exclusive but act in concerted mechanism to impart various biological efficiency to trehalose in stress alleviation and preservation of macromolecules [12].

## 13.4 Trehalose Production

In 1954, the first attempt of conversion via chemical transformation led to synthesis of trehalose as well as its anomers, but it was difficult to conduct it at industrial scale. Hence a transition was observed for extracellular production of trehalose based on enzymatic and as well as microbial conversion. The trehalose synthase based pathway has presented itself to be convenient as well as cost effective. Although enzymatic methods are efficient for trehalose production, microbe mediated methods are preferred for commercial production. Therefore, in addition to available enzymatic methods, trehalose accumulation was studied for economical production with *Corynebacterium E. coli* and *P.freudenreichii*. Optimization for enhanced production additionally included standardization of media, pH and temperature. Simple constituents like maltose as well as complex carbohydrates (extracts of rice, corn and kudzu root) have been used for lab scale production (2–5L). Here in the next section we discuss the details of both methods and organisms involved in these processes [8].

### 13.4.1 Enzymatic Conversion to Trehalose

TSase was identified first in *Pimelobacter* sp. R48, followed by discovery in many species like *Thermomonospora curvata, Thermus aquaticus, Picrophilus torridus, Pseudomonas stutzeri, Pseudomonas putida, Arthrobacter auresce, Enterobacter hormaechei, Meiothermus ruber*, to name a few [13]. The enzymes thus detected were characterized and used for trehalose production. The crude enzyme preparation from *Plesiomonas*, MTA, consisted of maltose phosphorylase (MP), trehalose phosphorylase (TP), saccharogenic amylase (SA) showing 20.5U/ml of MP, 34.4U/ml of TP and 5.8U/ml of SA. Starch debranching enzymes (SDE), pullulanase and isoamylase, was extracted from Flavobacterium having activity of 11,200 U/ml. Dialyzed concentrate of MTA and SDE was used concurrently for conversion of liquefied potato starch(25% w/v) to trehalose. 51% of solid weight of starch was converted to trehalose upon usage of *Klebsiella* pullulanase was used as a debranching enzyme. Subsequently, the digest was concentrated under high pressure followed by conventional crystallization to obtain pure (99%) crystallized trehalose. The maltose syrup containing approximately 45% trehalose gave the best yield [14].

*Plesiomonas* MP and TP were co-adsorbed on anion-exchange resin to facilitate the simultaneous activity of the enzyme system. Immobilized enzyme (as compared to non immobilized enzyme) retained 96% of its activity when the incubation conditions were maintained at 55 °C for 1 H at pH 7. The maximum yield (productivity rate 252 mg trehalose productivity rate mg/h/g) obtained was about 60% of solid mass with a flow rate of 1.0 $h^{-1}$ of the space velocity (SV) at 20% (w/v) maltose concentration [14]. Rice hydrolysate have been used in a 5 L fermentor system using a recombinant thermoacidophilic archaea, *Picrophilus torridus*, TSase (PTTS) system [15]. The PTTS system tends to use maltose as substrate and had high catalytic efficiency (60% trehalose bioconversion potential yield). The conversion rate at 30 °C was estimated to be 64.63 ± 4.05% with a yield of 94% and 92% purity. Kudzu root starch was identified and its conversion rate was calculated considering four factors namely, amount of dextrose, β-amylase, pullulanase and saccharification time. Maximum trehalose conversion rate of 70.6 % was acquired using saccharificaiton condition of 32.5U β-amylase and 1.25 U pullulanase per gram of starch.

Discovery and optimization of efficient TSase usage for conversion process is yet to reach a satisfactory scale-up level. Attempts are also been made to increase the half life of TSase with corresponding augmented yield by using immobilized enzyme in reactors. Thermostability of the enzyme also become a critical factor as the hydrolysis of the starting material, generally starch, is easily mediated at higher temperatures. The first attempt was made using enzymes in tandem immobilized in K-carrageen, namely, maltose phosphorylase (*Lactobacillus brevis*) and trehalose phosphorylase (*Euglena gracilis*). There was a slight shift in temperature (5 °C) with no change in optimal pH. The yield associated with the immobilized form was calculated to be 64%.

Using Eupergit C250L with 11 U/gm of support (equivalent to 92% of immobilization yield), a considerable shift in the optimum temperature in bioreactor was obtained (45°–65 °C) with 10 times reusability potential. The construct was stable at 70 °C for 16 days. A continuous flow reactor (25ml bed volume; 0.25 ml $min^{-1}$ rate and 50 g $L^{-1}$ maltose) produced a maximum yield of 42% trehalose [6]. *Thermus thermophilus* TSase was immobilized on glutaraldehyde treated chitosan; The enzyme post-immobilization maintained the same optimal pH and marked shift of temperature by 4 °C. Additionally, an increase of half life from 5.7 days to 6.3 days at 70 °C with 82% activity retention after 150 days at 40 °C [16].

Cross-linked polystyrene divinylbezene-based adsorbent with a metal chelator was used for immobilization of TSase (5.39 mg $g^{-1}$,) and a conversion yield of 64.1% was obtained within 6 H. Bound enzyme retained 80% residual activity even after 24 cycles. A residual activity of approximately 80% was maintained after 24 cycles in a repeated-batch operation [17]. Magnetic iron nanoparticles have also been used for immobilization (with highest yield of 10 mM with specific activity 0.134 U/g) and the construct retained 82% of its activity after twelve repeated reactions. The separation of nanoparticles were easily mediated by external magnetic field [18].

Another method encompasses a simultaneous release of extracellular enzyme and its subsequent adsorption on a substratum for efficient resuse. A recombinant trehalose synthase from *Picrophilus torridus* DSM 9790 was tagged with *B. subtilis* signal peptides and chitin-binding domains (ChBD) from *Bacillus circulans* WL-12 chitinase A1. The construct expressed in *Bacillus subtilis* was subsequently used in bioreactor containing chitin as an additive. The recombinant TreS was released extracellularly and

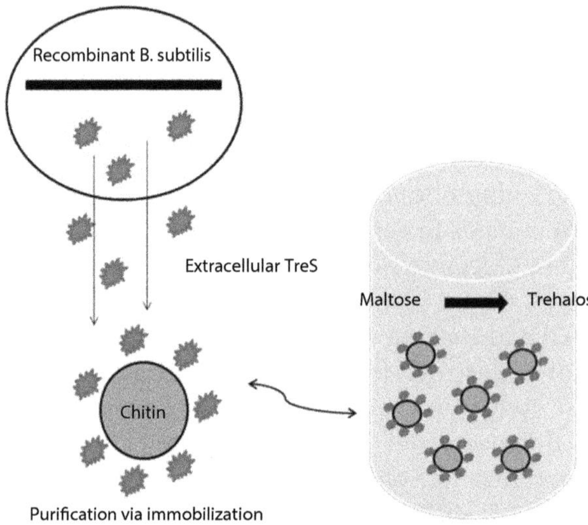


**Figure 13.2** Simultaenous secretion, purification and immobilization protocol for trehalose production: Permeabilized cells release recombinant TreS (containing chitin binding domain) in the solution and gets adsorbed on chitin beads. Immobilized enzyme tends to catalyze the conversion of maltose to trehalose.

subsequently adsorbed on chitin (Figure 13.2). In situ immobilized TreS gave a conversion yield up to 69.07% and 50% residual activity even after 21$^{st}$ use. The immobilized constructs tend to sustain 55% adsorption capacity even after being reused 53 times [19].

### 13.4.2 Microbe Mediated Fermentation

#### *13.4.2.1 Propionibacteria Mediated Conversion*

In 1958, trehalose was identified as a metabolite produced by three species of *Propionibacteria*. In 2004, a study was conducted by manipulating growth parameters of this bacteria and its effect on trehalose production. Change in pH from 7 to 4.7 led to doubling of the yield per gram of cell protein (200 to 400 mg). Additionally changing the concentration of NaCl and maintaining an aerobic condition (50% air saturation, 24 microM $O_2$, pH 7.0) increased the yield to 400 mg $g^{-1}$ of cell protein. *P. freudenreichii* ssp. *shermanii* NIZO B365 was also shown to produce similar levels of trehalose in skimmed milk. Thereafter, a study indicated that maintenance of anaerobic condition for a 24 hr cycle resulted in accumulation followed by degradation of trehalose in anaerobic sequencing batch reactor (ASBR), although within 1 hr time frame 40% of the carbohydrate source had been converted to trehalose [20]. Another osmomutant of *P. freudenreichii* ssp. *shermanii* managed to produce yields up to 391 mg/g of biomass and 90 mg/g of substrate consumed in the presence of KCl using crude glycerol as substrate [5]. A Taguchi method based study conducted for *P. acidipropionici* DSM 20273 using parameters like lactose concentration, pH, temperature and agitation rate gave a yield of 595 mg/L (pH 7; 30 °C; agitation rate 100 rev/min). Under these conditions, the highest yield of trehalose was produced [21].

### 13.4.2.2 *Micrococcus* Mediated Conversion

Trehalose was first reported in the micrococcus species in 1980. A variant of *Micrococcus* was found to accumulate commendable amount of extracellular trehalose. An optimum pH 6.0 at 32 °C for growth phase with concomitant increase to 34 °C for production phase lead to generation of 40 mg/ml trehalose from 100 mg $ml^{-1}$ glucose in 5 days in a 5 L jar fermentor [22]. Post optimization, the group obtained a US patent for the production parameters (US 5447856 A). A mutant strain of *M. luteus* (HS-208) demonstrated a production capacity of 6.547 g/l trehalose with conversion yield of 32.7% [23]. Later, in 2006, a cost-effective methodology using permeabilized *Micrococcus* cells was designed by Zhou and his team. Permeabilized *Micrococcus* QS412 cells were used for production whereby a yield of 188 mg/gm of wet cells after 12 hours was obtained. With batch reactions these cells could be reused 16 times and after 8 days, the yield enhanced to 2.5 g/g wet cells. [24].

### 13.4.2.3 *Brevibacterium* Mediated Conversion

*Brevibacterium*, a cornyeform non-pathogenic bacterium belongs to actinomycetes class of microorganisms. Biosynthesis of trehalose by *Brevibacterium flavum* using triose phosphate isomerase activity using malto-oligosaccharides as starting substrates was traced way back in 1985. A novel strain *Brevibacterium* sp. SY361 was used for optimal production of trehalose using glucose (4% V/V) as carbon source and peptone as nitrogen source (1 g/L) with the 180 rev/min and aeration rate of 0.15 vvm (volume air/ volume medium/min) at 28–32 °C and pH 6 with an incubation time of 60 H. Optimized parameters gave a maximum trehalose of 12.2 mg/ml with a conversion rate of 58.4%.

A US patent (US5484714) elaborates a method for production of trehalose by industrial mass production. *Brevibacterium lactofermentum* ATCC13869, *Brevibacterium flavum* ATCC14067 or *Brevibacterium divaricatum* ATCC21642 could be used in presence 10% saccharide and 2% ammonium sulphate as nitrogen source with an incubation temperature range of 31.5 C and pH 7. Using sucrose as a carbon source gave a accumulation of 1.25 gm/dL in a back and forth shaker after 48 H.

### 13.4.2.4 Fungal Strain Mediated Fermentation

An intensive study on *Saccharomyces cerevisiae* strain lead to identification of some parameters which could result in elevated production by yeast strains. For example, a glucose rich nitrogen deficit medium accumulated trehalose at higher level towards the end of the logarithmic phase. Additionally higher yield could be obtained, while using maltose or galactose as carbon source. While no appreciable disaccharide accumulation occurred during most of the logarithmic phase, a rapid synthesis took place during the final stages. Correlated studies confirmed that ATP supply by oxidative phosphorylation may not be the only criteria regulating trehalose metabolism. This was further mathematically substantiated by an experiment in fed batch reactor with starving carbon or nitrogen conditions. Results indicated trehalose accumulation to be 0.13 gm/gm biomass under carbon or nitrogen starvation. Using cassava starch (20 g/L) as a substrate, a yield of 25.8 g per 100 g of *Saccharomycopsis fibuligera* A11 dry weight in a 2 L fermentation shake flask. Crystalline trehalose was purified thereafter [25].

A one pot procedure for trehalose production and purification involved simultaneous integrated enzyme production, trehalose biotransformation, and bioremoval in one bioreactor setup, thus simplifying the traditional procedures used for trehalose production. A recombinant strain was generated by fusing TSase from *Picrophilus torridus* with an anchor gene, YlPir1 followed by insertion in *Yarrowia* yeast. The trehalose yield increased to 73% under optimal conditions. Following transformation from maltose, ethanol formed by a *S. cerevisiae* strain was removed by distillation, and hence high-purity trehalose could be retrieved from the fermentation broth [26]

#### *13.4.2.5 Recombinant Strains in Trehalose Production*

*Pseudomonas* sp is an omnipresent bacteria each having their unique specificity against trehalose production. Genome mining was done to identify the efficient strains of Pseudomonas and the selected gene (TreS from *P. stutzeri A1501*) were cloned in *Escherichia coli.* A permeabilized cell suspension of recombinant *E.coli* was used for trehalose production using 150 g /L maltose in CKBB buffer (Citrate, Phosphate, Barbitone and borate). After 4 h of biotransformation, the concentration of trehalose was 92.2 g/L with a productivity of 23.1 g/H and a yield of 61.5%. [13]

Combination of whole cell catalysis and autoinduction system was used for trehalose production using recombinant *E.coli.* The cells were permeabilized and recombinant TreS was induced by lactose present as carbon source. Glucose present as a by-product was used as an alternative carbon source by the recombinant *E.coli.* Batch cultures maintained a yield of upto 83.2 ± 5.0 % at the $10^{th}$ batch [27].

Synergetic carbon utilization mechanism (SynCar) couples usage of two carbon sources, for example both glucose and xylose whereby phosphoenol pyruvate (PEP) is generated by PEP-dependent phosphotransferase system (PTS) to promote non-catabolic use of one source. As lignocellulosic sugars contains both glucose and xylose in large percentage, SynCar system was used in production of trehalose. A recombinant strain of *Escherichia coli* was used to obtain a titer of 5.55 g/L in shake flask experiments. The conversion efficiency of substrate to trehalose was found to be fourfold higher than that of the original strain 30 [28]. Curli display technique uses a biofilm immobilized β-amylase (BA) and an intracellular expression of TreS within same *E.coli* cells. This bi-sequential conversion enabled production of 59% yield from starch with the consistency of 89 % cell viability even after 8 continuous rounds of biocatalysis [19].

### 13.4.3 Purification and Detection of Trehalose in Fermentation Process

Conversion of maltose to trehalose via enzymatic methods results in accumulation of high concentration of by product, glucose and residual maltose, which becomes a bottleneck in its downstream processing (Figure 13.3). An immobilized enzyme system (Glucoamylase on Eupergit® C and CM Sepharose) facilitated the conversion of maltose to glucose followed by subsequent oxidation to gluconic acid. Separation of gluconic acid and trehalose is done by ionic exchange process followed by crystallization leading to 94.2% recovery yield, hence simultaneous TreS mediated isomerization and glucose oxidase mediated conversion was found potential for the production and purification of trehalose on industrial scales [29].

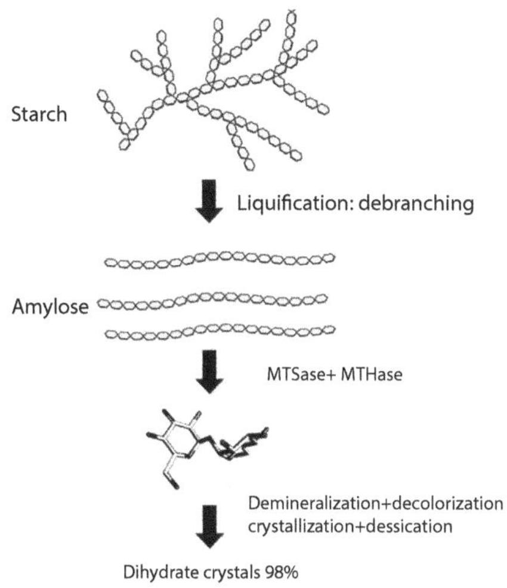


**Figure 13.3** Production protocol used by Hayashibara company (Okayama, Japan).

## 13.5 Application of Trehalose

Trehalose Market is classified based on application in food, pharmaceutical agriculture and cosmetics grade. Key players in trehalose production at the industrial level are Meihua Group,

Hayashibara, Lianmeng Chemical, Visionbio Technology, Sigma-Aldrich, Santa Cruz Biotechnology, Pfanstiehl and Sinozyme Biotechnology. Within a time frame of two years (1997–1999) demand-based sale of trehalose increased by approximately 15,000 tons. Global consumption augmented by 2–4 tonnes from 2010 to 2014. Hayashibara Ltd. alone tends to achieve a target of 31,500 tons/year using isomerase and hydrolase enzymes giving a yield of 85% and purity of 98%. Valued at 215.33 million USD in 2016 it is expected to reach 284.69 million USD by the end of 2022, growing at a compound growth rate of 4.76% between 2016 and 2022. Some analogues are also been synthesized to meet the ever increasing demand [21, 30]. In the next section, we review the various applications of trehalose in different areas.

### 13.5.1 Role of Trehalose in Food Industries

Trehalose has been found to act as a food preservative and tends to maintain the taste, visual and aromatic appeal of edibles. This property is attributed to its nonreactive nature and its ability to retent water. Thus, demonstrating its efficiency for increasing the shelf life of starch containing food products as well as with amino compounds such as amino acids or proteins (Figure 13.4a). In addition, trehalose protects phospholipids, proteins and gels from damage by freezing and drying. Thus it projects excellent restitution from freezing and drying and its utilization in frozen and dry food products preservation is heightened (Figure 13.4b). Described as "sugar of life", addition of 0.5% to 5% by weight has been found in almost over 8,000 food applications. The "just made freshness" has been maintained in products present in bakery, snack, confection, meat and seafood, fruit and vegetable processing, and beverages.

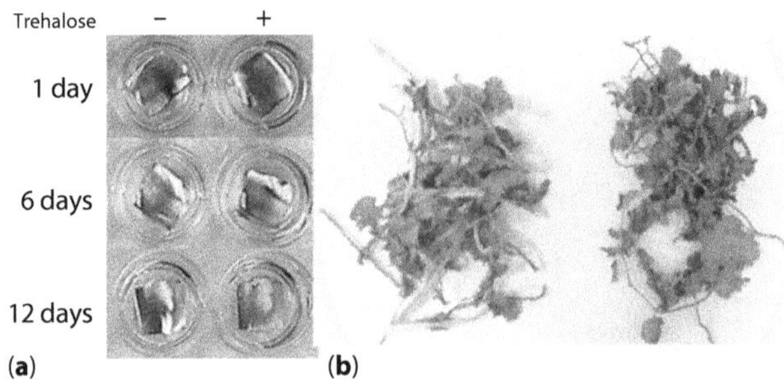


**Figure 13.4** (a) Denaturation profile of protein patch monitored by TMB assay, in presence and absence of trehalose. Secondary trehalose coating preserved the protein structure. (b) Change in crystalline growth in food preserved minimizes the destructiveness in freezing in minimum. [26, 35].

Whenever there is a phase transition from liquid to crystalline water, there is an associated damage to the cellular micro-architecture. Solution chemistry involves the study of water structure breaker and maker effect and trehalose forms one of the major molecules of consideration. Heat of solution of sugars is directly correlated with efficiency as cryoprotectant. Trehalose as an efficient cryoprotectant with $\Delta H_{sol}$ (kJ/mol) = 19.4 tends to disrupt the water structure around any molecule replacing it by membrane fluid interface and at the same time tends to maintain hydrated environment upon drying [31]. A study conducted for washed beef meat using trehalose as a cryoprotectant exhibited stabilisation of myofibrillar proteins during storage. With increasing concentration of trehalose significantly lesser loss of myofibrillar protein solubility, thermal shifts in temperatures of myosin and actin and increase of enthalpies of myosin and actin transitions was observed [32].

Trehalose itself is bland, non-toxic dietary sugar, which does not change significantly the flavor of foods to which it is added but has been found to suppresses unfavorable bitter/astringent and salty taste, and suggestions have been made that trehalose may mediate encapsulation of "tastants" and thus restrict the accessibility to the receptors [33]. This "masking" property of trehalose has been utilized to improve taste of food products. Aromatic appeal of edibles is another feature which is maintained by trehalose. When added to edibles (Strawberry apricot and pear puree and cubes) before being kept for preservation, it has been efficient (better than foam-mat drying) in preventing loss of volatiles from food, thus responsible for the maintaining the characteristic flavors of fresh fruits [34, 35].

A stabilizing agent created using trehalose-pullulan combination embedded with phages was layered on butcher paper and tested as antibacterial coating. Stand-alone utilization of trehalose/ pullulan led to decrease in efficiency as a coating but pullulan-trehalose mixture was able to inhibit *Lysteria monocytogenes* up to three months. The glass transistion property of trehalose had a synergistic effect on phage stability and hence was found to limit bacterial contamination [26].

### 13.5.2 Role of Trehalose in Cosmetics and Pharmaceutics

Anhydrobiosis is the ability of plants and animals to withstand prolonged periods of desiccation. The "isomorphous" with the dihydrate crystalline sugar tends to form gel preventing disruption of internal cell organelles by effectively preserving them in position and reversibly absorbs water to produce the di-hydrate conformation and supplementing the sample with water. Trehalose is able to protect cultures of *E. coli* and *B. thuringiensis* when the bacteria were rehydrated immediately after drying. However, only trehalose exhibited the protective ability ability on exposure to light and air at 60% relative humidity [36].

Drug formulation with protein composition like Adcetris, Blincyto, Inflectra, Voraxaze need to be sufficiently stable over shelf life at the intended storage condition, especially when they are lyophilized in a final drug product manufacturing step. It has been observed the protectant should be stable over a wide pH range and temperature and should have optimal water retention capability. Trehalose averts the detrimental effects on target because of its non-reactive nature and near optimal glass transition temperature. Formulation at lower pH is also possible with trehalose. Even after 1 hour in solution at pH 3.5 >99% activity was retained trehalose remains compared to ~0% sucrose remaining in the same conditions. There is an increase in stability and solubility of minerals (particularly magnesium and calcium) in suspensions. An injectable grade (Trehalose SG Nagase Group) containing lower levels of endotoxin tends to maintain the protein based therapeutic formulations. Storage of protein based formulations was found to be when an optimal range of trehalose-antibody (w/w) ratio, 0.2–2.4, was maintained. This ratio was adept of stabilizing antibodies formulations during long-term frozen storage; even for fast cooled (>100 °C/min) processes [37].

The disaccharide is capable of storing the dry liposomes in its original confirmation with fusion. As the sugar constitutes both inside and outside of the bilayer, efficiency of trapped solute is observed in rehydrated vesicles freeze-dried with 1.8 g trehalose $gm^{-1}$ dry phospholipid [8]. Subsequently liposome-based preservation has been studied in microbes as well as mammalian cells. As an essential component of cosmetics, it acts as a moisturizer and protecting agent that protects skin and hair from dehydration under extremely dry conditions. A peel-off facial mask containing myoinositol and trehalose-loaded liposomes (as activators of cutaneous autophagy) could improve cosmetic appearance adult female acne (AFA) levels in skin within 60 days [9, 38].

## 13.6 Conclusions

Trehalose is an ideal ingredient adding new life to existing food and beverage brands. Its unique properties have made it an indispensible molecule with an ever-increasing demand in the market. The Asia-Pacific region dominated the market in 2014, followed by North America. The market in the Asia-Pacific region is projected to grow significantly in terms of both production and consumption. China was one of the major markets for low-intensity sweeteners and accounted for a major share in the Asia-Pacific low-intensity sweeteners market in 2014. Over the years technological advances have added an impetus for the enhanced production of trehalose. Identification of novel metabolic pathway components in extremophiles has provided exciting avenues for engineering recombinants with higher production efficiency for meeting the growing demands of trehalose production in a sustainable manner.

## References


1. Z.I. Ahmad, J.R. Alden, and M.D. Montague, *Microbiology*, Vol. 121, p. 483, 1980.
2. J.C. Argüelles, *Journal of Molecular Evolution*, Vol. 79, p. 111, 2014.
3. G. Iturriaga, R. Suárez, and B. Nova-Franco, *International Journal of Molecular Sciences*, Vol. 10, p. 3793, 2009.
4. A. Patist, and H. Zoerb, *Colloids and Surfaces B: Biointerfaces*, Vol. 40, p. 107, 2005.
5. T. Shimada, J. Zilles, L. Raskin, and E. Morgenroth, *Water Research*, Vol. 41, p. 4721, 2007.
6. Y.J. Cho, O.J. Park, and H.J. Shin, *Enzyme and Microbial Technology*, Vol. 39, p. 108, 2006.
7. B.D. Connolly, L. Le, T.W. Patapoff, M.E. Cromwell, J.M. Moore, and P. Lam, *Journal of pharmaceutical sciences*, Vol. 104, p. 4170, 2015.
8. C.M. Figueroa, and J.E. Lunn, *Plant Physiology*, Vol. 172, p. 7, 2016.
9. G. Fabbrocini, C. Capasso, M. Donnarumma, M. Cantelli, M. Le Maître, G. Monfrecola, and E. Emanuele, *Journal of Cosmetic Dermatology*, Vol. 39, p. 1473, 2017.
10. M.A. Mensink, H.W. Frijlink, K. van der Voort Maarschalk, and W.L. Hinrichs, *European Journal of Pharmaceutics and Biopharmaceutics*, Vol. 114, p. 288, 2017.
11. S. Ohtake, and Y.J. Wang, *Journal of Pharmaceutical Sciences*, Vol. 100, p. 2020, 2011.
12. L. Jiang, X. Song, Y. Li, Q. Xu, J. Pu, H. Huang, C. Chao Zhong, *ACS Catalysis*, Vol. 8, p. 1837, 2018.
13. Y. Zhou, Q. Yuan, H. Gao, and R. Ma, *Journal of Molecular Catalysis B: Enzymatic*, Vol. 43, p. 137, 2006.
14. Z. Zheng, Y. Xu, Y. Sun, W. Mei, J. Ouyang, *PLOS ONE*, Vol. 10, p. e0140477, 2015.
15. P.T. Chen, C.J. Chiang, Y.T. Chen, H.C. Lin, C.H. Liu, Y.P. Chao, and J.F.Shaw, *Journal of Agricultural and Food Chemistry*, Vol. 60, p. 6063, 2012.
16. H. Kizawa, J.I. Miyazaki, A. Yokota, Y. Kanegae, K.I. Miyagawa, and Y.Sugiyama, *Bioscience, Biotechnology, and Biochemistry*, Vol. 59, p. 1522, 1995.
17. X. Yan, L. Zhu, Y. Yu, Q. Xu, H. Huang, and L. Jiang, L. *Journal of Agricultural and Food Chemistry*, Vol. 66, p. 1444, 2018.
18. S. Paul, and S. Paul, *The Journal of Physical Chemistry B*, Vol. 119, p. 1598, 2015.
19. J.K. Kaushik, and R. Bhat, *Journal of Biological Chemistry*, Vol. 278, p. 26458, 2003.
20. S. Suethao, B. Innawong, S. Sirisansaneeyakul, W. Vanichsriratana, and P. Parakulsuksatid, *Journal of Kasetsart*, Vol. 49, p. 726, 2015.
21. T.T.Wu, S.C. Lin, and J.F. Shaw, *Process Biochemistry*, Vol. 46, p. 1481, 2011.
22. D. KoVačeVić, and K. MaStaNJeVić, *Czech Journal of Food Sciences*, Vol. 29, p. 15, 2011.
23. M. Walmagh, R. Zhao, R. and T. Desmet, *International Journal of Molecular Sciences*, Vol. 16, p. 13729, 2015.
24. R. Ruhal, R. Kataria, R. and B. Choudhury, *Microbial Biotechnology*, Vol. 6, p. 493, 2013.
25. U. Breuer, and H. Harms, *Yeast*, Vol. 23, p. 415, 2006.
26. S. Magazu, and F. Migliardo, *Quality Assurance and Safety of Crops & Foods*, Vol. 2, p. 56, 2010.
27. M. Yoshida, N, Nakamura, and K. Horikoshi, *Enzyme and Microbial Technology*, Vol. 22, p. 71, 1998.
28. Y. Wu, J. Wang, X. Shen, J. Wang, Z. Chen, X. Sun, and Y. Yan, *Biotechnology and Bioengineering*, Vol. 115, p. 785, 2018.
29. S.W. Chang, P.T. Liu, L.C. Hsu, C.S. Chen, and J.F. Shaw, *Food Chemistry*, Vol. 134, p. 1745, 2012.
30. M.K. O'Neill, B.F. Piligian, C.D. Olson, P.J. Woodruff, and B.M. Swarts, *Pure and Applied Chemistry*, Vol. 89, p. 1223, 2017.

31. Poplinger, I. Shumilin, and D. Harries, *Food Chemistry*, Vol. 237, p. 1209, 2017.
32. N. Li, H. Wang, L. Li, H. Cheng, D. Liu, H. Cheng, and Z. Deng, *Journal of Agricultural and Food Chemistry*, Vol. 64, p. 6179, 2016.
33. M. Sajadi, F. Berndt, C. Richter, M. Gerecke, R. Mahrwald, and N.P. Ernsting. *The Journal of Physical Chemistry Letters*, Vol. 5, p. 1845, 2014.
34. V. Leung, A. Szewczyk, J. Chau, Z. Hosseinidoust, L. Groves, H. Hawsawi, and C.D. Filipe, *ACS Biomaterials Science & Engineering*, 2017.
35. D. Komes, T. Lovrić, and K.K. Ganić, *LWT-Food Science and Technology*, Vol. 40, p. 1578, 2007.
36. J.H. Crowe, *Membrane Hydration* Springer International Publishing, 2015.
37. C. Henry, S.W. Bledsoe, A. Siekman, A. Kollman, B.M. Waters, R. Feil, M. Stitt, and L.M. Lagrimini, *Journal of Experimental Botany*, Vol. 65, p. 5959, 2014.
38. R.J. Mancini, J., Lee, and H.D. Maynard, *Journal of the American Chemical Society*, Vol. 134, p. 8474, 2012.

# 14

# Production of Yeast Derived Microsomal Human CYP450 Enzymes (Sacchrosomes) in High Yields, and Activities Superior to Commercially Available Microsomal Enzymes

**Ibidapo Stephen Williams and Bhabatosh Chaudhuri***

*CYP Design Ltd, The Innovation Centre, 49 Oxford Street, Leicester, LE1 5XY, UK*

***Abstract***

Human CYP450 enzymes, naturally bound to livers' endoplasmic reticular (ER) membranes, are used for drug-metabolism studies and commercially produced from genetically engineered insect/bacterial cells. They are temperature-labile and must be stored at –80 °C. Unlike insect/bacterial cells, the intracellular environment of baker's yeast and human cells is similar. Here we show that six major human liver CYP enzymes (1A2/2C9/2C19/2D6/2E1/3A4), isolated from baker's yeast as ER-bound microsomes (Sacchrosomes™) in yields of 120–325 nmol/L, are 0.5–25 fold more active than commercial enzymes. $IC_{50}$ values for Sacchrosome inhibition by known CYP inhibitors are similar to those published using human-liver-microsomes (HLMs). Moreover, freeze-dried Sacchrosomes can be stored at +21 °C for at least 6 months without losing enzymatic activity. These results suggest that yeast cells, cheap and easy to grow, could provide an ideal environment for bulk production of highly active human CYP microsomal enzymes so that they could be used more widely than they are now.

***Keywords:*** Baker's yeast, human cytochrome P450, CYP450 activity, drug metabolism, freeze-drying, microsomal enzyme

## 14.1 Introduction

### 14.1.1 Cytochrome P450 (CYP) Enzymes in Humans

Chemical compounds that are usually harmful and toxic to the human organism are referred to as xenobiotics. The main oxidative enzyme system involved in xenobiotic metabolism, a process which makes humans safe from the toxicity of xenobiotics, is the family of haem-containing intracellular membrane-bound cytochrome P450 (CYP)

**Corresponding author*: bchaud00@gmail.com

Saurabh Saran, Vikash Babu, and Asha Chaubey (eds.) High Value Fermentation Products, Volume 1, (323–346) © 2019 Scrivener Publishing LLC

enzymes. They also help in removing potential xenobiotic carcinogens from the body [1]. Beside CYPs, enzymes like alcohol dehydrogenase, xanthine oxidase and the amine oxidases are also involved in xenobiotic metabolism, albeit having less important roles than the CYPs [2]. In addition to playing a part in xenobiotic metabolism, there are CYP enzymes within the human body that participate in the syntheses of diverse chemicals such as steroids, prostanoids, eicosanoids and fatty acids [3]. These CYPs use endogenous chemicals as natural substrates for such chemical syntheses.

Mutations in *CYP* genes are responsible for several human diseases resulting from enzymatic deficiencies caused by the mutations [1]. Certain CYPs have functionally important roles in the brain and are known to partake in vascular autoregulation [4]. There are other CYPs which activate exogenous pro-carcinogenic chemicals to carcinogens and thereby promote incidences of breast and lung cancers [5].

The CYPs that predominantly exist in the liver, the main organ where drug and toxin removal occur [6], are differentially expressed. Besides the liver, CYPs are also found in extrahepatic tissues such as the small intestine, lung [7], kidney, skin, heart and the brain [8–9].

### 14.1.2 Human Cytochrome P450 Enzymes and their Role in Drug Metabolism

Human CYPs participate in solubilising (i.e., metabolizing) a pharmaceutical drug (i.e., a medicine), which is imbibed, for excretion from the human body. Hence, they are known as drug metabolizing enzymes [1]. In the human liver there are at least 12 distinct CYP enzymes. At present, it appears that only six isoenzymes from families CYP 1, 2 and 3 are involved in the oxidation (i.e., metabolism) of most (i.e., ~95%) of the marketed pharmaceutical drugs. They are CYP1A2, CYP2C9, CYP2C19, CYP2D6, CYP2E1 and CYP3A4. These CYPs account for about 70% of the human hepatic CYPs. They are naturally bound to the liver endoplasmic reticular (microsomal) membranes. The CYP3 isoenzymes account for approximately 30% of the total CYPs present in the liver.

Drug metabolism research in pharmaceutical industry involves accurately (a) determining the potential of a drug and its metabolites to inhibit CYP enzymes and (b) identifying the metabolites of a drug which are formed through CYP-mediated biotransformation reactions. The degree of success in drug metabolism research is one of the crucial measures of the approval/rejection of a drug by regulatory authorities. However, drug metabolism studies take place only in the pre-clinical – clinical interface which is quite late in the drug discovery process for a failure to occur, which unfortunately happens quite often. Drug metabolism studies are not performed any earlier primarily because CYP enzymes are extremely costly and are difficult to use because of their inherent instability at room temperature. The experiments presented in this chapter attempt to address these issues.

The main aim of the work was to create a new set of tools that would facilitate the drug discovery process in general but, more specifically, allow these reagents to be used by chemists/biologists in early pre-clinical phase, where they are most needed, by trying to overcome the major hurdles of CYP's costs and stability. It is acknowledged that the human CYP enzymes are responsible for diverse sets of chemical reactions which occur in stereospecific, stereoselective, regiospecific and regioselective manner. Hence, they can play more important roles in drug discovery if they could be used more readily [10–11].

### 14.1.3 Requirement of Activating Proteins to Form Functional Human CYP Enzymes

A CYP enzyme requires a CYP reductase (CPR) for it to be active [12]. A CPR performs the function of an oxidoreductase. Both CYPs and CPR are bound to the intracellular endoplasmic reticular (ER) membranes. The presence of a CPR is essential for membrane-bound CYP activity. It seems another ER-bound protein, cytochrome *b5*, is required for the activity of some CYPs. However, cytochrome *b5*'s role in CYP activity is unclear [13]. In this chapter, a modified human CPR, which is less toxic than the wild-type human CPR and is more efficient in coupling with human CYP proteins [14], has been used for the expression of active human CYP enzymes.

### 14.1.4 Use of Yeast Biased Codons for the Syntheses of Human Cytochrome P450 Genes

There are 61 codons which code for the 20 amino acids that are essential for protein syntheses. It is known that the highly expressed genes in an organism use specific codons. This would indicate that there is codon bias for genes that produce abundant proteins. Hence, codons used by highly expressed genes in baker's yeast were used to create genetic sequences for expression of human CYP enzymes [15]. These codon-optimised human *CYP* genes, in theory, should provide high expression in yeast. To assess the assumption that yeast-biased codons would provide better expression of human CYPs, expression of 'synthetic' *CYP* genes was compared with the expression of 'native' *CYP* genes, isolated from a human liver cDNA library. Both synthetic and native genes were expressed from an episomal 2µ-based plasmid (Figure 14.1) which replicates extra-chromosomally [16]. It was clearly shown that, per constant number of yeast cells, the synthetic *CYP* genes produced far more CYP protein than the native *CYP* genes (unpublished observations).

### 14.1.5 Expression of Human CYP Genes in Baker's Yeast from an Episomal Plasmid

CYP protein levels may vacillate when human *CYP* genes are expressed from an episomal 2µ-based plasmid (Yep) within yeast cells which already contain a modified *CPR* (*CYP reductase*) gene expression cassette integrated on a chromosomal locus [14]. The vacillation of the copy number of a Yep depends on how toxic the human CYP protein is for the host yeast organism in which it is being expressed. This happens because the autonomously replicating sequence based on the yeast 2µ-plasmid allows for fluctuation of the plasmid's copy number depending on the cells' ability to cope with the levels of expression of a relatively toxic protein produced from the gene borne on the 2µ episomal plasmid.

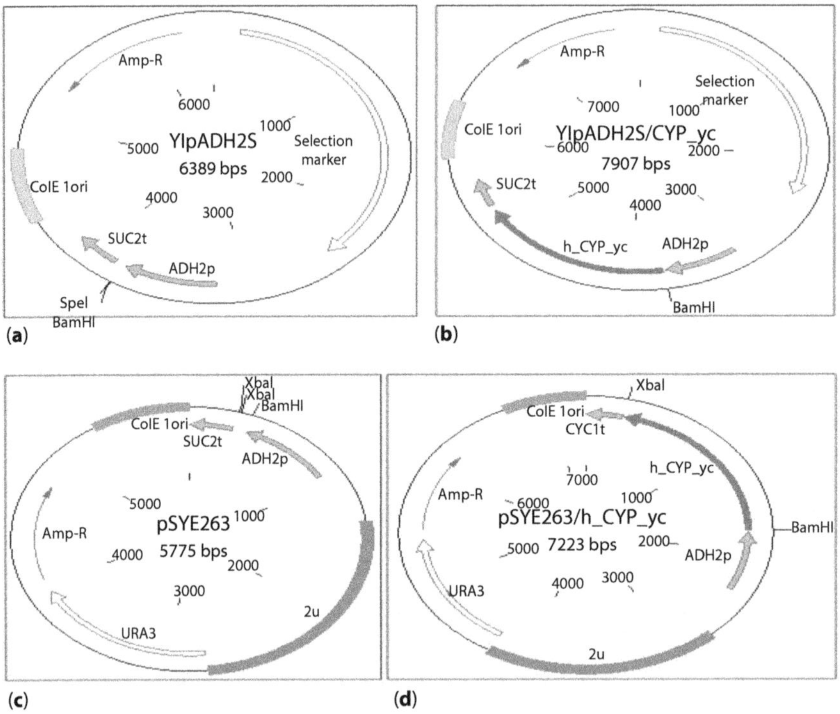


**Figure 14.1** (a) Shows a generic basic plasmid, YIpADH2S, used for cloning of human *CYP* genes, chemically synthesized using yeast-biased codons, that would allow integration of a *CYP* gene expression cassette into the yeast genome. It contains the ethanol-inducible yeast *ADH2* promoter (ADH2p) and the *SUC2* transcription termination signal (SUC2t). The 'colE1ori' sequence allows replication of the plasmid in bacterial cells for the plasmid to be amplified for isolation of plasmid DNA. The 'Selection Marker' could be any one of the functional yeast *ADE2, LEU2, HIS3, TRP1* or *URA3* genes which would permit auxotrophic selection of the plasmid during its integration into the genome of the yeast strains, YY7 and YAB79, derived from the strain W303–1a (ATCC 208352) which bears non-functional alleles *ade2, leu2, his3, trp1* and *ura3*. The synthetic human *CYP* genes were cloned as *BamHI-XbaI* fragments at the *BamHI, SpeI* sites of the integration plasmids bearing different auxotrophic markers. (b) Shows the map of a generic integration plasmid after cloning of a *CYP* gene downstream of the ADH2p in the basic integration plasmid, YIpADH2S. (c) Shows a basic episomal, 2μ-plasmid, pSYE263, that bears a functional *URA3* gene. The human *CYP* genes, chemically synthesized using yeast-biased codons, were cloned as *BamHI-XbaI* or *BamHI-XhoI* fragments at either the *BamHI, XbaI* or the *BamHI, XhoI* sites of pSYE263. (d) Shows the map of a generic episomal plasmid after cloning of a *CYP* gene downstream of the ADH2p in the basic episomal plasmid, pSYE263. The term 'CYP_yc' represents a *CYP* gene synthesized using yeast-biased codons.

It should also be pointed out that cells containing episomal plasmids can only be maintained, within yeast cells, when they are grown in selective synthetic minimal media [16]. However, production of a heterologous protein in yeast, using a minimal selective medium, would become a futile exercise if very little protein (in our case, CYP enzyme) would be obtained because of poor cell growth. Instead, complete full medium, which does not select for the presence of the plasmid bearing the heterologous *CYP* gene, could be used for short-term expression from a Yep, over a period of only 24 to 48 h. This is because, in full, non-selective complete medium, there would be huge loss of plasmid from the cell population after a few generations of growth, implying that cells would continue to grow in the absence of a plasmid that bears the heterologous (*CYP*) gene.

### 14.1.6 Expression of Human CYP Genes in Baker's Yeast from Integrative Plasmids

In order to avoid plasmid loss and allow growth of yeast cells in complete full medium, we have established a generic technology that allows chromosomal integration of expression cassettes bearing *CYP* genes, synthesized using yeast-biased codons, to create highly stable yeast strains that can grow in complete medium over weeks.

Expression of heterologous CYP proteins from genomic copies of *CYP* genes, integrated into the yeast genome via homologous recombination [17], would be far 'more stable' than expression from an episomal vector because

- There would not be any copy number fluctuations – the copy number would depend solely on how many copies of the *CYP* gene expression cassette are introduced into different chromosomal loci, and
- Once the *CYP* gene expression cassettes have stably been integrated into one or more of yeast's chromosomes, the cells could be grown in fully nutritious complete YPD (yeast extract-peptone-dextrose) medium, without any selection.

This would permit cells containing integrated copies of *CYP* gene expression cassettes to be used routinely for

- High-throughput screening of potential CYP inhibitors using whole cells, and also
- Production of active CYP enzymes on a large scale.

### 14.1.7 The *ADH2* Promoter for Production of Human CYP Enzymes in Baker's Yeast

The promoter from the *ADH2* gene (ADH2p; Saccharomyces Genome Database; YMR303C), derived from one of the two alcohol dehydrogenase genes present in yeast, was used for expression of human CYP enzymes in yeast. It was isolated by the polymerase chain reaction (PCR), with the help of a high-fidelity polymerase, using the genomic DNA of the wild-type yeast strain S288C (ATCC 204508) as a template.

For expression of a human CYP enzyme, the yeast cells contained either chromosomally integrated copies of *CYP* gene expression cassettes on the genome or an extrachromosomal episomal plasmid that encodes a *CYP* gene. In both cases, the *CYP* genes used for expression were synthesized using yeast-biased codons and were cloned downstream of the ADH2p.

The ADH2p is repressed in medium containing glucose and is induced in the presence of ethanol. Exhaustion of glucose, through steady conversion of glucose to ethanol, gradually induces the *ADH2* promoter and allows the yeast cells to slowly adapt to the overexpression of the toxic, heterologous human CYP enzymes which have powerful oxidoreductive properties [18]. The transcription termination signal that was used for expression in yeast of human *CYP* genes, with yeast-biased codons, was the *SUC2* terminator (SUC2t). It was isolated from the *SUC2* gene of yeast (Saccharomyces Genome Database; S000001424) by PCR using genomic DNA isolated from the strain S288C.

Generic *CYP* gene expression plasmids, integrative and episomal, used for transformation of the two yeast strains which already contain (a) a modified human *CPR* gene (yeast strain, YY7) or (b) a modified human *CPR* gene and the human *cytochrome b5* gene (yeast strain, YAB79) are depicted in Figure 14.1.

### 14.1.8 Growth of Yeast Cells Containing Integrated Copies of CYP Gene Expression Cassettes, Driven by the *ADH2* Promoter, for Production of CYP Enzymes

Yeast cells containing stably integrated *CYP* gene expression cassettes are grown in complete YPD medium (containing 1.5% glucose) to obtain a high cell density of ~100, measured at a wavelength of 600 nm ($OD_{600}$), in baffled shake flasks over a period of 54 h. This occurs without any plasmid loss. Such a high cell density was achieved over 3 days through repetitive replenishment with fresh medium, every 18 h. The time period could easily have been extended to obtain a higher number of cells. Since the promoter used for gene transcription is ADH2p, the duration of growth of yeast cells could be as long as the cells can divide within a defined enclosed space (i.e., a shake flask or a fermentor). Stoppage of cell growth, because of aging (aged cells being incapable of vigorous cell division), could take more than a fortnight, perhaps longer, depending on the conditions used for fermentation. Moreover, in order to prevent cells to become fully saturated with ethanol whereby they cannot grow any further, cells can be replenished with fresh medium when using shake flasks, or could be grown under fed-batch fermentation conditions in fermentors where exhausted medium could be drained away with fresh medium being drip-fed continuously.

In summary, although in the experiments that follow a time period of 54 h has been used for growth, yeast cells containing integrated copies of *CYP* gene expression cassettes could potentially be grown, over many more generations, for continuous production of CYP proteins of interest. Depending on the levels of required production, the yeast cells could be grown in a small scale in baffled shake flasks or in a larger scale in fermentors merely by feeding yeast cells continuously with nutrients present in complete cell culture medium [19].

## 14.2 Amounts of Microsomal CYP Enzyme Isolated from Yeast Strains Containing Chromosomally Integrated CYP Gene Expression Cassettes are far Higher than Strains Harbouring an Episomal Expression Plasmid Encoding a CYP Gene

### 14.2.1 Preparation of Microsomal CYP Enzymes

Yeast strains, each of which contain two copies of human *CYP* genes, downstream of the *ADH2* promoter, integrated into different chromosomal loci of the genome of the yeast strain YY7 or YAB79 were used for preparation of microsomal CYP enzymes. For comparison's sake, strains YY7 or YAB79 transformed with episomal 2μ-plasmids, encoding human *CYP* genes, were used for preparation of microsomal CYP enzymes. YY7 is a derivative of strain W303–1a (ATCC 208352) with an integrated copy of the modified human CYP450 reductase gene at its *LEU2* genomic locus [14]. YAB79 is a derivative of strain YY7 with an integrated copy of the *cytochrome b5* gene at its *TRP1* genomic locus.

Yeast strains (harbouring integrated copies of a *CYP* gene or an episomal plasmid bearing a *CYP* gene) from frozen stocks were revived in 500 mL Erlenmeyer baffled flasks containing 100 mL YPD medium (peptone 20 $gL^{-1}$, yeast extract 10 $gL^{-1}$, glucose 15 $gL^{-1}$, pH 6.0). The flasks were shaken at 200 rpm for 24 h at 28 °C. Three consecutive 400 mL YPD pre-cultures, starting with the 100 mL pre-culture inoculum, were then grown for production of high biomass in 2 L baffled flasks at 30 °C for 18 h. The cells were harvested, after every 18 h, before inoculation into a new 2 L baffled flask containing 400 ml YPD medium. The process, with the recombinant yeast cells grown for a total time period of 54 h, yielded 400 mL of cells with a final $OD_{600}$, of ~100 (a total of ~$1 \times 10^{11}$ cells). These cells were used for liberation of membrane-bound (microsomal) CYP proteins.

The cells, suspended in Buffer-A (4M sorbitol, 1M Tris-HCl pH 7.5, 0.5M EDTA pH 8.0, 0.1M AEBSF, 0.1M DTT), were lysed using a cell disruptor (Constant Systems Basic Z model with continuous processing head) at a pressure of 22.5 ksi. The cell lysates were collected from the cell disruptor and were centrifuged at 4,500 g for 20 min at 4 °C using a Sorvall bench-top centrifuge to remove all cell debris and broken cell membranes. The resulting supernatants were then transferred into chilled 30 ml tubes and centrifuged at 35,000 g for 15 min at 4 °C in a high-speed centrifuge (Beckman Coulter Avanti J-20XP) to remove cell debris, nuclei, peroxisomes, lysosomes and mitochondria. Centrifugation was repeated twice.

The supernatants, contained in a beaker, were diluted with ice cold Buffer-B (0.6M sorbitol, 60 mM Tris-HCl pH 7.6, 1 mM EDTA pH 8, 0.4 mM AEBSF, 0.04 mM DTT). The beaker was stirred at 1–2 rpm at 4 °C. NaCl was added gently to give a final concentration of 0.125 M. Ice-cold 50% PEG solution was added drop-wise (1–2 drops per second) from a separating funnel, whilst gently stirring the mix, to precipitate the microsomal (ER-bound) CYP enzymes [20]. Aliquots of the enzymes were frozen at –80 °C or were used directly for freeze-drying. Thawed enzymes from the –80 °C freezer or freeze-dried enzymes, re-suspended in ultrapure water, were used to perform the comparative analyses described in the Sections that follow.

### 14.2.2 Measurement of the Amounts of Functional CYPs in Microsomes Isolated from Baker's Yeast

The amounts (i.e. nanomoles) of functional CYP enzymes in the microsomes, obtained from 400 ml YPD cell culture, were ascertained via CO-difference spectroscopy [21]. The assay is based on the fact that, when carbon monoxide (CO) reacts with the ferrous form of a functionally active CYP, a complex is formed that produces a spectrum with a wavelength maximum at ~450 nm. The cytochrome P450 (CYP) concentration present in the yeast microsomes (Sacchroosmes™) was calculated from the difference between the absorbance at 450 nm (peak) and 490 nm, using an extinction coefficient of 91 $mM^{-1}cm^{-1}$. When CO reacts with an inactive (i.e., misfolded, non-functional) CYP, the CO-inactive CYP complex absorbs at a wavelength maximum of 420 nm. The protein cytochrome b5 also absorbs at 420 nM. The cytochrome b5 peak is seen in the microsomal enzymes, Sacchrosomes, prepared from the YAB79 strain.

### 14.2.3 Production of Functional Human CYP1A2 Microsomal Enzyme from Baker's Yeast

The results in Figure 14.2 show that cells containing 2 integrated copies of the human *CYP1A2* gene expression cassette produce around 70 nanomoles of functional CYP1A2 enzyme from cells grown over 3 days (54 h) from 400 ml of YPD culture, the yield per litre being ~175 nanomoles. The amount obtained, over the same time period, from the *CYP1A2* gene expressed from an episomal 2μ-plasmid was ~5 nanomoles (i.e. 12.5 nmol/L). The Western blot, in Figure 14.2 (D), confirms the presence of the CYP1A2 protein and shows a clear difference in the amounts of CYP1A2 protein produced within the cells of the three different strains that contain *CYP1A2* gene as: (i) 2 integrated copies, (ii) 1 integrated copy, and (iii) encoded by a 2u-plasmid.

### 14.2.4 Production of Functional Human CYP3A4 Microsomal Enzyme from Baker's Yeast

The results in Figure 14.3 show that cells containing 2 integrated copies of the human *CYP3A4* gene expression cassette produce ~48 nanomoles of functional CYP3A4 enzyme from cells grown over 3 days (54 h) from 400 ml of YPD culture, the yield per

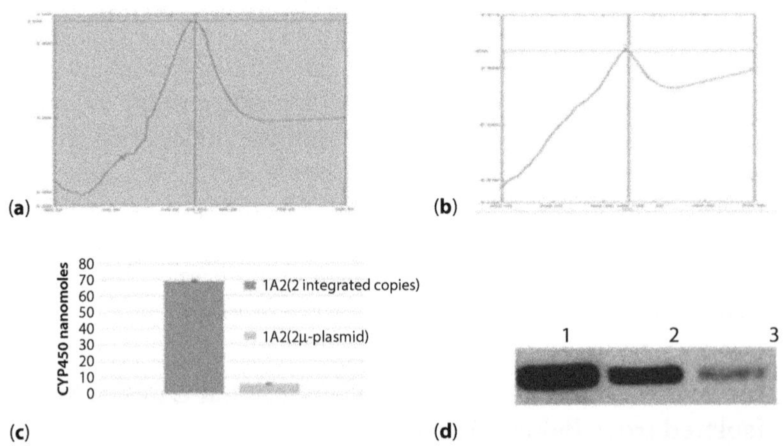


**Figure 14.2** (a), (b): Functional CYP450 content obtained from the strains YY7::1A2(2 integrated copies) (a) and YY7::1A2(2μ-plasmid) (b) was determined by CO difference spectroscopy using 1 and 3 mg of microsomal protein (21). (c) The amount of CYP450 (in nanomoles) obtained from the 2 strains is represented as bar plots. The data represent mean ± S.D. of three independent experiments. (d) Western blot using 10 μg of total protein, obtained after lysis of the three strains YY7::1A2(2 integrated copies) (lane 1), YY7::1A2(1 integrated copy) (lane 2) and YY7::1A2(2μ-plasmid) (lane 3) was probed with a CYP1A2 specific antibody (Santa Cruz Biotechnology, sc-53241).

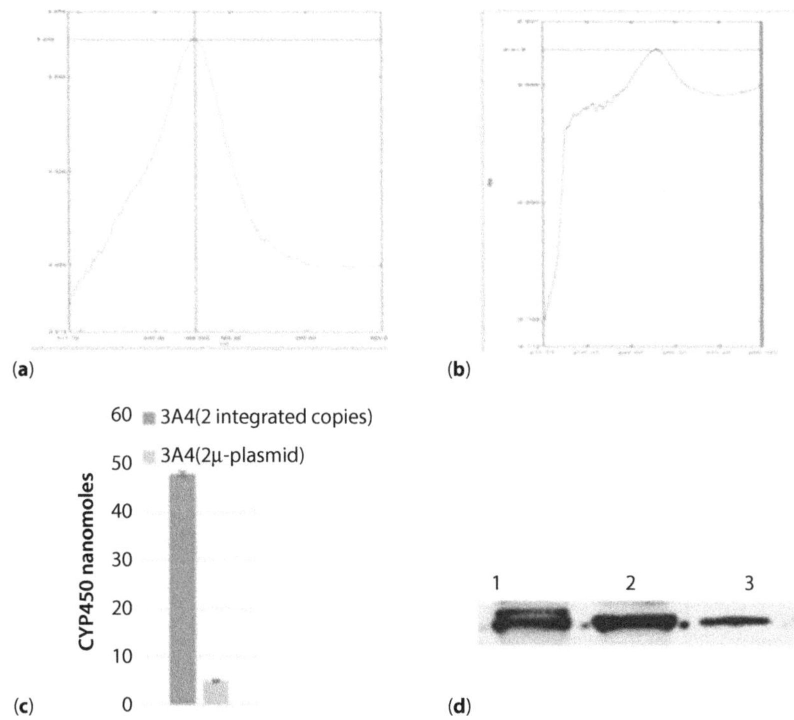


**Figure 14.3** (a), (b) Functional CYP450 content obtained from the strains YAB79::3A4(2 integrated copies) (a) and YAB79::3A4(2μ-plasmid) (b) was determined by CO difference spectroscopy using 1 and 5 mg of microsomal protein (21). (c) The amount of CYP450 (in nanomoles) obtained from the 2 strains is represented as bar plots. The data represent mean ± S.D. of three independent experiments. (d) Western blots where 10 μg of total protein, obtained after lysis of the three strains YAB79::3A4(2 integrated copies) (lane 1), YAB79::3A4(1 integrated copy) (lane 2) and YAB79::3A4(2μ-plasmid) (lane 3), was probed with a CYP3A4 specific antibody (Santa Cruz Biotechnology, Cat no:sc-27639).

litre being ~120 nanomoles. The amount obtained from the *CYP3A4* gene expressed, over the same time period, from an episomal 2μ-plasmid was ~4 nanomoles (i.e. 10 nmol/L). The Western blot, in Figure 14.3 (D), confirms the presence of the CYP3A4 protein and shows differences in the amounts of CYP3A4 protein produced within the cells of the three different strains containing *CYP3A4* gene as: (i) 2 integrated copies, (ii) 1 integrated copy, and (iii) encoded by a 2u-plasmid.

### 14.2.5 Production of Functional Human CYP2D6 Microsomal Enzyme from Baker's Yeast

The results in Figure 14.4 show that cells containing 2 integrated copies of the human *CYP2D6* gene expression cassette produce ~52 nanomoles of functional CYP2D6($Val^{374}$) [22] enzyme from cells grown over 3 days (54 h) from 400 ml of YPD culture, the yield per litre being ~130 nanomoles. The Val374 variant of CYP2D6 is the enzyme that is sold worldwide by Corning. The amount obtained from the *CYP2D6* gene expressed, over the same time period, from an episomal 2μ-plasmid was ~4 nanomoles (i.e. ~10 nmol/L). The Western blot, in Figure 14.4 (D), confirms the presence of the CYP2D6

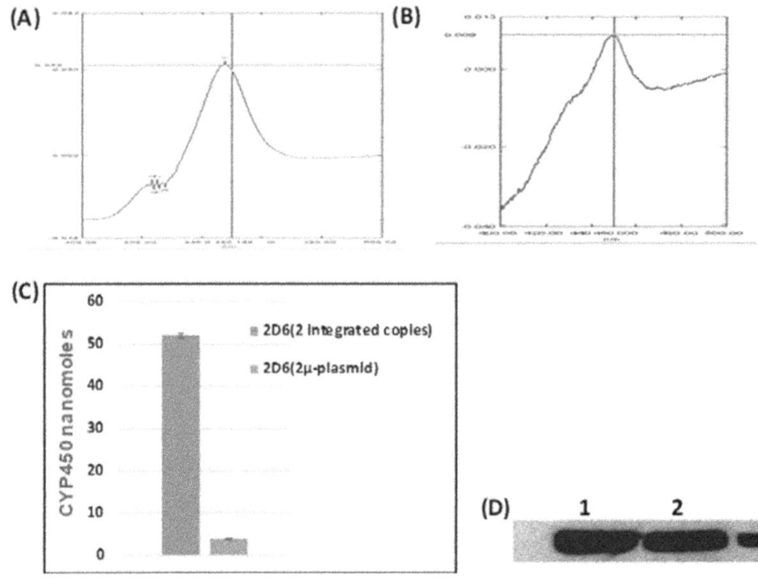


**Figure 14.4** (a), (b): Functional CYP450 content obtained from the strains YY7::2D6(2 integrated copies) (a) and YY7::2D6(2μ-plasmid) (b) was determined via CO difference spectroscopy using 1 and 3 mg of microsomal protein (21). (c) The amount of CYP450 (in nanomoles) obtained from the two strains is represented as bar plots. The data represent mean ± S.D. of three independent experiments. (d) Western blots where 10 μg of total protein, obtained after lysis of the three strains YY7::2D6(2 integrated copies) (lane 1), YY7::2D6(1 integrated copy) (lane 2), and YY7::2D6(2μ-plasmid) (lane 3), was probed with a CYP2D6 specific antibody (Santa Cruz Biotechnology, Cat no:sc-130366).

protein and shows differences in the amounts of CYP2D6(Val$^{374}$) protein produced within the cells of the three different strains containing *CYP2D6* gene as: (i) 2 integrated copies, (ii) 1 integrated copy, and (iii) encoded by a 2u-plasmid. Similar results have been obtained with five other variants, including single nucleotide polymorphs (SNPs), of CYP2D6.

### 14.2.6 Production of Functional Human CYP2C19 Microsomal Enzyme from Baker's Yeast

The results in Figure 14.5 show that cells containing 2 integrated copies of the human *CYP2C19* gene expression cassette produce nearly 70 nanomoles of functional CYP2C19 enzyme from cells grown, over 3 days (54 h), in 400 ml of YPD culture, the yield per litre being ~175 nanomoles. The amount obtained from the *CYP2C19* gene expressed, over the same time period, from an episomal 2μ-plasmid was ~4 nanomoles (i.e., ~10 nmol/L). The Western blot, in Figure 14.5 (D), confirms the presence of the CYP2C19 protein and shows differences in the amounts of CYP2C19 protein produced within the cells of the three different strains containing *CYP2C19* gene as: (i) 2 integrated copies, (ii) 1 integrated copy, and (iii) encoded by a 2u-plasmid.

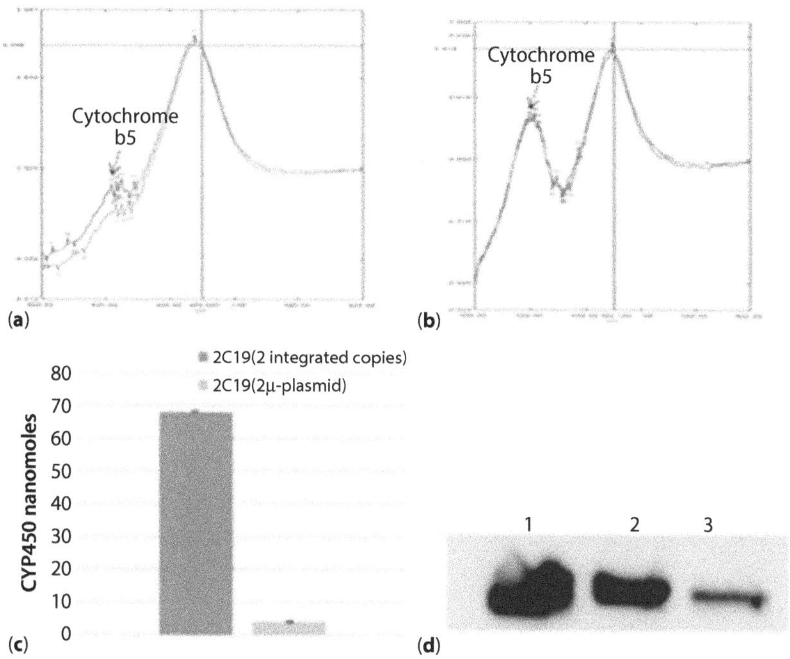


**Figure 14.5** (a), (b): Functional CYP450 content obtained from the strains YAB79::2C19(2 integrated copies) (a) and YAB79::2C19(2μ-plasmid) (b) was determined by CO difference spectroscopy using 1 and 5 mg of microsomal protein (21). (c) The amount of CYP450 (in nanomoles) obtained from the two strains is represented as bar plots. The data represent mean ± S.D. of three independent experiments. (d) Western blots where 10 μg of total protein from the strains YAB79::2C19(2 integrated copies) (lane 1), YAB79::2C19(1 integrated copy) (lane 2) and YAB79::2C19 (2μ-plasmid) (lane 3) were probed with a CYP2C19 specific antibody (Santa Cruz Biotechnology, Cat no:sc-25581).

### 14.2.7 Production of Functional Human CYP2C9 Microsomal Enzyme from Baker's Yeast

The results in Figure 14.6 show that cells containing 2 integrated copies of the human *CYP2C9* gene expression cassette produce around 130 nanomoles of functional CYP2C9 enzyme from cells grown over 3 days (54 h) in 400 ml of YPD culture, the yield per litre being 325 nanomoles. The amount obtained from the *CYP2C9* gene expressed, over the same time period, from an episomal 2μ-plasmid was ~5 nanomoles (i.e. ~12.5 nmol/L). The Western blot, in Figure 14.6 (D), confirms the presence of the CYP2C9 protein and shows differences in the amounts of CYP2C9 protein produced within the cells of the three different strains containing *CYP2C9* gene as: (i) 2 integrated copies, (ii) 1 integrated copy, and (iii) encoded by a 2u-plasmid.

### 14.2.8 Production of Functional Human CYP2E1 Microsomal Enzyme from Baker's Yeast

The results in Figure 14.7 show that cells containing 2 integrated copies of the human *CYP2E1* gene expression cassette produce around 56 nanomoles of functional CYP2E1

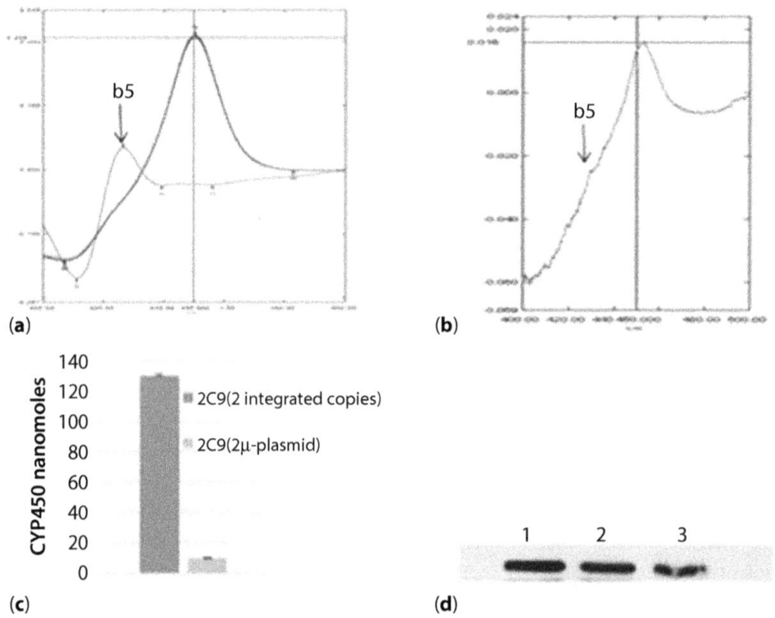


**Figure 14.6** (a), (b): Functional CYP450 content obtained from the strains YAB79::2C9(2 integrated copies) (a) and YAB79::2C9(2µ-plasmid) (b)] were determined by CO difference spectroscopy using 1 and 3 mg of microsomal protein (21). (c) The amount of CYP450 (in nanomoles), obtained from the two strains is represented as bar plots. The data represent mean ± S.D. of three independent experiments. (d) Western blots where 10 µg of total protein from the strains YAB79::2C9(2 integrated copies) (lane 1), YAB79::2C9(1 integrated copy) (lane 2) and YAB79::2C9(2µ-plasmid) (lane 3) were probed with a CYP2C9 specific antibody (Santa Cruz Biotechnology, Cat no:sc-374421).

enzyme from cells grown over 3 days (54 h) from 400 ml of YPD culture, the yield per litre being ~140 nanomoles. The amount obtained from the *CYP2E1* gene expressed, over the same time period, from an episomal 2µ-plasmid was ~5 nanomoles (i.e. ~12.5 nmol/L). The Western blot, in Figure 14.7 (D), confirms the presence of the CYP2E1 protein and shows differences in the amounts of CYP2E1 protein produced within the cells of the three different strains containing *CYP2E1* gene as: (i) 2 integrated copies, (ii) 1 integrated copy, and (iii) encoded by a 2u-plasmid.

### 14.2.9 Comments on the Production of Human CYP Enzymes from Baker's Yeast

Such high amounts (120 nmol/L to 325 nmol/L) of human CYP enzymes produced from yeast, as has been shown in this Section, have not been reported before [23]. The amounts reported could have been much higher if the cell culture was extended further beyond 54 h. Usually, large fermentors (30 L or larger) have been used for production of recombinant human CYPs from yeast. The amounts that have been reported from 30 L fermentors are far less than obtained from a 0.4 L yeast cell culture [23].

The yields of 120 nmol/L to 325 nmol/L of the six major human CYP enzymes, produced in yeast and commonly used for drug metabolism studies, compares well with

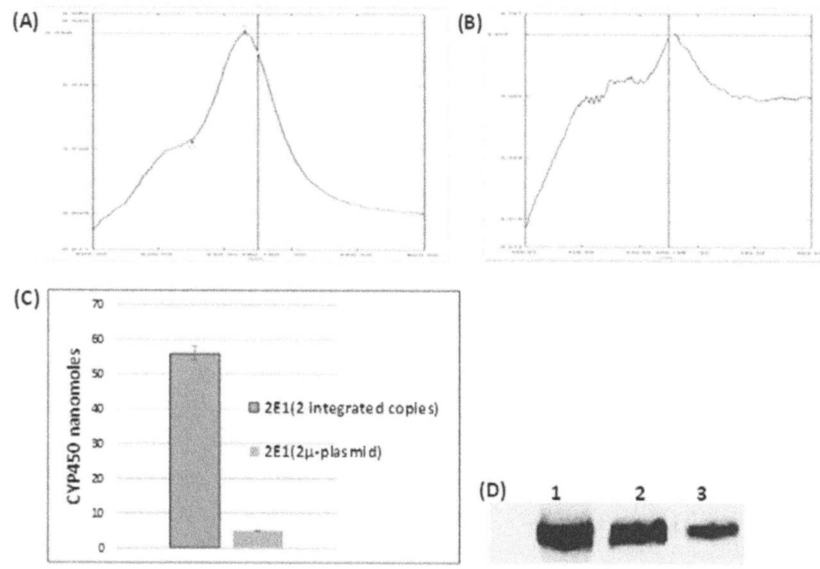


**Figure 14.7** (a), (b): Functional CYP450 content obtained from the strains YAB79::2E1(2 integrated copies) (a) and YAB79::2E1(2µ-plasmid) (b) were determined by CO difference spectroscopy using 1 and 3 mg of microsomal protein (21). (c) The amount of CYP450 (in nanomoles) obtained from the two strains is represented as bar plots. The data represent mean ± S.D. of three independent experiments. (d) Western blots where 10 µg of total protein from the strains YAB79::2E1(2 integrated copies) (lane 1), YAB79::2E1(1 integrated copy) (lane 2) and YAB79::2E1(2µ-plasmid) (lane 3) were probed with a CYP2E1 specific antibody (Santa Cruz Biotechnology, Cat no:sc-374421).

the available bacterial production systems of 'human' CYPs where, to the best of our knowledge, yields have always been less than 100 nmol/L for the human enzymes [24]. However, bacterial cells, being prokaryotes, suffer from the drawback that they do not contain the intracellular ER compartment. Hence, the human *CYP* genes must drastically be modified for bacterial expression that would allow association of the produced CYP proteins with bacterial cells' periplasmic membranes. The yields from insect cell systems are usually very low compared to bacterial expression. Insect cell production always depends on large stirred-vessel fermentors [25].

The yeast system described here relies on chromosomal integration of two copies of *ADH2* promoter driven human *CYP* gene expression cassettes. The genes consist of cDNA molecules that were chemically synthesized using yeast-biased codons. Genome-integrated copies allow recombinant cells to be grown continuously in complete YPD medium for generations without loss of genetic information. Use of the ethanol-inducible *ADH2* promoter allows continuous growth of cells in fresh YPD medium containing glucose. After complete exhaustion of glucose at ~12 h, the genes are induced in ethanol for ~6 h before the medium containing glucose, is replenished, every 18 h of cell growth. It should be noted that the modified, less toxic CPR, that was used for expression also plays a crucial role in the production of functional human CYP enzymes and their activities [14].

## 14.3 Comparison of CYP Enzyme Activity of Yeast-Derived Microsomes (Sacchrosomes) with Commercially Available Microsomes Isolated from Insect and Bacterial Cells

### 14.3.1 Fluorescence-based Assays for Determining CYP Enzyme Activities in Isolated Microsomes

Only a few chemical molecules have been found to be useful as selective probes (i.e., as inhibitors and substrates) of each human CYP enzyme. Analysis of CYP enzyme inhibition and identification of metabolites that are formed, upon reaction of a CYP enzyme with its substrate, usually involves the use of techniques such as high-pressure liquid chromatography (HPLC) and liquid chromatography in tandem with mass spectrometry (LC-MS). These methodologies are slow and enormously labour intensive.

Alternatively, non-fluorogenic substrates can be used for rapid screening of libraries of chemical compounds to identify a compound's potential to inhibit CYP enzymes. These substrates do not fluoresce on their own but form fluorescent products upon reacting with specific CYP enzymes. Historically, the very early non-fluorogenic substrates had limited aqueous solubility and showed poor metabolite formation. They also gave high background and low signal to noise ratio which would not have been appropriate for screening of CYP enzyme inhibitors. Later on, the *O*-alkyl derivatives of resorufin, fluorescein, 7-hydroxycoumarins and other molecules were discovered to study CYP enzyme mediated formation of fluorescent metabolites [26–28].

Very few other chemical substrates are selective for human CYP enzymes. Their products are usually non-fluorescent. They are useful for measuring individual CYP enzyme activities in human liver microsomes where there are 12–15 CYP enzymes present [29]. However, non-selective substrates, which form fluorescent products, are quite appropriate for measuring recombinant microsomal (membrane-bound) CYP enzyme activities and their inhibition since they manifest only individual CYP enzyme activity.

Fluorescence-based assays were used to compare the activities of microsomal CYP enzymes (Sacchrosomes™), isolated from yeast, with the CYP enzymes that are available commercially and produced in insect cells (by Corning and Invitrogen) and in bacterial cells (by Cypex). Enzymes were briefly incubated with non-fluorogenic substrates, for 5 min, for initiation of formation of the fluorescent product (i.e., metabolite). The kinetics of the enzymatic reaction (i.e., rate of product formation) was measured over a period of time. The product formed was quantified by measuring its fluorescence on a plate reader [30, 31].

The non-fluorogenic substrates, that were used for these comparative studies, and the fluorescent products that were formed are listed in Table 14.1.

### 14.3.2 Comparison of Enzyme Activity of CYP1A2 Sacchrosomes with Commercially Available CYP1A2 Microsomes Isolated from Insect and Bacterial Cells

Results in Figure 14.8 show that CYP1A2 Sacchrosomes, isolated from recombinant yeast cells, are ~50% (i.e., 0.5-fold) more active than the best commercially available

**Table 14.1** The non-fluorescent substrates that were used and the fluorescent products formed for measurement of CYP enzyme activities.

| CYP | Substrate | Metabolite formed |
|---|---|---|
| 1A2 | CEC (3-cyano-7-ethoxycoumarin) | CHC (3-cyano-7-hydroxycoumarin) |
| 2C9 | MFC (7-methoxy-trifluoromethylcoumarin) | HFC (7-hydroxytrifluoromethylcoumarin) |
| 2C19 | CEC (3-cyano-7-ethoxycoumarin) | CHC (3-cyano-7-hydroxycoumarin) |
| 2D6 | AMMC (3-[2-(N,N-diethyl-N-methylamino) ethyl]-7-methoxy-4-methylcoumarin) or EOMCC (7-Ethyloxymethyloxy-3-cyanocoumarin) | AHMC (3-[2-(N,N-diethyl-N-methylammonium)ethyl]-7-hydroxy-4-methylcoumarin) or 7_HCC (7-hydroxy-3-cyanocoumarin) |
| 2E1 | MFC (7-methoxy-trifluoromethylcoumarin) | HFC (7-hydroxytrifluoromethylcoumarin) |
| 3A4 | DBF (dibenzylfluorescein), BFC (7-benzoyloxy-4-trifluoromethyl coumarin) or BOMCC (3-cyano-7-(benzyloxymethoxy)-coumarin) | Fluoroscein or HFC (7-hydroxytrifluoromethylcoumarin) |

enzyme, Baculosome, available from Invitrogen (Thermo-Fisher). Baculosomes are isolated from insect cells after infection with recombinant baculoviruses bearing *CYP* gene expression cassettes.

### 14.3.3 Comparison of Enzyme Activity of CYP2C9 Sacchrosomes with Those of Commercially Available CYP2C9 Microsomes from Insect and Bacterial Cells

Results in Figure 14.9 show that CYP2C9 Sacchrosomes are 6-fold better in activity than the best commercially available enzyme, Bactosome, available from Cypex. Bactosomes are isolated from recombinant bacterial cells which do not contain endoplasmic reticular membranes. Instead, CYP proteins are allowed to secrete so that they can be bound to bacterial periplasmic membranes. In order for secretion to occur, the *CYP* and *CPR* genes must be devoid of their membrane anchor. Hence, the human genes used for secretion are not the same as the wild type genes.

### 14.3.4 Comparison of Enzyme Activity of CYP2C19 Sacchrosomes with Those of Commercially Available CYP2C19 Microsomes from Insect and Bacterial Cells

Results in Figure 14.10 show that CYP2C19 Sacchrosomes are 25-fold better in activity than the other three commercially available enzymes available from Corning, Cypex and Invitrogen. Like Invitrogen's enzymes, Corning's enzymes are also produced from insect cells infected with recombinant baculoviruses.

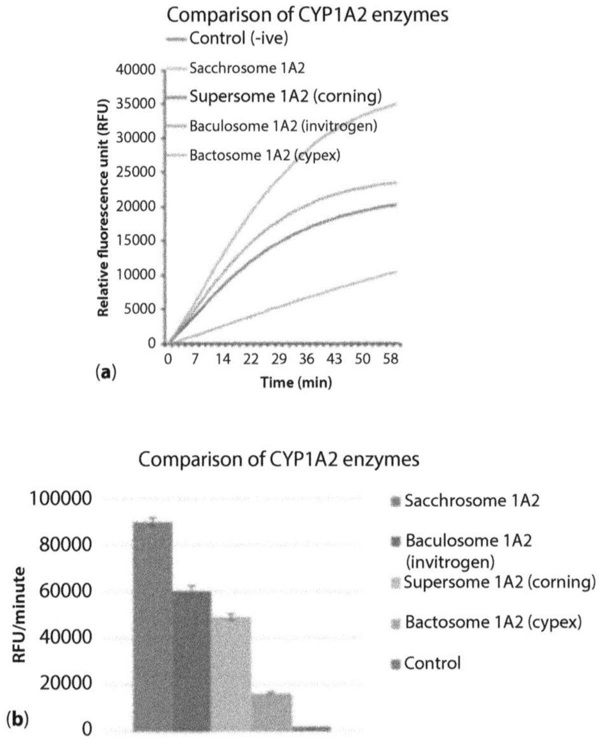


**Figure 14.8** 1 picomole of CYP1A2 microsomes, isolated from yeast (Sacchrosomes), insect cells [Baculosomes (Invitogen); Supersomes (Corning)], and bacterial cells [Bactosome (Cypex)], compared in fluorescence-based enzymatic assays. The 'Control' was microsomes obtained from yeast cells containing an empty plasmid bearing no human *CYP* gene. Substrate concentration used was 4 μM CEC (3-cyano-7-ethoxycoumarin) for each assay. The data represent mean ± S.D. of three independent experiments. (a) Shows the kinetics of the enzymatic reactions over a period of 60 min. (b) Shows the comparison of activities at the single time point of 60 min. 'RFU' represents 'relative fluorescence units'.

### 14.3.5 Comparison of Enzyme Activity of CYP2D6 Sacchrosomes with Those of Commercially Available CYP2D6 Microsomes from Insect and Bacterial Cells

Results in Figure 14.11 show that CYP2D6 Sacchrosomes are at least 2-fold better in activity than the other three commercially available enzymes available from Corning, Cypex and Invitrogen.

### 14.3.6 Comparison of Enzyme Activity of CYP3A4 Sacchrosomes with Those of Commercially Available CYP3A4 Microsomes from Insect and Bacterial Cells

Results in Figure 14.12 show that CYP3A4 Sacchrosomes are ~2-fold better in activity than the enzyme (Supersome) available from Corning.

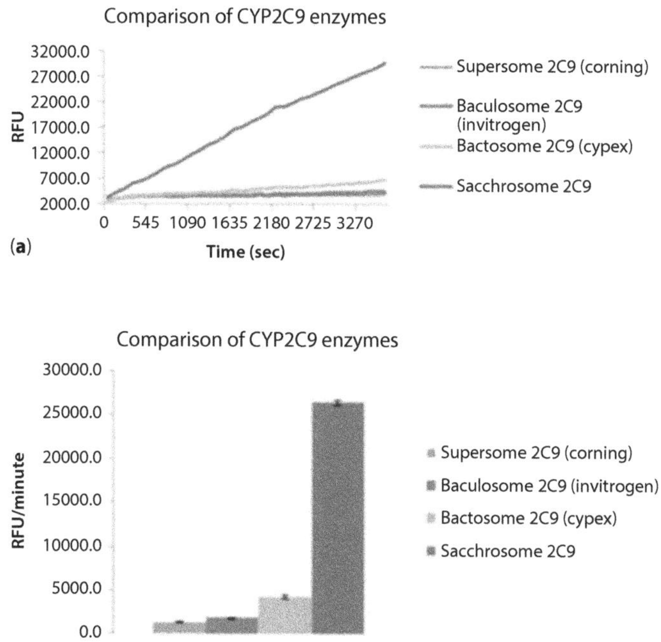


**Figure 14.9** 1 picomole of CYP2C9 microsomes, isolated from yeast (Sacchrosome), insect cells [Baculosome (Invitogen); Supersome (Corning)], and bacterial cells [Bactosome (Cypex)], compared in fluorescence-based enzymatic assays. Substrate concentration used was 10 μM 7-MFC (7-Methoxy-4-trifluoromethylcoumarin) for each assay. The data represent mean ± S.D. of three independent experiments. (a) Shows the kinetics of the enzymatic reaction over a period of 60 min. (b) Shows the comparison of activities at the single time point of 60 min. 'RFU' represents 'relative fluorescence units'.

### 14.3.7 Comparison of Enzyme Activity of CYP2E1 Sacchrosomes with One of the Commercial CYP2E1 Microsomes Available from Insect Cells

Results in Figure 14.13 show that CYP2E1 Sacchrosomes are at least ~20-fold better in activity than the enzyme (Supersome) available from Corning.

## 14.4 $IC_{50}$ Values of Known CYP Inhibitors Using Sacchrosomes, Commercial Enzymes and HLMs

The ability of known CYP enzyme inhibitors to inhibit different microsomal enzymes, obtained from yeast, insect and bacterial cells, were compared to see if the enzymes share similar active sites. This was ascertained by performing inhibition experiments with CYP-Sacchrosomes and then comparing the results obtained with the results that have been published using the commercial enzymes and human liver microsomes which contain a conglomeration of 12–15 CYPs. The $IC_{50}$ values obtained with the CYP-Sacchrosomes (Table 14.2) would indicate that yeast-produced human CYP enzymes closely resemble the 3D geometry of the active sites of the CYP enzymes that exist in the human liver (Table 14.2).

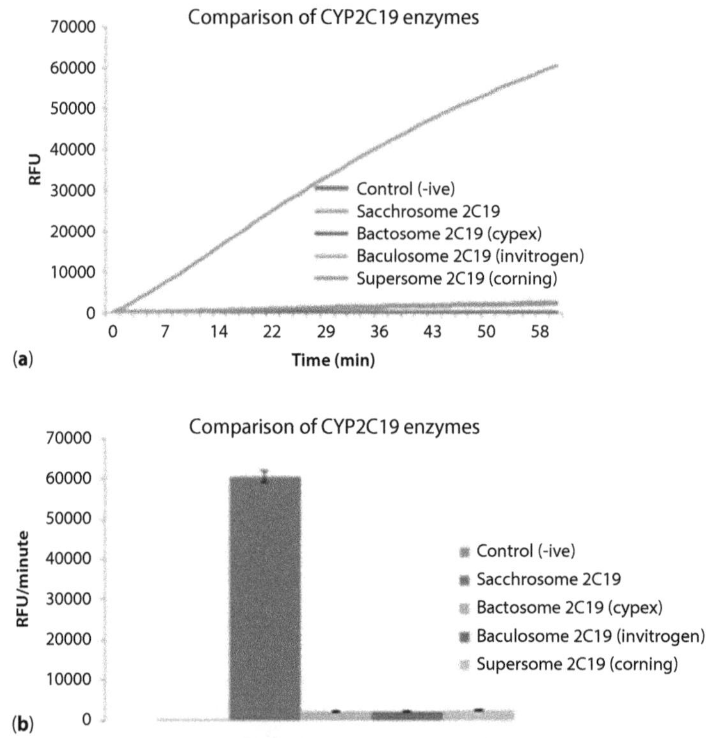


**Figure 14.10** 1 picomole of CYP2C19 microsomes, isolated from yeast (Sacchrosomes), insect cells [Baculosomes (Invitogen); Supersomes (Corning)], and bacterial cells [Bactosomes (Cypex)], compared in fluorescence-based enzymatic assays. Substrate concentration used was 4 µM CEC (3-Cyano-7-ethoxycoumarin) for each assay. The data represent mean ± S.D. of three independent experiments. (a) Shows the kinetics of the enzymatic reaction over a period of 60 min. (b) Shows the comparison of activities at the single time point of 60 min. 'RFU' represents 'relative fluorescence units'.

## 14.5 Stabilisation of Sacchrosomes through Freeze-drying

The recombinant cytochrome P450 enzymes (CYPs) are inherently unstable. Manufacturers of recombinant CYP enzymes (Corning, Invitrogen-Thermo Fisher and Cypex) recommend storage at very low temperatures (i.e., minus 80 °C) in their datasheets. They also mention that their enzymes are sensitive to freeze-thawing. Corning and Invitrogen provide recombinant CYPs made in insect cells and Cypex in bacteria.

Sacchrosomes (i.e., yeast-derived CYP enzymes) are naturally more stable than the Corning, Invitrogen and Cypex enzymes at both 4 °C and 20 °C (i.e., room temperature). Sacchrosomes maintain activity at these temperatures for 3 days (unpublished observations) whereas the commercial enzymes lose most of their activity within 15 h at 20 °C.

There is a necessity for the microsomal CYP enzymes, manufactured from different organisms, to be more stable so that they could be used more widely in chemical reactions that include biotransformations. The freeze-drying (i.e., lyophilisation) technology has been used to stabilise recombinant CYP-Sacchrosomes.

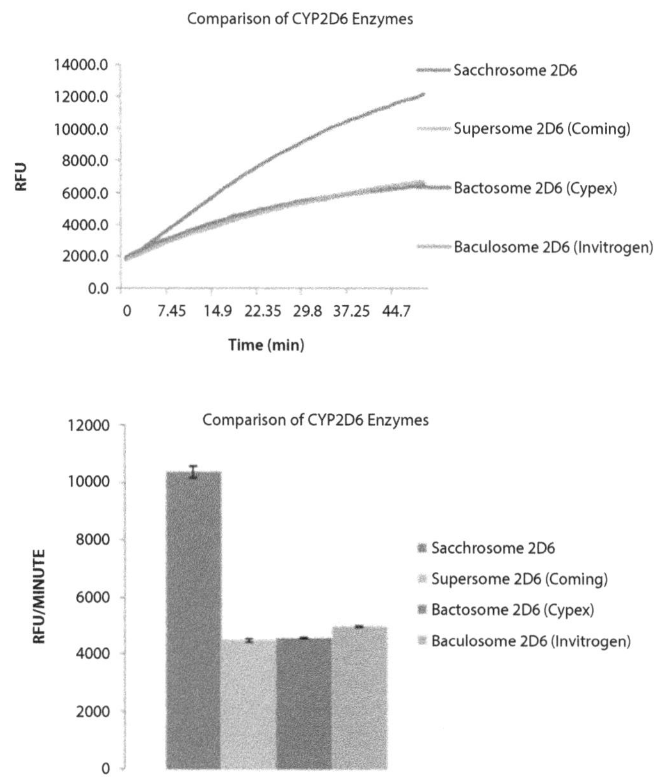


**Figure 14.11** 1 picomole of CYP2D6 microsomes, isolated from yeast (Sacchrosomes), insect cells [Baculosomes (Invitogen); Supersomes (Corning)], and bacterial cells [Bactosomes (Cypex)], compared in fluorescence-based enzymatic assays. Substrate concentration used was 4 μM EOMCC (7-Ethyloxymethyloxy-3- cyanocoumarin) for each assay. The data represent mean ± S.D. of three independent experiments. (a) Shows the kinetics of the enzymatic reaction over a period of 60 min. (b) Shows the comparison of activities at the single time point of 50 min. 'RFU' represents 'relative fluorescence units'.

1–2 picomoles of Sacchrosomes (e.g., in 3 μl), per well of a 96-well or a 384-well microtiter plate, were diluted, for example, in 7 μl of 100 mM potassium phosphate buffer, pH7.5, containing 20% w/v sucrose. The 10 μl volume of diluted Sacchrosomes were pipetted in duplicate in wells of the microtiter plate before sealing in a plastic bag which was placed at minus 80 °C for 1 h. The contents of the microtiter plate were then lyophilised in a Heto Dry Winner machine (model CT/DW60E), fitted with a BOC-Edwards vacuum pump (model # RV5), for ~2 h.

After drying, the plates were vacuum sealed (using Orved Eco Vacuum Pro) in polyamide bags that are expected not to leach plasticisers. Sealing the plates in bags should also prevent the ingress of water vapour so that the microsomes remained dry. The sealed microtiter plates were stored in a dark drawer, for 2–6 months, to prevent any potential photo-degradation. The temperature of the laboratory was constantly at 20 °C–21 °C.

The freeze-dried microsomes were resuspended in 10 μl of ultrapure water. 90 μl of a standard reaction buffer, containing a non-fluorogenic substrate that forms a fluorescent product upon reaction with CYP enzyme, was then added and the progression of

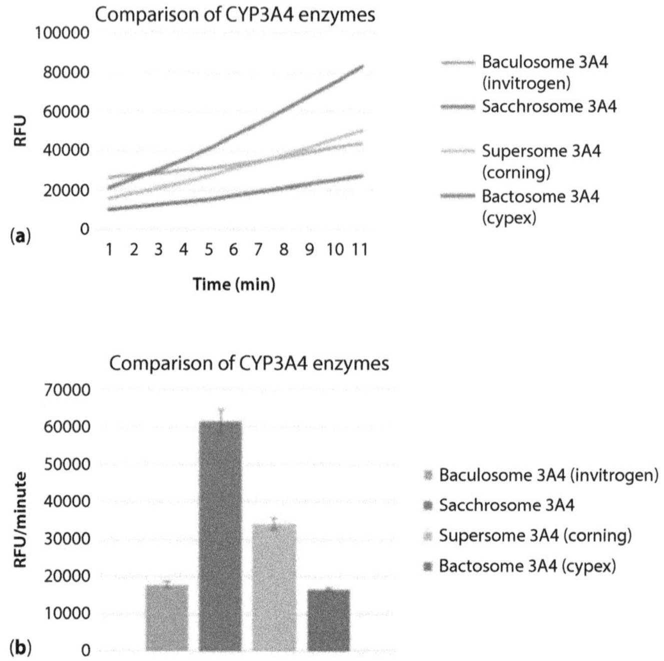


**Figure 14.12** 1 picomole of CYP3A4 microsomes, isolated from yeast (Sacchrosomes), insect cells [Baculosomes (Invitogen); Supersomes (Corning)], and bacterial cells [Bactosomes (Cypex)], compared in fluorescence-based enzymatic assays. Substrate concentration used was 10 μM BOMCC (3-cyano-7-(benzyloxymethoxy)-coumarin) for each assay. The data represent mean ± S.D. of three independent experiments. (a) Shows the kinetics of the enzymatic reaction over a period of 60 min. (b) Shows the comparison of activities at the single time point of 12 min. 'RFU' represents 'relative fluorescence units'.

the reaction was monitored. The results, at the end of reaction (50 min), are presented in Figure 14.14.

Following these observations, experiments have been conducted with freeze-dried Sacchrosomes in 6 and 24-well plates showing that Sacchrosomes can be lyophilised on any convenient plastic surface so that they could be used for further experimentation in chemical reactions and biotransformations.

## 14.6 Conclusions

The human cytochrome P450 (CYP) enzymes, that exist in the liver, play key roles in pharmaceutical drug metabolism. They are naturally bound to the intracellular ER membranes. For drug metabolism studies, microsomal ER-bound human CYPs are isolated from recombinant insect or bacterial cells mainly because these cells produce high activity enzymes. Unfortunately, insect cells lack a distinct ER whereas, in bacteria, ER is completely absent. Since the intracellular architecture of baker's yeast and human cells is alike, it was thought that human CYPs produced in yeast would closely resemble those present in human cells. Here we have described the isolation from the yeast *Saccharomyces cerevisiae* of six human CYP enzymes (1A2, 2C9, 2C19, 2D6, 2E1,

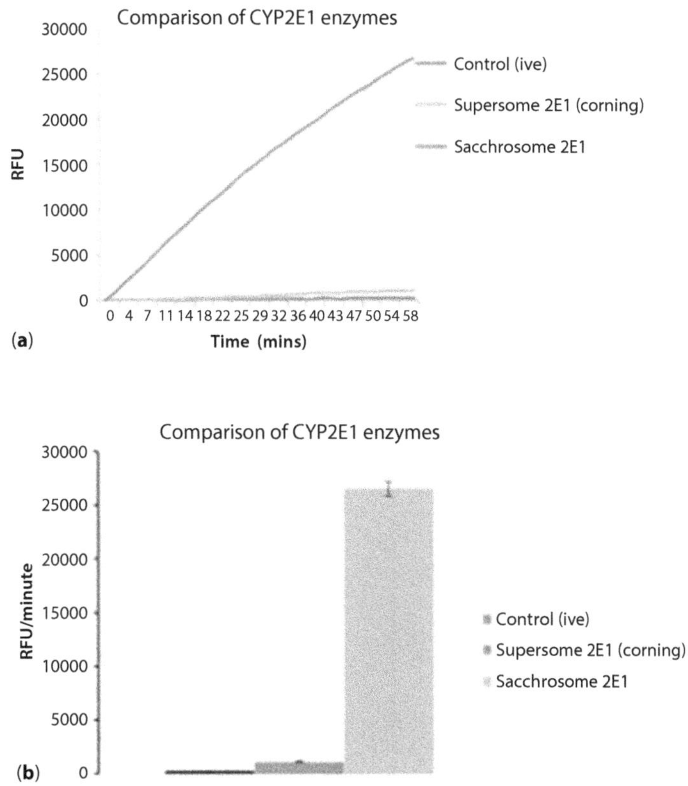


**Figure 14.13** 1 picomole of CYP2E1 microsomes, isolated from yeast (Sacchrosomes) and insect cells [Supersomes (Corning)], compared in fluorescence-based enzymatic assays. Substrate concentration used was 10 μM 7-MFC (7-Methoxy-4-trifluoromethylcoumarin) for each assay. The data represent mean ± S.D. of three independent experiments. (a) Shows the kinetics of the enzymatic reaction over a period of 60 min. (b) Shows the comparison of activities at the single time point of 60 min. 'RFU' represents 'relative fluorescence units'.

3A4) which have major roles in drug metabolism, in appreciable amounts. We have shown that the human CYP enzymes produced in yeast are, in general, far more active than the enzymes which are commercially available at the moment.

Since yeast cells are easy to grow and cheap to cultivate like bacteria, and the conditions for growth and cultivation being far less involved than insect cells, one could foresee a decrease of CYP enzyme prices with the possible commercial implementation of the yeast expression system, described in this Chapter. The current list price of ½ nanomole of Baculosomes (human microsomal CYP enzymes isolated form insect cells, available from Invitrogen-Thermo Fisher) is £130-£180. The listed prices of Supersomes (human microsomal CYP enzymes also isolated form insect cells and available from Corning) and Bactosomes (human microsomal CYP enzymes isolated form bacterial cells, available from Cypex) are similar. The availability of a less expensive and highly active form of recombinant human CYP enzymes, in a stable freeze-dried form, could open up new possibilities. For example, human CYP enzymes made in baker's yeast could be used very early in preclinical drug metabolism research. CYP enzymes, freeze-dried

**Table 14.2** Comparison of IC50 values of known CYP enzyme inhibitors using Sacchrosomes, commercial enzymes and HLMs. The column that depicts the data for Sacchrosomes (column 4) represent mean ± S.D. of three independent experiments.

| CYP enzyme | Inhibitor | HLMs (human liver microsomes) | Sacchrosomes (Baker's Yeast) | Supersomes (Corning; Insect) | Bactosomes (Cypex; Bacteria) | Baculosomes (Invitrogen; Insect) |
|---|---|---|---|---|---|---|
| 1A2 | ANF | 0.026[32] | 0.03 ± 0.002 | 0.0081–0.28[33] | 0.0032[34] | 0.06[35] |
| 1A2 | Furafylline | 0.6–0.7336 | 0.7 ± 0.12 | 5.22[33], 6[37] | 43.2–62.95[38] | 1[39] |
| 2D6 | Qunidine | 0.004[40] | 0.0045 ± 0.0003 | 0.002–0.2237 | 0.048[34] | 0.004[39] |
| 2C9 | Sulfaphenazole | 0.194[34] | 0.185 ± 0.013 | 0.11–1.23 33,37 | 0.2138–1.234 | 0.03[39]–0.31[35] |
| 2C19 | Ticlopidine | 0.23[i] | 0.25 ± 0.08 | 0.04 33–2.741 | 0.52[38] | 0.2[39] |
| 3A4 | Ketoconazole | 0.004[36] | 0.0027 ± 0.0012 | 0.165[34] | 0.21, 0.065[34] | 0.004[39]–0.13[35] |

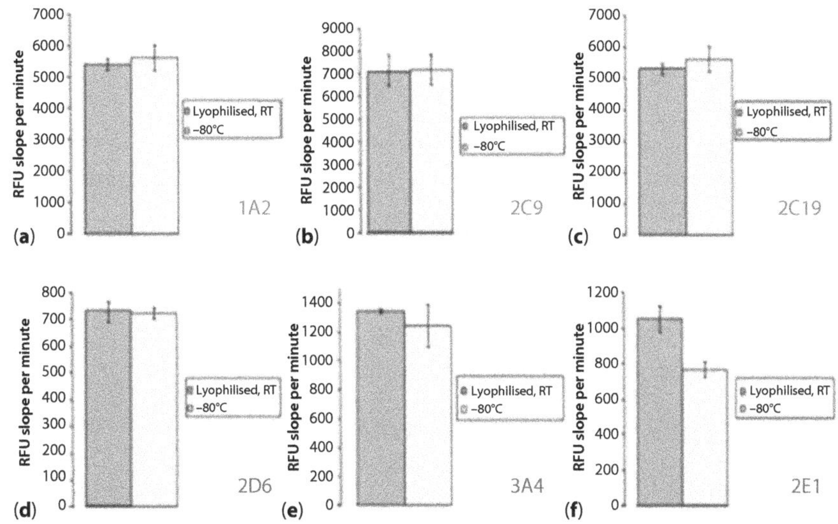


**Figure 14.14** Shows the comparative activity of lyophilised (i.e., freeze-dried) Sacchrosomes and Sacchrosomes taken directly from the minus 80 °C freezer. The activities were measured using fluorescence assays, as described in Section 1.4. The assays were performed exactly in the same manner under identical conditions. The lyophilised samples had been kept at room temperature in sealed bags for 6 months before performing the enzymatic activities. 'RFU' represents 'relative fluorescence units'.

on diverse plastic surfaces, could routinely be used for stereospecific, stereoselective, regiospecific and regioselective organic syntheses [18]. With their ease of use, they could also be employed for quick identification and characterization of metabolites of pharmaceuticals in clinical trials [32]. This has not been possible before because of the costs and intricacies involved in handling thermo-labile human CYP enzymes which was fearful for scientists with a non-biological background. Current global market size of recombinant CYPs, which are used at present solely at the preclinical-clinical interface, is thought to be at least $250 million. This could potentially increase manifold with the advent of yeast-produced enzymes and their regular use as facile tools in multiple new areas of research.

## References

1. D.W. Nebert, K. Wikvall, and W.L. Miller, *Philosophical Transactions of the Royal Society of London. Series B, Biological sciences*, Vol. 368, p. 20120431, 2013.
2. M. Strolin Benedetti, R. Whomsley, and Baltes E, *Expert Opinion on Drug Metabolism & Toxicology*, Vol. 2, p. 895, 2006.
3. Lewis DF, *Pharmacogenomics*, Vol. 5, p. 305, 2004.
4. V. Ravindranath, and H.W. Strobel, *Expert Opinion on Drug Metabolism & Toxicology*, Vol. 9, p. 551, 2013.
5. C. Rodriguez-Antona, and M. Ingelman-Sundberg, *Oncogene*, Vol. 25, p. 1679, 2006.
6. D.R. Nelson, L. Koymans, T. Kamataki, J.J. Stegeman, R. Feyereisen, D.J. Waxman, M.R. Waterman, O. Gotoh, M.J. Coon, R.W. Estabrook, I.C. Gunsalus, and D.W. Nebert, *Pharmacogenetics*, Vol. 6, p.1, 1996.

7. X. Ding, and L.S. Kaminsky, *Annual Review of Pharmacology and Toxicology*, Vol. 43, p. 149, 2003.
8. C.R. Rieder, R.B. Parsons, N.J. Fitch, A.C. Williams, and D.B. Ramsden, *Neuroscience Letters*, Vol. 278, p. 177, 2000.
9. F.J. Gonzalez, and S. Kimura, *Archives of Biochemistry and Biophysics*, Vol. 409, p. 153, 2003.
10. H. Renault, J.E. Bassard, B. Hamberger, and D. Werck-Reichhart, *Current Opinion in Plant Biology*, Vol. 19, p. 27, 2014.
11. E.M. Gillam, and M.A. Hayes, *Current Topics in Medicinal Chemistry*, Vol. 13, p. 2254, 2013.
12. D.S. Riddick, X. Ding, C.R. Wolf, T.D. Porter, A.V. Pandey, Q.Y. Zhang, J. Gu, R.D. Finn, S. Ronseaux, L.A. McLaughlin, C.J. Henderson, L.Zou, and C.E. Flück, *Drug Metabolism and Disposition: the Biological Fate of Chemicals*, Vol. 41, p. 12, 2013.
13. C. Barnaba, K. Gentry, N. Sumangala, and A. Ramamoorthy, *F1000Research*, Vol. 6, p. 662, 2017.
14. B. Chaudhuri, *Patent Publication Number*, WO2007129050 A2, 2007.
15. T.E. Quax, N.J. Claassens, D. Söll, and J. van der Oost, *Molecular Cell*, 59, p. 149, 2015.
16. C. Compagno, A. Tura, B. M. Ranzi, L. Alberghina, and E. Martegani, *Biotechnology Progress*, Vol. 9, p. 594, 1993.
17. T.M. Joska, A. Mashruwala, J.M. Boyd, and W.J. Belden. *Journal of Microbiological Methods*, Vol.100, p. 46, 2014.
18. F.P. Guengerich, and A.W. Munro, *Journal of Biological Chemistry*, Vol. 288, p. 17065, 2013.
19. L. Mears, S.M. Stocks, G. Sin, K.V. Gernaey, *Journal of Biotechnology*, Vol. 245, p. 34, 2017.
20. R.A. McLellan, M. Oscarson, M. Hidestrand, B. Leidvik, E. Jonsson, C. Otter, and M. Ingelman-Sundberg, *Archives of Biochemistry and Biophysics*, Vol. 378, p. 175, 2000.
21. F.P. Guengerich, M.V. Martin, C.D. Sohl, and Q. Cheng, *Nature Protocols*, Vol. 4, p.1245, 2009.
22. P. Rowland, F.E. Blaney, M.G. Smyth, J.J. Jones, V.R. Leydon, A.K. Oxbrow, C.J. Lewis, M.G. Tennant, S. Modi, D.S. Eggleston, R.J. Chenery, and A.M. Bridges, *Journal of Biological Chemistry*, Vol. 281, p. 7614, 2006.
23. Y. Yabusaki, *Cytochrome P450 Protocols*, Totowa, New Jersey, Humana Press Incorporated, 1998.
24. C.M. Jenkins, I. Pikuleva, and M.R. Waterman, *Cytochrome P450 Protocols*, Totowa, New Jersey, Humana Press Incorporated, 1998.
25. S.R. Hood, G. Shah, and P. Jones, *Cytochrome P450 Protocols*, Totowa, New Jersey, Humana Press Incorporated, 1998.
26. L.H. Cohen, M.J. Remley, D. Raunig, and A.D. Vaz, *Drug Metabolism and Disposition: the Biological Fate of Chemicals*, Vol. 31, p. 1005, 2003.
27. V.P. Miller, D.M. Stresser, A.P. Blanchard, S. Turner, and C.L. Crespi, *Annals of the New York Academy of Sciences*, Vol. 919, p. 26, 2000.
28. P. Szymański, M. Markowicz, and E. Mikiciuk-Olasik, *International Journal of Molecular Sciences*, Vol. 13, p. 427, 2012.
29. A. Lahoz, M.T. Donato, J.V. Castell, and M.J. Gómez-Lechón, *Current Drug Metabolism*, Vol. 9, p. 12, 2008.
30. M.U. Mohd Siddique, G.J. McCann, V. Sonawane, N. Horley, I.S. Williams, P. Joshi, S.B. Bharate, V. Jayaprakash, B.N. Sinha, and B. Chaudhuri, *Organic & Biomolecular Chemistry*, Vol. 14, p. 8931, 2016.
31. M.U. Mohd Siddique, G.J. McCann, V. Sonawane, N. Horley, I.S. Williams, P. Joshi, S.B. Bharate, V. Jayaprakash, B.N. Sinha, and B. Chaudhuri, *European Journal of Medicinal Chemistry*, Vol. 130, p. 320, 2017.
32. L. Rey, *Freeze-Drying/Lyophilization of Pharmaceutical and Biological Products*, New York, CRC Press, 2010.

# 15

# Artemisinin: A Potent Antimalarial Drug

**Alok Malaviya[1*], Karan Malhotra[2], Anil Agarwal[1] and Katherine Saikia[3]**

*[1]Department of Chemistry, CHRIST (Deemed To Be University), Hosur Road, Bengaluru, (India)*
*[2]National Centre for Biological Sciences, Tata Institute of Fundamental Research, GKVK Campus, Bellary Road, Bengaluru, (India)*
*[3]Department of Life Sciences, CHRIST (Deemed To Be University), Hosur Road, Bengaluru, (India)*

***Abstract***
Artemisinin is known to be a potent antimalarial drug which is naturally obtained from the plant *Artemisia annua* L. Malaria is a global health problem with nearly 1.2 billion people at high risk. In 2001, WHO recognised artemisinin based combination therapies (ACTs), as the front-line drugs to fight against malaria and therefore, artemisinin is the most effective anti-malarial drug. It appears to be a safe drug with no adverse reactions or noticeable side effects, even for pregnant women. However, access to ACTs by malarial patients, especially in poor countries, is inadequate due to high volatility in price, unpredictable demand and low yield from *A. annua*. The huge gap in demand and supply has motivated researchers to explore artemisinin production in alternative systems like bacteria, yeast and tobacco. Scientists have been successful in producing this wonder molecule in heterologous hosts. Challenges associated with large-scale production and drug resistance against artemisinin has also been discussed to present a comprehensive picture of artemisnin production, application and limitations.

***Keywords:*** *Artemisia annua*, artemisinin, malaria, terpenoid

## 15.1 Introduction

The Chinese herb *Artemisia annua L.* (also known as sweet wormwood) has traditionally been used in China (for over 2,000 years) for the treatment of malaria as well as a source of essential oil to be used in the perfume industry. Application of *A. annua* as a herbal tea to treat malaria has been recorded in "Compedium of Materia Medica" by Li Shenzhen in 1596 [1]. Artemisinin is the main bioactive compound of *A. annua*, which was first extracted and purified by Prof. Youyou Tu and her team between 1969 and 1972 [2]. For this discovery, the 2015 Nobel Prize in Physiology and Medicine was awarded to Youyou Tu.

**Corresponding author*: alokkumar.malaviya@christuniversity.in

Saurabh Saran, Vikash Babu, and Asha Chaubey (eds.) High Value Fermentation Products, Volume 1, (347–364) © 2019 Scrivener Publishing LLC

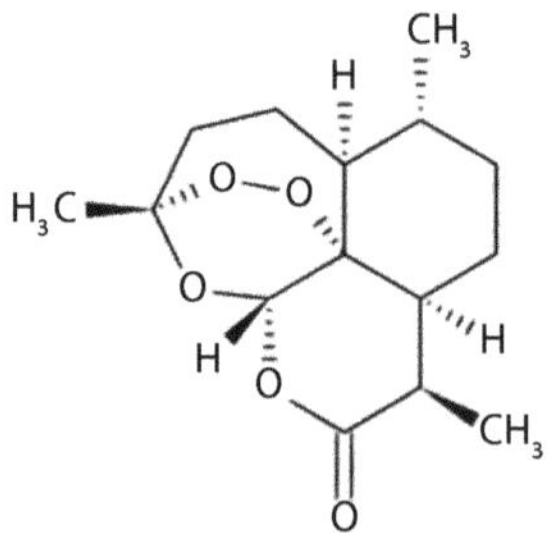


**Figure 15.1** Chemical structure of artemisinin.

Artemisinin is a sesquiterpenoid lactone with a rare endoperoxide bridge (Figure 15.1) which is considered as the most potent antimalarial drug. Following its recognition by WHO in 2001, artemisinin-based combination therapies, commonly referred to as ACTs, are the frontline drugs in the fight against malaria [3]. It kills the young circulating parasites and is highly effective against both sexual and asexual stages of parasites [4–6]. It is more effective than traditional quinine-based drugs for treating severe malaria as it reduces the number of parasites that mature and invade blood capillaries while traditional drugs do not prevent sequestration [7]. However, artemisinin has a very short half-life and is rapidly metabolized by human cytochrome enzymes, thereby resulting in high rates of recrudescence [8]. Utilizing artemisinin as a monotherapy has shown high rates of recrudescence and may result in reduced drug sensitivity [9]. Therefore, artemisinin derivatives are always used in combination with another drug with a longer half-life [10].

Artemisinin has also been found to demonstrate cytotoxicity against cancer cells and schistosomiasis [11–13]. However, ACTs are not readily available to patients in malaria endemic regions due to the low yield of artemisinin from the native plant. The plant yield is highly dependent on (i) variation in genetic backgrounds [14, 15], (ii) cultivation conditions [16], (iii) plant tissue and harvesting periods [17, 18], and (iv) supply of mineral nutrients [19, 20], which severely limits the production of artemisinin from its natural source. Preparation of this compound through chemical synthesis is also highly expensive [21, 22]. Further, price is also affected by shortage of substrate and unpredictable demand [23]. Due to mentioned challenges, people have started looking towards alternative heterologous production platforms so as to meet the demands of artemisinin. Some of these strategies have been discussed in Section 15.4 of this chapter.

## 15.2 Biosynthesis of Artemisinin in *Artemisia annua* and Pathways Involved

Artemisinin belongs to the terpenoid family, one of the most important families of plant secondary metabolites. Terpenoids plays some of the important roles in plant system e.g., plant hormones (Gibberellins, ABA and cytokines), attracts pollinators and helps in defense against insects [1]. Biosynthesis of all terpenoids including artemisinin occurs via the universal isoprenoid precursors – isopentenyl pyrophosphate (IPP)

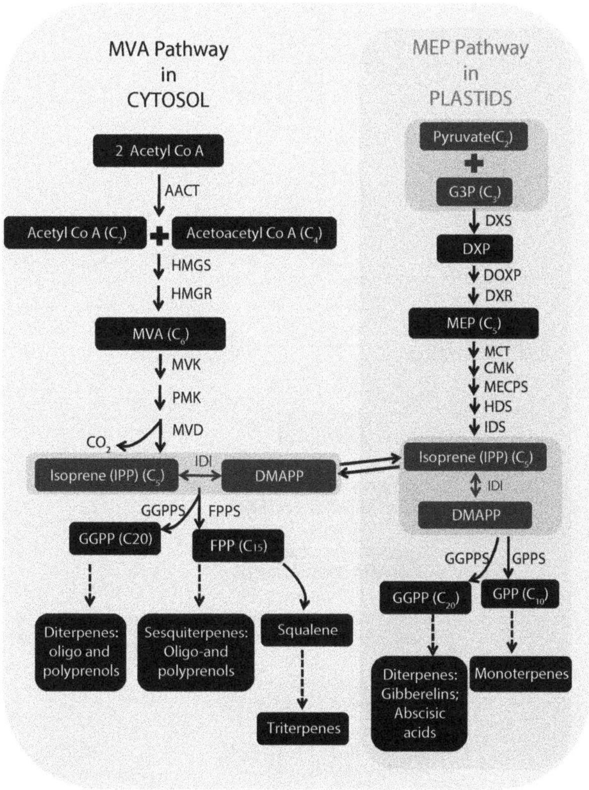


**Figure 15.2** The mevalonic acid (MVA) and the 2-C-methyl-D-erythritol 4-phosphate (MEP) pathway for terpene biosynthesis. AACT, acetoacetyl-CoA thiolase; HMGS, HMG-CoA synthase; HMGR, HMG-CoA reductase; MVA, mevalonic acid; MEP, 2-C-methyl-D-erythritol 4-phosphate; MVK, mevalonate kinase; PMK, phophomevalonate kinase; MVD, mevalonate diphosphate decarboxylase; DMAPP, dimethyl diphosphate; FPP, farnesyl diphosphate; FPPS, farnesyl diphosphate synthase; GGPP, geranyl diphosphate; GGPPS, geranyl diphosphate synthase; DXS, deoxy-d-xylulose-5-phosphate synthase; DXR, 1-deoxy-D-Xylulose-5-phosphate reductoisomerase; MCT, 2-C-methyl-d-erythritol 4-phosphate cytidylyltransferase; CMK, 4-(cytidine 5'-diphospho)-2-C-methyl-d-erythritol kinase; MECPS, 2-C-methyl-d-erythritol 2,4-cyclodiphosphate synthase; HDS, (E)-4-hydroxy-3-methylbut-2-enyl diphosphate synthase; IDS, isopentyl diphosphate synthase; IDI, isopentyl diphosphate isomerase. (adapted from Wang B, (29).

and dimethylallyl pyrophosphate (DMAPP). These isoprenoid precursors are formed by two pathways in plants (Figure 15.2) – (i) Mevalonic acid pathway (cytosol) and, (ii) 2-C-methyl-D-erythritol 4-phosphate (MEP) pathway (plastids).

Artemisinin biosynthesis takes place in the cytosolic compartment of *A. annua* which are then accumulated within the apical glandular secretory trichomes of *A. annua* [24, 25]. A putative pathway for artemisinin synthesis in *A. annua* has been presented in Figure 15.3. Artemisinin is formed by the condensation of (1'-4) of IPP and DMAPP through a series of chemical reactions. Conjugation of IPP and its isomer dimethylallyl pyrophosphate (DMAPP) leads to the formation of acyclic prenyl diphosphate farnesyl diphosphate (FPP) In the next step, FPP is cyclized to amorpha-4,11-diene with the help of amorphadiene synthase (ADS) [26].

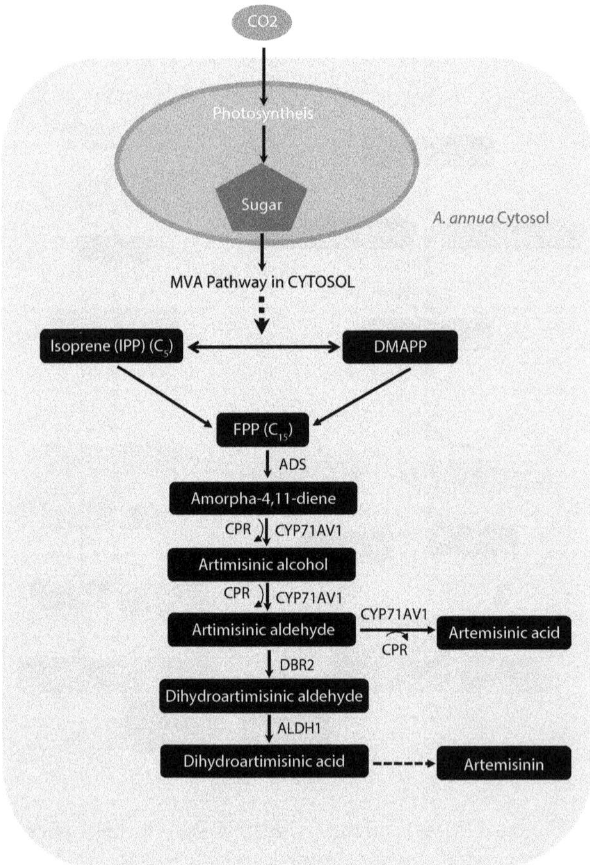


**Figure 15.3** The artemisinin biosynthesis pathway in *A. annua*. Broken lines indicates the involvement of more than one step. Dihydroartimisinic acid to artemisinin conversion takes place in presence of light and oxygen; IPP, isopentyl diphosphate; FPP, farnesyl diphosphate; CYP71AV, amorphadiene oxidase; CPR, cytochrome P450 reductase; DBR2, artemisinic aldehyde double-bond reductase; ALDH1, aldehyde dehydrogenase; IDI, isopentyl diphosphate isomerase. [adapted from Wang B (29].

Amorphadiene is the first dedicated precursor of the artemisinin biosynthetic pathway. The successive oxidation steps beyond amorphadiene to artemisinic acid is catalyzed by cytochrome P450 artemisinic alcohol to artemisinic aldehyde and artemisinic acid [27]. In the next step, Alcohol Dehydrogenase I (ADHI) helps in the oxidation of artemisinic alcohol to aldehyde. From artemisinic aldehyde, there are two alternative proposed pathways which leads to artemisinin biosynthesis – (i) via artemisinic acid and the other (ii) via dihydroartemisinic acid (DHAA). Aldehyde Dehydrogenase I (ALDHI) helps with oxidation of artemisinic aldehyde to artemisinic acid. Artemisinic aldehyde is then converted to dihydroartemisinic aldehyde with the help of artemisinic aldehyde Δ11 [13] reductase (DBR2). This compound then gets oxidised to dihydroartemisinic acid (DHAA) with the help of ALDHI which is then converted non-enzymatically to artemisinin [28, 29]. Many reports suggests DHAA as the immediate precursor to artemisinin and its bioconversion occurs spontaneously in presence of light and oxygen [30, 31].

## 15.3 Yield Enhancement Strategies in *A. annua*

The effectiveness of artemisinin towards early stages of malarial parasites with almost no reports of resistance makes this sesquiterpenoid molecule the most effective antimalarial. Following WHO recommendation, the price and availability of artemisinin have fluctuated substantially from US$600 per kg in early 2012 to under $300 per kg in late 2013 [32]. The Semi-Synthetic project led by Prof. Jay Keasling and his group has led to the development of an alternative route which could supplement the existing supply of artemisinin [33]. However, *A. annua* still continues to remain the main source for the commercial supply of artemisinin to endemic regions. Ever since the discovery of medicinal properties of *A. annua*, several research groups have worked on different strategies to improve its yield.

Initial efforts to enhance the yield of artemisinin were mainly focused on manipulating the phytohormones levels in hairy root cultures of *A. annua* [34–36]. Ferreira and Janick [34] found a strong correlation between artemisinin content and numbers of roots in *A. annua*. They even observed a reduction in the artemisinin content at concentrations of phytohormones known to reduce rooting. Later, Smith and co-workers [35] studied the effect on hairy roots of *A. annua* at different concentrations of gibberellic acid ($GA_3$). At $GA_3$ concentrations of 28.9 μM, both root growth (~25% increase) and artemisinin concentration was detected in maximum amount. However, production of artemisinin from undifferentiated cultures of *A. annua* was very low [37–39]. Later, production of artemisinin through suspension cultures of *A. annua* was explored by the addition of precursor (mevalonate lactone) and elicitor (methyl jasmonate, MeJA) [40]. Through an integrated bioprocess approach, artemisinin titers reached up to ~ 110 mg/L [40] and 97 mg/L in suspension and cell cultures of *A. annua* respectively [41]. A similar positive effect on the yield of artemisinin and its precursors was observed by exogenous application of MeJA and salicylic acid on whole plant of *A. annua* [42, 43]. However, the yield of artemisinin through *in vitro* approach is quite variable with almost no artemisinin in undifferentiated callus cultures of *A. annua* [44]. Moreover, endeavors to enhance artemisinin through the use of elicitors and hormones in suspension cultures of *A. annua* have met with limited success. Interestingly, studies done on the different ecotypes of *A. annua* have demonstrated large variations in their overall artemisinin content [45, 46]. Selective breeding of superior strains of *A. annua* thus appears to be a feasible option for developing hybrids with high levels of the drug artemisinin. This approach has resulted in artemisinin content up to ~ 1.4% on dry weight basis in the famous F1 hybrid, Artemis [46]. Transcriptomic analysis of Artemis by Graham and co-workers [47] has led to the identification of key genes that are responsible for artemisinin yield and development of glandular trichomes (GTs). Extensive field trials have led to development of a quantitative trait loci (QTL) map that helps in identifying key traits affecting artemisinin yield that may be used for developing hybrids with commercial value. Such a strategy would be a great boost for restoring the confidence of farmers in the endemic regions [48]. Similar efforts were done by Indian researchers from Central Institute of Medicinal and Aromatic Plants (CIMAP), Lucknow [49, 50] to develop the hybrid variety "CIM-Arogya" with yields comparable to Artemis.

Besides plant breeding, development of metabolic engineering tools has allowed researchers to manipulate the artemisinin biosynthetic pathway genes of *A. annua*. Numerous reports suggest that artemisinin concentration reaches maximum levels at the time of flowering [17, 51]. To investigate this, Wang and co-workers [52] expressed the flowering promoting factor1 (*fpf1*) from *Arabidopsis thaliana* into *A. annua* under constitutive promoter using agroinfection. Although transgenic lines expressing *fpf1* flowered 20 days earlier, no significant differences in artemisinin content were observed between the transformed and untransformed plants. It is well known that artemisinin is accumulated in the GTs of *A. annua* [24] and its concentration is dependent on the density of glandular trichomes. Attempts to increase the density of GTs using external stress have met with limited success [53]. Previous reports have demonstrated that expression of β-glucosidase in tobacco chloroplasts [54] has boosted the biomass as well as GTs. Transgenic *A. annua* expressing *bgl1* from *Trichoderma reesei* were observed with increased trichome density of up to ~ 66% in flowers and ~ 20% in leaves of transgenic lines [55]. This was accompanied by an increase in artemisinin content in both leaves (~1.4%) and flowers (~2.56%).

Another approach to improve artemisinin content in the native plant is to block a competing pathway. FPP which acts as precursor for sesquiterpenoids, is also utilized by metabolic pathways leading to formation of sterols, dolichols and ubiquinone [56]. Squalene synthase (SS) is one enzyme that diverts FPP flux towards sterols. Suppression of SS in *A. annua* using antisense approach resulted in drastic reduction in SS transcript along with an increase in transcript level of AACPR, CYP71AV1 and CPR and thus artemisinin biosynthesis [57]. In a completely different approach, Chen and co-workers [58] introduced the cDNA encoding farnesyl diphosphate synthase (FDS) from cotton into the nuclear genome of *A. annua* under the control of a constitutive promoter. This resulted in more than two-fold enhancement in the content of artemisinin than in the control plants. A similar strategy involved overexpressing *hmgr* from *Catharanthus* and *ads* gene from *A. annua* resulted in more than seven-fold improvement in the yield of artemisinin [59]. Interestingly, when only *hmgr* was expressed, it resulted in only a marginal change of artemisinin content.

## 15.4 Artemisinin Production Using Heterologous Hosts

Metabolic engineering offers a viable alternative strategy for expressing complete metabolic pathways into foreign hosts [60, 61]. Heterologous systems are routinely being used as highly productive platforms for the production of various industrially valuable metabolites. On the same line, various terpenoid and plant secondary metabolite pathways have been engineered into a variety of microbial hosts like *Escherichia coli*, *Saccharomyces cerevisiae* or in model plants like tobacco and mosses. In the following sections we will briefly discuss the various approaches used for the heterologous production of artemisinin.

### 15.4.1 Microbial Engineering

The low yield of the sesquiterpenoid antimalarial drug artemisinin from the Chinese herb *A. annua* shifted the focus of researchers towards the use of synthetic biology tools to engineer heterologous hosts with genes of artemisinin biosynthetic pathway. Initial attempts to biosynthesize the terpenoid precursors in *E. coli* mainly focused on accumulating the isoprenoid precursors IPP/DMAPP by manipulating the mevalonate or the MEP pathways. Martin and co-workers [60] expressed the genes of the yeast mevalonate pathway into *E. coli* in an attempt to bypass the feedback inhibition associated with the native MEP pathway in *E. coli*. However, accumulation of isoprenoid precursors using this strategy severely impaired the growth of *E. coli*. Co-expression of a synthetic amorphadiene synthase gene (ADS) in *E. coli* not only resulted in the production of amorphadiene (~24 μg caryophyllene equivalent/ml) but also minimized the adverse affect on the growth of *E. coli*. Mass spectrometry studies identified 3-hydroxy-3-methyl-glutaryl-CoA (HMGCoA), a non-native metabolite to *E. coli* responsible for this toxicity [62]. Substituting the yeast HMGR with a type II HMGR from *Staphylococcus aureus* resulted in amorphadiene titres upto 25 g/L using a novel fermentation strategy [63]. However, commercial production of artemisinin would essentially require conversion of the olefinic precursor amorphadiene to an oxidised metabolite artemisinic acid using plant cytochrome P450s [64]. Expressing eukaryotic P450s in prokaryotic machinery like *E. coli* is not feasible at industrial level [65]. Therefore, yeast was chosen as a platform for future microbial engineering strategies for the semi-synthetic production of artemisinin.

Ro and co-workers [66] were the first to demonstrate the biosynthesis of artemisinic acid up to ~ 100 mg/L in *S. cerevisiae* using a three-step engineering strategy. This involved overexpressing the key enzymes of the farnesyl diphosphate (FPP) biosynthetic pathway, engineering ADS gene from *A. annua* and cloning cytochrome P450 (CYP71AV1) along with *A. annua* redox partner cytochrome P450 oxidoreductase (CPR). To enhance FPP production, they first overexpressed 3-hydroxy-3-methyl-glutaryl-coenzyme A reductase (tHMGR) and ERG20 (encoding FPP synthase). Also, down regulation of ERG9 (encoding for squalene synthase) allowed to minimize the utilization of FPP towards the sterol biosynthetic pathway. Initial metabolic engineering studies on artemisinic acid biosynthesis were done on *S. cerevisiae* strain S288C [67]. Unlike S288C, CEN.PK2 is an industrial friendly strain [68] and was used for achieving high-level biosynthesis of artemisinic acid. Using CEN.PK2 as the base strain, Westfall and co-workers [69] demonstrated amorphadiene titers up to 40 g/L by using an inexpensive carbon source. Here, they overexpressed each enzyme of the MEV pathway including ERG20. However, this approach did not result in a concomitant increase in the titers of artemisinic acid. Interestingly, CEN.PK2 expressing CYP71AV1 and CPR1 experienced severe oxidative stress owing to poor coupling between these two enzymes which was not observed in CEN.PK2 strains expressing pathway only up to amorphadiene [61]. As a result, both CYP71AV1 and CPR1 were expressed under strong inducible-promoters on a multi-copy plasmid. Co-expression of CYP71AV1 and cytochrome b5 [70] along with ALDH1 and ADH from *A. annua* resulted in the final CEN.PK2 strain producing artemisinic acid titers up to ~ 25 g/L which eliminates any toxic intermediate that is formed during the fermentative process [61]. Commercial production of artemisinin beyond artemisinic acid is done using a photochemical process developed by Sanofi in specialized bioreactors [71].

### 15.4.2 Plant Metabolic Engineering

Although the semi-synthetic approach [61] using yeast achieved high-titers of artemisinic acid, complete biosynthesis of artemisinin using microbial hosts has never been achieved till date. Moreover, conversion of artemisinin acid to artemisinin requires an environmental unfriendly chemical process. Also, the biosynthesis of artemisinin is limited to GTs in *A. annua* which makes suspension cultures not a viable commercial alternative. Endeavours to improve artemisinin yields in *A. annua* have met with limited success [35, 40, 41] and rely mostly on genetic approaches of breeding high yielding varieties [47]. With the advancement in genetic engineering tools, it is possible to transfer genes of the artemisinin biosynthetic pathway to model plant systems with high-leafy biomass and faster life-cycle. Previous reports suggest that tobacco serves as an ideal host for rapid genetic manipulations and for the production of recombinant therapeutic proteins [72–74].

A group of researchers from the Netherlands [75] were the first to characterize the terpene synthase *ads* responsible for the biosynthesis of amorphadiene, the first precursor of artemisinin biosynthetic pathway. Using a constitutive promoter, amorphadiene was expressed in the nuclear genome of tobacco plants. However, expression of amorphadiene in transgenic lines were very low (~ 2 ng/g dry weight) which may be attributed to the limited pools of FDP in cytosol.

A major breakthrough in terpenoid biosynthesis was achieved when biosynthesis of terpenoids was explored using different subcellular compartments [76]. Interestingly, subcellular targeting of both *ads* and farnesyl diphosphate synthase (*fps)* gene to chloroplasts resulted in much higher accumulation of amorphadiene (> 25µg) than when *ads* was engineered in cytosol [77]. Later, van Herpen and co-workers [78] attempted to biosynthesize artemisinic acid by combinatorial expression of FPP, HMGR, ADS and CYP71AV1 genes and subcellular targeting. Surprisingly, untargeted mass spectrometry studies revealed glycosylation of the end-product to artemisinic acid-12-β-diglucoside by endogenous glycosyl transferases of tobacco with the overall yield of ~ 39.5µg/g FW. Nevertheless, these studies also showed that FPP [a precursor to amorphadiene] is available in different cellular compartments and could be channeled towards enhanced amorphadiene and artemisinin biosynthesis. Biosynthesis of advance stage artemisinin precursors in tobacco provided a major boost to heterologous expression studies. A parallel study explored the biosynthesis of artemisinin in tobacco using the nuclear genome engineering [79]. In this study, expression of ADS only led to the biosynthesis of amorphadiene. However, co-expression of ADS and CYP71AV1 resulted in formation of artemisinic alcohol and minimal amount of amorphadiene. Strangely, no accumulation of artemisinic acid was observed in the transgenic tobacco lines. It is well known that CYP71AV1 is responsible for the conversion of amorphadiene to artemisinic acid via artemisinic aldehyde and artemisinic alcohol [66, 80]. This was confirmed by *in vivo* metabolic experiments where wildtype tobacco leaves were supplied with artemisinic aldehyde and dihydroartemisinic aldehyde. GC-MS of both feeding experiments showed accumulation of aldehyde and artemisinic alcohol but no artemisinic acid which suggests that either CYP71AV1 is not active or the internal environment favor reduction to alcohols. To reconfirm this, wildtype tobacco was again transformed with binary vectors harboring *aldh1* and *dbr2* in addition to *fps*, *ads* and *cyp71av1*. Here also, only artemisinic alcohol and dihydroartemisinic alcohol was detected but no corresponding acids. Overall, it may be concluded from this study that the cytosolic environment does not favor formation of sesquiterpenoid aldehyde and acids. Also, future studies on biosynthesis of artemisinin precursors in tobacco should consider compartmentalization of the biosynthetic pathway.

Farhi and co-workers [81] were the first to demonstrate the complete biosynthesis of artemisinin in *Nicotiana tabacum*. Using a synthetic biology approach, they engineered the tobacco nuclear genome with five *A. annua* genes of and a deregulated form of HMGR (tHMGR) from yeast. However, yield of artemisinin in these lines were low (~7 μg/g dry-weight) which was possibly due to limited pool of isoprenoid precursors IPP/DMAPP. Despite extensive metabolic engineering and pathway reconstruction strategies, the biosynthesis of artemisinin in alternative plant hosts remained dramatically low [75]. Most of these studies mainly focused on engineering nuclear genomes with targeting intermediate metabolites to different subcellular compartments. However, biosynthesis of artemisinin precursors through the chloroplast genome was never explored. Using the biolistic approach, Saxena *et al.* [82] expressed 12 genes of MEV and artemisinic acid biosynthetic pathways in the chloroplast genome of tobacco. The mevalonate pathway genes were expressed under the 16s-ribosomal rRNA promoter whereas the artemisinic acid pathway was under the chloroplast psbA promoter [83]. Surprisingly, the transplastomic lines displayed stunted phenotype which was attributed to the excess gene-load on a single organelle. Further, the authors reasoned out that the aberrant phenotype in transgenic tobacco may be due to an imbalance of isoprenoid precursors which are utilized by diverse metabolic pathways [84]. Nonetheless, artemisinic acid was detected in transformed tobacco plants which were confirmed using mass spectrometry and NMR studies. Based on these assumptions, a rationalized transformation strategy was designed where the biosynthesis of artemisinin was explored in tobacco using both chloroplast and nuclear genome [85]. In comparison to the earlier approach [82], a balanced compartmentalized metabolic engineering was done through which an excess IPP flux was first achieved by engineering the exogenous yeast mevalonate pathway in the chloroplast genome of tobacco [85]. Using a binary vector, five genes (FPP, farnesyl pyrophosphate; ADS, amorphadiene synthase; AACPR, *Artemisia annua* cytochrome P450 reductase; amorpha-4,11-diene C-12 oxidase; CYP71AV1, and DBR2, double bond reductase2) of the artemisinin biosynthetic pathway till DHAA was introduced into the nuclear genome of tobacco. Among these, AACPR, CYP71AV1 and DBR2 genes were targeted to chloroplast while ADS was targeted to mitochondria using cox4. As discussed earlier in Section 2, the conversion of DHAA to artemisinin is a spontaneous photo-oxidation process and does not require additional enzymes [31]. High-light conditions and post-harvest process assisted this bioconversion in the double-transformed tobacco lines. Thus, using a dual transformation strategy, complete biosynthesis of artemisinin was achieved up to ca. 0.8 mg/g dry weight. Yield of artemisinin and its precursors in various heterologous plant production platforms are presented in Table 15.1.

A similar approach was also reported by Fuentes *et al.* [86] demonstrating the biosynthesis of artemisinic acid in tobacco using combinatorial super transformation of transplastomic recipient lines (COSTREL). Here, they first engineered the key enzymes viz. ADS, FPP, CYP71AV1 and CPR in different orientations into the chloroplast genome of tobacco to biosynthesize artemisinic acid. Reports [61, 87] have identified CYB5, ADH1, ALDH1 enzymes that may help in driving the flux towards the biosynthesis of artemisinic acid. These three genes along with DBR2 and 1-deoxy-D-xylulose-5-phosphate

**Table 15.1** Yield of artemisinin and its precursors in heterologous plant production platforms.

| Plant species | Gene delivery method and organelle involved | Yield of artemisinin/precursors | Reference |
|---|---|---|---|
| *Physcomitrella patens* | Protoplast transformation; nuclear genome | 0.21 mg/g DW artemisinin | [88] |
| *Nicotiana tabacum* L. cv. Petit Havana SR1 | Biolistic and agroinfection; chloroplast and nuclear genome | 0.8 mg/g DW artemisinin | [85] |
| *Nicotiana tabacum* cv. Petit Havana | Biolistic; chloroplast and nuclear genome | 0.12 mg/g FW artemisinic acid | [86] |
| *Nicotiana tabacum* L. cv. Petit Havana SR1 | Biolistic; chloroplast genome | 0.1 mg/g FW artemisinic acid | [82] |
| *Nicotiana tabacum* cv. Samsun | Agroinfection; nuclear genome | 6.8 μg/g DW artemisinin | [81] |
| *Nicotiana benthamiana* | Agroinfection; nuclear genome | 39.5 μg/g FW artemisinic acid-12-β-glucoside | [101] |

reductoisomerase (DXR) from *Synechocystis* were mixed in equal proportions and delivered simultaneously using gene-gun to transplastomic plants expressing artemisinic acid. Large-scale metabolite screening of COSTREL plants identified tobacco plants biosynthesizing artemisinic acid up to ~ 120 mg/g on dry weight basis.

Recent reports [88, 89] describe the biosynthesis of artemisinin in a non-vascular haploid bryophyte, *Physcomitrella patens*, using an *in vivo* assembly [90] of genes from artemisinin biosynthetic pathway. Isolated *P. patens* protoplasts transformed with ADS gene under a constitutive promoter, Ubiquitin1 from *Zea mays* and selected using geneticin (G418), resulted in a high amorphadiene biosynthesis (~ 0.2 g/L). The best line was subsequently engineered with downstream enzymes CYP71AV1 and ADH1 linked by a linker polypeptide (LP4) from *Impatiens balsamina*. Similar screening of transgenic lines was repeated again and was finally transformed with DBR2 and ALDH1. The expression of CYP71AV1-LP4/2A-ADH1 and DBR2-LP4/2A-ALDH1 were driven by actin promoter from rice and *Arabidopsis* ubiquitin promoter respectively. Molecular studies confirmed the site-specific integration of all these five genes. Continuous aeration and high-illumination phase during the cultivation conditions assisted in the bioconversion of DHAA to artemisinin which was in agreement with the earlier published reports [81, 85]. Transformed *P. patens* peaked maximum artemisinin concentration (ca. 0.2 mg/g DW) after 3 days of cultivation while maximum accumulation (~2.5 mg) was observed on the $12^{th}$ day. Thus, *P. patens* offer a commercially viable alternative for scaling up artemisinin production without the need for chemical synthesis.

## 15.5 Spread of Artemisinin Resistance

The development and spread of drug resistant malaria parasites is proving to be one of the greatest challenges to malaria control today. Resistance can be identified when there is a significant delay in the time it takes to clear malaria parasites from infected patients, indicating the declining efficacy of the drug. *Plasmodium falciparum*, the deadliest species of malaria parasite, is responsible for the vast majority of the mortality and morbidity associated with malaria infection and has developed resistance to nearly all antimalarial drugs currently in use.

Artemisinin-based combination therapies (ACTs, introduced in the 1990s) are the most effective drugs available to treat malaria parasite infection [91]. Although these treatments are working well in many parts of the world, there is a serious concern that malaria parasites are once again developing widespread resistance to this vital treatment. ACTs have recently been found to be losing efficacy in Southeast Asia [92]. Artemisinin resistance was first reported along the Thailand-Cambodia border in 2008 and has continued to spread in the region to four countries: Cambodia, Myanmar, Thailand and Vietnam [93]. Recently, Lu *et al.* [94] have reported an artemisinin-resistant strain of *P. falciparum* that was contracted in Africa, in addition to the widespread resistance to artemisinin in many countries in Southeast Asia.

There are many factors that are thought to have contributed to the emergence and spread of artemisinin resistance in the Greater Mekong sub-region of Southeast Asia [95]. This artemisinin resistance, defined by a delayed parasite clearance time, has been associated with several genetic mutations. Genetic mutations help the parasite to hide from our immune system or by making the parasite resistant to a particular drug. One of the most significant genetic mutations (C580Y) identified in the resistant parasite genome is on chromosome 13, in a gene called kelch13 giving rise to *Kelch13* mutant *P. falciparum* parasites (PfKelch13) [96, 97]. These dominant artemisinin-resistant parasites are developing resistance to the ACT partner drugs, and are causing high rates of malaria treatment failure.

Another important factor is thought to be the use of oral artemisinin alone (as a monotherapy) in place of the World Health Organization (WHO) recommended ACTs. Failure to complete treatment with artemisinin monotherapy can leave the patient with malaria parasites in their blood, which can then be transmitted by mosquitoes to other people. Other contributing factors are the use of substandard and counterfeit anti-malarial drugs and the difficulty of controlling malaria within migrant populations.

In response to this threat, WHO launched an emergency plan of action to tackle artemisinin resistance in the Greater Mekong Sub-region. To develop an effective strategy to combat malaria, it is crucial to understand the genetic factors that determine how drug resistance emerges and spreads.

## 15.6 Challenges in Large-Scale Production

Artemisinin is the first-choice of treatment for malaria; however, when administered as a single compound, this allows malaria parasites to become drug-resistant. As a result, WHO has recommended ACTs with two other drugs, mefloquine and chlorproguanil, in order to avoid artemisinin resistance. But the supply of artemisinin is not stable or sufficient and as a result treatment remains expensive. Artemisinin is generally derived from the plant *A. annua* in very small quantities. Artemisinin has a complex chemical structure (Figure 15.1) with multiple chiral centres making it very difficult and not economically feasible to synthesize chemically.

Zhu and Cook [98] reported an efficient and cost-effective approach for the synthesis of artemisinin. They developed mild, complexity-building reaction cascades that allowed the use of readily available, affordable cyclohexenone as the key starting material (Figure 15.4).

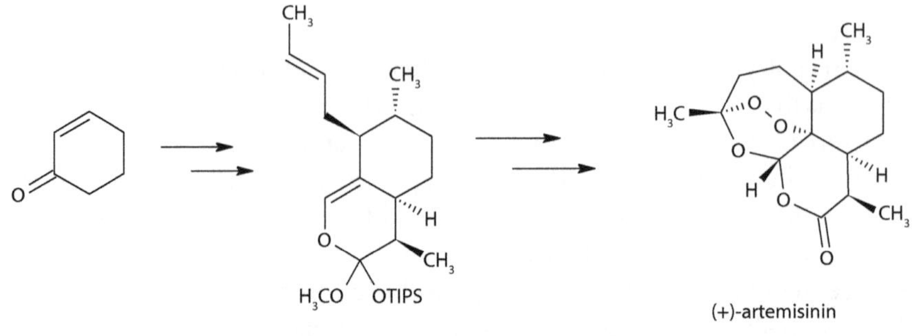


**Figure 15.4** Synthetic plan for synthesis of (+)-artimisin.

Later, Turconi *et al.* [99] from Sanofi-Aventis have reported a new commercial-scale alternative manufacturing process to produce a complementary source of artemisinin to add-on the plant-derived supply. Their major breakthrough was the discovery of a highly efficient biosynthetic route to artemisinic acid (A) in engineered *S. cerevisiae* [61]. The biosynthetic artemisinic acid was then converted into semi-synthetic artemisinin using diastereoselective hydrogenation followed by photooxidation as pivotal steps (Figure 15.5). This commercial synthesis delivers 365 kg batches of artemisinin starting from 600 kg of A and is expected to produce 60 tons per annum. This process was accepted by Prequalification of Medicines Programme (PQP) in 2013 as a first source of nonplant-derived-artemisinin in industrial scale from Sanofi production facility in Garessio, Italy.

Ikram *et al.* [88] have recently demonstrated that artemisinin can be rapidly produced by genetically engineered moss at an industrial scale. The researchers introduced five genes responsible for biosynthesizing the precursor of artemisinin, dihydroartemisinic acid, into the moss *Physcomitrella patens* using multiple DNA fragments.

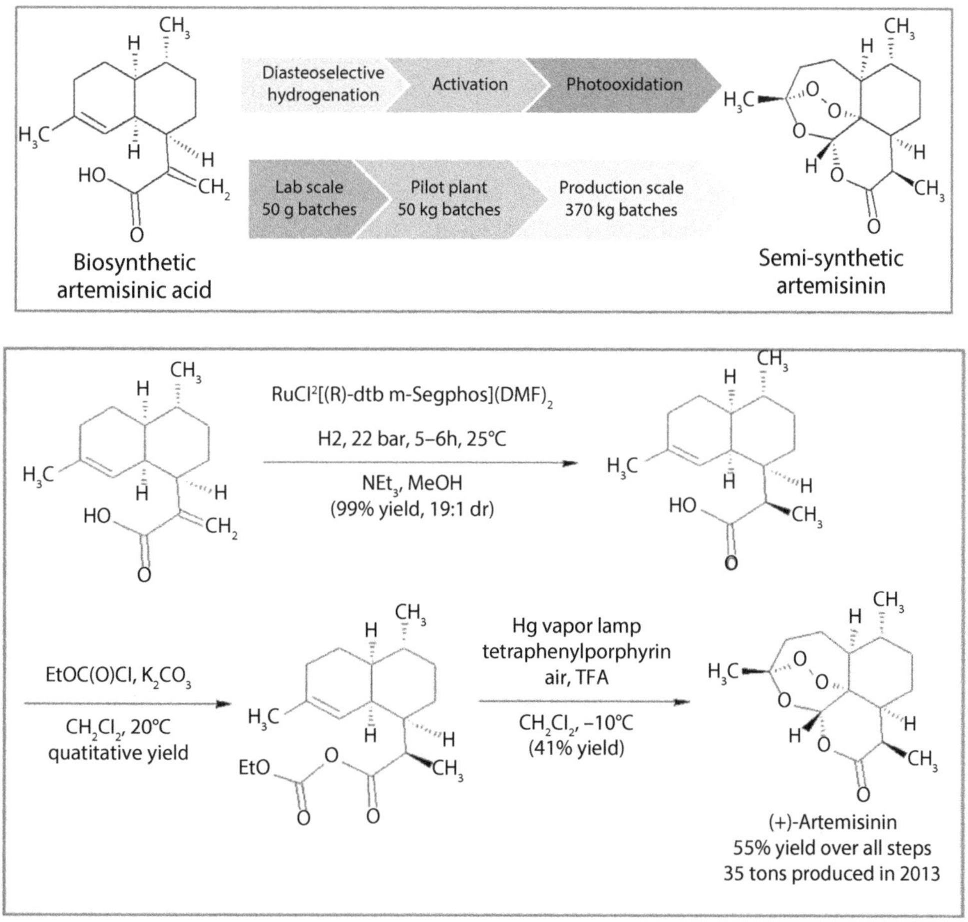


**Figure 15.5** Top - Synthetic plan for production scale synthesis of artemisinin. Bottom - Synthetic scheme for the scale up of artemisinin.

## 15.7 Future Prospects

Resistance to antimalarial drugs is one of the biggest problems in malaria control. As the threat of antimalarial drug resistance grows, there is increasing pressure to sustain the efficacy of existing treatments and develop alternative treatments, as well as put in place preventative measures such as bednets. Scaling up prevention and control interventions and implementing all of WHO's recommendations require considerable financial resources, long-term political commitment, and strong cross-border cooperation. Governments of endemic countries also need to take targeted regulatory measures to remove oral artemisinin-based monotherapies from markets, along with antimalarials that do not meet international quality standards.

Endemic countries outside Greater Mekong Sub-region and in particular on the African continent, where malaria took an estimated 438,000 lives in 2015 [100] also need to identify additional resources to fully implement WHO recommendations to prevent the emergence of artemisinin resistance. One of the most urgent challenges is to strengthen pharmaceutical market regulation and remove oral artemisinin-based monotherapies from markets around the world once and for all. Recent studies looking at the genome of the malarial parasite could help scientists understand how drug resistance has evolved and develop the tools needed to keep it in check. The development of a malaria vaccine could be the most successful way not only to curb the spread of malaria but to completely eradicate it. Although there is currently no effective malaria vaccine on the market, enormous progress has been made in the last 10 years toward developing malaria vaccines. Currently, more than a dozen vaccine candidates are in clinical development and the most advanced one is from GlaxoSmithKline Biologicals' RTS, S, which is in Phase III clinical trial, and in the near future we can expect to have a cost-effective and easily available artemesinin-based preventive healthcare solution for malaria.

## References

1. B. Wang, "Metabolic engineering of biosynthesis and sequestration of artemisinin". PhD Thesis, Wageningen University, 2016.
2. C. Faurant, *Parasite*, Vol. 18, p. 215, 2011.
3. D. Dalrymple, *Social History of Medicine*, Vol. 26, p. 799, 2013.
4. P.Q. Chen, G.Q. Li, X.B. Guo, K.R. He, Y.X. Fu, L.C. Fu, and Y.Z. Song, *Chinese Medical Journal*, Vol. 107, p. 709, 1994.
5. N. Kumar, and H. Zeng, *Parasitol*, Vol. 76, p. 214, 1990.
6. T.S. Skinner, L.S. Manning, W.A. Johnston, and T.M. Davis, *International Journal of Parasitology*, Vol. 26, p. 519, 1996.
7. H.M. McIntosh, and P. Olliaro, *Cochrane Database of Systematic Review*, Vol. 2, p. CD000256, 2000.
8. U.S.H. Svensson, and M. Ashton, *British Journal of Clinical Pharmacology*, Vol. 48, p. 528, 1999.
9. S.R. Meshnick, T.E. Taylor, and S. Kamchonwongpaisan, *Microbiological Reviews*, Vol. 60, p. 301, 1996.

10. T.M. Davis, H.A Karunajeewa, and K.F. llett, *The Medical Journal of Australia.*, Vol. 182, p. 181, 2005.
11. T. Efferth, *Current Drug Targets,* Vol. 7, p. 407, 2006.
12. T. Efferth, F. Herrmann, A. Tahrani, and M. Wink, *Phytochemistry Reviews,* Vol. 18, p. 959, 2011.
13. J. Utzinger, S.H. Xiao, M. Tanner, and J. Keiser, *Current opinion in investigational drugs,* Vol. 8, p. 105, 2007.
14. D.J. Charles, J.E. Simon, K.V. Wood, and P. Heinstein, *Journal of Natural Products*, Vol. 53, p. 157, 1990.
15. D.C. Jain, A.K. Mathur, M.M. Gupta, A.K. Singh, R.K. Verma, and A.P. Gupta, *Phytochemistry,* Vol. 43, p. 993, 1996.
16. E. Van Geldre, A. Vergauwe, E. Van den Eeckhout, *Plant Molecular Biology,* Vol. 33, p. 199, 1997.
17. J.F. Ferreira, J.E. Simon, and J. Janick, *Planta Medica*, Vol. 61, p. 167, 1995.
18. J.C. Laughlin, *Transactions of the Royal Society of Tropical Medicine & Hygiene,* Vol. 88, p. 21, 1994.
19. M.J. Davies, C.J. Atkinson, C. Burns, J.G. Woolley, N.A. Hipps, R.R. Arroo, N. Dungey, T. Robinson, P. Brown, and I. Flockart, *Annals of Botany,* Vol. 104, p. 315, 2009.
20. M. Davies, C. Atkinson, C. Burns, R. Arroo, and J. Woolley, *Industrial Crops and Products,* Vol. 34, p. 1465, 2011.
21. M. Enserink, *Science*, Vol. 307, p. 33, 2005.
22. N.J. White, *Science,* Vol. 320, p. 330, 2008.
23. J.M. Kindermans, J. Pilloy, P. Olliaro, M. Gomes, *Malaria Journal,* Vol. 6, p. 125, 2007.
24. M.E. Olsson, L.M. Olofsson, A.L. Lindahl, A. Lundgren, M. Brodelius, and P.E. Brodelius, *Phytochemistry*, Vol. 70, p. 1123, 2009.
25. M.V. Duke, R.N. Paul, H.N. Elsohly, G. Sturtz, and S.O. Duke, *International Journal of Plant Sciences,* Vol. 155, p. 365, 1994.
26. H.J. Bouwmeester, T.E. Wallaart, M.H. Janssen, B. van Loo, B.J. Jansen, M.A. Posthumus, C.O. Schmidt, J.W. de Kraker, W.A. Knig, and M.C. Franssen, *Phytochemistry*, Vol. 52, p. 843, 1999.
27. K. Cankar, A. van Houwelingen, D. Bosch, T. Sonke, H. Bouwmeester, J. Beekwilder, FEBS Letters, Vol. 585, p. 178, 2011.
28. H.M. Ting, B. Wang, A.M. Ryde, L. Woittiez, T. Herpen, F.W. Verstappen, C. Ruyter-Spira, J. Beekwilder, H.J. Bouwmeester, and A. Krol, *New Phytologist,* Vol. 199, p. 352, 2013.
29. K. Yang, R.S. Monafared, H. Wang, A. Lundgren, and P.E. Brodelius, *Plant Molecular Biology,* Vol. 88, p. 325, 2015.
30. G.D. Brown, and L.K. Sy, *Tetrahedron,* Vol 60, p. 1139, 2004.
31. J.F. Ferreira, and D.L. Luthria, *Journal of Agricultural and Food Chemistry,* Vol. 58, p. 1691, 2010.
32. Artemisinin imports into India, http://www.a2s2.org/, 2017.
33. C.J. Paddon, and J.D. Keasling, *Nature Reviews Microbiology*, Vol. 12, p. 355, 2014.
34. J.F. Ferreira, and J. Janick, *Plant Cell, Tissue and Organ Culture,* Vol. 44, p. 211, 1996.
35. T.C. Smith, P.J. Weathers, and R.D. Cheetham, *Plant,* Vol. 33, No. 1, p. 75, 1997.
36. P.J. Weathers, G. Bunk, and M.C. McCoy, *In Vitro Cellular and Developmental Biology-Plant,* Vol. 41, p. 47, 2005.
37. X.C. He, M.Y. Zeng, G.F. Li, and Z. Liang, *Acta Botanica Sinica,* Vol. 25, p. 87, 1983.
38. G.J. Kudakasseril, L. Lam, and E.J. Staba, *Planta Medica*, Vol. 53, p. 501, 1987.
39. J. Jha, T.B. Jha, and S.B. Mahato, *Current Science*, Vol. 57, p. 344, 1988.
40. A. Baldi, and V.K. Dixit, *Bioresource Technology,* Vol. 99, No. 11, p. 4609, 2008.

41. A. Baldi, and V.K. Dixit, *Current Trends in Biotechnology and Pharmacology*, Vol. 2, p. 341, 2008.
42. G.B. Pu, D.M. Ma, J.L. Chen, L.Q. Ma, H. Wang, G.F. Li, H.C. Ye, and B.Y. Liu, *Plant Cell Reports,* Vol. 28, No. 7, p. 1127, 2009.
43. H. Wang, C. Ma, Z. Li, L. Ma, H. Wang, H. Ye, G. Xu, and B. Liu, *Industrial Crops and Products,* Vol. 31, No. 2, p. 214, 2010.
44. D.P. Fulzele, A.T. Sipahimalani, and M.R. Heble, *Phytotherapy Research*, Vol. 5, No. 4, p. 149, 1991.
45. A. Singh, R.A. Vishwakarma, and A. Husain, *Planta Medica,* Vol. 54, No. 5, p. 475, 1998.
46. N. Delabays, X. Simonnet, and M.D.A.D. Gaudin, *Current Medicinal Chemistry*, Vol. 8, No. 15, p. 1795, 2001.
47. I.A. Graham, K. Besser, S. Blumer, C.A. Branigan, T. Czechowski, L. Elias, I. Guterman, D. Harvey, P.G. Isaac, A.M. Khan, T.R. Larson, Y. Li, T. Pawson, T. Penfield, A.M. Rae, D.A. Rathbone, S. Reid, J. Ross, M.F. Smallwood, V. Sequra, T. Townsend, D. Vyas, T. Winzer and D Bowles, *Science.*, Vol. 327, No. 5963, p. 328, 2010.
48. www.york.ac.uk/org/cnap/artemisiaproject/pdfs/AEconference-report-web.pdf, accessed online 1 Dec., 2017
49. S. Paul, S.P. Khanuja, A.K. Shasany, M.M. Gupta, M.P. Darokar, D. Saikia, and A.K. Gupta, *Planta Medica*, Vol. 76, No. 13, p. 1468, 2010.
50. S. Paul, S.P. Khanuja, and M.M. Gupta, *Industrial Crops and Products*, Vol. 56, p. 67, 2014.
51. J.H. Woerdenbag, P. Niesko, V.W. Uden, T. Elco Wallaart, C.A Beekman, and B.C Lugt, *Pharmacy World and Science*, Vol. 16, No. 4, p. 169, 1994.
52. H. Wang, L. Ge, H.C. Ye, K. Chong, B.Y. Liu, and G.F. Li, *Planta Medica,* Vol. 70, No. 4, p. 347, 2004.
53. A. Kjaer, K. Grevsen, and M. Jensen, *AoB Plants,* p. pls018, 2012.
54. S. Jin, A. Kanagaraj, D. Verma, T. Lange, and H. Daniell, *Plant Physiol*ogy, Vol. 155, p. 222, 2011.
55. N.D. Singh, S. Kumar, and H. Daniell, *Plant Biotechnology Journal,* Vol. 14, No. 3, p. 1034, 2016.
56. S. Takahashi, Y. Yeo, B.T. Greenhagen, T. McMullin, L. Song, J. Maurina-Brunker, R. Rosson, J.P. Noel, and J. Chappell, *Biotechnology and Bioengineering,* Vol. 97, No. 1, p. 170, 2007.
57. R.Y. Yang, L.L. Feng, X.Q. Yang, L.L. Yin, X.L. Xu, and Q.P. Zeng, *Planta Medica,* Vol. 74, No. 12, p.1510, 2008.
58. D.H. Chen, H.C. Ye, and G.F. Li, *Plant Science*, Vol. 155, No. 2, p. 179, 2000.
59. P. Alam, and M.Z. Abdin, *Plant Cell Reports*, Vol. 30, No. 10, p. 1919, 2011.
60. V.J. Martin, D.J. Pitera, S.T. Withers, J.D. Newman, and J.D. Keasling, *Nature Biotechnology*, Vol. 21, p. 796, 2003.
61. C.J. Paddon, P.J. Westfall, D.J. Pitera, K. Benjamin, K. Fisher, D. McPhee, M.D. Leavell, A. Tai, A. Main, D. Eng, D.R. Polichuk, K.H. Teoh, D.W. Reed, T. Treynor, J. Lenihan, M. Fleck, S. Bajad, G. Dang, D. Dengrove, D. Diola, G. Dorin, K.W. Ellens, S. Fickes, J. Galazzo, S.P. Gaucher, T. Geistlinger, R. Henry, M. Hepp, T. Horning, T. Iqbal, H. Jiang, L. Kizer, B. Lieu, D. Melis, N. Moss, R. Regentin, S. Secrest, H. Tsuruta, R. Vazquez, L.F. Westblade, L. Xu, M. Yu, Y. Zhang, L. Zhao, J. Lievense, P.S. Covello, J.D. Keasling, K.K. Reiling, N.S. Renninger, and J.D. Newman, *Nature*, Vol. 496, p. 528, 2013.
62. D.J. Pitera, C.J. Paddon, J.D. Newman, and J.D. Keasling, *Metabolic Engineering,* Vol. 9, p. 193, 2007.
63. H. Tsuruta, C.J. Paddon, D. Eng, J.R. Lenihan, T. Horning, L.C. Anthony, R. Regentin, J.D. Kaesling, N.S. Renninger, and J.D. Newman, *PLoS ONE,* Vol. 4, No. 2, p. e4489, 2009.

64. C.M. Bertea, J.R. Freije, H. van der Woude, F.W. Verstappen, L. Perk, V. Marquez, J.W. De Kraker, M.A. Posthumus, B.J. Jansen, A. de Groot, M.C. Franssen, and H.J. Bouwmeester, *Planta Medica*, Vol. 71, No. 1, p. 40, 2005.
65. M.C. Chang, R.A. Eachus, W. Trieu, D.K. Ro, and J.D. Keasling, *Nature Chemical Biology*, No. 3, p. 274, 2007.
66. D.K. Ro, E.M. Paradise, M. Ouellet, K.J. Fischer, K.L. Newman, J.M. Ndungu, K.A. Ho, R.A. Eachus, T.S. Ham, J. Kirby, M.C.Y. Chang, S.T. Withers, Y. Shiba, R. Sarpong, and J.D. Kaesling, *Nature*, Vol. 440, p. 940, 2006.
67. R.K. Mortimer, and J.R. Johnston, *Genetics*, Vol. 113, p. 35, 1986.
68. J.P. Van Dijken, J. Bauer, L. Brambilla, P. Duboc, J.M. Francois, C. Gancedo, M.L.F. Giuseppin, J.J. Heijnen, M. Hoare, H.C. Lange, E.A. Madden, Niederberger, J. Nielsen, J.L. Parrou, T. Petit, D. Porro, M. Reuss, N. van Riel, M. Rizzi, H.Y. Steensma, C.T. Verrips, J. Vindelov, and J.T. Pronk, *Enzyme and Microbial Technology*, Vol. 26, p. 706, 2000.
69. P.J. Westfall, D.J. Pitera, J.R. Lenihan, D. Eng, F.X. Woolard, R. Regentin, T. Horning, H. Tsuruta, D.J. Melis, A. Owens, S. Fickers, D. Diola, K.R. Benjamin, J.D. Kaesling, M.D. Leavell, D.J. McPhee, N.S. Renninger, J.D. Newman, and C.J. Paddon, *Proceedings of the National Academy of Sciences*, Vol. 109, p. E111, 2012.
70. H. Zhang, S.C. Im, and L. Waskell, *The Journal of Biological Chemistry*, Vol. 282, p. 29766–29776, 2007.
71. Sanofi en France - Accueil, Sanofi.fr, http://www.sanofi.fr/l/fr/fr/layout.jsp?cnt=F4BEF56E-3727-40C9-A6A2-29894774AD12, 2017
72. H. Daniell, S. Chebolu, S. Kumar, M. Singleton, and R. Falconer, *Vaccine*, Vol. 23, No. 15, p. 1779, 2005.
73. J.M. Staub, B. Garcia, J. Graves, P.T. Hajdukiewicz, P. Hunter, N. Nehra, V. Paradkar, M. Schlittler, J.A. Carroll, L. Spatola, and D. Ward, *Nature Biotechnology*, Vol. 18, No. 3, p. 333, 2000.
74. R. Tremblay, D. Wang, A.M. Jevnikar, and S. Ma, *Biotechnology Advances*, Vol. 28, No. 2, p. 214, 2010.
75. T.E. Wallart, H.J. Bouwmeester, J. Hille, L. Poppinga, N.C. Maijers, *Planta*, Vol. 212, No. 3, p. 460, 2001.
76. I.F. Kappers, A. Aharoni, T.W. Van Herpen, L.L. Luckerhoff, M. Dicke, and H.J. Bouwmeester, *Science*, Vol. 309, No. 5743, p. 2070, 2005.
77. S. Wu, M. Schalk, A. Clark, R.B. Miles, R. Coates, and J. Chappell, *Nature Biotechnology* Vol. 24, No. 11, 2006.
78. T.W. van Herpen, K. Cankar, M. Nogueira, D. Bosch, H.J. Bouwmeester, and J. Beekwilder, *PLoS One*, Vol. 5, No. 12, p. e14222, 2010.
79. Y. Zhang, G. Nowak, D.W. Reed, and P.S. Covello, *Plant Biotechnology Journal*, Vol. 9, No. 4, p. 445, 2011.
80. K.H. Teoh, D.R. Polichuk, D.W. Reed, G. Nowak, and P.S. Covello, *FEBS Letters*, Vol. 580, p. 1411, 2006.
81. M. Farhi, E. Marhevka, J. Ben-Ari, A. Algamas-Dimantov, Z. Liang, V. Zeevi, O. Edelbaum, B. Spitzer-Rimon, H. Abeliovich, and B. Schwartz, *Nature. Biotechnology*, Vol. 29, p. 1072, 2011.
82. B. Saxena, K. Subramaniyan, K. Malhotra, N.S. Bhavesh, S.D. Potlakayala, and S. Kumar, *Journal of Biosciences*, Vol. 39, p. 33, 2014.
83. D. Daniell, and B.A. McFadden, *Proceedings of the National Academy of Sciences*, Vol. 84, p. 6349, 1987.
84. J.L. Goldstein, and M.S. Brown, *Nature*, Vol. 343, p. 425, 1990.

85. K. Malhotra, M. Subramaniyan, K. Rawat, M. Kalamuddin, M.I. Qureshi, P. Malhotra, *Molecular Plant*, Vol. 9, p. 1464, 2016.
86. P. Fuentes, F. Zhou, A. Erban, D. Karcher, J. Kopka, and R. Bock, *eLife*, Vol. 5, p. 13664, 2016.
87. J.B. Schenkman, and I. Jansson, *Pharmacology & Therapeutics*, Vol. 97, No. 2, p. 139, 2003.
88. N.K.B. Ikram, A. Beyraghdar Kashkooli, A.V. Peramuna, A.R. Van Der Krol, H. Bouwmeester, and H.T. Simonsen, *Frontiers in Bioengineering and Biotechnology*, Vol. 5, No. 47, doi: 10.3389/fbioe.2017.00047, 2017.
89. N.K.B. Ikram, and H.T. Simonsen, *Frontiers in Plant Science*, Vol. 8, No. 1966, doi: 10.3389/fpls.2017.01966, 2017.
90. B. King, K. Vavitsas, N. Ikram, J. Schrøder, L. Scharff, J. Bassard, B. Hamberger, P. Jensen, and H. Simonsen, *Scientific Reports*, Vol. 6, p. 31261, 2016.
91. Malaria Consortium, Malariaconsortium.org, http://www.malariaconsortium.org, 2017.
92. J. Wang, C. Xu, Z.R. Lun, and S.R. Meshnick, *Trends in Pharmacological Sciences*, Vol. 38, No. 6, p. 506, 2017.
93. WHO status reports on artemisinin and ACT resistance, World Health Organization, http://www.who.int/malaria/areas/drug_resistance/updates/en/, 2017.
94. F. Lu, R. Culleton, M. Zhang, A. Ramaprasad, L. von Seidlein, H. Zhou, G. Zhu, J. Tang, Y. Liu, W. Wang, Y. Cao, S. Xu, Y. Gu, J. Li, C. Zhang, Q. Gao, D. Menard, A. Pain, H. Yang, Q. Zhang, and J. Cao, *The New England Journal of Medicine*, Vol. 376, p. 991, 2017.
95. C.H. Sibley, *Science*, Vol. 347, No. 6220, p. 373, 2015.
96. F. Ariey, B. Witkowski, C. Amaratunga, J. Beghain, A.C. Langlois, N. Khim, S. Kim, V. Duru, C. Bouchier, L. Ma, P. Lim, R. Leang, S. Duong, S. Sreng, S. Suon, C.M. Chuor, D.M. Bout, S. Ménard, W.O. Rogers, B. Genton, T. Fandeur, O. Miotto, P. Ringwald, J. Le Bras, A. Berry, J.C. Barale, R.M. Fairhurst, F. Benoit-Vical, O. Mercereau-Puijalon, and D. Ménard, *Nature*, Vol. 505, No. 7481, p. 50–55, 2014.
97. M. Imwong, K. Suwannasin, C. Kunasol, K. Sutawong, M. Mayxay, H. Rekol, F.M. Smithuis, T.M. Hlaing, K.M. Tun, R.W. van der Pluijm, R. Tripura, O. Miotto, D. Menard, M. Dhorda, N.P.J. Day, N.J. White, and A.M. Dondorp, *The Lancet Infectious Diseases*, Vol. 17, No. 5, p. 491, 2017.
98. C. Zhu, and S.P. Cook, *Journal of the American Chemical Society*, Vol. 134, No. 33, p. 13577, 2012.
99. J. Turconi, J. Griolet, R. Guevel, G. Oddon, R. Villa, A. Geatti, M. Hvala, K. Rossen, R. Göller, and A. Burgard, *Organic Process Research and Development*, Vol. 18, No. 3, p. 417, 2014.
100. https://ourworldindata.org/malaria/, 2017.
101. T.W.J.M. van Herpen, K. Canker, M. Nogueira, D. Bosch, H.J. Bouwmeester, J. Beekwilder, *PLOS One*, Vol. 5, No. 12, p. e14222, 2010.

# 16

# Microbial Production of Flavonoids: Engineering Strategies for Improved Production

**Aravind Madhavan[1], Raveendran Sindhu[1,2*], KB Arun[1], Ashok Pandey[3], Parameswaran Binod[2] and Edgard Gnansounou[4]**

*[1]Rajiv Gandhi Center for Biotechnology, Thiruvananthapuram, Kerala, (India)*
*[2]Microbial Processes and Technology Division, CSIR-National Institute for Interdisciplinary Science and Technology (CSIR-NIIST), Thiruvananthapuram, Kerala (India)*
*[3]CSIR- Indian Institute for Toxicology Research (CSIR-IITR), 31 MG Marg, Lucknow, (India)*
*[4]Ecole Polytechnique Federale de Lausanne, ENAC GR-GN, GC A3, Station 18, CH-1015, Lausanne, (Switzerland)*

***Abstract***
Flavonoids are natural plant-derived compounds which have immense applications in pharmaceutical industry and have attracted the attention of both the industry and researchers involved in all fields of biology and medicine. Microbial production systems can act as efficient cell factories for flavonoid production due to their versatile genetic and metabolic capabilities which can produce a wide variety of flavonoids. Moreover, metabolic engineering and synthetic biology techniques are used to rewire the metabolic pathways of flavonoid production for enhancing the product yield. This chapter is presented as an update on the various engineering approaches to the microbial production of natural or synthetic flavonoids in heterologous microbial hosts.

***Keywords:*** Flavonoids; synthetic biology; microbial cell factory; metabolic engineering

## 16.1 Introduction

Flavonoids are natural compounds known to be produced by terrestrial plants. They include a large number of naturally occurring compounds synthesised from the phenylpropanoid metabolism, which has originated in plants to produce a number of flavonoid structures. Structurally, they have the C6-C3-C6 carbon framework. Flavonoids are of greater interest due to their health-nourishing benefits and their vast applications in the pharmaceutical and food industries. The pharmaceutical properties of these compounds have been associated with the lower risks of major devastating

**Corresponding author*: sindhurgcb@gmail.com

Saurabh Saran, Vikash Babu, and Asha Chaubey (eds.) High Value Fermentation Products, Volume 1, (365–380) © 2019 Scrivener Publishing LLC

diseases including heart diseases, cancer, inflammation and several other major degenerative diseases [1, 2].

The extraction of these flavonoids from plants does not produce in the required amounts which are needed to meet the increasing commercial market demands. The isolation, separation and purification of these naturally occurring compounds using conventional extraction and purification processes, especially regarding those required for marketing purpose are limited due to the low amounts of these phyto-compounds in plants, compared to large amount of plant biomass. Moreover, these extraction processes are time and money consuming, wasting natural resources and not environmentally safe due to the usage of a huge amount of solvents during the extraction and purification processes. Thus the microbial production of flavonoids has attracted researchers as an alternative, cost-effective production platform and thus a substantial increase of availability levels.

In 2015, the flavonoid market was estimated as US$410 million. The major share of the 2015 US market is possessed by application of flavonoids in functional food. The use of anthocyanin, an important flavonoid and coloring agent in the food & beverage industry, was the major driving factor of the market in this segment. Even though it is easy to consume flavonoids in their native form, changing food habits of consumers have led to their utilization in the form of various functional food products.

This chapter gives an overview of flavonoids and health benefits, the impact of modern tools and techniques of metabolic engineering synthetic biology on flavonoid production.

## 16.2 Flavonoids

Flavonoids belong to the cluster of polyphenolic compounds which specifically possess a benzo-pyrone structure [3]. Plants are the ubiquitous source of these resourceful compounds. In plants, flavonoids are synthesized by phenyl propanoid pathway [4], mainly in response to microbial infection; draw attention of insects for pollination; controlling cell growth [5], and many more. Rutin was the first isolated flavonoid in 1930, and later on a line of flavonoid compounds have been identified continuously [6]. Approximately, 9,000 flavonoids have been identified from plants and microbes [7]. Flavonoids play a vital role in human health due to their potent biological activities. However, chemical nature of flavonoids determined the bioavailability and biological activities in humans, and in turn the flavonoids chemical nature of confides on their structural class, degree of hydroxylation and polymerization, functional group substitutions and conjugations [8].

## 16.3 Flavonoid Chemistry and Classes

Basic flavonoid structure (Figure 16.1) contains a fifteen-carbon skeleton with two benzene rings (I & II) connected through a pyrane ring (III). Flavonoids are broadly classified into six groups - chalcones, flavones, flavonols, flavandiols, anthocyanins, and proanthocyanidins or condensed tannins. Aurones and isoflavonoids are specialized

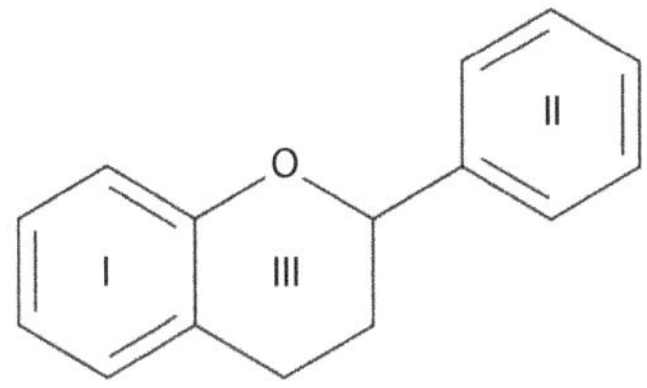


**Figure 16.1** Basic flavonoid structure.

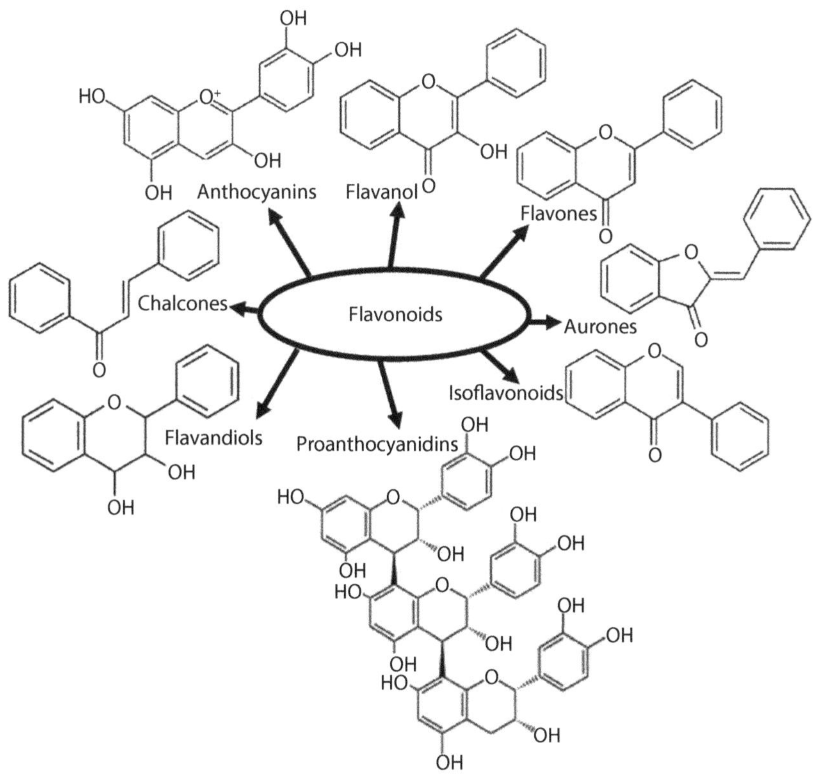


**Figure 16.2** Basic structures of different classes of flavonoids.

flavonoids found in some species [9, 10]. The structures of the main classes of flavonoids are shown in Figure 16.2 and common examples are shown in Table 16.1. The different classes of flavonoids vary in the oxidation level and III ring substitution outline, where as each compounds inside a class differ in the substitution of I and II rings. In nature, flavonoids exist as a glycones, glycosides, and methylated forms.

## 16.4 Health Benefits of Flavonoids

Flavonoids are one of the most vital phytochemicals in diets with diverse bioactivity. They are naturally occurring antioxidants and are reported to have various biological

**Table 16.1** Different classes of flavonoids with examples.

| Flavonoids | Examples |
|---|---|
| Chalcones | Butein, Licochalcone, Xanthohumol, Isobavachalcone |
| Flavones | Apigenin, Luteolin, Chrysin |
| Flavanols | Quercetin, Kaempferol, Myricetin |
| Flavandiols | Catechin, Epicatechin, Epigallocatechin, Gallocatechin |
| Anthocyanins | Apigenidin, Cyanidin |
| Proanthocyanidins | Pycnogenol, Procyanidins, Prodelhinidins, Propelargonidins |
| Aurones | Aureusidin, Hispidol, Leptosidin, Sulfuretin, 4,5,6 trihydroxyaurone |
| Isoflavonoids | Genistein, Daidzein |

activities like anti-inflammatory, anticancer, anti-diabetic, etc. Researchers have given great attention to elucidate the potential of flavonoids and have gathered enormous data which highlights the benefits of flavonoids in the prevention and management of life-style associated diseases - cancer [11–13], diabetes [14–16] and cardiovascular diseases [17–19]. Flavonoids are proven to scavenge free radicals and its consumption enhances the plasma antioxidant capacity which in turn strengthen capillary [20], reduces hyper-tension and thereby reduces the risk of developing heart problems [21]. The protec-tive effects against heart diseases can also be attributed to prevention LDL oxidation and platelet aggregation property of flavonoids [22]. Flavonoids exert their anticancer potential in all the stages of cancer progression due to their ability to induce cell cycle arrest, apoptosis [23] and prevent metastasis of cells. Diabetes research has focused on flavonoids and has proved that flavonoids can stimulate the insulin secretion from the existing β-cells of pancreas, decrease blood glucose level, increase glucose uptake and improve glucose homeostasis [24]. Thus flavonoids possess enormous biological effect which makes them an active research target.

Even though flavonoids are an essential part of human diet, they are not produced by the human system. Plants are the only source through which human acquired fla-vonoids. It is estimated that approximately 100–1000 milligrams of flavonoids were ingested by humans via diet per day [25]. Only a part of these are absorbed by the gastrointestinal tract and are further degraded by bacteria present in the colon. Most flavonoids are bound to one or two sugar moieties, that is, they usually exist as glyco-sides particularly in 'β' form [26]. Due to this particular chemical nature, they undergo widespread metabolism and fast excretion occurs. Hence, the bioavailability of flavo-noids is comparatively less when compared to other plant-based active compounds. So to meet the daily requirement of flavonoids dietary supplements are being used widely.

## 16.5 Flavonoid Biosynthesis in Microorganism

The industrial production of flavonoids increased significantly due to the enormous health benefits of flavonoids on humans. The microbial production of flavonoids is gaining more importance as the conventional methods of flavonoid production like plant extraction and chemical synthesis fails to meet the rising requirement of flavonoids effectively [27]. The low concentration of flavonoids in plants and the complex chemical structure of flavonoids respectively form the major hurdle in plant extraction and chemical synthesis of flavonoids. Advanced techniques in metabolic and protein engineering as well as synthetic biology prevail over the margins and challenges of microbial production system for flavonoids.

Unless and until there is an environmental or developmental stimuli plants will not produce flavonoids, which limits the extraction of flavonoids from plant tissues. The chemical synthesis includes usage of organic solvents and extreme reaction conditions; often fails to undergo modifications that affect the quality of flavonoids [28]. These challenges force the research groups to drag their focus on the microbial production of flavonoids utilizing metabolic engineering and synthetic biology tools. With the advancements in molecular engineering tools, scientists have generated very vigorous cell factories that can manufacture different types of compounds of both natural and non-natural origin [29]. *E. coli*, *Saccharomyces cerevisiae*, *Streptomyces venezuelae* [30, 31, 32] and *Phellinus igniarius* are some of the prokarytoes and eukaryote that have been implemented for flavonoids manufacturing [33]. In addition to these organisms, many other species *Actinobacteria* [34, 35] and oleaginous microorganisms [36] have also been reported to be suitable for flavonoid production.

As mentioned earlier, phenylpropanoid pathway is employed by the plants for the synthesis of flavonoids. The pathway triggers by starting the conversion of phenylalanine to cinnamic acid which is catalysed by phenylalanine ammonia lyase. Cinnamic acid is converted to 4-coumaric acid by the action of cinnamate-4-hydroxylase. After that 4-coumaric acid is modified to 4-coumaroyl-CoA by 4-coumarate:coenzyme A ligase further step by step condensation results in naringeninchalcone, which is the precursor molecule for many flavonoids [37]. However in microbial system the heterologous expression imparts specific characteristics. *E. coli* was engineered to assembled genes of phenylpropanoid for production of flavanones from different amino acid precursor's tyrosine and phenyl alanine [38]. Phenylalanine ammonia lyase from *Rhodotorula rubra* has the freedom to select either phenyl alanine or tyrosine as substrate to start the flavonoid synthesis [39]. Since some microorganisms lack the cytochrome P-450 hydroxylase, the system bypasses cinnamate-4-hydroxylase step [40]. Another characteristic is that *Streptomyces coelicolor* cinnamate/coumarate: CoA ligase can directly transfer CoA from malonyl CoA to both cinnamic acid and 4-coumaric acid [38].

De novo synthesis of precursors is another strategy adapted for efficient flavonoid production. Na *et al.* [41] have reported to develop a modified *E. coli* strain which produces 2 g/L of tyrosine by modulating expression of certain target genes using small synthetic regulatory RNAs. Similar methods have been applied to engineer the yeast *S. cerevisiae* to enhance the yield of tyrosine and phenylalanine [42, 43, 44]. Malonyl-CoA, another important precursor for flavonoid production, can be increased by introducing a recombinant malonateor acetate assimilation pathway [45, 46]; by enhancing intracellular content of acetyl-CoA [47]; and by deleting competing pathways to channel carbon towards malonyl-CoA [48]. In case of *S. cerevisiae* over expression of native acetyl CoA carboxylase is followed to increase malonyl-CoA precursor puddle [49, 50].

## 16.6 Engineering of Flavonoid Biosynthesis Pathway

Number of gene engineering is required to control the metabolic flux in microorganisms during recombinant flavonoid production. Modern genetic engineering techniques makes the control and manipulation of biological systems easier by developing a series of different expression vectors for the controlled expression of multiple genes involved in a biosynthetic pathway [51]. All genetic manipulation inside the host cell requires vectors. Several components of vectors can be engineered, for example: Promoter, terminator, origin of replication, ribosome binding sites to ensure the high product titre [52]. Several synthetic vectors have been designed and they allow tight control of artificially introduced biosynthetic pathways and this will lead to high production titre of various products. The e Path Brick vector system has been established for the production of flavanone naringnin [53]. In this work, the pseudo-operon structure of gene (desired gene was under the individual control of independent promoter while the mRNA with the same terminator) containing multiple gene containing pathway for naringenin production and malonyl-CoA synthesis led to the highest naringenin production from allgene combinations tested (36.7 $mgL^{-1}$ without ACC encoding genes and ~80 $mgL^{-1}$with ACC genes). Zhao *et al.* [54] reported catechin production using ePath Brick vector system with a yield of 910.9 $mgL^{-1}$from 1.0 $gL^{-1}$of eriodictyol in batch culture. Another study reported that seven NDP-sugar biosynthesis genes were heterologously expressed in monocistronic form under the separate T7 control element and transcriptional terminator for enhanced production of flavonol glycosides. Further experimental validation of these expressions constructs were done by bioconversion of flavonols and was found to be highly successful to achieve the yield of 392.6 $mgL^{-1}$of fisetin 3-*O*-glucoside and 342 $mgL^{-1}$ of fisetin 3-*O*rhamnoside. These cassettes were tested for the production of other flavanols (quercetin, kaempferol, myricetin) [55, 56]. A bio-sensor based expression cassettes have been described to easily detect different flavonoid compounds at the mono cell level by introducing genetically engineered transcriptional regulators from *Herbaspirillum seropedicae* and *Bacillus subtilis* [57]. Silencing of pathways that deactivate carbon from the flavonoid pathway including lipid synthesis genes was accomplished by the use of RNAi technology and produced high intracellular titre of malonyl-CoA in *E. coli.* The engineered *E. coli* with naringenin and resveratrol biosynthetic genes from plants produced 1.7 fold high resveratrol biosynthesis (268.2 $mgL^{-1}$) and 1.53 fold increased naringenin (91.31 $mgL^{-1}$) compared to the standard control ones [58].

## 16.7 Metabolic Engineering Strategies

Flavonoids are the most important plant secondary metabolites that have been the target of metabolic engineering for decades. Production of metabolically modified flavonoids is reported by several researchers from microbial sources such as *Escherichia coli*, *Saccharomyces cerevisiae*, and *Streptomyces* species. Several previous experimental studies have shown that imbalances within metabolic production pathways resulted in the unwanted accumulation of intermediary metabolites in multiple-gene controlled pathways [51]. To address this serious limitation, the combination of metabolic and different protein engineering techniques has opened several opportunities for the optimization of multi-gene controlled pathways [59, 60]. For example, Zhao *et al.* (2015) evaluated 18 different gene combinations to reduce the dihydroquercetin production and could achieve a production titre of 130 mg/L of catechin from eriodictyol. In another study of naringenin biosynthesis, multiple copies of the CHS gene were inserted to remove the bottleneck downstream of coumaric acid, resulted in enhanced production of naringenin production [61]. Similarly, various promoter optimization and gene copy number optimization were done to balance different upstream and downstream control elements to produce a maximum yield of 1.02 g/L taxadiene [62]. The pinosylvin production was increased through the in vivo evolution of stilbene synthase, which also resulted in the decreased production of cinnamic acid [63].

Four modified *S. cerevisiae* strains were developed by introducing flavonoid biosynthetic pathway genes from various organisms. *PAL*gene and *CPR*gene was isolated from *Populus* hybrid, *C4H, 4CL, CHS, CHI,* isoflavone synthase, flavanone 3-hydroxylase, and flavonol 3′-hydroxylase was obtained from *Glycine max*, resveratrol synthase from *Vitis vinifera*cv. *Soultanina*, and *FLS* from *Symphytum tuberosum* were expressed for the production of stilbenes, isoflavones, flavanones and flavonols. The modified yeast strains were specific for a particular flavonoid compound from the precursor phenyl alanine. The engineered yeast strains produced 0.29mgL$^{-1}$ of *trans*-resveratrol, 8.9 mgL$^{-1}$ of naringenin, 0.1 mgL$^{-1}$ of genistein, 0.9 mgL$^{-1}$ of kaempferol and 0.26 mgL$^{-1}$ of quercetin [64, 29]. In another study, eriodictyol was heterologously produced from L-tyrosine in recombinant *E. coli. TAL* gene from *R. glutinis*, *4CL* from *P. crispum*, *CHS* from *P. X hybrida*, and *CHI* gene from *M. sativa* were used for engineering along with ACC genes from *C. glutamicum* and *acs* gene from *E. coli* to synthesise naringenin. Naringenin was further transformed by hydroxylation with a fusion protein of truncated *F3′H* from Gerbera hybrid and *CPR* from *Catharanthus roseus* to eriodictyol. Upon transformation of the host strain by the inactivation of acetate kinase the modified strain produced 5.70 mgL$^{-1}$of ediodictyol. The production was enhanced upto 107 mgL$^{-1}$when L-tyrosine was supplemented exogenously and *ACC* genes were over expressed in a high copy number plasmid [65].

Recently an *E. coli* strain was developed which produces high amount tyrosine which is required for the production of flavonoids from simple carbon sources (high titres of tyrosine of about 400 mgL$^{-1}$). This strain was further modified by introducing plasmids with a codon optimized *TAL* and 4-coumarate 3-hydroxylase (*C3H/sam5*) from *Saccharothrix espanaensis*, and an O-methyltransferase gene from *A. thaliana*. The engineered strain produced 974 mgL-1, 150 mgL$^{-1}$, and 196 mgL$^{-1}$ of *p*-coumaric acid, caffeic acid, and ferulic acid, respectively, from glucose [66]. This group also studied about the pathway for methylated resveratrol production [67]. Recently engineered flavonoid compounds produced in microbial hosts are listed in the Table 16.2.

## 16.8 Applications of Synthetic Biology in Flavonoid Production

The main advantage of synthetic biology is the possibility of writing the gene code and its customization for the production of various products like natural molecules

**Table 16.2** Flavonoid compounds produced in microbial hosts.

| Flavonoids | Source | Host | Precursor | Yield (mg/ml) | Ref. |
|---|---|---|---|---|---|
| Naringenin | *A.thaliana* | *S. cerevisiae* | Glucose | 112.9 | [42] |
| Naringenin | *M. satira and R. trifoli* | *E. coli* | Glucose | 84 | [63] |
| Resveratol | *R. sphaeroides* | *S. cerevisiae* | Glucose | 1.06 | [7] |
| Kampferol | *Capparis spinosa* | *E. coli* | Naringenin | | [68] |
| Quercetin | *Capparis spinosa* | *E. coli*<br>*S. cerevisiae* | P. coumaric acid<br>Phenyl alanine | .05 | [69]<br>[64] |
| Genistein | *Glycine max* | *E.coli*<br>*S. cerevisiae* | Naringenin<br>Naringenin | 16.2 | [70] |
| Quercetin | *A. thaliana* | *E.coli* | | 65 | [71] |
| Pterostilbene | *A. thaliana* | *E. coli* | | 33.6 | [72] |

or biofuels. Natural products are metabolites produced by plants, animals and microorganisms with wide range of structural and functional activities [73]. Most of these natural molecules are under characterised and in the coming decades the advent of new technologies will address the current challenges and close the gap [74]. The availability of large numbers of sequenced genomes, new computational tools, proteomics, transcriptomics and metabolomics data helps to unlock the diversity of undiscovered natural molecule [75].

Different pathways for alcohols, lipids, terpenoids, polyketides, have been constructed synthetically by introducing individual gene parts from different organism and transferred in to *E. coli* host [76]. This approach can be implemented for several other molecules. The atremisin production is a classic example for the success of synthetic biology approach [77]. Another pioneering work in the area of synthetic biology was the synthetic pathway design for D-hydroxy phenyl glycine- the building block for the side chain of penicillins and cephalosporins [78]. They constructed the synthetic pathway by incorporating hydoxymandelate synthase and a hydroxyl mandelate oxidase from *Streptomyces coilicolor* and *Amycolatopsis orientalis* and hydroxyl phenyl glycine aminotransferase from *Pseudomonas putida*. Several other approaches attempted the rewiring of transcriptional units and by introducing enzymes from different organisms led to the improved production of fuels and chemicals.

The widely accepted application of synthetic biology in health sector is the microbial production of different pharmaceutical like artemisinin, the antimalarial drug. This drug is now produced large quantities in several heterologous hosts like *E. coli* and *Saccharomyces cerevisiae* [77]. Last few years witnessed dramatic increase in the production of pharmaceuticals with the help of synthetic biology tools like taxol [79], farnesene [65] and many others [80, 81]. The heterologous expression of the huge molecular synthetic gene assembly for pharmaceutically important natural products is demonstrated in *E. coli* [82]. They could express relatively huge protein, a 654 k Da non-ribosomal peptide synthase for valinomycinin active form using genome engineering tools. Shao *et al.* [83] has introduced the refactoring technique of synthetic biology for activating repressed biosynthetic pathway by deleting the native regulatory control elements of the spectabilin cluster of *Streptomyces orinoci* and replaced with strong promoters and obtained an improved production of the product. Recently power of synthetic biology has been utilized in producing a series of 74 novel compounds using a combinatorial genetics approach in baker's yeast [84].

The key importance of synthetic biological approach to synthetic pathway design is several possible roots of different pathway are considered for the production of a single compound. Some computational tools are designed for the alteration of the metabolic regulatory pathway by using gene knockouts or by the introduction of new catalytically active enzymes. Others involved in finding entirely new metabolic pathway other than the first one [85]. One important tool for metabolic pathway prediction is from metabolite to metabolite (FMM) – freely available software [86]. This includes the compilation of the KEGG map and KEGG ligand data to obtain a combined pathway. Cipher of Evolutionary Design (CiED) was introduced and was used for strain development and design using a previously generated *E. coli* metabolic model [87]. This software allowed the reconstruction and modification of existing *E. coli* metabolic models and resulted in the development of a new *E. coli* metabolic model with increased carbon flux towards different cofactors. Based on this CiED computational tool, deletion of citrate cycle genes *sdhCDAB* encoding for succinate dehydrogenase and citrate lyase, the amino acid transport, acetaldehyde dehydrogenase, along with over-expressions of Coenzyme A and malonyl-CoA synthetic pathways, developed an improvedhost genotype for the production of flavanones. The manipulated *E. coli* strain lacking *sdhA*, *adhE*, *brnQ* and *cite* and harbouring flavanone biosynthesis pathway genes (*Pc4CL2*, *PhCHS*, and *MsCHI*) and high expression of cofactor and other assimilation pathways produced 270 $mgL^{-1}$ of naringenin and 150 $mgL^{-1}$ of eriodictyol [88]. This tool was further used to develop *E. coli* metabolic models generating improved reduced nicotinamide adenine dinucleotide phosphate (NADPH). This modified strain was used for leucoanthocyanidin and (+)-catechin biosynthesis in which NADPH is an essential cofactor requirement. The engineering of host metabolism to redirect the metabolic flux in favour of the newly introduced biosynthetic pathway is extremely important. The newly introduced pathway normally depends on the cellular machinery for the generation of initial building blocks as well as the necessary cofactors like ATP, NADH and NADPH [89]. Cells normally sense the increased demand for all these factors and normalise their internal metabolism. Sometimes this imposes an extra burden on the cell and results in stressed condition and leads to decreased expression of introduced biosynthetic gene cluster [90]. To resolve all these problems, several stoichiometric – metabolic models are applied for considering the whole metabolic network and for studying the effect of host cell metabolism during the introduction of new biosynthetic gene cluster [91].

Twenty-two modified *E. coli* strains were generated which was deficient in various genes and the development of *E. coli* BL21 deficient in phospholipase A or lysophospholipase L, glucose-6-phosphate isomerase and phosphoenolpyruvate carboxylase could produce 817 $mgL^{-1}$ leucocyanidin while using *DuLAR* from *D. uncinatum* and supplementing 5 mM of dihydroquercetin. *E. coli* strain was modified by the silencing of fumerase coding gene - *fumC* and succinyl-CoA synthetase coding gene- *sucC* and over-expression of *accABCD*, *pgk*-phosphoglycerate kinase gene, and *PDC* complex-pyruvate dehydrogenase complex coded by *aceEF* and *lpdA* genes resulted in enhanced (4 fold) intracellular malonyl–CoA than wild type strain. This modified strain was further modified by the introduction of additional genes for naringenin biosynthesis (*Pc4CL2*, *PhCHS* and *MsCHI*) resulted in 474.2 mgL-1 of naringenin [48]. The same strain was also used for the production of 1.6 g/l of resveratrol by introducing *A. thaliana At4CL-2* and *V. vinifera VvSTS* genes [92].

## 16.9 Post-modification of Flavonoids

Flavonoids with diverse structure are created by extensive enzymatic post-modification processes resulting in the production therapeutically or nutraceutical novel compounds. Post-modification of flavonoids was achieved by utilizing fungal mycelia, streptomyces and plant cells by simple biotransformation method [93–95, 90]. The indigenous co-factors and enzymes present within these actinomycetes and eukaryotic cells with post modification efficacy, efficiently modifies the supplied flavonoids. Classical flavonoids modification reactions are carried out by O-methyltransferases (OMT), glycosyltransferases (GT), cytochrome P450 and prenyltransferases [96–99]. In addition to these foremost post-modifications, flavonoids are being modified by dehydroxylation [100], deglycosylation [101, 100], hydrogenation [102], halogenation [103], and sulfation [104]. Glycosylation results in the transfer of an activated sugar nucleotide diphosphate sugar to an acceptor molecule. This process is carried out GTs and is widely used for the flavonoid modification. Flavonoids are post-translationally methylated by S-adenosyl-Lmethionine-dependent caffeoyl-CoA OMTs. By the aid of biotransformation prenylated flavonoids [conjugated with lipophilic prenyl side-chain (5-carbon (dimethylallyl), 10-carbon (geranyl) or 15-carbon (farnesyl)] are produced which are of great therapeutic importance [105]. The chemical synthesis of modified flavonoids is difficult and always has limited production. Hence, the biotransformation approach for the post-modification of flavonoids holds a critical role, as it reduces the difficulty in post-modification simultaneously with better production rate.

## 16.10 Purification of Flavonoids

It is necessary to isolate flavonoids from microbial culture in pure form as these flavonoids are manufactured for its biological activity. Advanced chromatographic techniques are widely implemented for the isolation and purification of microbial flavonoids. Fingerprint analyses and highly sensitive molecular biology assays make sure of the purity of desired compound and it substantially reduces the labour and expenditure concerned in purification process. Purification of natural products starts from prehistoric age from the production of ethanol and thereafter dyes used for coloring [106]. Penicillin, the first drug, discovery draws the attention of the scientific community in isolation of microbial natural products [107]. Different groups had studied on penicillin isolation which overcomes the hurdles and finally results in large-scale penicillin production.

Penicillin is extracted by organic solvents using continuous counter-current extraction method. Counter-current chromatography works on differential solubility of two immiscible solvents. Conocandin is an active antibiotic isolated from *Hormococcus conorum* using the same strategy adopted for penicillin [108]. Droplet counter-current chromatography is another method widely used for isolation of natural products. Gibberellin A58, a plant growth regulator, is isolated from *Gibberella fujikuroi* by this droplet counter method [109]. High speed counter-current chromatography separates desired compound from microbial extracts. This method is far better than earlier mentioned methods and is widely used in isolation of natural compounds [110–113]. Compounds like cephamycin A [114], gualamycin [115], and zaragozic acid A [116] are isolated by employing liquid ion exchange method. Calphostin D [117], tubelactomicin A [118], teicoplanin A 2 -2 [119] and pentostatin [120] are isolated and purified by liquid solid chromatography.

Advanced bioengineering techniques revolutionized the strain development process and the tailor-made strains are efficient enough to produce a large variety of biologically important compounds of plant origin in large scale. Meantime, the advancements in chromatographic techniques make sure the final compound we isolated from microbes is pure enough for human use.

## 16.11 Conclusion

Flavonoids are a large and diverse group of phyto-compounds with fascinating compounds that display extraordinary antioxidant activity. The biological production pathways of flavonoids production have been extensively studied and well characterized. Furthermore, a huge number of flavonoids compounds have been isolated and characterised from plants and some of which are recognised as highly potent lead compounds with various pharmaceutical applications. In addition, flavonoids and their derivatized compounds have also been used in the food industry, nutraceuticals, and the cosmetic industry. The current commercial market of flavonoids production is largely dependent upon their plant sources, which is largely influenced by season expensive and time-consuming extraction and purification process. Thus introduction of recombinant flavonoid biosynthetic pathways form plants in heterologous microbial host are an alternative method of production. The advent of recent genetic engineering and system/synthetic biology tools has created the potential for the biosynthesis of a variety of flavonoids and their synthetic derivatives from well-established production hosts such as *E. coli*, *Streptomyces* and *S. cerevisiae*. However, low yield of flavonoids in microbial production hosts is one of the challenging factors for industrial and commercial production. The identification of current bottlenecks, optimization of multigene flavonoid biosynthetic pathways, and rewiring of metabolic networks are to be addressed.

## Acknowledgements

One of the authors, Raveendran Sindhu acknowledges DST for sanctioning a project under DST-WOS-B scheme. Parameswaran Binod and Raveendran Sindhu acknowledge EPFL, Lausanne for visiting fellowship. Aravind Madhavan and K B Arun acknowledges DST-SERB for National Post-Doctoral Fellowship.

## References

1. A. Gholami, N. De Geyter, J. Pollier, S. Goormachtig, and A. Goossens, *Natural Products Reports*, Vol. 31, p. 356, 2014.
2. F. Ververidis, E. Trantas, C. Douglas, G. Vollmer, G. Kretzschmar, and N. Panopoulos, *Biotechnology Journal*, Vol. 2, p. 1214, 2007.
3. L. H. Yao, Y. M. Jiang, J. Shi, F. A. Tomas-Barberan, N. Datta, R. Singanusong, and S. S. Chen, *Plant Foods for Human Nutrition*, Vol. 59(3), p. 113, 2004.
4. M. L. Falcone Ferreyra, S. P. Rius, and P. Casati, *Frontiers in Plant Sciences*, Vol. 3, p. 222, 2012. http://doi.org/10.3389/fpls.2012.00222
5. S. Kumar, and A. K. Pandey, *The Scientific World Journal*, Vol. 2013, p. 162750, 2013.
6. E. J. Middleton, *Advances in Experimental Medicine & Biology*, Vol. 439, p. 175, 1998.
7. Y. C. Wang, S. Chen, and O. Yu, *Applied Microbiology Biotechnology*, Vol 91, p. 949, 2011.
8. E. H. Kelly, R. T. Anthony, and J. B. Dennis, *Journal of Nutritional Biochemistry*, Vol. 13(10), p. 572, 2002.
9. B. Winkel-Shirley, The biosynthesis of flavonoids, in *The Science of Flavonoids*, Grotewold E., (Ed.). pp. 71–95, Springer, New York, 2006.
10. E. Miadoková, *Interdisciplinary Toxicology*, Vol. 2(4), p. 211, 2009.
11. T. Tanaka, H. Makita, K. Kawabata, H. Mori, M. Kakumoto, K. Satoh, *et al.* Carcinogenesis, Vol. 18, p. 957, 1997.
12. F. Casagrande, and J. M. Darbon, *Biochemical Pharmacology*, Vol. 61, p. 1205, 2001.
13. V. P. Androutsopoulos, S. Mahale, R. R. Arroo, G. Potter, *Oncology Reports*, Vol. 2, p. 1525, 2009.
14. S. Srinivasan, and L. Pari, *Chemico-Biological Interactions*, Vol. 195, p. 43, 2012.
15. R. Dhanya, K. B. Arun, H. P. Syama, P. Nisha, A. Sundaresan, T. R. Santhosh Kumar, and P. Jayamurthy, *Food Chemistry*, Vol. 158, p. 546, 2014.
16. F. Hajiaghaalipour, M. Khalilpourfarshbafi, and A. Arya, *International Journal of Biological Sciences*, Vol. 11, p. 508, 2015.
17. I., Huk, V., Brovkovych, J., NanobashVili, G., Weigel, C., Neumayer, L., Partyka, *et al.* *British Journal of Surgery*, Vol. 85(8), p. 1080, 1998.
18. D. J. Son, M. R. Cho, Y. R. Jin, S. Y. Kim, Y. H. Park, S. H. Lee, *et al. Prostaglandins, Leukotrienes and Essential Fatty Acids*, Vol.71, p. 25, 2004.
19. R. R. Holt, L. Actis-Goretta, T. Y. Momma, and C. L. Keen, *Journal of Cardiovascular Pharmacology*, Vol. 47, p. S187, 2006.
20. K. Chow, and I. Kramer, *All the Tea in China*. San Francisco: China Books and Period, Inc. 1990.
21. G. G. Duthie, S. J. Duthie, and J. A. M. Kyle, *Nutrition Research Reviews*, Vol. 13, p.79, 2000.
22. L. H. Yao, Y. M. Jiang, J. Shi, F. A. Tomas-Barberan, N. Datta, R. Singanusong, and S. S. Chen, *Plant Foods for Human Nutrition*, Vol. 59(3), 113, 2004.
23. R. VidyaPriyadarsini, R. SenthilMurugan, S. Maitreyi, K. Ramalingam, D. Karunagaran, and S. Nagini, *European Journal of Pharmacology*, Vol. 649(1–3), p. 84, 2010.
24. C. M. Agapakis, *ACS Synthetic Biology*, Vol. 3, p. 121, 2014.
25. R. Vinayagam, and B. Xu, *Nutrition & Metabolism*, Vol. 12, p. 60, 2015.

26. A. S. Aherne, and N. M. Obrien, *Nutrition*, Vol. 18, p. 75, 2002.
27. G. Williamson, Common features in the pathways of absorption and metabolism of flavonoids. In: Meskin, M.S., Bidlack, R.W., Davies, A.J., Lewis, D.S., Randolph, R.K., eds. Phytochemicals: Mechanisms of Action, pp. 21–33. Boca Raton, CRC Press, 2004.
28. T. Delmulle, S. L. De Maeseneire, and M. De Mey, Challenges in the microbial production of flavonoids. *Phytochemistry Reviews*, Vol. 17(2), p. 229, 2018.
29. S. Mavel, B. Dikic, S. Palakas, P. Emond, I. Greguric, A. G. de Gracia, F. Mattner, M. Garrigos, D. Guilloteau, and A. Katsifis, *Bioorganic & Medicinal Chemistry*, Vol. 14, p. 1599, 2006.
30. R. P. Pandey, ParajuliPrakash, A. G. Koffas Mattheos, and S. J. Kyung, *Biotechnology Advances*, Vol. 34(5), p. 634, 2016.
31. S. R. Park, J. A. Yoon, J. H. Paik, J. W. Park, W. S. Jung, Y-H Ban, E. J. Kim, Y. J. Yoo, A. R. Han, and Y. J. Yoon, *Journal of Biotechnology* Vol. 141(3–4), p. 181, 2009.
32. S. R. Park, J. H. Paik, M. S. Ahn, J. W. Park, Y. J. Yoon, *Journal of Microbiology and Biotechnology*, Vol. 20(9), p. 1295, 2010.
33. S. R. Park, M. S. Ahn, A. R. Han, J. W. Park, and Y. J. Yoon, *Journal of Microbiology and Biotechnology*, Vol. 21(11), p. 1143, 2011.
34. B. A. Pfeifer, and C. Khosla, *Microbiology and Molecular Biology Reviews*, Vol. 65(1), p. 106, 2001.
35. R. E. de Lima Proco´pio, I. R. da Silva, M. K. Martins, J. L. de Azevedo, and J. M. de Arau´jo, *Brazilian Journal of Infectious Diseases*, Vol. 16(5), p. 466, 2012.
36. A. Beopoulos, J. Cescut, R. Haddouche, J. L. Uribelarrea, C. Molina Jouve, J. M. Nicaud, *Progress in Lipid Research*, Vol. 48, p. 375, 2009.
37. B. Weisshaar, and G. I. Jenkins, *Current Opinion in Plant Biology*, Vol. 1, p. 251, 1998.
38. E. I. Hwang, M. Kaneko, Y. Ohnishi, and S. Horinouchi, *Applied Environmental Microbiology*, Vol. 69(5), p. 2699, 2003.
39. J. Kyndt, T. Meyer, M. Cusanovich, and J. Van Beeumen, *FEBS Letters*, Vol. 512, p. 240, 2002.
40. D. Pompon, B. Louerat, A. Bronine, and P. Urban, *Methods in Enzymology*, Vol. 272, p. 51, 1996.
41. D. Na, S. M. Yoo, H. Chung, H. Park, J. H. Park, and S. Y, Lee, Nature Biotechnology, Vol. 31(2), p. 170, 2013.
42. F. Koopman, J. Beekwilder, B. Crimi, A. van Houwelingen, R. D. Hall, D. Bosch, A. J. van Maris, J. T. Pronk, and J. M. Daran, *Microbial Cell Factories*, Vol. 11(1), p. 155, 2012.
43. M. A. H. Luttik, Z. Vuralhan, E. Suir, G. H. Braus, J. T. Pronk, and J. M. Daran, *Metabolic Engineering*, Vol.10(3–4), p.141, 2008.
44. A. Rodriguez, K. R. Kildegaard, M. Li, I. Borodina, and J. Nielsen, *Metabolic Engineering*, Vol. 31, p. 181, 2015.
45. E. Leonard, Y. Yan, Z. L. Fowler, Z. Li, C. G. Lim, K. H. Lim, and M. A. G. Koffas, *Molecular Pharmaceutics*, Vol. 5(2), p. 257, 2008.
46. I. Miyahisa, M. Kaneko, N. Funa, H. Kawasaki, H. Kojima, Y. Ohnishi, and S. Horinouchi, *Applied Microbiology and Biotechnology*, Vol. 68(4), p. 498, 2005.
47. Y. Krivoruchko, V. Zhang, Y. Siewers, J. Chen, Nielsen, *Metabolic Engineering*, Vol. 28, p. 28, 2015.
48. P. Xu, S. Ranganathan, Z. L. Fowler, C. D. Maranas, and M. A. G. Koffas, *Metabolic Engineering*, Vol. 13(5), p. 578, 2011.
49. S.Shi, Y. Chen, V. Siewers and J. Nielsen, mBio Vol 5(3), p. 1, 2014.
50. S. Y. Lee, and H. U. Kim, *Nature Biotechnology*, Vol. 33(10), p. 1061, 2015.
51. P. Xu, Q. Gu, W. Wang, L. Wong, A.G. Bower, C.H.Collins, and M.A. Koffas. *Nature Communications*, Vol. 4, p. 1409, 2013.

52. C. M. Agapakis, *ACS Synthetic Biology*, Vol. 3(3), p. 121, 2014.
53. D. Endy, and T. F. Knight, *Journal of Biological Engineering*, Vol. 2, p. 5, 2008
54. S. Zhao, J. A.Jones, D. M. Lachance, N. Bhan, O. Khalidi, S. Venkataraman, *Metabolic Engineering*, Vol. 28, p. 43, 2015.
55. A. K. Chaudhary, J. W. Park, Y. J. Yoon, B. G. Kim, J. K. Sohng, *Biotechnol Letters* Vol. 35, p. 285, 2013.
56. P. Parajuli, R. P. Pandey, N. T. Trang, A. K. Chaudhary, J. K. Sohng, *Microbial Cell Factories*, Vol. 14, p. 76, 2015.
57. S. Siedler, S. G.Stahlhut, S. Malla, J. Maury, A. R. Neves, *Metabolic Engineering*, Vol. 21, p. 2–8, 2014.
58. Y. Yang, Y. Lin, L. Li, R. J. Linhardt, Y. Yan, *Metabolic Engineering*, Vol. 29, p. 217, 2015.
59. J. W. Lee, *et al.*, *Nature Chemical Biology*, 8, p.536, 2012.
60. J. Wu, G. Du, J. Zhou, and J. Chen, *Metabolic Engineering*, 16, p. 48–55, 2013.
61. F. Koopman, *et al.*, *Microbial Cell Factories*, Vol. 11, p. 43–53, 2012.
62. P. K. Ajikumar, *et al.*, *Science*, Vol. 330, p. 70–74, 2010.
63. C. N. S. Santos, M. Koffas, and G. Stephanopoulos, *Metabolic Engineering* 13, p. 392, 2011.
64. E. Trantas, N. Panopoulos, and F. Ververidis. *Metabolic Engineering*, Vol. 11, p. 355, 2009.
65. F. Zhu, X. Zhong, M. Hu, L. Lu Z. Deng and T. Liu, (2014). *Biotechnology and Bioengineering*, Vol. 111, p. 1396, 2014.
66. S. Y. Kang, O. Choi, J. K. Lee, B. Y. Hwang, T. B. Uhm, and Y. S. Hong, *Microbial Cell Factories*, Vol. 11, p. 153, 2012.
67. S. Y. Kang, J. K. Lee, O. Choi, C. Y. Kim, J. H. Jang, B. Y. Hwang, *et al. BMC Biotechnology*, Vol. 14, p. 67, 2014.
68. P. Xu, A. Vansiri, N. Bhan, M. A. Koffas, *ACS Synthetic Biology*, Vol. 1, p. 256, 2012.
69. E. Leonard, Y. Yan, M. A. Koffas, *Metabolic Engineering*, Vol. 8, p. 172, 2006.
70. D. H. Kim, B. G. Kim, N. R. Jung, and J. H. Ahn. *Journal of Microbiology and Biotechnology*, Vol. 2(12), p. 1612, 2009.
71. D. G. An, S. M. Yang, B. G. Kim, J. H. Ahn, *Journal of Industrial Microbiology and Biotechnology*, Vol. 43(6), p. 841, 2016.
72. K. T. Heo, S. Kang, and Y. Hong, *Microbial Cell Factories*, Vol. 16, p. 30, 2017.
73. A. S. Khalil, and J. J. Collins, *Nature Reviews Genetics*, Vol. 11, p. 367, 2010.
74. J. D. Keasling, *ACS Chemical Biology*, Vol 3, p. 64, 2008.
75. Y. H. Wang, K. Y. Wei, and C. D Smolke, *Annual Review of Chemical and Biomolecular Engineering*, 4, p. 69, 2013.
76. V. Singh, *Gene*, Vol. 535, p. 1, 2014.
77. C. J. Paddon, P. J. Westfall, D. J. Pitera, K. Benjamin, K. Fisher, D. McPhee, M. D. Leavell, A. Tai, A. Main, and D. Eng, *Nature*, Vol. 496, p. 528, 2013.
78. U. Müller, F. van Assema, M. Gunsior, S. Orf, S. Kremer, D. Schipper, A. Wagemans, C. A. Townsend, T. Sonke, R. Bovenberg, and M. Wubbolts, *Metabolic Engineering*, Vol. 8, p. 196, 2006.
79. M. Jiang, G. Stephanopoulos, and B. A. Pfeifer, *Applied Microbiology and Biotechnology*, Vol. 94, 841, 2012.
80. C. M. Immethun, A. G. Hoynes-O'Connor, A. Balassy, and T. S. Moon, *Frontiers in Microbiology*, 4, 2013.
81. T. Wriessnegger, and H. Pichler, *Progress in Lipid Research*, Vol. 52, p. 277, 2013.
82. J. Jaitzig, J. Li, R. D. Süssmuth, and P. Neubauer, *ACS Synthetic Biology*, Vol. 3, p. 432, 2014.
83. Z. Shao, G. Rao, C. Li, Z. Abil, Y. Luo, and H. Zhao, Refactoring the silent spectinabilin gene cluster using a plug-and-play scaffold. *ACS Synthetic Biology*, Vol. 2, p. 662, 2013.
84. J. Klein, J. R. Heal, W. D. O. Hamilton, T. Boussemghoune, T. O. Tange, F. Delegrange, G. Jaeschke, A. Hatsch, and J. Heim, *ACS Synthetic Biology*, Vol. 3, p. 314, 2014.

85. K. L. J. Prather, and C. H. Martin, *Current Opinion in Biotechnology*, Vol. 19, p. 468, 2008.
86. M. H. Medema, R. Raaphorst, E. Takano, and R. Breitling, *Nature Reviews Microbiology*, Vol. 10, p. 191, 2012.
87. J. L. Reed, T. D. Vo, C. H. Schilling, B. O. Palsson, *Genome Biology*, Vol. 4, p. R54, 2003.
88. Z. L. Fowler, W. W. Gikandi, M. A. Koffas, *Applied Environment Microbiology*, 75, p. 5831, 2009.
89. V. Chubukov, L. Gerosa, K. Kochanowski, and U. Sauer, *Nature Reviews Microbiology*, Vol. 12, p. 327, 2014.
90. S. G. Wu, L. He, Q. Wang, and Y. J. Tang, *Microbial Cell Factories*, Vol. 14, p. 39, 2015.
91. Z. A., King, C. J., Lloyd, A. M. Feist, and B. O. Palsson, *Current Opinion in Biotechnology*, Vol. 35, p. 23, 2015.
92. N. Bhan, P. Xu, O. Khiladi, M. A. Koffas, *Chemical Engineering Science*, Vol. 103, p. 109, 2013.
93. K. Shimoda, N. Kubota, K. Taniuchi, D. Sato, N. Nakajima, and H. Hamada, *Phytochemistry*, Vol. 71, p. 201, 2010.
94. J. H. Kim, *Journal of Biotechnology* Vol. 150, p. 347, 2010.
95. B. F. da Silva, and E. Rodrigues-Fo, *Journal of Molecular Catalysis B: Enzymatic*, Vol. 67, p. 184, 2010.
96. R. K. Ibrahim, A. Bruneau, and B. Bantignies, *Plant Molecular Biology*, Vol. 36, p. 1, 1998.
97. J. H. Kim, B. G. Kim, J. H. Ko, Y. Lee, H. G. Hur, Y. Lim, and J. H Ahn, *Plant Sciences*, Vol. 170, p. 897, 2006.
98. M. A. Schuler, and D. Werck-Reichhart, *Annual Review of Plant Biology*, 54, p. 629, 2003.
99. B. Winkel-Shirley, *Plant Physiology*, Vol. 126, p. 485, 2001.
100. S. Lin, Q. Zhu, L. Wen, B. Yang, G. Jiang, H. Gao, F. Chen, and Y. Jiang, *Food Chemistry*, Vol. 145, p. 220, 2014.
101. J. Y. Liu, H. S. Yu, B. Feng, L. P. Kang, X. Pang, C. Q. Xiong, Y. Zhao, C. M. Li, Y. Zhang, and B. P. Ma, *Chinese Journal of Natural Medicine*, Vol. 11, p. 684, 2013.
102. M. J. C. Corrêa, F. M. Nunes, H. R. Bitencourt, F. C. Borges, G. M. S. P. Guilhon, M. S. P. Arruda, A. M. R. Marinho, A. S. Santos, C. N. Alves, D. S. B. Brasil, and L. S. Santos, *Journal of Brazilian Chemical Society*, Vol. 22, p. 1333, 2011.
103. W. S. Xiang, J. Zhang, J. D. Wang, L. Jiang, B. Jiang, Z. D. Xiang, and X. J. Wang, *Journal of Agriculture Food Chemistry*, Vol. 58, p.1933, 2010.
104. K. C. F. Araujo, E. M. M. B. Costa, F. Pazini, M. C. Valadares, and V. de Oliveira, *Food and Chemical Toxicology*, Vol. 51, p. 93, 2013.
105. B. Botta, A. Vitali, P. Menendez, D. Misiti, and G. DelleMonache, *Current Medicinal Chemistry*, Vol. 12, p. 717, 2005.
106. S. Olov, Isolation of microbial natural products. In: Satyajit D. S. and Lutfun N. (eds.), *Methods in Molecular Biology*, Vol. 864, p. 393, 2012.
107. D. A. Dias, S. Urban, and U. Roessner, *Metabolites*, Vol. *2*(2), p. 303. 2012.
108. J. M. Mueller, H. Fuhrer, J. Gruner, W. Voser, Roback, *Helvetica Chimica Acta*, Vol. 59, p. 2506, 1976.
109. D. B. Stierle, A. A. Stierle, B. Ganser, *Journal of Natural Products*, Vol. 60, p. 1207, 1997.
110. K. Hostettmann, A. Marston, Natural products isolation of droplet counter current chromatography. In: Mandava NB, Ito Y (eds) *Counter current chromatography – theory and practice, Chromatographic science series* 44. Marcel Dekker Inc, New York, pp. 465–492, 1988.
111. J. McAlpine, Separation by high-speed counter current chromatography. In: Cannell RJP (ed.) *Natural products isolation*. Humana, Totowa, NJ, pp 247–260. 1998.
112. M. J. Dawson, J. E. Farthing, P. S. Marshall, R. F. Middleton, M. J. O'Neill, A. Shuttleworth, C. Stylli, R. M. Tait, P. M. Taylor, H. G. Wildman, A. D. Buss, D. Langley, M. V. Hayes, *Journal of Antibiotics*, Vol. 45, p. 639, 1992.

113. S. Hillebrand, E. C. Montilla, N. Koehler, P. Winterhalter, *Agro Food Industry Hi-Tech*, Vol. 20, p. 52, 2009.
114. T. W. Miller, R. T. Goegelman, R. G. Weston, I. A. Putter, F. J. Wolf, *Antimicrobial Agents and Chemotherapy*, Vol. 2, p. 132, 1972.
115. K. Tsuchiya, S. Kobayashi, T. Harada, T. Kurokawa, T. Nakagawa, N. Shimada, K. Kobayashi, *Journal of Antibiotics*, Vol. 48, p. 626, 1995.
116. J. D. Bergstrom, M. M. Kurtz, D. J. Rew, A. M. Amend, J. D. Karkas, R. G Bostedor, V. S. Bansal, C. Dufresne, F. L. Van Middlesworth, O. D. Hensens, J > M. Liesch, D. L. Zink, K. E. Wilson, J. Onishi, J. A. Milligan, G. Bills, L. Kaplan, M. Nallin Omstead, R. G. Jenkins, L. Hunag, M. S. Meinz, L. Quinn, R. W. Burg, Y. L. Kong, S. Mochales, M. Mojena, I. Martin, F. Pelaez, M. T. Diez, A. W. Alberts, Proceedings of *National Academy of Sciences USA*, Vol. 90, p. 80, 1993.
117. E. Kobayashi, K. Ando, H. Nakano, T. Iida, H, Ohno, M, Morimoto, T Tamaoki, *Journal of Antibiotics,* Vol. 42, 1470,1989.
118. M. Igarishi, C. Hayashi, Y. Homma, S. Hattori, N. Kinoshita, M. Hamada, T. Takeuchi, *Journal of Antibiotics*, Vol. 53, p. 1096, 2000.
119. T.-W. Kang, B.-T. Choi, G.-S. Choi, Y.-R. Choi, S.-H. Hwang, Method for purifying teicoplanin A2. US Patent Appl. 20040024177, 2004.
120. J. C. French, C. R. Edmunds, P. Mc Donnell, H. D. H. Showalter Process for purifying pentostatin. US Patent 5,463,035, 1995.

# 17

# Astaxanthin: Current Advances in Metabolic Engineering of the Carotenoid

**Manmeet Ahuja, Jayesh Varavadekar, Mansi Vora, Piyush Sethia, Harikrishna Reddy and Vidhya Rangaswamy***

*High Value Chemicals, Breakthrough R & D, Reliance Industries Limited. Reliance Corporate Park. Thane Belapur Road, Ghansoli, Navi Mumbai, (India)*

***Abstract***
Astaxanthin is a valuable carotenoid, naturally produced by some yeast and algae. It has significant biological properties. The molecule is available in synthetic and natural form. The increasing market demand and preference for biological astaxanthin provides an impetus for exploration in this area. This chapter discusses the recent progress made in production and fermentation from the natural producers *Phaffia rhodozyma* and *Hematococcus pluvialis*. This chapter also covers in detail the current state of the art in metabolic engineering of organisms including plants, bacteria, algae and yeasts to produce astaxanthin biologically.

***Keywords:*** Astaxanthin, carotenoid, biological production, metabolic engineering

## 17.1 Introduction

Over the last decade, the family of carotenoids has received a lot of interest due to their potential health benefits in decreasing the risk of diseases relating to heart and eye, and cancer [1, 2]. Carotenoids or terpenoids are a class of more than 750 naturally occurring pigments which are produced by plants, algae as well as some bacteria and fungi; the majority of which contain eight isoprene units with a C40 skeleton [3, 4]. Carotenoids are mainly classified into two groups: carotenes (α-carotene, β-carotene and lycopene) and xanthophylls (zeaxanthin, cantaxanthin and astaxanthin) [5].

What is Astaxanthin? Astaxanthin, often referred to as "the king of carotenoids," is a red pigmented xanthophyll which imparts an attractive red colour to shrimp, lobsters and other crustaceans fishes such as salmon and redsnapper and also to the feathers of certain birds [6]. Astaxanthin is naturally synthesized by some bacteria, algae, yeast and plants. Shrimps and fish cannot synthesize their own astaxanthin and get the pigment through algae which is ingested by these organisms and which gets accumulated in their fatty tissue [7].

*Corresponding author*: Vidhya.Rangaswamy@ril.com

Saurabh Saran, Vikash Babu, and Asha Chaubey (eds.) High Value Fermentation Products, Volume 1, (381–400) © 2019 Scrivener Publishing LLC

### 17.1.1 Structure of Astaxanthin

The xanthophyll astaxanthin has a molecular weight of 596.84 g/mol with a molecular formula $C_{40}H_{52}O_4$. The difference between carotenes and xanthophylls is that the latter contain oxygen groups in addition to carbon and hydrogen moieties. The oxygen can be present either in the form of carbonyl or hydroxyl groups or in a combination of both, which is the case in astaxanthin [8]. The 3' 3' position of the β-ionone ring of astaxanthin contains two asymmetric carbons on either end. Astaxanthin occurs in stereoisomers, geometric isomers, free and esterified forms [9]. The structure of free astaxanthin is as follows (Figure 17.1).

### 17.1.2 Natural vs. Synthetic Astaxanthin

Natural sources of astaxanthin include bacteria from the genus *Paracoccus, Brevundimonas* and *Agrobacterium* [10], many species of algae including *Haematococcus pluvialis, Ostreococcus,* and *Micromonas* [11, 12], yeast such as *Xanthophyllomyces dendrorhous* [13] and flowers of plants such as *Adonis aestivalis* [14]. Amongst these sources, *Haematococcus pluvialis* is the highest accumulator of natural astaxanthin. *Haematococcus* is a fresh water green microalgae found in pools around the world. Under stressful conditions such as low nutrients, excessive light etc., it changes from its regular green vegetative form to a red cyst-form. This stage allows accumulation of astaxanthin [15]. This unique characteristic provides an efficient method for high-scale industrial production of astaxanthin. The first industrial-scale production of the compound from algae was developed in the late 1990s[16]. In nature, the stereoisomers 3S 3'S and 3R 3'R are the ones most commonly found and *Haematococcus* biosynthesizes the 3S 3'S form of astaxanthin. Another good producer of astaxanthin is the yeast, *X. dendrorhous.* Unlike *Haematococcus,* this organism produces the 3R 3'R isomeric form of astaxanthin. It produces astaxanthin *via* the mevalonate pathway and strain variations have been shown to increase yields [17].

Astaxanthin can be produced by synthetic and biological means, although the former dominates (>95%) the commercial market. In the synthetic market, BASF and Hoffman-La Roche are the leading producers of astaxanthin. Synthetic astaxanthin product contains different isomers of the compound than those normally found in nature. Synthesis mainly occurs from petrochemicals using various strategies such as hydroxylation of cantaxanthin or isomerization of lutein to zeaxanthin and then

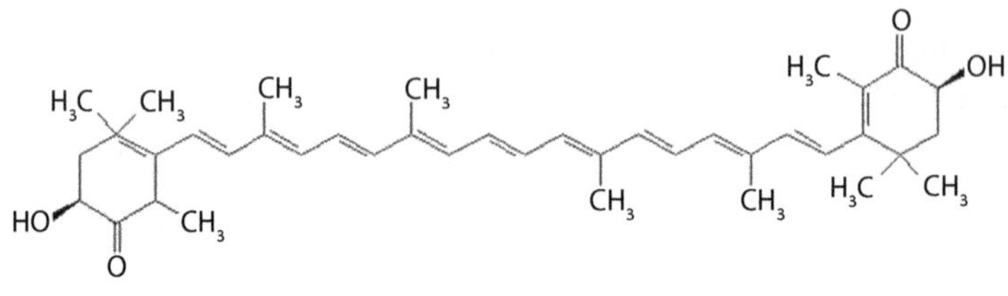


**Figure 17.1** Chemical structure of astaxanthin.

oxidation to astaxanthin or a Wittig reaction of two C15-phosphonium salts with a C10-dialdehyde, which is the most commonly used method [18–22].

### 17.1.3 Uses and Market of Astaxanthin

Astaxanthin is primarily used as a food supplement for humans and aquaculture. Since astaxanthin is the principal carotenoid pigment of several aquatic animals, it is considered to be of great importance in aquaculture. The red coloration imparted to aquaculture feeding on food rich in astaxanthin makes it aesthetically attractive, thereby rendering a competitive advantage to maintain high market demand. This pigment is used for imparting the pinkish-red coloration to the flesh of salmon, trout, shrimp, lobsters and other bred species which results in better quality and acceptance in the consumer market [16, 23, 24]. Astaxanthin has also been approved by the U.S. Food and Drug Administration (FDA) for use as a food colour additive. Naturally produced astaxanthin has been given generally recognised as safe (GRAS) status so it can also be sold as a dietary supplement. Astaxanthin products are available in the form of capsules, soft gels, tablets, powder, cream, energy drinks and oil extracts. As a dietary supplement, astaxanthin has been shown to have beneficial effects in anti-ageing, immune system boosting, anti-inflammatory, anti-cancer and as an anti-diabetic [25–29]. Astaxanthin is a very potent antioxidant able to cross the blood-brain barrier and act as a neuroprotective agent [30, 31]. Astaxanthin is considered as one of the best antioxidants for ROS-burst-mediated acute diseases like avian influenza infection and ischemia-reperfusion [32]. It is an anti-inflammatory therapeutic agent used to prevent the risk of cardiovascular diseases, shown to possess preclinical anti-tumor efficacy [26, 33–37]. On a different note, low doses of astaxanthin have been shown to improve memory of the BALB/c mice experimentally [32]. It has been shown to protect the skin from UV rays acting as a sunscreen [38, 39]. A patent, granted in 2001 for improvement of muscle function in rats, horses and humans by the aid of astaxanthin as a measure to target muscle related disorders [40, 41], claims to extend this function further to apply astaxanthin supplementation for physically intensive sports. For a more detailed study of the basic applications of astaxanthin Lorenz and Cysewskihave compiled a more comprehensive review [16].

Owing to the projected commercial rise in demand for astaxanthin based healthcare and cosmetic applications, several corporate organizations including Alga technologies Ltd. (Israel), Asta Real Co. Ltd. (Japan), BASF SE (Germany), Beijing Ginkgo Group (China), Cyanotech Corporation (US), Divis Neutraceuticals (India), DSM NV (the Netherlands), E.I.D. Parry Ltd. (India), Fenchem Biotek Ltd. (US), Igene Biotechnology Inc. (US), Nutrex Hawaii (US), Supreme Biotechnologies Ltd. (New Zealand), and Yunnan Alphy Biotech Company Ltd. (China) are involved in production and supply of the molecule. Natural astaxanthin from *Haematococcus pluvialis* is being produced and marketed, by Alga technologies, Israel, under the brand name Asta Pure®, by Asta Real Co., Ltd. Japan, under the brand name Asta Real® and astavita®. The USD 555 million worldwide astaxanthin market (as in 2016) is estimated to cross over USD 800 million by 2022 as per the prospective market analysis studies reported in May 2017 (http://www.marketsandmarkets.com/Market-Reports/astaxanthin-market-162119410.html, as accessed on 14$^{th}$ September, 2017).

## 17.2 Pathway of Astaxanthin

Astaxanthin biosynthesis pathway has been elucidated in a number of the natural producers including bacteria, fungi and plants. This section covers some of the differences in the carotenogenic pathway components and routes leading to the different carotenoid products.

### 17.2.1 Bacteria

Some marine bacteria belonging to the genus *Paracoccus* and *Brevundimonas* as well as *Agrobacterium aurantiacum* are known natural producers of astaxanthin [10]. The bacterial pathway for the production of astaxanthin uses glyceraldehyde 3' phosphate (GAP) and pyruvate as the precursors for the production of DMAPP and IPP *via* the methyl erythritol phosphate (MEP) route. DMAPP and IPP are finally converted into β-carotene in a series of steps catalysed by the *crt* cluster of genes [42]. The *crtE* gene catalyses the first conversion of DMAPP and IPP to GGPP. Phytoene, which is the first colourless carotene, is produced by the condensation of two GGPP molecules catalysed by the enzyme CRTB which is a phytoene synthase. The phytoenedesaturase gene *crt* converts phytoene to lycopene which is an important intermediate in carotene biosynthesis. The final step in the synthesis of β-carotene is catalysed by the enzyme lycopene β-cyclase CRTY[43].

All-trans-β-carotene is converted to astaxanthin *via* a series of hydroxylation and ketolation steps that generate a number of intermediates including zeaxanthin, cantaxanthin and adonixanthin amongst others. In bacteria, only two enzymes CRTZ/BHY (Beta-carotene hydroxylase) and CRTW/BKT (Beta-carotene ketolase) are capable of carrying out these reactions [44]. Extensive structural and functional analysis has been carried out to elucidate the biosynthetic pathway of *A. aurantiacum*. One such research conducted by *Misawa et al.* isolated a cluster of five *crt* genes (*crtW*, *crtZ*, *crtY*, *crtB* and *crtI*) from *A. aurantiacum*. The *crtZ* gene catalyses the hydroxylation reaction from β-carotene to zeaxanthin *via* cryptoxanthin. The *crtW* gene catalyses the conversion of the methylene to keto groups to synthesize canthaxanthin from β-carotene *via* echinenone. The products of these two genes were found to have low substrate specificity. Hence, a number of intermediates of astaxanthin such as adonixanthin, phoenicoxanthin (adonirubin), canthaxanthin, 3*-hydroxyechinenone, and 3-hydroxyechinenone are also produced as shown in Figure 17.2 [43].

### 17.2.2 Algae

The green microalga *H. pluvialis* is recognised as one of the primary natural source of astaxanthin in the aquatic food chain. This species has an exceptional ability to accumulate large amounts of astaxanthin under stress conditions [11]. Like bacteria, algae also use the MEP pathway for the production of IPP and DMAPP. The conversion of astaxanthin from IPP and DMAPP in algae is almost similar to that of bacteria. In certain algae such as *Ostreococcus* and *Micromonas*, two classes of phytoene synthase enzymes were found whereas certain other green algae only possess one copy of the enzyme [12]. Interestingly, cyanobacteria cannot naturally produce 3S, 3'S astaxanthin

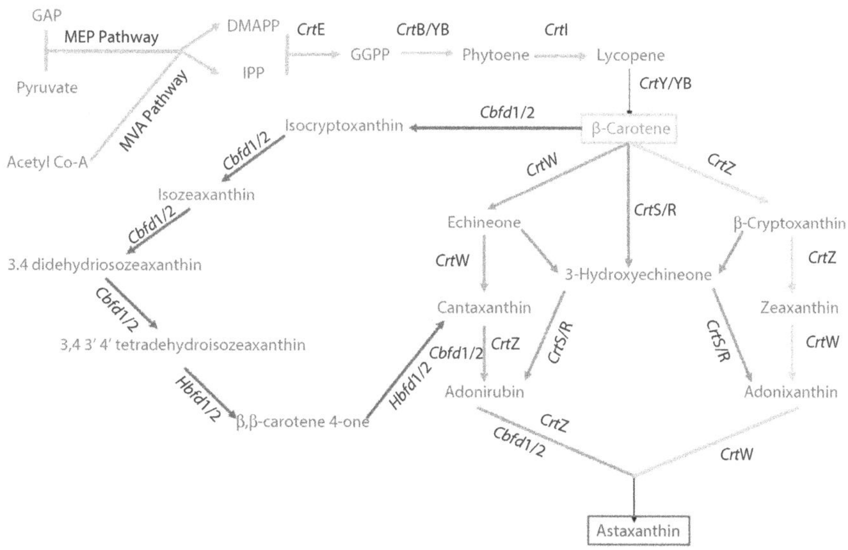


**Figure 17.2** Pathway of astaxanthin production in bacteria, algae, yeast and plants. The naturally occurring conversion of astaxanthin from precursor molecules dimethylallyl pyrophosphate (DMAPP) and isopentenyl pyrophosphate (IPP) is illustrated. Genes are shown next to the reaction catalyzed by the encoded enzyme (*crtE*: Prenyltransferase, *crtB*: Phytoene synthase, *crtI*: Phytoenedesaturase, *crtYb*:, Bifunctional lycopene cyclase/phyotene synthase, *crtY*: Lycopene-cyclase, *crtZ*: β-Carotene hydroxylase (3,3'-beta-ionone ring hydroxylase), *crtW*: β-Carotene ketolase (4,4'-beta-ionone ring ketolase), *crtS*: Astaxanthin synthase, *crtR*: β-carotene hydroxylase, *cbfd*1/2-carotenoid β-ring 4-dehydrogenase, *hbfd*1/2-carotenoid 4-hydroxy-β-ring-4-dehydrogenase.

but they share the reaction sequences catalysed by enzymes such as β-carotene ketolase (CRTO) and β-carotene hydroxylase (CRTR) which, however, show no sequence homology to related bacterial or plant enzymes [45]

### 17.2.3 Yeast

The heterobasidiomycetus yeast *X. dendrorhous* produces the 3R, 3'R form of astaxanthin which is readily absorbable and hence considered more advantageous. This organism produces astaxanthin and all its intermediates via the mevalonate pathway. There are certain differences in the conversion of IPP and DMAPP to β-carotene as compared to the bacterial pathway [13]. The conversion of GGPP to phytoene is catalysed by the CRTYB enzyme. This phytoene synthase is a bifunctional enzyme that has both phytoene synthase and lycopene cyclase activities. Thus, this enzyme also synthesises β-carotene from lycopene. In *Xanthophyllomyces*, the role of the ketolase and hydrolase required to convert β-carotene to astaxanthin is covered by a single enzyme, astaxanthin synthase (CRTS) which belongs to the cytochrome P450 family of proteins [46]. Another enzyme CRTR which also belongs to the same family, has been shown to have an auxillary role to CRTS in *X. dendrorhous*, providing the necessary electrons for substrate oxygenation. It has been shown that without CRTR, β-carotene cannot be converted to astaxanthin, thus proving its importance in the pathway [47].

### 17.2.4 Plants

The mono and di-esters of astaxanthin have been found in plants but only in the flower petals of *Adonis aestivalis*. The enzymes involved in the biosynthesis of astaxanthin from *Adonis* differ significantly from those observed in other organisms. IPP and DMAPP are condensed to form geranylgeranyl pyrophosphate (GGPP). Two molecules of GGPP are further condensed to form 15-cis-phytoene [48]. In plants, as well as in cyanobacteria, the 15-cis-phytoene is converted to lycopene in a multi-step reaction which involves the action of three enzymes: Phytoenedesaturase (PDS), ζ-carotene desaturase (ZDS) and prolycopene isomerase (CRTISO). The bacterial CRTI enzyme however can perform the three reactions from 15-cis-phytoene to lycopene all by itself. This characteristic of CRTI has ensured its use in metabolic engineering for the production of carotenoids. Lycopene is further converted to β-carotene by lycopene β-cyclase (LCY-b) encoded by lcy-b [48]. In higher plant *Adonis aestivalis*, β-carotene is converted to astaxanthin in three steps catalysed by two enzymes. The first step after the synthesis of β-carotene is its conversion to isocryptoxanthin catalysed by the enzyme CBDF1/2. The isocryptoxanthin in turn is converted to isozeaxanthin by the enzyme CBDF1/2. These enzymes also catalyse the conversion to 3,4-didehydroisozeaxanthin and 3,4,3'4' tetradehydroisozeaxanthin. The latter is the substrate for the enzymes HBDF1 and HBDF2 which convert it to β-β-carotene-4-one and subsequently cantaxanthin. The final conversion of cantaxanthin to astaxanthin is again catalysed by CBDF1 and CBDF2 *via* adonirubin [14]. The two enzymes act synergistically (unlike bacterial CRTZ and CRTW) to result in complete conversion of β-carotene to astaxanthin (Figure 17.2). However, BKTs are present only in a few bacteria and green algae.

## 17.3 Challenges/Current State of the Art in Fermentation/Commercial Production

Biological or Natural astaxanthin can be produced using the algae *H. pluvialis* or the yeast *X. dendrorhous* but the algal process is preferred owing to its high astaxanthin content. Astaxanthin has been known to naturally accumulate in basidiomycetous yeast *X. dendrorhous* [49]. Several genetic modifications have been attempted in the wild type yeast strain (ATCC 96594) to result in a better yielding culture, as has been discussed in the section below. This has led to the development of standardized production strategies that have been patented for commercial purposes [50–53].

The process is divided into two phases – an initial growth phase followed by a production phase. During growth phase, media conditions are optimized such that the cells are cultivated to accumulate maximum possible biomass. As the culture approaches stationary phase, growth slows down and simultaneously astaxanthin production kicks in. The switch from growth phase to production phase plays a crucial role in obtaining high titers. It is best achieved by inducing stress which hinders cell growth and diverting all cellular resources towards synthesis of secondary metabolites including carotenoids like astaxanthin. Schewe *et al.* have explored a pH shift based mechanism wherein they found that pH 3.5 during the production phase results in yield improvement, accumulating 0.7 g astaxanthin per kg of culture broth [54]. The last step in the pathway involves conversion of β-carotene to astaxanthin catalyzed by astaxanthin synthase. It has been demonstrated that rapid synthesis of β-carotene requires high oxygen levels for its complete oxidation to astaxanthin failing which the ratio of astaxanthin to other carotenoids is reduced. Switch to pH 3.5 possibly slows down the catalytic activity of astaxanthin synthase and allows enough oxygen to be available for complete oxidation and leads to an improved ratio. The authors of another study have used the glucose consumption rate (GCR) feeding method to achieve 20% higher astaxanthin production rate in the production phase [52]. Another successful strategy for improving yields of astaxanthin produced by *X. dendrorhous* to 5000 ppm (425 mg/L) has been external supplementation of glutamate as demonstrated in a patent by de la Fuente *et al.* [51].

It has also been shown in a separate study that the availability of dissolved oxygen plays an important role in astaxanthin synthesis and affects the proportion of astaxanthin accumulated among the various carotenoids [55]. A study by de la Fuente *et al.* in 2010 has shown that astaxanthin production specificity is significantly increased (4.7 mg per g) by using white or ultra violet light [56]. The authors hypothesize that this might be due to generation of active oxygen molecules in response to particular wavelengths resulting in stimulation of carotenogenesis. Gassel *et al.* adopted a combination of classical mutagenesis approach along with directed metabolic engineering and increased precursor supply to achieve efficient synthesis of astaxanthin (9 mg per g) [57].

The techno-economic modelling illustrates a tough fight for biological processes to overcome the advantages of chemical process of astaxanthin synthesis. However, the environment-friendly nature of the biological process and a futuristic foresight outweighs all the merits of chemical process [22]. Despite varied efforts the relative yield of astaxanthin that could be obtained from *X. dendrorhous* is lower than 0.1% g/g dry biomass in comparison to 6% g/g dry biomass obtained from the microalga *H. pluvialis. H. pluvialis* is grown in photobioreactors or in enclosed ponds and astaxanthin accumulation is induced by subjecting the cells to nutrient/environmental stress. Streamlined production processes have been designed for *H. pluvialis* and commercialized [58, 59]. The fresh water green microalgae is also cultivated in 2 phases as the aforementioned yeast. The first phase emphasizes building up of biomass by using optimal growth conditions and is known as the green stage because of accumulation of chlorophyll. This is immediately followed by the production phase wherein priority shifts from growth to accumulation of astaxanthin and other secondary metabolites. A sudden shift in growth condition leads to peculiar morphological changes in cellular structure and contents depending on the extent of stress exerted by the surrounding microenvironment. For instance, increase in cell volume along with thickened cell wall is distinct in case of sulfur starvation induced shift from green phase to red phase [60]. The microscopic morphology and ultra-structural changes occurring in the cellular structure during different phases of growth/production have been markedly described in a review by Shah *et al.* [61].

Adverse environmental conditions like nutrient (phosphate/iron/sulfur) starvation, temperature shift, light intensity and high salinity have been attempted in isolation as well as in combinations to persuade a productive shift from green phase to red phase [60, 62–65]. Recently, Sun *et al.* have proposed a novel strategy describing "staged cultivation" during the green phase [66]. The single green phase is subdivided in different sub-stages and the respective bottlenecks were addressed by using a combination of fed batch culture and replacement culture techniques along with the use of a mixture of red and white light in the ratio of 2:1. This resulted in accumulation of significantly higher amount of biomass before the red phase kicks in by means of any of the above mentioned stress-based induction strategies. Cheng *et al.* adopted a different strategy; they exposed the microalgal cells to 4000 Gy γ-ray irradiation to enhance growth rate and concurrently used high concentration of $CO_2$ to synergistically result in enhanced yield of astaxanthin (46 mg/L, 2.4 fold higher than wild type strains) [67].

## 17.4 Metabolic Engineering for Astaxanthin

The isoprenoid pathway is essential for the cell and serves as a source for many precursor molecules involved in the cellular metabolism. The primary rationale applied toward increasing the astaxanthin yield is to enhance the carbon flux through the isoprenoid pathway. This can be attempted by several strategies including overexpression of genes coding for enzymes involved in rate limiting steps such as *dxs* (1-deoxy-D-xylulose-5-phosphate synthase) and *idi* (Isopentenyl-diphosphate delta-isomerase), or introducing heterologous genes which code for enzymes with better activity and are less affected by the cell's own regulatory regime [68].

Metabolic engineering for astaxanthin production in organisms not known to produce this molecule naturally, has essentially involved incorporation of heterologous genes encoding for lycopene cyclase (*crtY*), carotene ketolase (*crtW*), and hydroxylase (*crtZ*). Of these three enzymes, the ketolase (CRTW) and the hydroxylase (CRTZ) are known to accept β-carotene as well as hydroxylated or ketolated products as substrate, resulting in the formation of different carotenoid intermediates reducing the final yield of astaxanthin (as mentioned in the previous section). Thus, a lot of efforts have been directed towards maximizing the astaxanthin yield by balancing/fine tuning of CRTZ and CRTW activities with respect to each other [69]. Another approach used is to engineer the enzyme to modify its activity, towards improving the ratio of astaxanthin versus the by-products. This is achieved by mutagenesis as was done in the case of β-carotene ketolase (CRTW) by Tao *et al.* [70].

### 17.4.1 Bacteria

Engineering attempts have been made in bacterial platforms such as *E. coli* [68]. Lemuth and others, for example, have made use of the λ-Red recombineering technique for creating a plasmid free *E. coli* strain having *crtE*, *crtB*, *crtI*, *crtY*, and *crtZ* from *Pantoeaananatis* and *crtW*148 (NpF4798) from *Nosto cpunctiforme* PCC 73102 stably inserted into the chromosome along with an enhanced expression of the native *E. coli* genes *idi* and *dxs*. Further, to enhance expression levels of genes coding for enzymes in the rate limiting steps of the isoprenoid pathway, an additional copy of *idi* was integrated under the control of a PTAC promoter and the promoter of the native *dxs* was changed to strong bacteriophage promoter T5. The resultant *E. coli* strain produced astaxanthin as the sole carotenoid with yields of 1.4 mg/g CDW (Cell Dry Weight).

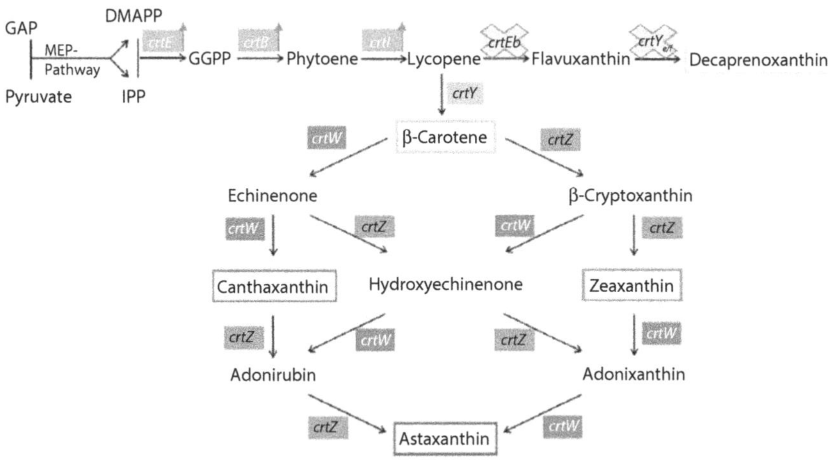


**Figure 17.3** Scheme of C40 cyclic carotenoid biosynthesis in recombinant *C. glutamicum*. The biosynthesis of C40 cyclic carotenoids derived from precursor molecules dimethylallyl pyrophosphate (DMAPP) and isopentenyl pyrophosphate (IPP) is illustrated. Genes are shown next to the reaction catalyzed by the encoded enzyme (*crtE*: Prenyltransferase, *crtB*: Phytoene synthase, *crtI*: Phytoenedesaturase, *crtEb*: Lycopene elongase, *crtYe/f*: C45/50 carotenoid "ε-cyclase, *crtY*: Lycopene β-cyclase, *crtZ*: β-Carotene hydroxylase (3,3'-beta-ionone ring hydroxylase), *crtW*: β-Carotene ketolase (4,4'-beta-ionone ring ketolase). Endogenous genes are shown in grey boxes and their overexpression indicated by green arrows. Heterologous genes are highlighted in coloured boxes (Image reproduced from Henke *et al.* [71]).

Significant progress has been made in astaxanthin production by *Corynebacterium glutamicum,* a gram-positive bacterium known to naturally synthesize a yellow C50 carotenoid decaprenoxanthin. Henke *et al.* have worked on precursor supply involved in the astaxanthin pathway [71]. They sequentially designed a strain capable of producing β-carotene (precursor for astaxanthin) which served as a chassis strain upon which all the genetic modification necessary for astaxanthin production were made (Figure 17.3).

Firstly, conversion of lycopene to decaprenoxanthin was prevented by deletion of lycopene elongase encoding gene *crtEb* and ε-cyclase. Further the expression of prenyltransferase (*CrtE*), phytoene synthase (*CrtB*) and phytoenedesaturase (*CrtI*) encoding genes was enhanced by replacing the native promoter with a strong $P_{tuf}$ promoter resulting in an 8- fold improvement in lycopene titers. Changing the native promoter of *dxs* with a strong promoter ($P_{tuf}$ in this work) markedly improved (34%) the lycopene titer further to about 0.43 ± 0.02 mg/ g CDW. This was previously also noted in *E. coli* [68]. Heterologous expression of the lycopene -cyclase gene *crtY* (under the control of $P_{tuf}$) from *P. ananatis* in the same strain enabled it to produce β-carotene. Further to balance expression of carotene ketolase (*CrtW*) and carotene hydroxylase (*CrtZ*) encoding genes with respect to astaxanthin production, a combinatorial gene assembly was used. *CrtW* from *Byrsia aurantiaca* and *CrtZ* from *P. ananatis* were combined in an artificial operon under the control of the constitutive $P_{tuf}$ with various combinations of RBS, start codons and spacers (between RBS and start codons). The resulting theoretical translation initiation rates (TIR) varied between 14 to greater than 30,000.

The resulting library did not show a strict co-relation between TIR and astaxanthin production and other pigments such as canthaxanthin and β-carotene dominated the product profile. However, the authors found that a higher TIR for *crtW* led to higher astaxanthin. Furthering their efforts the authors screened *CrtW* and *CrtZ* from marine and non-marine bacteria and found that *C. glutamicum* strain overexpressing *CrtW* and *CrtZ* from *Fulvimarina pelagi* accumulated astaxanthin as main carotenoid to titers of 1.6 ± 0.3 mg/g CDW) with volumetric productivity of 0.4 mg/l/hr.

Milo *et al.* in their patent application WO 2014013489 A1 disclose a similar strategy wherein the expression of the *Icy-B* and the *crtW* is maintained at a greater level of expression of the *crtI* to result in higher titers of astaxanthin. Titers of up to 5 mg/g CDW in *E. coli* have been reported. To achieve this, the applicants used a well-characterized RBS sequence to regulate expression of the genes involved in the astaxanthin pathway [72].

### 17.4.2 Plants

Although IPP and DMAPP, the C5 building blocks for carotenoids can be synthesized via two separate pathways: the cytoplasmic mevalonate pathway and the plastid-located (or MEP) pathway in plants, it is known that MEP pathway has a larger contribution in the carbon flux that goes into carotenoids [73]. The advantages of using plants for the production of astaxanthin are manifold: a) conferring foods such as tomato, carrot and maize with the ability to accumulate astaxanthin makes them fortified, b) plants such as those of tomato and carrot are inherently good producers of carotenoids such as lycopene and β-carotene, and hence have well functional intrinsic pathways designed towards production of carotenoids, making genetic manipulations designed towards furthering of pathway to astaxanthin plausible.

Recently, in an attempt to make astaxanthin easily available for consumption, Bai *et al.* attempted reconstruction of the astaxanthin biosynthesis pathway in rice endosperm using endosperm-specific promoters, although astaxanthin could be eventually funnelled to the rice grains, simultaneous prevalence of canthaxanthin and adonirubin suggested a bottleneck in the pathway leading to astaxanthin [74]. There stands a need to enhance a hydroxylation reaction involved in the ketocarotenoid pathway which would lead to increased relative abundance of astaxanthin in corresponding rice grains. Similarly, genetic engineering was performed in maize genome to make it synthesize astaxanthin in the kernel endosperm [75]. This was achieved by introduction of a hydroxylase and a ketolase enzyme in the host genome. It was then crossed with a high oil-maize genotype to improve the storage capacity for lipophilic astaxanthin and tested as an alternate feed supplement to confer desired pigmentation to the flesh of rainbow trout [75, 76]. Farré *et al.* genetically engineered white maize by introducing BKT from *Chlamydomonas reinhardtii* and BHY from *Brevundimonas sp.* Strain SD212 along with overexpression of phytoene synthase and knock-down of lycopene ε-cyclase. The modifications resulted in accumulation of astaxanthin in the maize endosperm to about 10.76 ± 1.56 μg/g dw (dry weight). Astaxanthin being lipophilic, they further crossed the said astaxanthin transgenic line into a high oil- maize genotype in order to increase the astaxanthin content in the seed and found an increase in astaxanthin levels (16.77 ± 1.45 μg/g dw) (75).

Tomato is being considered as an excellent plant platform for the production of astaxanthin because of its inherent ability to produce lycopene and β-carotene. Huang *et al.*, 2013, used functional complementation in *E. coli* and identified β-carotene ketolase from *Chlamydomonas reinhardtii* and β-carotene hydroxylase from *H. pluvialis*, as the enzymes having the best cooperation in converting β-carotene to astaxanthin [77]. They then co-expressed the genes of the said enzyme pair in tomato which resulted in up-regulation of native carotenogenic genes and directed the carbon flux towards carotenoids. As a result the total carotenoids increased by 16-fold with astaxanthin levels reaching 3.12 mg/g and 16.1mg/g in leaves and fruit respectively without the plant growth or development being affected [77]. Jayaraj *et al.* have similarly used bacterial complementation in their work which involved introducing a β-carotene ketolase gene (from *H. pluvialis*) fused to pea Rubisco small sub-unit transit peptide [78]. The transit peptide helped target the enzyme to plastids in leaf and root tissues. Further different promoters were screened, and the double CaMV 35S was used as the promoter of choice resulting in astaxanthin accumulation of 34.7 (±5.86) μg/g FW and 91.6 (±10.8) μg/g FW (fresh weight) in leaves and roots respectively [77]. Like in tomato, the endogenous expression of β-carotene hydroxylases was up-regulated in carrot leaves and roots [78]. The *H. pluvialis* β-carotene ketolase gene was also transferred to tobacco under the regulation of the tomato *pds* (phytoene desaturase) promoter by Mann *et al.* which resulted in production of astaxanthin in the flowers of the plant [79].

### 17.4.3 Synechocystis

The photosynthetic cyanobacterium *Synechocystis* sp. PCC 6803 has also been engineered to produce astaxanthin [80]. Although *Synechocystis* does not naturally produce astaxanthin, it does produce carotenoids such as myxoxanthophyll, β-carotene, zeaxanthin, and echinenone. The engineered mutant *Synechocystis* strain had insertion of two genes, β-carotene di-ketolase gene *crtW148* from *Nostocpunctiforme* and an additional copy of the endogenous β-carotene hydroxylase gene *crtR* from *Synechocystis*. Further the open reading frame of the endogenous β-carotene mono-ketolase gene *crtO* was disrupted. The resultant strain showed an overall increase in carotenoid content by 178 ± 10% of that seen in wild type cells with astaxanthin titers reaching production rates of 1.11 ± 0.07 mg/l/day. Albers also controlled the expression of *crtW148* by studying and using promoters with varying strengths and found that using a $P_{SCA6-2}$ promoter (which is more active) for driving expression of crtW148 improved the carotenoid content by 150% to 197% w.r.t to having $P_{sigA}$ promoter driving the same gene construct.

### 17.4.4 Algae

*H. pluvialis*, the unicellular alga, is a natural source for astaxanthin production. There have been efforts directed towards increasing the production of this molecule from the organism [58, 81, 82]. Advances made in engineering the organisms are mentioned in this section. Steinbrenner and Sandmann were able to develop a transformation protocol wherein a modified phytoene desaturase was integrated into the nuclear genome of *H. pluvialis*, and observed accelerated accumulation of astaxanthin in one of the transformants upon stress induction by high light [83]. The unicellular green alga *Chlorella zofingiensis* could also be a promising alternative. Several modified phytoene desaturase were tested, PDS-L516F, which had a 33% higher desaturation activity resulting in a higher total carotenoid and astaxanthin yields, the yield increase in these transformants is possibly related to enhanced transcripts of other pathway genes involved in astaxanthin biosynthesis as well [84].

### 17.4.5 Yeast

Metabolic engineering of yeasts to produce astaxanthin has also garnered significant attention due to their desirable attributes such as high rate of growth and the capability to grow to high cell densities. In organisms known to produce astaxanthin or related carotenoids naturally, the work has been directed towards increasing the expression of native enzymes responsible for astaxanthin production, deleting genes known to code for enzymes of a competing pathway and fine tuning/adjusting the expression levels of all the enzymes in a cascade maximize production of astaxanthin [71]. Yeasts such as *X. dendrorhous* (also known as *Phaffia rhodozyma*) are known to accumulate astaxanthin where it represents about 85% of the total carotenoid content. Astaxanthin serves to protect the *X. dendrorhous* cells from oxidative damage caused by reactive oxygen species or peroxyl radicals [85, 86]. There are several comprehensive reviews available which cover aspects of fermentation process optimization from this organism [87] while strategies to increase astaxanthin via genetic engineering, random mutagenesis are also covered [17, 86]. There are several limitations to enhancing yields of this molecule in this organism primarily because of a lack of genetic background information and suitable tools for metabolic engineering. A more recent review [88] covers in detail the biotechnology-based and omics approaches made towards improvements in astaxanthin production. Some of the more relevant studies are being included in this section. Recently development of a system to delete diploid genes, in this case, CYP61 genes encoding for sterol desaturases involved in the ergosterol biosynthesis resulted in an enhancement of astaxanthin yields in *X. dendrorhous* [89]. This study corroborates the previous finding that a mutant over-producer of astaxanthin strain of *Phaffia rhodozyma*, termed MK19 had a much lower total fatty acid content with reduced fatty acid and ergosterol biosynthesis [90]. A *crtS*-overexpressing strain of *Phaffia rhodozyma* was constructed employing a recombinant plasmid with lower copy numbers. This led to the enhancement in other genes of the pathway and increased astaxanthin yields and other carotenoids [91]. Alternatively the over-expression of three MVA pathway enzymes acetoacetyl-CoA thiolase, HMG-CoA synthase and HMG-CoA reductase together with genes involved in β-carotene and astaxanthin synthesis showed an increase in production without any deleterious effects on cell growth [92].

*S. cerevisiae*, the extremely well-established chassis used for microbial production of various molecules is an attractive alternative. The initial study in this organism involved introduction of *crtI* and *crtY* from *X. dendrorhous* resulting in β carotene production. Enhancement of the native GGPP synthase increased the levels of β carotene; however, accumulation of only astaxanthin required the co-expression of both *CrtS* along with *CrtR* (cytochrome P450 reductase) from *X. dendrorhous.* On the other hand introduction of the bacterial *crtW* and *crtZ* (ketolase and hydroxylase) in the β carotene producing strain resulted in accumulation of astaxanthin and its intermediates [93]. While in the most recent study in *S. cerevisiae,* the precursor β carotene supply was increased by overexpressing a positive mutant of GGPP synthase (*CrtE*03M) along with three other rate-limiting enzymes tHMG1, CRTI and CRTYB. Directed evolution was employed to generate a triple mutant OBKTM (H165R/V264D/F298Y) [94]. Along with other strain improvements the authors were able to demonstrate astaxanthin production by the over-expression of OCrtZ and OBKTM and showed a yield of 8.10 mg/g DCW (47.18 mg/L) of (3S, 3'S)-astaxanthin. The same group had previously reported a yield of 4.7 mg/g DCW in the shake-flask cultures by introducing algal β-carotenoid hydroxylase (*crtZ*) and ketolase (*bkt*) genes in a β-carotene over-producer [95]. In each study, improvements have been demonstrated in yield by adopting strategies of gene copy number adjustment, codon optimization and cofactor supplementation. In other yeast species, maximum yield of 31 μg/g DCW astaxanthin was achieved in *Kluyveromyces marxianus* strain Sm23, following the adoption of the above-mentioned strain improvement techniques. By site-directed mutagenesis of the gene *Hpchyb* (β-carotene hydroxylase) from *H. pluvialis* to increase enzyme efficiency and half-life of the protein along with repeated-integration approach of *bkt* and the mutated *Hpchyb* into the strain, the authors were able to produce the 3S, 3'S-astaxanthin at 9972 μg/g DCW in a 5 L fermenter [96].The highest reported yields with *S. cerevisiae* so far have been 81.0 mg/L in 5L bioreactor [97]. The authors achieved this by introducing *CrtZ* from *Alcaligenes* sp. strain PC-1 and *CrtW* from *Brevundimonas vesicularis* into a β-carotene over-producing strain.

Of the other yeasts, astaxanthin production has been demonstrated in *Candida utilis*, wherein genes from *Erwinia uredovora* were codon optimized and expressed under the control of native constitutive promoters and terminators [98]. The other non-carotenoid producing yeast is *Yarrowia lipolytica* [59, 99] where heterologous genes from other organisms have been introduced for production of astaxanthin.

## 17.5 Future Prospects

The astaxanthin commercially available is primarily from chemical synthesis; the major drawback of this process is that the product is a mixture of stereoisomers [58] with a possibility of a carryover of the synthetic intermediates [81]. Despite progress made in current biological production techniques, the economics of the process are not able to compete with the synthetic one. The product available in the market from biological sources via fermentation is from *Phaffia* or *Haematococcus.* (https://www.gminsights.com/industry-analysis/astaxanthin-market as accessed on 12th September, 2017). Astaxanthin use in fisheries is slowly being out-scaled by its demand in the cosmetic and nutraceutical industry. The powerful anti-oxidants market demand is substantial and will only increase but the price for the naturally derived astaxanthin will only be competitive compared with the synthetic price once the process is more economical.

Organisms such as *E. coli* and *S. cerevisiae* are being extensively employed for the production of commodity as well as high-value chemicals. Adapting some of the approaches in use to the production of astaxanthin in the natural producers as well as its heterologous production in these work-horse organisms will indeed be promising. There are well-established tools for genetic manipulations of *E. coli* and *S. cerevisiae* for pathway modifications such as overexpressing rate limiting enzymes, introducing heterologous genes, creating multiple gene knockouts [100] and introducing multiple copies of genes into genome [101]. A tool kit has even been developed to program gene expression in *S. cerevisiae* by CRISPR/Cas9 technology [102].

Though recent advances have been made in metabolic engineering, high-throughput tools have not yet been developed for the known natural producers. Table 17.1 summarizes the yields achieved from the studies covered in the previous section. It is evident that the higher yields so far have only been possible with the known natural producers. There are several drawbacks, some of which include: carotenoids are naturally synthesised in the chloroplast of algae and engineering algae for astaxanthin may require nuclear, mitochondrial, cytoplasmic genetic modifications. Nuclear over-expression of genes results in lesser protein production, but has the advantage of post-translational modification. Some of the nuclear expressed proteins are transported to

**Table 17.1** Significant values of astaxanthin yield from studies reviewed.

| Strain | Method | Maximum yield/Productivity | Ref. |
|---|---|---|---|
| *Phaffia rhodozyma* | pH shift | 0.7 g/kg of culture broth | [54] |
| *X. dendrorhous* | External supplementation of glutamate | 425 mg/L | [51] |
| *X. dendrorhous* | Classical mutagenesis approach along with directed metabolic engineering | 9 mg/g | [57] |
| *X. dendrorhous* | White or ultra violet light | 4.7 mg/g | [56] |
| *H. pluvialis* | 4000 Gy γ-ray irradiation | 46 mg/L | [67] |
| *E. coli* | Metabolic Engineering | 1.4 mg/g CDW | [68] |
| *E. coli* | Metabolic Engineering | 5 mg/g CDW | [72] |
| *C. glutamicum* | Metabolic Engineering | 1.6 ± 0.3 mg/g CDW | [71] |
| *Maize* | Metabolic Engineering | 16.77 ± 1.45 μg/g CDW | [75] |
| *Tomato* | Metabolic Engineering | 3.12 mg/g and 16.1mg/g in leaves and fruit<br>34.7 (±5.86) μg/g FW and 91.6 (±10.8) μg/g FW in leaves and roots | [77] |
| *Synechocystis sp.* | Metabolic Engineering | 1.11 ± 0.07 mg/l/day | [80] |
| *X. dendrorhous* | Metabolic Engineering | 1.65 mg/L | [89] |
| *Phaffia rhodozyma* | Metabolic Engineering | 27.8 mg/L | [91] |
| *X. dendrorhous* | Metabolic Engineering | 2.45 mg/L | [106] |
| *S. cerevisiae* | Metabolic Engineering | 8.10 mg/g DCW | [94] |
| *S. cerevisiae* | Metabolic Engineering | 29 μg/g CDW | [93] |
| *K. marxianus* | Metabolic Engineering | 9972 μg/g DCW | [96] |
| *S. cerevisiae* | Metabolic Engineering | 81.0 mg/L | [97] |

chloroplast and mitochondria to perform their functions. In contrast, protein overproduction is relatively high in chloroplasts compared to nuclear expressed genes, but lack posttranslational modifications [103]. Carotenoids synthesis genes are encoded by the nucleus and the protein synthesis occurs in the cytoplasm and then transported to chloroplasts [104]. A genetic tool has been developed to overexpress *aadA* gene in chloroplast of *H. pluvialis* [105]. Even though pathways have been characterized for carotenoid synthesis, there is a lack understanding of control mechanisms and Knock in/Knock out of genes by homologous recombination is not possible in microalgae and plants. Apart from the development of tools there is a lot of research needed to understand and develop flux control, transcriptional regulators in *X. dendrorhous* and microalgae.

Novel strategies are persistently coming up from both academic institutions as well as corporate organizations, thus promising a greater chance of success towards developing a cost-effective and environment-friendly bioprocess. Intensive research work continues to push the boundaries to render it feasible for the biological process to take over the existing chemical process for astaxanthin production. A switch to biologically produced astaxanthin has the potential not only to provide high-quality feed supplement for aquaculture but also to serve as an excellent source of various demonstrated benefits to mankind.

## References

1. E.A. Johnson, and W.A. Schroeder, in *Biotechnology for Improved Foods and Flavors*, Vol. 637, ed. American Chemical Society, pp. 39–50, 1996.
2. E.J. Johnson, *Nutrition in Clinical Care: an official publication of Tufts University*, Vol. 5, p. 56, 2002.
3. P.D. Fraser, and P.M. Bramley, *Progress in Lipid Research*, Vol. 43, p. 228, 2004.
4. C. Sieiro, M. Poza, T. de Miguel, and T.G. Villa, *International Microbiology: The Official Journal of the Spanish Society for Microbiology*, Vol. 6, p. 11, 2003.
5. P. Bhosale, and P.S. Bernstein, *Applied Microbiology and Biotechnology*, Vol. 68, p. 445, 2005.
6. T. Matsuno, *Aquatic Animal Carotenoids*, Vol. 67, ed., 2001.
7. T. Maoka, *Marine Drugs*, Vol. 9, p. 278, 2011.
8. L.M.A.J. Seabra, and L.F.C. Pedrosa, *Revista de Nutrição*, Vol. 23, p. 1041, 2010.
9. R.R. Ambati, S.M. Phang, S. Ravi, and R.G. Aswathanarayana, *Marine Drugs*, Vol. 12, p. 128, 2014.
10. N. Misawa, *Marine Drugs*, Vol. 9, p. 757, 2011.
11. S. Boussiba, L. Fan, and A. Vonshak, *Methods in Enzymology*, Vol. 213, p. 386, 1992.
12. D. Tran, J. Haven, W.G. Qiu, and J.E. Polle, *Planta*, Vol. 229, p. 723, 2009.
13. M. Rodriguez-Saiz, J.L. de la Fuente, and J.L. Barredo, *Applied Microbiology and Biotechnology*, Vol. 88, p. 645, 2010.
14. F.X. Cunningham, Jr., and E. Gantt, *The Plant Cell*, Vol. 23, p. 3055, 2011.
15. M. Kobayashi, Y. Kurimura, T. Kakizono, N. Nishio, and Y. Tsuji, *Journal of Fermentation and Bioengineering*, Vol. 84, p. 94, 1997.
16. R.T. Lorenz, and G.R. Cysewski, *Trends in Biotechnology*, Vol. 18, p. 160, 2000.
17. I. Schmidt, H. Schewe, S. Gassel, C. Jin, J. Buckingham, M. Humbelin, G. Sandmann, and J. Schrader, *Applied Microbiology and Biotechnology*, Vol. 89, p. 555, 2011.

18. K.D. Bernhard, R.K.D. Muller, and R.D. Spruijtenburg, Process for the preparation of astaxanthine and intermediates in the astaxanthine synthesis, European Patent Office 0101597 A3, assigned to F. Hoffmann-La Roche & Co., November 27, 1985.
19. G.C. Schloemer, and J.L. Davis, Preparation of astaxanthin, US Patent 6376717 B2, assigned to Prodemex, S.A. de C.V., April 23, 2002.
20. E. Widmer, R. Zell, E.A. Broger, Y. Crameri, H.P. Wagner, J. Dinkel, M. Schlageter, and T.o. Lukáč, *Helvetica Chimica Acta*, Vol. 64, p. 2436, 1981.
21. K.D. Nguyen, *Chem Biomol Eng Publ Other Work*, 1, 11.
22. G. Panis, and J.R. Carreon, *Algal Research*, Vol. 18, p. 175, 2016.
23. G.R. Cysewski, and R. Todd Lorenz, in *Handbook of Microalgal Culture*, Blackwell Publishing Ltd, pp. 281–288, 2007.
24. K.C. Lim, F.M. Yusoff, M. Shariff, and M.S. Kamarudin, Reviews in Aquaculture, 0.p.1,2018. (DOI: 10.19080/AIBM.2018.09.555768).
25. P. Kidd, *A Journal of Clinical Therapeutic*, Vol. 16, p. 355, 2011.
26. C.J. Ying, F. Zhang, X.Y. Zhou, X.T. Hu, J. Chen, X.R. Wen, Y. Sun, K.Y. Zheng, R.X. Tang, and Y.J. Song, *Cellular and Molecular Neurobiology*, Vol. 35, p. 1027, 2015.
27. L. Zhang, and H. Wang, *Marine Drugs*, Vol. 13, p. 4310, 2015.
28. Y. Yoshihisa, T. Andoh, K. Matsunaga, M.U. Rehman, T. Maoka, and T. Shimizu, *PLoS ONE*, 11, e0152288, 2016.
29. R.G. Fassett, and J.S. Coombes, *Marine Drugs*, Vol. 9, p. 447, 2011.
30. P. Lobos, B. Bruna, A. Cordova, P. Barattini, J.L. Galaz, T. Adasme, C. Hidalgo, P. Munoz, and A. Paula-Lima, *Neural Plasticity*, 3456, p.783, 2016.
31. H. Wu, H. Niu, A. Shao, C. Wu, B.J. Dixon, J. Zhang, S. Yang, and Y. Wang, *Marine Drugs*, Vol. 13, p. 5750, 2015.
32. X. Zhang, L. Pan, X. Wei, H. Gao, and J. Liu, *Environmental Geochemistry and Health*, Vol. 29, p. 483, 2007.
33. P. Regnier, J. Bastias, V. Rodriguez-Ruiz, N. Caballero-Casero, C. Caballo, D. Sicilia, A. Fuentes, M. Maire, M. Crepin, D. Letourneur, V. Gueguen, S. Rubio, and G. Pavon-Djavid, *Marine Drugs*, Vol. 13, p. 2857, 2015.
34. J. Monroy-Ruiz, M.A. Sevilla, R. Carron, and M.J. Montero, *Pharmacological Research*, Vol. 63, p. 44, 2011.
35. P.R. Augusti, A. Quatrin, S. Somacal, G.M. Conterato, R. Sobieski, A.R. Ruviaro, L.H. Maurer, M.M. Duarte, M. Roehrs, and T. Emanuelli, *Journal of Clinical Biochemistry and Nutrition*, Vol. 51, p. 42, 2012.
36. L. Zhang, and H. Wang, *Marine Drugs*, Vol. 13, p. 4310, 2015.
37. X.L. Xue, X.D. Han, Y. Li, X.F. Chu, W.M. Miao, J.L. Zhang, and S.J. Fan, *Stem Cell Research & Therapy*, Vol. 8, p. 7, 2017.
38. K. Suganuma, H. Nakajima, M. Ohtsuki, and G. Imokawa, *Journal of Dermatological Science*, Vol. 58, p. 136, 2010.
39. E. Camera, A. Mastrofrancesco, C. Fabbri, F. Daubrawa, M. Picardo, H. Sies, and W. Stahl, *Experimental Dermatology*, Vol. 18, p. 222, 2009.
40. B. Capelli, U. Jenkins, and G. Cysewski, *Role of Astaxanthin in Sports Nutrition*, Vol., ed., 2013.
41. A. Lignell, Medicament for improvement of duration of muscle function or treatment of muscle disorders or diseases, US Patent 6245818 B1, assigned to Astacarotene AB, June 12, 2001.
42. A. Yokoyama, H. Izumida, and W. Miki, *Bioscience, Biotechnology, and Biochemistry*, Vol. 58, p. 1842, 1994.
43. N. Misawa, Y. Satomi, K. Kondo, A. Yokoyama, S. Kajiwara, T. Saito, T. Ohtani, and W. Miki, *Journal of Bacteriology*, Vol. 177, p. 6575, 1995.

44. P.D. Fraser, Y. Miura, and N. Misawa, *The Journal of Biological Chemistry*, Vol. 272, p. 6128, 1997.
45. C. Liang, F. Zhao, W. Wei, Z. Wen, and S. Qin, *International Journal of Biological Sciences*, Vol. 2, p. 197, 2006.
46. V. Alvarez, M. Rodriguez-Saiz, J.L. de la Fuente, E.J. Gudina, R.P. Godio, J.F. Martin, and J.L. Barredo, *Fungal Genetics and Biology: FG & B*, 43, 261–272, 2006.
47. J. Alcaino, S. Barahona, M. Carmona, C. Lozano, A. Marcoleta, M. Niklitschek, D. Sepulveda, M. Baeza, and V. Cifuentes, *BMC Microbiology*, Vol. 8, p. 169, 2008.
48. G. Giuliano, *Current Opinion in Plant Biology*, Vol. 19, p. 111, 2014.
49. A.G. Andrewes, H.J. Phaff, and M.P. Starr, *Phytochemistry*, Vol. 15, p. 1003, 1976.
50. L. De Boer, V. Hell, and A.J.J. Krouwer, Strains of *phaffia rhodozyma* containing high levels of astaxanthin and low levels of 3 hydroxy-3',4'-didehydro-β, Ψ-caroten-4-one (HDCO), US Patent 5648261, assigned to Gist-brocades, July 15, 1997.
51. J. De La Fuente Moreno, E.P. Cezon, B.D. Garcia, A.M. Rodriguez, C.S. Sanchez, M.R. Saiz, C.R. Otero, W. Cabri, and J.B. Fuente, Method of producing astaxanthin by fermenting selected strains of *xanthophyllomyces dendrorhous*, US Patent 0124032 A1, assigned to Vitatene S.A., June 9, 2005.
52. T. Hoshino, Y. Setoguchi, and Y. Takagi, Astaxanthin production using fed-batch fermentation process by *phaffia rhodozyma*, US Patent 0134734 A1, assigned to DSM IP Assets B.V., June 22, 2006.
53. G.K. Jacobson, S.O. Jolly, J.J. Sedmak, T.J. Skatrud, and J.M. Wasileski, Astaxanthin overproducing strains of *phaffia rhodozyma* method for their cultivation and their use in animal feeds, US Patent 6015684, assigned to Archer-Daniels-Midland Company, January 18, 2000.
54. H. Schewe, A. Kreutzer, I. Schmidt, C. Schubert, and J. Schrader, *Biotechnology and Bioprocess Engineering*, Vol. 22, p. 319, 2017.
55. W. Wang, and L. Yu, *Journal of Biosciences*, Vol. 64, p. 853, 2009.
56. J.L. de la Fuente, M. Rodriguez-Saiz, C. Schleissner, B. Diez, E. Peiro, and J.L. Barredo, *Journal of Biotechnolgy*, Vol. 148, p. 144, 2010.
57. S. Gassel, J. Breitenbach, and G. Sandmann, *Applied Microbiology and Biotechnology*, p. 98, 2014.
58. S. Boussiba, A. Vonshak, Z. Cohen, and A. Richmond, Procedure for large-scale production of astaxanthin from *haematococcus*, US Patent 6022701, assigned to Ben-Gurion University of the Negev, February 8, 2000.
59. S.K. Samanta, A. Banerjee, A. Paradkar, Q. Cheng, P.L. Sharpe, and Q.Q. Zhu, Astaxanthin production using a recombinant microbial host cell, US Patent 0167041 A1, assigned to E.I. Du Pont De Nemours And Company, June 18, 2015.
60. P. He, J. Duncan, and J. Barber, *Journal of Integrative Plant Biology*, Vol. 49, p. 447, 2007.
61. M.M. Shah, Y. Liang, J.J. Cheng, and M. Daroch, *Frontiers in Plant Science*, Vol. 7, p. 531, 2016.
62. A.R. Dominguez-Bocanegra, I. Guerrero Legarreta, F. Martinez Jeronimo, and A. Tomasini Campocosio, *Bioresource Technology*, Vol. 92, p. 209, 2004.
63. G. Markou, and E. Nerantzis, *Biotechnology Advances*, 31, 1532–1542, 2013.
64. Z. Zhang, B. Wang, Q. Hu, M. Sommerfeld, Y. Li, and D. Han, *Biotechnology and Bioengineering*, Vol. 113, p. 2088, 2016.
65. M. Harker, A.J. Tsavalos, and A.J. Young, *Bioresource Technology*, Vol. 55, p. 207, 1996.
66. H. Sun, B. Liu, X. Lu, K.W. Cheng, and F. Chen, *Bioresource Technology*, Vol. 233, p. 326, 2017.
67. J. Cheng, K. Li, Z. Yang, J. Zhou, and K. Cen, *Bioresource Technology*, Vol. 204, p. 49, 2016.
68. K. Lemuth, K. Steuer, and C. Albermann, *Microbial Cell Factories*, Vol. 10, p. 29, 2011.

69. V.M. Ye, and S.K. Bhatia, *Biotechnology Letters*, Vol. 34, p. 1405, 2012.
70. L. Tao, J. Wilczek, J.M. Odom, and Q. Cheng, *Metabolic Engineering*, Vol. 8, p. 523, 2006.
71. N.A. Henke, S.A. Heider, P. Peters-Wendisch, and V.F. Wendisch, *Marine Drugs*, Vol. 14, 2016.
72. R. Milo, L. Zelcbuch, and N. Antonovsky, Methods of production of products of metabolic pathways, US Patent 9719122 B2, assigned to Yeda Research and Development Co. Ltd., August 1, 2017.
73. M. Rodriguez-Concepcion, *Archives of Biochemistry and Biophysics*, Vol. 504, p. 118, 2010.
74. C. Bai, J. Berman, G. Farre, T. Capell, G. Sandmann, P. Christou, and C. Zhu, *Transgenic Research*, Vol. 26, p. 13, 2017.
75. G. Farre, L. Perez-Fons, M. Decourcelle, J. Breitenbach, S. Hem, C. Zhu, T. Capell, P. Christou, P.D. Fraser, and G. Sandmann, *Transgenic Research*, 25, 477–489, 2016.
76. J. Breitenbach, M. Nogueira, G. Farre, C. Zhu, T. Capell, P. Christou, G. Fleck, U. Focken, P.D. Fraser, and G. Sandmann, *Transgenic Research*, Vol. 25, p. 785, 2016.
77. J.C. Huang, Y.J. Zhong, J. Liu, G. Sandmann, and F. Chen, *Metabolic Engineering*, Vol. 17, p. 59, 2013.
78. J. Jayaraj, R. Devlin, and Z. Punja, *Transgenic Research*, Vol. 17, p. 489, 2008.
79. V. Mann, M. Harker, I. Pecker, and J. Hirschberg, *Nature Biotechnology*, Vol. 18, p. 888, 2000.
80. S.C. Albers, thesis, Colorado State University (Fort Collins, Colorado), 2016.
81. M.M.R. Shah, Y. Liang, J.J. Cheng, and M. Daroch, *Frontiers in Plant Science*, Vol. 7, p. 531, 2016.
82. H. Sun, Q. Kong, Z. Geng, L. Duan, M. Yang, and B. Guan, *Bioresource Technology*, Vol. 186, p. 67, 2015.
83. J. Steinbrenner, and G. Sandmann, *Applied and Environmental Microbiology*, Vol. 72, p. 7477, 2006.
84. J. Liu, Z. Sun, H. Gerken, J. Huang, Y. Jiang, and F. Chen, *Applied Microbiology and Biotechnology*, Vol. 98, p. 5069, 2014.
85. H. Visser, A.J. van Ooyen, and J.C. Verdoes, *FEMS Yeast Research*, Vol. 4, p. 221, 2003.
86. L. Ye, W. Xie, P. Zhou, and H. Yu, *ChemBioEng Reviews*, Vol. 2, p. 107, 2015.
87. M. Rodriguez-Saiz, J.L. Fuente, and J.L. Barredo, *Applied Microbiol Biotechnology*, p. 88, 2010.
88. J. Barredo, C. García-Estrada, K. Kosalkova, and C. Barreiro, *Journal of Fungi*, Vol. 3, p. 44, 2017.
89. K. Yamamoto, K.Y. Hara, T. Morita, A. Nishimura, D. Sasaki, J. Ishii, C. Ogino, N. Kizaki, and A. Kondo, *Microbial Cell Factories*, Vol. 15, p. 155, 2016.
90. L. Miao, S. Chi, Y. Tang, Z. Su, T. Yin, G. Guan, and Y. Li, *FEMS Yeast Research*, Vol. 11, p. 192, 2011.
91. S. Chi, Y. He, J. Ren, Q. Su, X. Liu, Z. Chen, M. Wang, Y. Li, and J. Li, *Microbial Cell Factories*, Vol. 14, p. 90, 2015.
92. K.Y. Hara, T. Morita, M. Mochizuki, K. Yamamoto, C. Ogino, and M. Araki, *Microbial Cell Factories*, p. 13, 2014.
93. K. Ukibe, K. Hashida, N. Yoshida, and H. Takagi, *Applied and Environmental Microbiology*, Vol. 75, p. 7205, 2009.
94. P. Zhou, W. Xie, A. Li, F. Wang, Z. Yao, Q. Bian, Y. Zhu, H. Yu, and L. Ye, *Enzyme and Microbial Technology*, Vol. 100, p. 28, 2017.
95. P. Zhou, L. Ye, W. Xie, X. Lv, and H. Yu, *Applied Microbiology and Biotechnology*, Vol. 99, p. 8419, 2015.
96. Y.-J. Lin, J.-J. Chang, H.-Y. Lin, C. Thia, Y.-Y. Kao, C.-C. Huang, and W.-H. Li, *Bioresource Technology*, 2017.

97. R. Wang, X. Gu, M. Yao, C. Pan, H. Liu, W. Xiao, Y. Wang, and Y. Yuan, *Frontiers of Chemical Science and Engineering*, Vol. 11, p. 89, 2017.
98. Y. Miura, K. Kondo, T. Saito, H. Shimada, P.D. Fraser, and N. Misawa, *Applied and Environmental Microbiology*, Vol. 64, p. 1226, 1998.
99. P.L. Sharpe, R.W. Ye, and Q.Q. Zhu, Carotenoid production in a recombinant oleaginous yeast, US Patent 0142082 A1, assigned to P.L. Sharpe, R.W. Ye, and Q.Q. Zhu, June 7, 2012.
100. S.I. Jensen, R.M. Lennen, M.J. Herrgard, and A.T. Nielsen, *Scientific Reports*, Vol. 5, p. 17874, 2015.
101. C. Ronda, J. Maury, T. Jakočiūnas, S.A. Baallal Jacobsen, S.M. Germann, S.J. Harrison, I. Borodina, J.D. Keasling, M.K. Jensen, and A.T. Nielsen, *Microbial Cell Factories*, Vol. 14, p. 97, 2015.
102. A. Reider Apel, L. d'Espaux, M. Wehrs, D. Sachs, R.A. Li, G.J. Tong, M. Garber, O. Nnadi, W. Zhuang, N.J. Hillson, J.D. Keasling, and A. Mukhopadhyay, *Nucleic Acids Research*, Vol. 45, p. 496, 2017.
103. J.A. Gimpel, V. Henriquez, and S.P. Mayfield, *Front Microbiology*, Vol. 6, p. 1376, 2015.
104. R. Leon, and E. Fernandez, *Advances in Experimental Medicine and Biology*, Vol. 616, p. 1, 2007.
105. C.L. Gutierrez, J. Gimpel, C. Escobar, S.H. Marshall, and V. Henriquez, *Journal of Phycology*, Vol. 48, p. 976, 2012.
106. K.Y. Hara, T. Morita, Y. Endo, M. Mochizuki, M. Araki, and A. Kondo, *Applied Microbiology and Biotechnology*, p. 98, 2014.

# 18

# Exploitation of Fungal Endophytes as Bio-Factories for Production of Functional Metabolites through Metabolic Engineering; Emphasizing on Taxol Production

**Sanjog Garyali[1]*, Puja Tandon[2], M. Sudhakara Reddy[3] and Yong Wang[1]**

[1]*Key Laboratory of Synthetic Biology, Shanghai Institutes for Biological Sciences, Chinese Academy of Sciences, Shanghai, (China)*
[2]*School of Environmental Science and Engineering, Shanghai Jiao Tong University, Shanghai, PR (China)*
[3]*Department of Biotechnology, Thapar University, Patiala 147004 (India)*

***Abstract***
Endophytic fungi represent a lesser explored resource of novel compounds that possess the capacity to synthesize various types of plant secondary metabolites. One of such metabolites is taxol (paclitaxel), a plant alkaloid and competent antineoplastic drug extracted from the bark of the *Taxus* plants. In the search for more effective ways to treat cancer, the taxol producing endophytic fungi have demonstrated high potential in increasing the efficiency of extraction and sustainable production of taxol and cancer treatment products. In the present chapter, we deliver an exhaustive outline of the current knowledge on the subject and considerable emphasis is given on exploiting the astounding diversity of the endophytic world for its pharmaceutical potential. Additionally, the acuity of modulation of the complete cellular network (regulation and modification) in place of only specific pathways resulting in the concept of "system metabolic engineering" approaches for enhancing the taxol production have been discussed.

***Keywords:*** Endophytic fungi, plant-microbe interaction, secondary metabolites, Taxus, taxol, metabolic engineering

## 18.1 Introduction

Drug discovery and its development consist of a long history that dates back to the early days of human civilization. The Mayans used endophytic fungi grown on roasted green corn to treat intestinal sickness [1]. Later, the Benedictine monks (800 AD) began to apply *Papaver somniferum* belonging to the poppy family (*Papaveraceae*) as

**Corresponding author*: sanjog_garyali@yahoo.com

Saurabh Saran, Vikash Babu, and Asha Chaubey (eds.) High Value Fermentation Products, Volume 1, (401–430) © 2019 Scrivener Publishing LLC

an anesthetic and analgesic agent. In 1928, Alexander Fleming found penicillin from the mold *Penicillium notatum*. Ever since the discovery of penicillin, the attention of pharmaceutical companies and research laboratories was focused more on fungi as a source of lead compounds. Since then, researchers have been engaged to search and discover new microbial metabolites with activity against plant, animals, and human pathogens.

Endophytes play a significant role in the daily life of human beings; besides, they may produce an overabundance of compounds with a wide range of properties, including antibiotics, antivirals, antimycotics, antioxidants, immunosuppressant, antidiabetic, insecticidal and anticancer agents [2–9]. The various secondary metabolites including active principles produced by endophytes possess unique structures and great bioactivities, representing a huge reservoir which offers an immense potential for exploitation for medicinal, agricultural and industrial uses [10–12].

The new technologies offer distinctive prospects for natural product screening which serves as the major strategies for drug discovery. Microorganisms act as an absolute, readily renewable, reproducible and inexhaustible source of enduring potential for pharmaceutics. It is hypothesized that almost 99% of the microorganisms are not discovered [13]. In recent years, the potential of endophytes as sources of novel bioactive compounds have been rigorously evaluated. The diversity and specialized habituation make these endophytes an exhilarating field of study in the search for new medicines or novel drugs.

Metabolic engineering can be applied for the transformation of microorganisms and plants in order to synthesize a range of valuable metabolites that are otherwise challenging or complex to produce from their natural sources or by their chemical synthesis [14–16]. The production and modulation of pre-existing pathways, novel metabolic enzymes, and biosynthetic pathways so as to optimize the synthesis of desired products could be effectively employed by the use of technological framework and theoretic approaches to synthetic biology, system biology and evolutionary engineering [17, 18]. The techniques involved in metabolic engineering were used to boost chemical production of molecular biological entities in the "post-generic era," leading to the establishment of more specific pathways for studying cellular metabolism [19]. This led to shifting the focus of metabolic engineers from conventional techniques such as overexpression and/or deletion of endogenous genes to controlling gene expression and modulation of regulatory pathways throughout the cell [20]. The chemical synthesis of certain valuable natural plant compounds is commercially infeasible as many of them possess chemical structures with multiple chiral centers. Therefore, in order to synthesize these compounds other approaches such as exploitation of native biological pathways *via* natural harvest (codeine and morphine), semi-synthesis (taxol), heterologous production or plant cell culture techniques (taxol) need to be applied [21–23].

In this chapter, we deliver an exhaustive outline of the current knowledge on the subject and considerable emphasis is put on exploiting the astounding diversity of the endophytic world for its pharmaceutical potential. Additionally, the acuity of modulation of the complete cellular network (regulation and modification) in place of only specific pathways resulting in the concept of "system metabolic engineering" approaches for enhancing the taxol production have been discussed.

## 18.2 Taxol: History and Clinical Impact

Functional metabolites of endophytic origin have displayed a considerable potential to impact the pharmaceutical sector [4, 10, 11, 24, 25]. Among the bioactive compounds of endophytic fungi origin described to date, the ones notable for their antitumor activity have drawn special attention, with paclitaxel (taxol) as the most striking example (Figure 18.1). It is one of the most successful and widely used anticancer drugs developed in the past five decades. This compound was first found in the bark of Pacific yew tree; *Taxus brevifolia* [26]. All the species of *Taxus* are known to produce taxol; a chemical substance of tetracyclic diterpene lactam which is highly efficient, has low toxicity and is a broad-spectrum natural anticancer drug [27, 28].

Paclitaxel represents a new class of antineoplastic agent as it has a distinctive mode of action. Unlike other antimicrotubule agents like podophyllotoxin, colchicine, vinca alkaloids and combretastatin, which impede microtubule assembly, paclitaxel stabilizes microtubules against depolymerization; it promotes the polymerization of microtubules but inhibits depolymerization [29–31]. This atypical stability blocks the cell's ability to disassemble the mitotic spindle during cell division; cells are blocked in the G2/M phase of the cell cycle [29, 32, 33] and this finally leads to cell death. Taxol underwent clinical trials in the 1980s and was approved by the Food and Drug Administration (FDA) in 1992 as a very important chemotherapeutic agent for treatment of a variety of cancers [16]. While the number of cancers treated with taxol is expanding, to date it has been chiefly used to treat metastatic carcinomas of the ovary, cell lung cancer and metastatic breast cancer as well as in second-line treatment of AIDS-related Kaposi's sarcoma [34–37]. It has also been investigated for the treatment of diseases not related to cancer such as Alzheimer's or Parkinsonism [38].

## 18.3 Endophytes

Endophytes are the organisms which spend whole or part of their life cycle colonizing inter and/or intracellularly within tissues of the host plant [39–41]. As per earlier

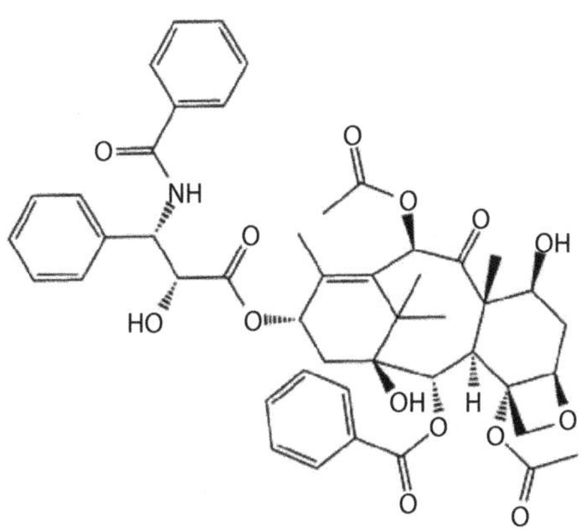


**Figure 18.1** Structure of Paclitaxel (Taxol).

research, these are defined as "the organisms which inhibit plant organs that at some time in their life colonize internal plant tissues without causing apparent harm to the host" [42, 43]. The term "endophyte" was firstly introduced in 1866 [44] and thereby included "Fungi and bacteria which, for all or part of their life cycle, invade the tissues of living plants and cause inapparent, asymptomatic infections entirely within the plant tissues but cause no symptoms of disease" [39, 45]. In literal translation, the word endophyte is derived from the Greek 'endon,' meaning within, and 'phyton,' meaning plant. These endophytes are found to be very important components of the plant micro-ecosystems [46]. They have been found to influence distribution, ecology, physiology, and biochemistry of host plants [47].

The plant endophytes spend the whole or part of their life cycle inside the healthy tissues of the host plants [10, 11, 48] (Figure 18.2). The first report describing these microbes antecede to the turn of the 19$^{th}$ and 20$^{th}$ century [49]. The most often encountered endophytes are from fungi; however, the existence of many endophytic bacteria has also been documented. Plant endophytic fungi have been found in each plant species examined and it is estimated that there are over one million fungal endophytes in nature [42, 43]. Endophytes represent the vast diversity of microbial adaptations which have developed in special environments. The diversity and distinctive habituation make these organisms important in the search for novel molecules. Endophytic fungi have been acknowledged as an important and novel resource of natural bioactive products. These endophytes produce a number of bioactive compounds for helping the host plant to ensure external biotic and abiotic stresses and benefit the host growth in return [50–52]. Some endophytic fungi have developed the potential to produce the same/similar bioactive compounds as those generated from the host plants and to develop a substitutable approach for efficiently synthesizing these scarce and valuable bioactive compounds [25, 46, 53]. Many medicinal plants have been reported to be a storehouse

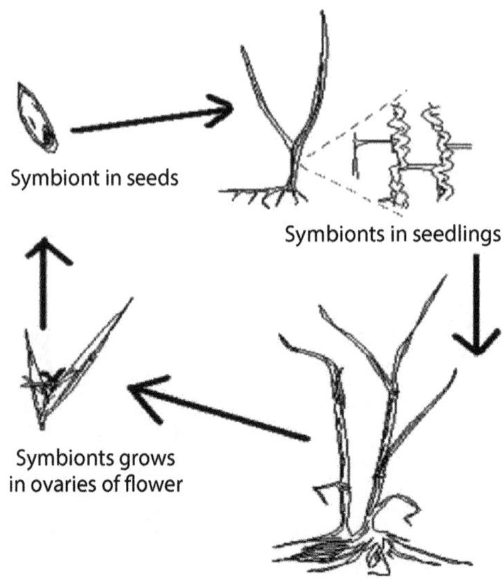


**Figure 18.2** Endophyte asexual life cycle [11].

of fungal endophytes with metabolites of pharmaceutical importance [54]. Several reviewers have shed light on natural bioactive products produced by these endophytes with potential applications in agriculture, medicine and food industry [4, 10, 11, 55].

### 18.3.1 Biodiversity of Endophytes

Many theoretical models and experimental tests described the important functions of diversity [56] including the enhancement of primary productivity, nutrient retention [57], nutrient flow [58], water availability and resistance to pathogen invasion [59, 60]. The diversity of endophytes is exhibited not only in the specificity of the hosts and their morphology, but also the type of benefits which they provide to the host [61]. Nowadays, endophytes have been isolated from diverse groups of plants ranging from large trees [62, 63], palms [64, 65], sea grasses [66] and even from lichens [67]. Most endophytes isolated to date belong to ascomycetes and their anamorphs; however, later several others were placed in basidiomycetes as well [68]. However, the colonization rate and the isolation rate of these fungi from plants vary considerably; some medicinal plants harbor more endophytes than others [69]. Some of the common endophytes not only prevailed in more plant hosts but also with high relative frequencies within the host. In contrast, some other endophytic fungi were perceived in only one given plant host [69–72].

### 18.3.2 Endophyte vs. Host Plant: the Relationship

Endophytes develop special mechanisms to penetrate and inhibit the host tissues in close association. They possess the exoenzymes compulsory to colonize their hosts and grow well in the apoplastic washing fluid of the host. To truly define the interaction between endophyte and the host plant seems to be quite a task [54, 72, 73]. Though a variety of relationships have been studied to exist between fungal endophytes and their host plants, varying from mutualistic or symbiotic to antagonistic or slightly pathogenic, based on a fine-tuned equilibrium between the demands of the invader and the plant response [55, 74–77]. It has been found that the concentrations of some plant defense metabolites are lower than in the control when the host is infected with a pathogen than with an endophyte [78]. There exists equilibrium between fungal virulence and plant defense. Schulz and co-workers documented an elegant hypothesis postulating this relationship to be a 'balanced antagonism' (Figure 18.3) [55, 78, 79].

The aforesaid perception could be interpreted as equilibrium under environmental, physiological and genetic control, resulting in fitness benefits for both partners. Any interruption in this balance, resulting from either a decrease in plant defense or an increase in fungal virulence, leads to the development of a disease. On the one hand, theoretically, fungal endophytes are considered as artists of phenotypic plasticity, capable of colonizing as endo-symbionts that inhabit cryptically and eventually sporulate as either parasites/pathogens and/or saprophytes. This creative variability suggests links to evolutionary potential. On the other hand, the possibility of secondary metabolites being a contribution of the mutualistic relationship with the endophytic partner cannot be excluded. A conflict of interests always persists at all stages of associations amongst endophytes and plant partners [80, 81]. However, for a deeper understanding of the

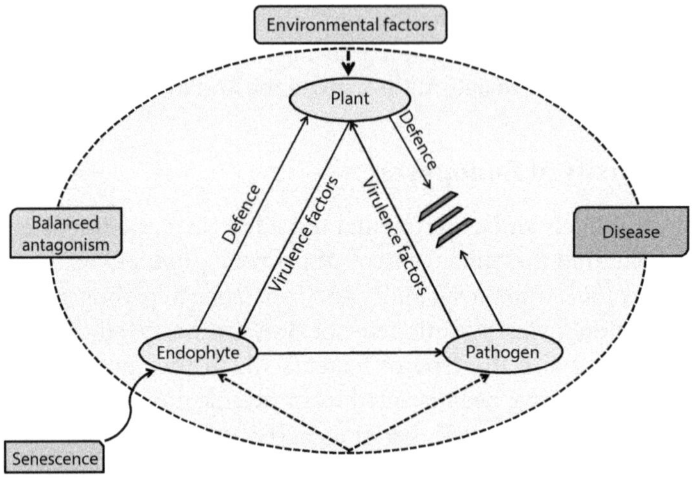


**Figure 18.3** Balanced antagonism hypothesis of endophyte vs. host plant relationship [55].

mechanism that balances virulence against the defense (hostility against hospitality), efforts should be put into the development of tools for non-invasive monitoring of sub-cellular activities during the establishment of symbiotic (mutual) interactions [77]. Endophytes synthesize metabolites in order to compete with epiphytes and/or pathogens to colonize the host and to regulate host metabolism in the balanced association.

Selection of host plant, screening, and utilization of potential endophytes involves studies on plant diversity, ethnobotany, and fungal taxonomy. The metabolic interactions of the endophyte with its host through mutualism may favor the synthesis of some similar secondary metabolites [82, 83]. Endophytes undergo long-term symbiotic relationships with their host plants and many of them may produce bioactive substances as part of these relationships. They exist in the same habitat, through long-term coexistence, and through direct contact, they might exchange genetic material [84, 85].

In order to adjust to the ecological environment, plants have developed several mechanisms to overcome microbial diseases including the production of several toxic substances. Some are available in healthy plants and some are synthesized during pathogenesis. Endophytes have a sturdy tolerance toward the host's unique metabolites. The detoxification of these highly bioactive defense compounds is an important transformation ability of many endophytes, which to a certain extent decides the colonization span of their hosts [85]. Biotransformation abilities of endophytes help in detoxification effects towards toxic metabolites produced by the host plant and production of some novel bioactive secondary metabolites [86–88]. Only with excellent biotransformation abilities, they can face the external environments directly. It is believed that the types of active compounds produced by endophytes have been much more than those produced by their host plants [85, 89]. The former has become an important origin of novel biologically active secondary metabolites with potential in different sectors of agriculture, medicine, and food. As several microorganisms have developed resistance to some of the current drugs, there has been a hunt for new drugs and endophytes have been found to produce several novel drugs like molecules. Improvement of existing drugs by modifying them with endophytes is one of the novel ways of exploiting these bioactive metabolites. Because of their effective biotransformation enzymes, endophytic fungi have been employed to change the three-dimensional conformation of compounds. Some researchers have attempted to use endophytes to obtain more active substances. Earlier studies reported that divergent metabolites could be obtained by using different types of fungi and those metabolite productions were stereoselective [90–93]. Utilization of endophytes for the region and stereoselective production of novel products allows us to procure novel compounds which cannot be synthesized by chemical ways. So, natural product drugs generated as microbial secondary metabolites exhibit a number of properties that make them excellent candidates for industrial processes [88, 94]. The endophytes in culture also offer a higher yield of secondary metabolites upon subjecting to strain improvement program [95, 96]. These factors encourage us to explore and study different groups of fungi from different biotopes in order to utilize their biotechnological promise. Documented plant species should also be evaluated from the point of their distribution and taxonomy and also for their chemical and microbial profile.

## 18.4 The Plausibility of Horizontal Gene Transfer (HGT) Hypothesis

Initially, the presence of bacterial genes in phagotrophic eukaryotes was supported by the 'you are what you eat hypothesis' [97, 98]. Contrary to this, the existence of bacterial genes in nonphagotrophic organisms (including members of the fungal kingdom), exhibited the involvement of mechanisms other than phagocytosis are responsible. While enthusiasts call it 'the essence of the phylogenetic process and the driving force in a new paradigm for evolution' [97, 98], skeptics delineate it as no more than one of many phylogenetic anomalies [99]. Horizontal (or lateral) gene transfer (HGT) is defined as the exchange and stable integration of genetic material between different strains or species [97, 98]. HGT differs from vertical gene transfer, which is the normal transmission of genetic material from parent to offspring. Often observed in prokaryotes and until recently HGT was supposed to be of limited importance to eukaryotes. However, pieces of evidence strongly suggest that in eukaryotic genome evolution (particularly in unicellular organisms) HGT plays an important role. The transfer of individual genes, gene clusters or entire chromosomes can have a significant influence on niche specification, disease emergence or shift in capabilities of metabolites. The fact that the fungal kingdom is one of the most densely sampled eukaryotic lineages and stands at the forefront of eukaryote comparative genomics in context to genomic sequencing, equips and encourages us to use fungi to study eukaryotic evolutionary mechanisms including HGT. The hypothesis of HGT with all associated controversies still seems to be quite fascinating. Two intriguing examples from endophytes have been well studied, and in both cases, the host and the invader concept revealed the occurrence of identical natural products in unrelated taxa. The potent cytotoxic agents, maytansinoids were first detected in *Maytenus serrata* [100]. In later investigations the occurrence of maytansinoids was detected not only from higher plants [101–104], but also in Gram-positive *Actinomycetes* [105, 106] and in mosses [107, 108]. In support, it can be assumed that through the course of evolution the biosynthesis of these natural products has been repeatedly developed. However, it is highly unlikely to believe, as the fact remains that approximately 48 genes are involved in the bacterial synthesis of maytansinoids [109]. Similarly, the aforementioned ubiquity of paclitaxel manifestation in yews as well as in taxonomically distant fungi raises questions. Therefore, HGT during the course of evolution seems a reasonable explanation that took place between taxonomically unrelated species, thus describing the distant distribution of the antineoplastic secondary metabolites.

Nevertheless, before supplicating horizontal gene transfer, alternative and equally possible explanations need to be thoroughly considered. In the case of maytansinoids, all affirmations seem to suggest that plant-associated microorganisms are ultimately producing these compounds. Maytansine, the unique parent compound, was absent in cell suspension cultures from *Maytenus buchananii* [110] as well as in callus cultures raised from *Maytenus wallichiana* [111] and *Putterlickia verrucosa* [112, 113]. This matches with the result obtained from an in-depth search for the unique gene involved in maytansinoid biosynthesis, mainly focused on 3-amino-5-hydroxybenzoic acid (AHBA) synthase, in *P. verrucosa*cell cultures. When subjected to a comprehensive PCR based homology screen, results were found negative, indicating that plants do not produce maytansinoids *ab initio* [72, 114]. However, an active role of the plant in an overall biosynthesis cannot be eliminated as the host converts a bacterially synthesized precursor into the final biologically active compound. Secondly, the fact remains that maytansine is only produced in response to a pathogen attack on the plant. Hence there is the possibility that plants may contain a bacterially produced precursor in biologically inactive form, which is converted into the potent final product only after a pathogen attack which triggers a response signal. Alternatively, the activation of maytansinoid precursor triggered by a plant signal in response to the pathogen aggression could be the reason for bacterial maytansinoid production [115, 116].

On the contrary, the production of paclitaxel by sterile cell suspension cultures of *Taxus* species provides adequate evidence supporting the synthesis of this diterpenoid to be a genuine feature of the yew host [16, 117–122]. Further validation to this was provided by the work of Croteau and his associates who successfully isolated paclitaxel biosynthetic genes of plant origin [123]. The taxadiene synthase (*ts*) gene with long N-terminal targeting sequence play a role in localization and processing in the plastids, clearly indicating that this gene is plant-derived rather than of a fungal origin [124–126]. Accordingly, an extensive PCR-based screening for microbial *ts*, *dbat* (encoding 10-deacetylbaccatin III-10-O-acetyltransferase) and *bapt* (encoding C-13 phenylpropanoyl side chain-CoA acyltransferase) using the designed PCR primer based on the conserved regions of these key genes of taxol biosynthetic pathway in yew provides essential evidence for the molecular blueprint of taxol biosynthesis being an inherent genetic trait of endophytic fungi [127–130]. Moreover, the biochemical detection of taxol production affords definitive proof for the presence of taxol biosynthetic pathway in endophytic fungi, supporting the fact that evolutionary trajectory of taxol gene cluster between microbial and plant origin might be coexisting.

Very meager incidences of eukaryote (non-fungal) to fungal HGT have been reported and current evidence suggests that rates of HGT into and between fungi are relatively low; therefore, reconstructing the fungal tree of life [FTOL] is a viable endeavor [131]. Furthermore, there is not much evidence yet to suggest that fungal HGT has been so rampant that it undermines a tree of life outlook, replacing it with a web of life hierarchy similar to what is observed in prokaryotes. To recapitulate the evidence for lateral gene transfer in eukaryotes remains largely anecdotal [132, 133]. Although tempting and attractive, the HGT hypothesis has yet to give a route to a more plausible alternative to postulate the endophyte-host co-evolution.

## 18.5 Endophytes as Biological Factories of Functional Metabolites

Diverse endophytic fungi inhabit the plants, representing an abundant resource of bioactive natural compounds with potential for exploitation in the pharmaceutical and agricultural field [78]. However, it is observed that most of the endophytic fungal diversity remains uncovered [69]. Many of the compounds produced by these fungi are biologically active and comprise alkaloids, flavonoids, steroids, terpenoids, peptides, polyketones, quinols, and phenols as well as some chlorinated compounds. Until 2003, approximately 4,000 secondary metabolites with biological activity have been reported from fungi [118, 135]. Most of these metabolites are produced by species of *Acremonium*, *Aspergillus*, *Fusarium*, and *Penicillium*. Another group of researchers isolated around 6,500 endophytic fungi and tested their biological potential [78]. They analyzed 135 secondary metabolites and found that 51% of bioactive compounds (38% for soil isolates) isolated from endophytic fungi were new natural products. These workers deduced that endophytic fungi are a good source of novel compounds and that "screening is not a random walk through a forest".

A large number of secondary metabolites have been extracted and characterized from different endophytic microbes [11, 88, 94, 134, 136]. In some cases, plant-associated fungi are able to make the identical bioactive metabolites as the host plant itself. One of the best examples of this is the discovery of phytohormones "gibberellins" in *Fusarium fujikuroi* in the early 1930s [137, 138]. Another crucial example is that of taxol (currently used as an anti-cancer agent) which was first found in the bark of the Pacific yew tree [26]. The discovery that taxol could be produced by endophytes of the Yew tree (*Taxus* sp.) lead to an outburst of endophyte studies worldwide on a number of other medicinal plants [3, 11, 88, 94, 139].

Fungal endophytes render important roles in the biosynthesis of secondary metabolites. The combination of inducing factors, including both plants and endophytic fungi increased the accumulation of secondary metabolites in plants and fungi, respectively [140–142]. Biosynthetic pathway studies unveil that plants and endophytic fungi have similar but distinct metabolic pathways for production of secondary metabolites [143]. Independent production of taxol by endophytic fungi has been shown by the isolation of the gene 10-deacetylbaccatin-III-10-O-acetyl transferase from the endophytic fungus *Cladosporiumcladosporiodes* MD2 isolated from *Taxus media* [140, 141]. This gene is involved in the taxol biosynthetic pathway and shares 99% identity with *T. media* and 97% identity with *Taxus wallichiana* var. *marirei*. A few studies showed that endophytes associated with non-taxol-producing plants (Yew species) have also been found to produce taxol. An endophytic taxol-producing fungus *Colletotrichum gloeosporioides* isolated from the leaves of a medicinal plant, *Justicia gendarussa*, produced a high amount of taxol [144]. While studies of another group of researchers revealed the endophytic association of *Colletotrichum* species as endophytes most frequently isolated from *T. mairei*, have not yet been reported as endophytes of *Taxus* though they have been reported as usual endophytes from other plants [65, 145–147]. Several other bioactive natural products of industrial importance have been found to be produced by different endophytic fungi [148–151].

## 18.6 Taxol Producing Endophytic Fungi

The hunt for plant-derived natural products as potential anticancer agents dates back to 1550 BC, but the scientific era of this search is much more recent, emerging in the 1950s with the discovery and development of the camptothecin, vinblastine, vincristine, and cytotoxic podophyllotoxins (Figure 18.4) [4, 137, 139, 152–156]. In comparison with

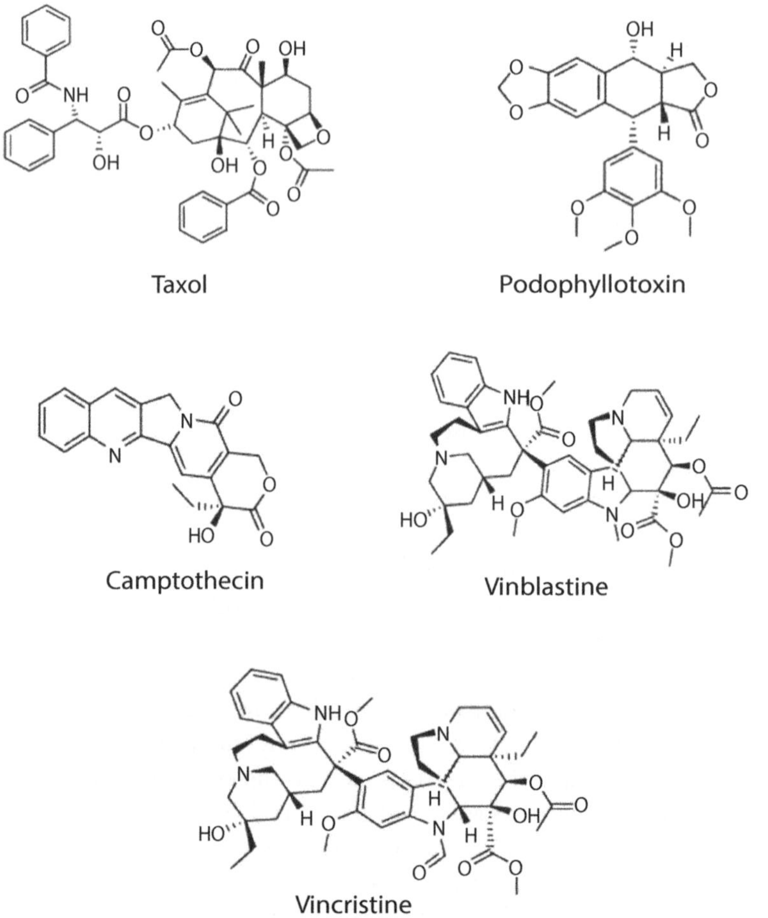


**Figure 18.4** Structures of some industrially relevant secondary metabolites used as anticancer compounds.

other antineoplastic agents, the binding dynamics of taxol with tubulin are peculiar. The former enhance microtubule disassembly whereas taxol promotes assembly and stabilization [157, 158]. Because of its unique ability to bind specifically to β-tubulin and its cytotoxicity at lower concentrations, taxol is being used for the treatment of several classical tumors. Taxol has been successfully employed in clinical treatments of advanced, progressive and drug-refractory ovarian cancer [34, 159] and breast cancer [36]. Presently, it is also used for the treatment of lung [160, 161], head and neck [162, 163], renal, prostate, colon, cervix, gastric and pancreatic cancers [159, 164, 165].

Production of a plant-based natural drug is incessantly not up to the desired level. Moreover, it is generated at a specific developmental stage or under specific environmental condition, stress or nutrient availability. Also, the plants may be very slow growing drawing several years to acquire a suitable growth phase for product accumulation and extraction. The same is the case with taxol, whose supply from the bark is limited (0.01–0.05 %) [166] because the plant is not abundantly found in nature [167], grows slowly taking several decades to increase a few inches in diameter [168, 169] and contains trace amounts of Paclitaxel (0.01% of dry weight of the bark) [60, 170]. Thus the supply of this potent antineoplastic agent was soon to become a matter of scarcity, as even early estimations indicated that the demand of paclitaxel for clinical treatment of cancer(s) might exceed 300 Kg, which would amount to 750,000 yew trees per year [171, 172] i.e.,1 Kg of paclitaxel is produced after extraction from 10,000 Kg of *Taxus* bark [173] and a cancer patient needs approximately 2.5–3.0 gm of paclitaxel for the treatment [174]. Additionally, extraction of taxol from yew trees requires a complex system and specific purification techniques using advanced and expensive technology. The supply catastrophe, as well as the ecological implications resulting in the plant-endangered distinction, prompted the search for the development of alternative resource and a potential strategy to fulfill the growing demand of the market and make the drug more widely available.

In the last 40 years, many effective approaches such as field cultivation, plant cell, and tissue culture, chemical synthesis for paclitaxel production have been developed and much progress has been achieved (Table 18.1) [149, 175–178]. However, these approaches for the production of taxol are not realistic as they come with a number of problems: they are time-consuming, product yield is lower and even the processes are non-economical.

After several years of continuous efforts, a novel paclitaxel-producing endophytic fungus *Taxomyces andreanae* was successfully discovered from the Pacific yew in 1993 [179]. This colossal finding firstly revealed that the plant endophytic fungi also had the potential to produce paclitaxel, giving us a novel and promising approach to producing this valuable compound. Since then many scientists throughout the globe have been expanding their interests in exploring fungal endophytes as potential candidates for producing paclitaxel. Comprehensive research for Paclitaxel-producing endophytic fungi from *Taxus* species as well as from other related plant species, microbial fermentation processes and genetic engineering for improving paclitaxel production has been developed and much advancement has been achieved during the past two decades. Table 18.2 shows varying yields of taxol production by various endophytic fungi. In order to deal with low taxol yield, current studies need to be focused on isolation and identification of high taxol-producing cell lines, the stable yield of taxol and optimization

**Table 18.1** Various biotechnological approaches for the production of taxol other than obtaining it from yew bark and other plant parts.

| Plant species | Culture type | Taxol yield | References |
|---|---|---|---|
| *Taxus cuspidata* | Callus | 0.020 % DW | [175] |
| *Taxus brevifolia* | *In vivo* | 0.01 % of dry weight of the bark | [170] |
| *Taxus media* | Cell suspension | 115.2 mg/L | [120] |
| *Taxus chinensis* | Elicitors (*Aspergillus niger)* | 2-fold increase to control | [12] |
| *Taxus x media* | Hairy root culture | Twice the amount of taxol than that in the bark of *T. brevifolia* | [162] |
| *Taxus chinensis* | Cell cultures in bioreactors | 612 mg/L | [149] |

of fermenting conditions. Strain improvement and optimization of the media of taxol-producing fungus *Fusarium maire* has been discussed previously [180, 181]. The establishment of efficient transformation system of taxol-producing endophytic fungus EFY-21 (*Ozonium* sp.) from *T. chinensis* var. *mairei* led to improved taxol production. Till now around 200 endophytic fungi belonging to more than 40 fungal genera from several different orders representing mostly Ascomycota and Deuteromycota, with only a few from Basidiomycota and Zygomycota, have been reported to produce taxol [72, 114]. Table 18.2 summarizes a list of taxol-producing endophytic fungi isolated from 1993 till 2017.

## 18.7 Molecular Basis of Taxol Production by *Taxus* Plants (Taxol Biosynthetic Pathway)

As a natural diterpenoid, taxol is formed exclusively from geranyl geranyl diphosphate (GGPP), which is synthesized from three isopentenyl diphosphate (IPP) molecules and the isomer dimethyl diphosphate (DMAPP) by the enzyme geranyl geranyl diphosphate synthase (Figure 18.5). This enzyme is of special interest as it leads to the formation of a branched point progenitor of a variety of diterpenoids and tetraterpenoids. According to previous research, the IPP involved in the biosynthesis of the taxane ring is formed by the plastidic route [182]. However, other studies [16, 183–185] have shown the involvement of the cytosolic pathway. A previous research study also suggested that cytosolic IPP could play a role in taxol production in the initial growth phase of *Taxus* cells [186]. Additionally, another research study reported that after supplementing *T. chinensis* cell suspensions with two inhibitors of metabolite translocation, suggested that translocation of IPP through the plastidic membrane only occurs during the late growth phase of the culture [187]. A recent study in *T. baccata* cell cultures showed that while taxol biosynthesis was blocked by the addition of fosmidomycin (an inhibitor of the plastidic

**Table 18.2** A list of taxol-producing endophytic fungi isolated from yew associated and non-associated plants (1993–2017).

| Endophytic fungus | Host | Taxol yield | References |
|---|---|---|---|
| *Taxomyces andreanae* | *Taxus brevifolia* | 24–50 ng/L | [179] |
| *Pestalotiopsis microspora* | *Taxus wallichiana* | 60–70 μg/L | [139] |
| *Pestalotiopsis microspora* | *Taxodium distichum* | 0.05–1.49 μg/L | [227] |
| *Monochaetia* sp. & *Fusarium lateritium* | *Taxus baccata* | 102 & 130 ng/L | [139] |
| *Alternaria* sp. & *P. microspora* | *Taxus cuspidata* | 157 & 268 ng/L | |
| *Pestalotiopsis microspora* | *Taxus wallichiana* | 500 ng/L | |
| *Pithomyces* | *Taxus sumatrana* | 95 ng/L | |
| *Pestalotia bicilia* | *Taxus baccata* | 1081 ng/L | |
| *Tubercularia* sp. | *Taxus chinensis* var. *mairei* | 185 μg/L | [149] |
| *Taxomyces* sp. | – | 2.3 μg/L | [187] |
| *Rhizoctonia* sp. | – | 1.43 μg/L | [55, 236] |
| *Phoma* | – | 32.93 μg/L | |
| *Botrytis* | – | 4.092 μg/L | |
| *Penicillium* | – | 8.24 μg/L | |
| *Trichoderma* | – | 19.586 μg/L | |
| *Mucor* | – | 1.08 μg/L | |
| *Chaetomium* | – | 21.1 μg/L | |
| *Alternaria alternate* | – | 84.5 μg/L | [11, 60] |
| *Ozonium* sp. | *T. chinensis* var. *mairei* | 4–18 μg/L | [180] |
| *Pestalotiopsis* | – | 8.5 μg/L | [94, 144] |
| *Fusarium* | – | 2.7 μg/L | [89, 55] |
| *Fusarium mairei* | – | 20 μg/L | [169, 18] |
| *Fusarium solani* | *Taxus celebica* | 1.6 μg/L | [55] |
| *Cladosporium cladosporioides* | *Taxus media* | 800 μg/L | [140] |
| *Fusarium solani*, Tax-3 | *Taxus chinensis* | 163.35 μg/L | [169, 89] |
| *Aspergillus niger* var. *taxi* | *Taxus cuspidata* | 273.46 μg/L | [129] |
| *Pestalotiopsis versicolor* | *Taxus cuspidata* | 478 μg/L | [204] |
| Ozonium sp. (EFY-21) | *T. chinensis* var. *mairei* | - | [72] |
| *Stemphylium sedicola* | *Taxus baccata* | 6.9 μg/L | [150] |
| *Guignardia mangiferae*, *Fusarium proliferatum* & *Colletotrichum gloeosporioides* | *Taxus x media* (Anglojap yew) | 720, 240 & 120 ng/L | [203] |
| *Fusarium redolens* | *Taxus baccata L. ssp. wallichiana* | 66 μg/L | [60] |

(*Continued*)

**Table 18.2** Cont.

| Endophytic fungus | Host | Taxol yield | References |
|---|---|---|---|
| *Fusarium tricinctum, Gibberella avenacea, Microdiplodia sp. and Paraconiothyrium brasiliense* | *Taxus baccata L. ssp. wallichiana* | 23.47, 11.03, 27.4 & 19.6 μg/L | [60, 169] |
| *Fusarium* | *Taxus wallichiana var. mairei* | 0.0153 mg/L | [145] |
| *Paraconiothyrium variabile & Epicoccum nigrum* | *Taxus baccata* | 1.75 and 1.32 μg/L | [181] |
| *Epicoccum nigrum* | *Taxus baccata* | 57.1±11.8μg/L | [181] |
| *Phoma medicaginis* | *Taxus wallichiana var. mairei* | 1.215 mg/L | [145] |

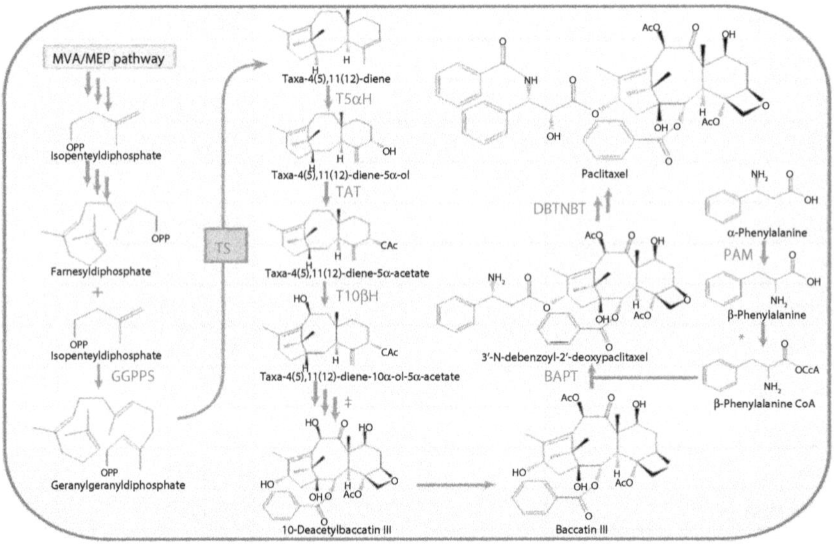


**Figure 18.5** Established biosynthetic pathway for Taxol production from *Taxus* species. Abbreviations: MVA: mevalonic acid; MEP: 2-C-methyl-D-erythritol-4-phosphate; GGPPS: geranylgeranyldiphosphate synthase; TS: taxa-4(5),11(12)-diene synthase that catalyzes the committed step of this pathway; T5αH: taxa-4(5),11(12)-diene-5α-hydroxylase; TAT: taxa-4(5),11(12)-diene-5α-ol-O-acetyltranseferase; T10βH: taxane-10β-hydroxylase; ‡: 'oxetane ring' formation and branch migration enzymes including taxane 2α-O-benzoyltransferase (T2BT or DBBT = debenzoyltaxane-2′-α-O-benzoyltransferase) as well as C-13 hydroxylation and steps taking pathway flux towards non-Taxol-type molecules; DBAT: 10-deacetylbaccatin III-O-acetyltransferase; BAPT: baccatin III 13-O-(3-amino-3-phenylpropanoyl) transferase; DBTNBT: 3′-N-debenzoyl-2′-deoxytaxol-N-benzoyltransferase which follows hydroxylation in the side chain by an unknown enzyme; PAM: phenylalanineaminomutase; *: β-phenylalanine coenzyme A ligase. Multiple arrows imply more than one biosynthetic step. The Taxol biosynthetic pathway is proposed to have about 20 different enzymatic steps in *Taxus* plants [72, 73].

pathway), it was also reduced by mevinolin (an inhibitor of the cytosolic pathway), indicating that both pathways could be involved in the biosynthesis of taxol [16, 185].

The first committed step of taxol biosynthesis is the cyclization of GGPP to the taxa-(4,5),[11, 12]-diene, a reaction catalyzed by taxadiene synthase (TS), a monomeric protein of 79 kDa (Figure 18.6). This enzyme was purified and characterized and the gene that codifies for TS has been cloned and functionally expressed in *E. coli* [88, 123, 188]. Afterwards, oxygen and acyl groups are added to the taxane core by oxygenation at multiple positions mediated by cytochrome P450 mono-oxygenases. The hydroxylation at the C5 position of the taxane ring by the enzyme cytochrome P450 taxadiene-5α-hydroxylase (T5αH) results in the formation of taxa-4(20),11(12)-dien-5α-ol, which is the second step in taxol biosynthesis [189]. T5αH is a protein of 56 kDa with an N-terminal sequence of insertion in the membrane of the endoplasmic reticulum. This enzyme, apart from its hydroxylating activity, also conditions the migration of the double bond from 4(5) to 4(20). Although these two metabolic steps, cyclization and hydroxylation; are slow, they do not seem to be rate-limiting in taxol biosynthesis [190, 191].

The next step in the pathway is catalyzed by a specific taxadiene-5α-ol-O-acetyl transferase (TAT) that acylates taxa-4(20),11(12)-dien-5α-ol at the C5 position to form taxa- 4(20),11(12)-dien-5α-yl-acetate. This enzyme is a protein of 50 kDa that bears no N-terminal organellar targeting information [16, 192]. The product of this reaction is then hydroxylated by taxoid 10β-hydroxylase (T10βH) at C10. T10βH is a P450-dependent monooxygenase cloned and functionally characterized in yeast [193]. Another CYP P450-dependent hydroxylase leading to the formation of taxa-4(20), 11(12)-dien-5α-13α-diol has been found [143, 194, 195]. The fact that this enzyme uses the same substrate as TAT, the taxa-4(20), 11(12)-dien-5α-ol suggests that taxol biosynthesis is not a linear pathway and that there are branch points that can lead to other related taxoids. It has been observed that this alternative step is especially frequent in cell cultures elicited with methyl jasmonate [16, 196]. The taxoid 14β-hydroxylase (T14βH) is responsible for the formation of taxa-4(20), 11(12)-dien-5α-acetoxy-10β-14β-diol [194, 195]. This enzyme does not use substrates already hydroxylated at the C13 position but only those hydroxylated at the C10 position, suggesting that T14βH cannot be involved in the production of taxol, which does not present any hydroxylation at the C14 position.

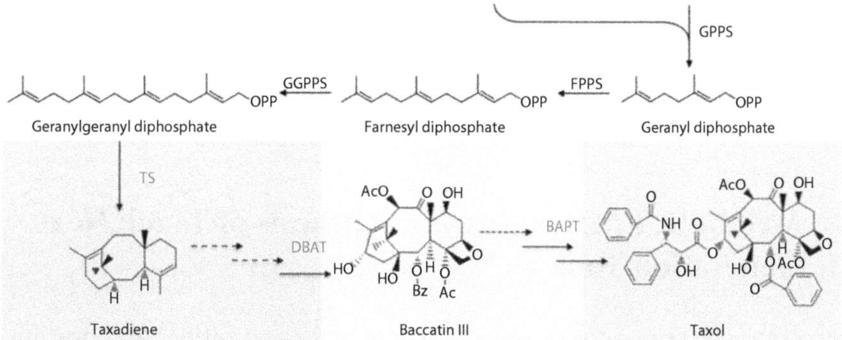


**Figure 18.6** Key genes involved in Taxol biosynthesis in *Taxus* species. Abbreviations: GGPPS: geranylgeranyldiphosphate synthase; TS: taxa-4(5),11(12)-diene synthase that catalyzes the committed step of this pathway; DBAT: 10-deacetylbaccatin III-O-acetyltransferase; BAPT: baccatin III 13-O-(3-amino-3-phenylpropanoyl) transferase. Multiple arrows imply more than one biosynthetic step [203].

The last steps in the taxol biosynthetic pathway, after the formation of taxa-4(20),11(12)-dien-5α,10β-diol 5-acetate, include several hydroxylations at the C1, C2, C4 and C7 positions, oxidation of C9 and epoxidation at the C4-C5 double bond. It is known that the hydroxylations are mediated by CYP P450 enzymes but not exactly in which order. Taking into account the oxidation frequency of the taxoids found in cell cultures, a probable sequence validated by phylogenetic analyses of previously cloned taxoid P450 oxygenases could be C5, C10, C2, C9, C13, C7 and finally C1 [197, 198]. However, rather than intermediates in taxol biosynthesis, some of these taxoids might be commodities of the *in vitro* cultures.

Although different mechanisms for the oxetane ring formation have been proposed [143, 191, 199], it is currently accepted that the process involves epoxidation of the 4(20) double bond followed by migration of the α-acetoxy group from the C5 to the C4 position together with the expansion of the oxirane to the oxetane group. It is possible that this step precedes hydroxylation at C1 in taxol biosynthesis, and in this case, the hypothetical polyhydroxylated intermediate would be a taxadiene-hexaol rather than a heptaol hydroxylated at C1 [178, 191]. The enzyme that epoxidates the C4-C20 double bond has not yet been functionally characterized and the expansion of the oxirane-to-oxetane ring is also an incompletely known step. After the formation of the hypothetical polyhydroxylated precursor by the activity of the enzyme 2α-O-benzoyl transferase (DBT), a protein of 50 kDa, the next compound obtained is 10-deacetylbaccatin III. Another identified transacetylation reaction in the taxol biosynthetic pathway involves hydroxylation at the C10 position of the 10-DAB (10-deacetylbaccatin III), which is catalyzed by the enzyme 10-deacetyl-baccatin III-10-O-acetyl transferase (DBAT). It leads to the formation of a diterpene intermediate, baccatin III, using 10-DAB and acetyl CoA as substrates (Figure 18.6).

An essential step in the taxol biosynthesis is the esterification of the C13 hydroxyl group of baccatin III with the β-phenylalanoyl-CoA side chain. The side chain is obtained from the amino acid β-phenyalanine by the action of phenylalanine aminomutase (PAM) [16, 200]. An unknown ester CoA ligase probably activates the compound so it can bind to baccatin III. The enzyme that catalyzes the conjugation of the β-phenylalanoyl-CoA side chain to baccatin III is C-13-phenylpropanoyl-CoA transferase (BAPT), yielding the compound 3′-N-debenzoyl-2′-deoxytaxol (Figure 18.6). This compound, by the action of an unknown CYP P450-dependent hydroxylase that hydroxylates the C2′ position and the enzyme 3′-N-debenzoyl-2′-deoxytaxol N-benzoyl transferase (DBTNBT) that conjugates benzoyl-CoA to 3′-N-debenzoyl-2′-deoxytaxol, yields taxol as the final compound. This enzyme can be exploited to improve the production of taxol in genetically engineered systems [201–203].

## 18.8 Metabolic Engineering for Synthesis of Taxol: Next Generation Tool

Nearly twenty-five years ago the employment of biological entities for boosting chemical production received attention with the conception of metabolic engineering [204]. The next two decades empowered by the endless developments in molecular biology methodologies referred to as the 'post-generic era', lead to accumulation of huge volume of the gene, protein and metabolic data, decreasing the overall cost of oligonucleotide production and an establishment of more specific/precise pathways of studying cellular metabolism. This led to shifting the focus of metabolic engineers from perturbing single pathways to modifying the entire cell thereby generating the concept of 'system metabolic engineering' [204, 205]. The focus of metabolic engineering 'toolbox' has rapidly moved from conventional techniques such as overexpression and/or deletion of endogenous genes to controlling gene expression and modulation of regulatory pathways throughout the cell [20]. The *in vivo* and *in vitro* techniques have been used to diversify plant secondary metabolites that possess potential pharmacological application in order to generate desired compounds libraries [206]. Plants serve as a perfect system for metabolic engineering as they can be cultivated economically and with relative ease. Their (plant) cell factories can be exploited massively in order to scale-up the production of high valued secondary metabolites such as antineoplastic drug Taxol, synthesized from *Taxus* sp. but in minute amount [19]. As taxol consists of complex molecular architecture, therefore, chemical production of it is challenging. Wide range of techniques such as taxol cell culture, microbial metabolic engineering and heterologous yew plant (*Taxus* sp.) metabolic engineering have been exploited by scientists throughout the world in order to streamline an approach in protecting the limited resources of *Taxus sp*. and make it more economical [21, 22, 198].

### 18.8.1 Plant Cell Culture

Plant cell cultures, comprising hairy root cultures and cell suspension cultures, through metabolic engineering can offer a probable production alternative in comparison to traditional cultivation methods [207, 208]. The production of medicinally important Phytochemicals *via* cell cultures can be optimized utilizing minimal growing space and manipulated with the solicitation of elicitors and other stresses accordingly in order to exploit the synthesis of targeted metabolites [19, 209]. Plant cell cultures systems for *Taxus* sp. are one of the most extensively studied and commercially successful culture systems for the production of paclitaxel [208, 210, 211]. In spite of extensive metabolic engineering efforts to obtain high yields of paclitaxel still, the *Taxus* sp. cells demonstrate low and variable productivity of metabolite. For example, genetic transformation of mature *T. brevifolia* and *T. baccata* by a wild strain of *Agrobacterium tumefaciens* and hairy root culture of *Taxus* sp. by *A. rhizogenes* resulted in low yield [212–214].

However, when the hairy root culture of *Taxus* sp. was elicited with methyl jasmonate (100 μM) the accumulation of paclitaxel was almost doubled as compared to as obtained from the bark of *T. brevifolia* [214]. It has been previously reported that methyl jasmonate induced the activity of geranylgeranyl diphosphate synthase (GGPS) and taxane accumulation in *Taxus baccata* cell suspension cultures [215]. However, methyl jasmonate was effective to elicitate the *Taxus* sp. suspension culture in order to synthesize paclitaxel but to only varying degrees.

The yield of paclitaxel from *Taxus* sp. cultures could be optimized and variability regulated after exploiting the rational engineering strategies involved in regulation of gene expression. These researchers after analyzing the biosynthetic pathway after elicitation with methyl jasmonate suggested the presence of additional molecular regulatory points outside the paclitaxel biosynthesis. They compared gene expression between cultures accumulating varying levels of paclitaxel in order to comprehend the interactions amongst the paclitaxel biosynthetic pathway and other molecular pathways so as to propose effective targets for metabolic engineering [216]. Thus at present, only few reports are available related to use of plant cell cultures for metabolic engineering because of the complications involved firstly in unfolding the precise biosynthetic pathways of some secondary metabolites and secondly the precise sequence of specific rate limiting steps involved for *Taxus* sp. cell systems [215, 217]. Further attempts for paclitaxel synthesis engaging metabolic engineering have been carried out involving cloning of specific enzymes such as taxadiene synthase, GGPS and taxa-4(20),11(12)-dien-5-ol-O-acetyl transferase [218, 219]. The increase in the yield of paclitaxel could be improved by the use of antisense technology wherein the over-expressing genes controlled the slow steps and suppressed the synthesis of unwanted metabolites thereby enhancing the titers of desired toxoids [215].

The production of new drugs, for example, 9-dihydrotaxol and/or 10-deacetyl-9-dihydrotaxol directly in order to synthesize additional precursors and rare toxoids can be employed in future for the transgenic redirection of the key pathway in *Taxus* sp. cell cultures [219]. The *Taxus* sp. suspension cell cultures were commercially utilized for paclitaxel production (100,000 different metabolites) as early as 1994 by Phyton (Ithaca, NY, USA) and ESCA genetic (San Carlos, CA, USA) in 1994 and in 2001 by Samyang Genex (Taejon, South Korea) [215, 220, 221]. These large-scale successful industrial applications of suspension cultures of *Taxus* sp. have triggered future research pathways and new biosynthetic insights into Taxane production.

### 18.8.2 Microbial Metabolic Engineering for Synthesis of Taxol and Its Precursors

The employment of genomics and post-genomics strategies could suggest insights for molecular regulation, so as to enhance together with the productivity in addition to the selectivity of secondary products. The knowledge of key metabolic targets (genes and enzymes) within a biosynthetic pathway is essential to manipulate the pathway and enhance its secondary metabolites [216]. After redesigning of the biosynthetic pathway in an individual host, microorganisms (microbial cell factories) can synthesize highly functionized compounds by utilizing simple carbon sources without the production of toxic byproducts or high energy depletion. Microbial cell factories can synthesise a lot of secondary natural metabolities, for example isoprenoids or terpenoids [222]. Terpenoids are built from phosphorylated five carbon units (isoprene) that are mostly cyclized, rearranged and modified so as to synthesize their highly bioactive compounds. Paclitaxel is a type of diterpene that is formed after combination of two isoprene units [17]. Microorganisms such as, *Escherichia coli* (superior and robust host) and *Saccharomyces cerevisiae* (large cell size and better resistance) have been mostly employed as host because they synthesise a lot of precursors and intermediates through their central metabolic processes [14, 19, 222].

In 2001, using metabolically engineered *E. coli*, the first successful biosynthesis of taxadiene (1.3 mg per liter), a key intermediate of paclitaxel production, was reported [19, 223]. It was enzymatically extracted from isopentenyl diphosphate using genetically modifying *E. coli* by the overexpression genes encoding isopentenyl diphosphate isomerase, geranylgeranyl diphosphate synthase (GGDP) and taxadiene synthase (TXS; 3 enzymes) of terpenoid pathway [223]. In a latter study, a multivariate-modular approach was used for metabolic engineering the terpenoid pathway using genetically modifying *E. coli* that was effective in increasing the taxadiene production upto 1 gram per liter [17]. These researchers divided the taxadiene biosynthesis pathway into two modules, i.e., *E. coli* tailored an upstream isoprenoid module and a heterologous downstream terpenoid module (Figure 18.7). In upstream module, taxol and other

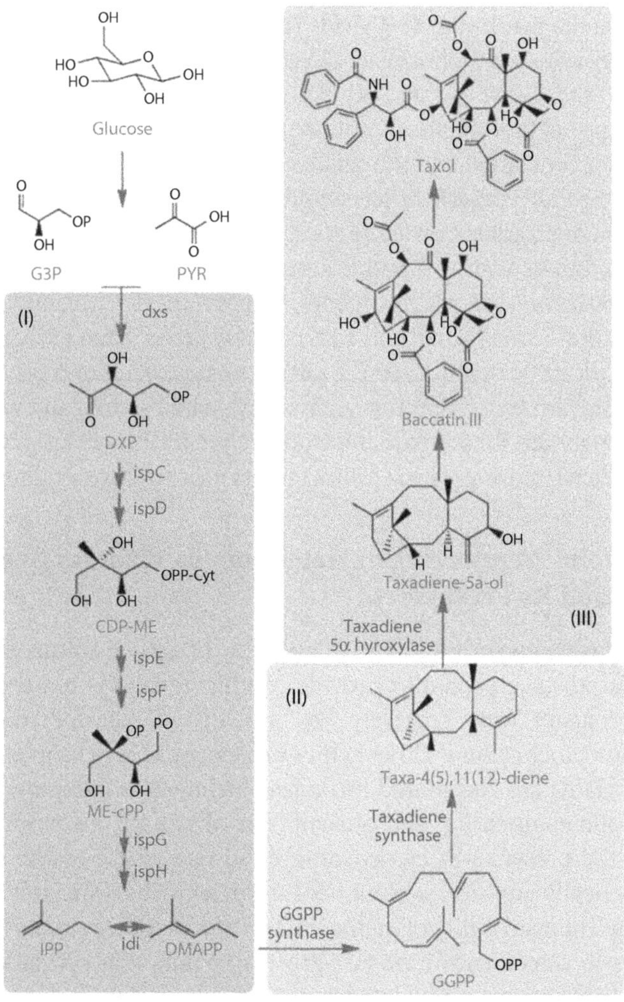


**Figure 18.7** Optimization of Isoprenoid pathway for taxol biosynthesis. **(I)** Native upstream methylerythritol-phosphate (MEP) module that generates IPP and DMAPP. Enzymatic bottlenecks have been targeted (*dxs, idi, ispD*, and *ispF*). MEP isoprenoid pathway is initiated by the condensation of the precursor's glyceraldehyde-3 phosphate (G3P) and pyruvate (PYR) from glycolysis. **(II)** A synthetic operon of downstream genes GGPP synthase and taxadiene synthase was constructed for taxol biosynthesis. The taxol pathway bifurcation starts from the isoprenoid precursors IPP and DMAPP to form GGPP, and then taxadiene. **(III)** Taxadiene undergoes stereospecific oxidations, acylations, and benzoylation to form the late intermediate Baccatin III and, after assembling the side chain, taxol [206].

isoprenoids can be extracted after the synthesis of isopentenyl pyrophosphate (IPP) and dimethylallyl pyrophosphate (DMAPP) from methylerythritol-phosphate (MEP). However, downstream module was rebuilt using *E. coli* and *S.cerevisiae* collectively and they inturn overexpressed the upstream module enzymes to produce taxol (10 mg/liter) via three enzymes utilized in terpenoid pathway [17, 222, 223]. These two modules were further optimized in order to obtain maximum production of taxol and decrease the accumulation of indole that acted as the inhibitor of the terpenoid pathway. The next step of paclitaxel biosynthesis was then engineered *via* P450-mediated 5a-oxidation of taxadiene to taxadien-5a-ol [17, 206].

In another study, an *in silico* modelling used different hosts and carbon sources to evaluate and compare the maximum IPP yields (theoretical) and the thermodynamic properties of the 1-deoxylulose 5-phosphate (DXP) pathway and the MVA pathway. These researchers found that *E.coli* and its DXP pathway was most effective in synthesis of IPP with almost 870 mg per liter (highest amount in heterologous host) extraction of taxadiene [224]. They further concluded that the genetic manipuled DXP pathway was highly effectively in enhancing the synthesis of terpenoids in the heterologous host. Another study for the production of taxadiene varied specific inducible promoters (T7, Trc, and T5) and the cellular backgrounds within K and B derived *E. coli* strains. They used T7 promoter system for the upstream MEP pathway genes, whereas the T7, Trc, and T5 promoters for downstream plasmid-borne GGDP and TXS pathway genes. After evaluation of the results the K-derived *E.coli* strain synthesized 2.5 fold more taxadiene as compared to B-derived *E.coli* strain. A study on transcriptomic analysis evaluated significany variability in pyruvate metabolism between these two *E.coli* strains, thereby establishing the discrepancy in synthesis of taxadiene and suggesting future targets for metabolic engineering [222, 225].

### 18.8.3 Metabolic Engineering in Heterologous Plant for Synthesis of Taxol and Its Precursors

Enhancing paclitaxel biosynthesis by manipulating *Taxus* cell culture systems and engineered microbial hosts appeared to provide promising results but the yields obtained from both approaches were relatively low and insufficient for routine production [226, 227]. Plant cytochrome P450 enzymes are essential for complete biosynthesis of the taxol. Thus plants are considered more feasible hosts as compared to microorganisms for metabolic engineering and biosynthesis of natural medicinal plant products (Taxol) due to the presence of cytochrome P450-related biosynthetic genes [19]. In order to metabolically engineer a plant, its system must be amenable to stable genetic transformations. The availability of inefficient transformation systems for *Taxus* sp. generated pathways to screen substitute fast-growing plants that can be easily genetically manipulated for production of taxol and/or its precursors [19, 22, 226]. In 2004, the first committed step for isoprenoid biosynthesis was explored after metabolic engineering of *Arabidopsis thaliana*. The constitutive expression of a recombinant *Taxus baccata* taxadiene synthase (TXS) gene (lacking plastid-localized sequence) fused to a C-terminal histidine tag in A. *thaliana* resulted in taxadiene synthesis with growth retardation and decreased accumulation of carotenoids in transgenic plants. The decreased accumulation and interference of the photosynthetic pigments indicated the relocation of GGDP pool in these transgenics.

In another study, using transgenic tomato fruit the effective production of TXS led to redirection of photosynthetic pigment (carotenoid) metabolism. In tomato fruit the production and use of GGDP is exclusively used to synthesize the coloured pigments (carotenoids) that in high levels are localized in plastids. The expression of TXS gene in this fruit line lost the ability to utilize GGDP for pigment production; however, almost 160 mg of taxadiene (pure) was extracted from 1 kg of frozen fruit of tomato [228]. The transgenic moss *Physcomitrella patens* was used to synthesize TXS gene extracted from *Taxus brevifolia* using a ubiquitin promoter. Compared to other transgenic plants utilized to produce TXS, no growth inhibition due to alteration of diterpenoid metabolic pools or deleterious phenotypic consequences was exhibited in the transgenic moss [229]. As secondary metabolities the moss can synthesise two types of diterpenoids (ent-kaurene and 16-hydroxykaurane) which together constitute 0.2% of its fresh weight. After knocking out of the moss bifunctional diterpene synthase these two diterpenoids can be eliminated, which will enhance the production of taxadiene [229, 230] providing future possibilities of metabolically engineering *Physcomitrella patens* for the large scale synthesis of Taxol and its precursors.

## 18.9 Future Perspectives

Fungi are a divergent group of organisms that are single-celled, multinucleated and multicellular filamentous. They mostly live as parasites or symbionts; however, to date the diversity of very few species has been examined. By the twentieth century, several fungal species possessing simplest eukaryotic models with incomparable benefits were documented in order to examine them in a better way through molecular biology, genetics and biochemistry approaches [231, 232]. Fungal fermentation has led to development of innumerable primary and secondary products such as antibiotics, amino acids, enzymes, vitamins, polymers, etc. [233, 234]. In the near future, fungi will also be exploited to understand and compare their exceptional life systems within higher organisms, for example, life cycle assessment, defenses against foreign invasion, modifications in response to harsh environments, senescence and finally death [231, 232].

The symbiotic fungi co-exist and mutually exchanges raw material with its host. With the passage of time the fungal physiological and biochemical features become more specific and the biosynthesis of its secondary metabolites more virtual [235]. Only after centuries of hunt/mining upon taxol producing endophytic fungi from diverse hosts, the concept of synthesizing taxol and its precursors via large-scale fermentation has been conceived. In order to boost the production and discovery of novel drugs and increase the yield of microbial processes, molecular manipulations have been carried out using mutational techniques [233]. To date, numerous studies report the use of gene cloning for taxol synthesis in *Taxus* sp. [16, 236, 237]. This has led to better understanding of the biosynthetic pathway involved in taxol synthesis in *Taxus* sp. [125]. In addition, fungi are extremely employable in carrying out molecular manipulations using biotransformation processes [233, 234]. After deep analysis of endophytic fungi it becomes easier to clone genes involved in taxol biosynthesis pathway and from that further information of its chemical structure and enzymatic reactions can be deduced. The speedy advancements made in genomics/proteomics and metabolomics in addition to clear understanding of the synthesis pathway and metabolism had made possible biotransformation of microbial strains with more accuracy. The reported taxol biosynthetic pathway engineered in *E. coli* [238–241] and *S.cerevisiae* [14, 16] utilizing metabolic engineering approaches have resulted in engineered taxadiene synthesis. However, engineering of taxol biosynthetic pathway beyond taxadiene led to total decrease in yield of paclitaxel or its precursors. For example, the lower yield was obtained for the seven consecutive gene transfers in *S. cereviseae*, recorded to be the highest number of steps involved in taxol pathway, engineered to heterologous host so far [239].

Nevertheless, instead of these unsuccessful ventures for taxol production, still researchers were able to enhance the yields of taxadiene by approximately hundred milligrams per liter along with directed transformation of taxanes intermediates using microbial enzymes have provided some routes that are worth exploring. The use of heterologous hosts along with specific optimization strategies and tools for intracellular compartment optimization, storage and efflux modulation and control of pathway regulatory elements might aid in enhancing the overall yield of paclitaxel. A clear understanding of the molecular mechanisms involved in biosynthesis and regulation of taxol is a prerequisite for conducting all such attempts. Enhancing paclitaxel biosynthesis by manipulating *Taxus* cell culture systems and engineered microbial hosts appeared to provide promising results but the yields obtained from both approaches were relatively low and insufficient for routine production [226]. The plants were found to be feasible hosts as compared to microorganisms for metabolic engineering and biosynthesis of natural medicinal plant products (taxol). However, more work needs to be done in order to obtain stable genetic transformations in plants. After unfolding the biosynthetic pathway of *Taxus* sp., it seems to be obvious that it consists of a lengthy and complex pathway, engineering and integration of which is difficult in heterologous hosts. Moreover, it is challenging to transform genes involved in the entire pathway and in addition to that regulation of paclitaxel synthesis surrounding epigenetic modulation and signaling cross-talk is evidently also a lesser explored topic.

The fungi require simpler culture medium in comparison with animal or plant cells that need biological reactors for their production. Furthermore, fungi can overcome low yield synthesis issues with limited starting raw that are mostly faced during extraction of compounds from plants. In addition, the raw material sources are also affordable with lower production costs. Earlier metabolic studies have evaluated that after optimizing fermentation conditions along with addition of precursors, elicitors and/or inhibitors can also aid in intensifying biosynthetic quantities of fungal taxol volume [242, 243]. The transgenic strains require a comparatively simple nutrient growth medium for culturing in large-scale industrial fermenters; thus their extracellular proteins secretion could be easily exploited. Taxol extracted from fungi at large scale using industrial fermenters have huge prospects in attracting wide market range along with social benefits. Even though the synthesis of paclitaxel from endophytic fungi seems cumbersome, still it presents a viable and long-term future target.

## Acknowledgements

The author wishes to thank Dr. M.S. Reddy and Dr. Anil Kumar, Department of Biotechnology, Thapar University, Patiala, India, and Dr. Yong Wang, Shanghai Institute for Biological Sciences, Chinese Academy of Sciences, Shanghai, China, for their full support and guidance throughout the writing of the chapter.

## References

1. E.J. Chrystal, S.K. Wringley, R. Thomas, N. Nicholson, and M. Hayes, *Biodiversity: New Leads for the Pharmaceutical and Agrochemical Industries*. Royal Society of Chemistry, 2007.
2. G. Strobel, *Current Opinion in Microbiology*, Vol. 9(3), p. 240, 2006.
3. G. Strobel, *Natural Products from Endophytic Microbes: Historical Perspectives, Prospects, and Guidance*, in *Chemical Biology of Natural Products*, CRC Press. p. 189, 2017.
4. G. Strobel and B. Daisy, *Microbiology and Molecular Biology Reviews*, Vol. 67(4), p. 491, 2003.
5. N.P. Keller, G. Turner, and J.W. Bennett, *Nature Reviews Microbiology*, Vol. 3(12), p. 937, 2005.
6. A.H. Aly, A. Debbab, J. Kjer, and P. Proksch, *Fungal Diversity*, Vol. 41(1), p. 1, 2010.
7. A.H.Aly, A. Debbab, and P. Proksch, *Applied Microbiology and Biotechnology*, Vol. 90(6), p. 1829, 2011.
8. V.C. Verma, S.K. Gond, A. Kumar, R.N. Kharwar, and G. Strobel, *Microbial Ecology*, Vol. 54(1), p. 119, 2007.
9. A. Porras-Alfaro, and P. Bayman, *Annual Review of Phytopathology*, Vol. 49(1), p. 291, 2007.
10. S.S. El-Hawary, A.M. Sayed, M.E. Rateb, W. Bakeer, S.F. AbouZid, and R. Mohammed, *Natural Product Research*, Vol. 31(21), p. 2568, 2017.
11. R.X. Tan and W.X. Zou, *Natural Product Reports*, Vol. 18(4), p. 448, 2017.
12. H.W. Zhang, Y.C. Song, and R.X. Tan, *Natural Product Reports*, Vol. 23(5), p. 753, 2006.
13. K.E.R. Davis, S.J. Joseph, and P.H. Janssen, *Applied and Environmental Microbiology*, Vol. 71(2), p. 826, 2005.
14. B Engels, P. Dahm, and S. Jennewein, *Metabolic Engineering*, Vol. 10(3), p. 201, 2008.
15. S. Howat, B. Park, I.S.Oh, Y.W.Jin, E.K.Lee, and G.J. Loake, *New Biotechnology*, Vol. 31(3), p. 242, 2014.
16. C. McElroy, and S. Jennewein, *Biotechnology of Natural Products*, p. 145, 2018
17. S.P.M. Ventura, B.P.Nobre, F. Ertekin, M. Hayes, M. Garciá-Vaquero, F. Vieira, and A.M.F. Palavra,*19 - Extraction of value-added compounds from microalgae A2 - Gonzalez-Fernandez, Cristina*, in *Microalgae-Based Biofuels and Bioproducts*, R. Muñoz, Editor. Woodhead Publishing. p. 461, 2018.
18. X. Yu, L. Chen, and W. Zhang, *Frontiers in Microbiology*, Vol. 6(56), 2015.
19. S. Kundu, S. Jha, and B. Ghosh, *Metabolic Engineering for Improving Production of Taxol*, in *Transgenesis and Secondary Metabolism*, S. Jha, Editor. Springer International Publishing: Cham. p. 463, 2017.
20. K.E.J. Tyo, K. Kocharin, and J. Nielsen, *Current Opinion in Microbiology*, Vol. 13(3), p. 255, 2010.
21. V. De Luca, V. Salim, S.M. Atsumi, and F. Yu, *Science*, Vol. *336*(6089), p. 1658, 2012.

22. S.A. Wilson, and S.C. Roberts, *Current Opinion in Biotechnology*, Vol. 26, p. 174, 2014.
23. M. Nogueira, E.M. Enfissi, J. Almeida, and P.D. Fraser, *Current Opinion in Biotechnology*, Vol. 49, p. 80, 2018.
24. G. Strobel, B. Daisy, U. Castillo, and J. Harper, *Journal of Natural Products*, Vol. 67(2), p. 257, 2004.
25. A.A.L. Gunatilaka, *Journal of Natural Products,* Vol. 69(3), p. 509, 2006.
26. M.C. Wani, H.L. Taylor, M.E. Wall, P. Coggon, and A.T. McPhail, *Journal of the American Chemical Society*, Vol. 93(9), p. 2325, 1971.
27. G.I. Georg *et al.*, *ChemInform*, Vol 26(17): p. 1, 2010.
28. S.P. Chavan, *Synthetic studies towards taxol and development of synthetically useful methodology*. 2017. (2006)
29. P.B. Schiff, J. Fant, and S.B. Horwitz, *Nature,* Vol. 277(5698), p. 665, 1979.
30. S. Horwitz, L. Lothstein, J.J. Manfredi, W. Mellado, J. Parness, S.N. Roy, and R. Zeheb, *Annals of the New York Academy of Sciences*, Vol. *466*(1), p. 733, 1986.
31. H. Lin, *et al.*, *Phytotherapy Research,* 2018.(L.Lin)
32. P.B. Schiff, and S.B. Horwitz, *Proceedings of the National Academy of Sciences,* Vol. 77(3), p. 1561, 1980.
33. M.J. Khandekar, A.S. Banks, D. Laznik-Bogoslavski, J.P. White, J.H. Choi, L. Kazak, and P.R. Griffin, *Proceedings of the National Academy of Sciences*, Vol. 115(3), p. 561, 2018.
34. W.P. McGuire, E.K. Rowinsky, N.B. Rosenshein, F.C. Grumbine, D.S. Ettinger, D.K. Armstrong, and R.C. Donehower, *Annals of internal medicine*, Vol. *111*(4), 273,1989.
35. E.K. Rowinsky, L.A. Cazenave, and R.C. Donehower, *JNCI: Journal of the National Cancer Institute*, Vol. *82*(15), p. 1247,1990.
36. F.A. Holmes, R.S. Walters, R.L. Theriault, A.U. Buzdar, D.K. Frye, G.N. Hortobagyi, and M.N. Raber, *JNCI: Journal of the National Cancer Institute*, Vol. *83*(24), p. 1797, 1990.
37. F. Li, B. Pan, N. Liang, Z. Chang, Y. Zhou, L. Wang, and B. Xing, *Environmental Pollution*, Vol. *227*, p. 49, 2017.
38. T. Austin, L. Qiang, and P. Baas, *Mechanisms of Neuronal Microtubule Loss in Alzheimer's Disease*, in *Neuroprotection in Alzheimer's Disease*. Elsevier. p. 59, 2017.
39. D. Wilson, *Oikos,* Vol 73(2): p. 274, 1995.
40. A.V. Sturz, and J. Nowak, *Applied Soil Ecology,* Vol. 15(2), p. 183, 2000.
41. M. Sarasan, J. Puthumana, N. Job, J. Han, J.S. Lee, and R. Philip, *Journal of Microbiology and Biotechnology,* 2017.
42. O. Petrini, *Fungal Endophytes of Tree Leaves*, in *Brock/Springer Series in Contemporary Bioscience.* Springer New York, p. 179, 1991.
43. M. Ibrahim, T.N. Sieber, and M. Schlegel, *Fungal Ecology,* Vol. 29, p. 10, 2017.
44. A. Bary, *Morphologie und physiologie der pilze, flechten und myxomyceten*. 1866: W. Engelmann.
45. H. Nisa, A.N. Kamili, I.A. Nawchoo, S. Shafi, N. Shameem, and S.A. Bandh, *Microbial Pathogenesis*, Vol. 82, p. 50, 2015.
46. X. Zhou, H. Zhu, L. Liu, J. Lin, and K. Tang, *Applied Microbiology and Biotechnology*, Vol. *86*(6), p. 1707, 2010.
47. J.A. Osorio, C.J. Crous, Z.W. De Beer, M.J. Wingfield, and J. Roux, *Fungal biology*, Vol. 121(4), 361, 2017.
48. C.W. Bacon, and D.M. Hinton, *Microbial Endophytes: Future Challenges*, in *Advances in Endophytic Research.*, Springer India. p. 441, 2013.
49. E.M. Freeman, *Biological Sciences,* Vol. 196(214–224), p. 1, 1904.
50. S. Firáková, M. Šturdíková, and M. Múčková, *Biologia,* Vol. **62**(3), 2007.
51. R.J. Rodriguez, J.F. White Jr, A.E. Arnold, and A.R.A. Redman, *New Phytologist*, Vol. 182(2), p. 314, 2009.

52. M. Jia, L. Chen, H.L. Xin, C.J. Zheng, K. Rahman, T. Han, and L.P. Qin, *Frontiers in Microbiology*, Vol. *7*, p. 906, 2016.
53. J. Liu, X. Wang, H. Pu, S. Liu, J. Kan, and C. Jin, *Carbohydrate Polymers*, Vol. 157, p. 1113, 2017.
54. L. Padhi, Y.K. Mohanta, and S.K. Panda, *Journal of Advanced Pharmacy Education & Research,* Vol. 3(3), p. 152, 2013.
55. B. Schulz, and C. Boyle, *Mycological Research*, Vol. 109(6), p. 661, 2005.
56. M. Nadeem, *Phytochemistry,* Vol. 60(6), p. 627, 2002.
57. D. Tilman, C.L. Lehman, and K.T. Thomson, *Proceedings of the National Academy of Sciences,* Vol. 94(5), p. 1857, 1997.
58. B.J. Cardinale, M.A. Palmer, and S.L. Collins, *Nature,* Vol. 415(6870), p. 426, 2002.
59. J.M. Levine, J.M. and C.M. D'Antonio, *Oikos,* Vol. 87(1), p. 15, 1999.
60. S. Garyali, A. Kumar, and M.S. Reddy, *Annals of Microbiology,* Vol. 64(3), p. 1413, 2014.
61. C.W. Bacon, and J. White, *Microbial Endophytes*. 2000: CRC Press.
62. S.K. Lee, S.K. Lee, H. Bae, S.T. Seo, J.K. Lee, *Mycobiology*, Vol. 1;42(4), p. 331, 2014.
63. L.C. Cline, Z. Song, G.A. Al-Ghalith, D. Knights, and P.G. Kennedy, *New Phytologist,* Vol. 216(3), p. 629, 2017.
64. J.E. Taylor, K.D. Hyde, and E.B.G. Jones, *New Phytologist,* Vol. 142(2), p. 335, 1999.
65. J. Fróhlich, K.D. Hyde, and O. Petrini, *Mycological Research,* Vol. 104(10), p. 1202, 2000.
66. P. Alva, EHC. McKenzie, S.B. Pointing, *Do sea grasses harbour endophytes?* Fungal Diversity Research Series, 2002.
67. H. Liu, T. Li, Y. Ding, Y. Yang, Z. Zhao, *Journal of Plant Interactions,* Vol 12(1), p. 264, 2017.
68. Y. Mishra, *et al.*, *Journal of Microbogy and Biochemistry*, Vol 8, p. 2, 2014.
69. R.E. Jalgaonwala, B.V. Mohite, and R.T. Mahajan, *Journal of Microbiology and Biotechnology Research,* Vol 1(2), p. 21, 2017.
70. A.E. Arnold, Z. Maynard, and G.S. Gilbert, *Mycological Research,* Vol 105(12), p. 1502, 2001.
71. L. Bettucci, R. Alonso, and S. Tiscornia, *Mycological Research,* Vol 103(4), p. 468, 1999.
72. S. Kusari, C. Hertweck, and M. Spiteller, *Chemistry & Biology,* Vol 19(7), p. 792, 2012.
73. S. Kusari, S. Singh, and C. Jayabaskaran, *Trends in Biotechnology,* Vol 32(6), p. 304, 2014.
74. S. Perotto, M. Girlanda, and E. Martino, *Ericoid mycorrhizal fungi: some new perspectives on old acquaintances*, in *Diversity and Integration in Mycorrhizas,* Springer Netherlands. p. 41, 2002.
75. A.E. Arnold, *Fungal Biology Reviews,* Vol 21(2–3), p. 51, 2007.
76. M. Möller, and E.H. Stukenbrock, *Nature Reviews Microbiology,* Vol 15(12), p. 756, 2017.
77. K.H. Kogel, P. Franken, and R. Hückelhoven, *Current Opinion in Plant Biology,* Vol 9(4), p. 358, 2006.
78. B. Schulz, C. Boyle, S. Draeger, A.K. Römmert., *Mycological Research,* Vol 106(9), p. 996, 2002.
79. C. Kosawang, *et al.*, *Fungal Biology,* 2017.
80. S.E. Smith, and D. Read, Elsevier. p. 117, 2008.
81. K. Elhindi, *et al.*, *Soil Science and Plant Nutrition*, p. 1, 2017.
82. V. Priti, B.T. Ramesha, S. Singh, G. Ravikanth *Current Science,* Vol 97(4), p. 477, 2009.
83. S. Kusari, S.P. Pandey, and M. Spiteller, *Phytochemistry,* Vol 91, p. 81, 2013.
84. S. Ponnusamy, S. Zinjarde, and A. RaviKumar, Elsevier. p. 251, 2017.
85. Y. Wang, and C.-C. Dai, *Annals of Microbiology,* Vol 61(2), p. 207, 2010.
86. M. Zikmundova, *et al.*, *Applied and Environmental Microbiology,* Vol 68(10), p. 4863, 2002.
87. M. Saunders, and L.M. Kohn, *New Phytologist,* Vol 182(1), p. 229, 2009.
88. M.V. Tejesvi, and A.M. Pirttilä, Springer. p. 213, 2018,
89. F. Feng, J. Ge, Y. Li, J. Cheng, J. Zhong, *Journal of Agricultural and Food Chemistry,* Vol 65 (6), p. 1131, 2017.

90. K.B Borges, W.D.S. Borges, M.T. Pupo, *Journal of Pharmaceutical and Biomedical Analysis,* Vol 46(5): p. 945, 2008.
91. A. Agusta, S. Maehara, K. Ohashi., *Chemical & Pharmaceutical Bulletin,* Vol 53(12), p. 1565, 2005.
92. M. Verza, NS Arakawa, NP Lopes, MJ Kato, *Journal of the Brazilian Chemical Society,* Vo 20(1), p. 195, 2009.
93. Y.K. Zheng, *et al., Annals of Microbiology,* Vol 66(2), p. 529, 2016.
94. M.V. Tejesvi, *et al., Canadian Journal of Microbiology,* Vol 54(9), p. 769, 2008.
95. M.A. Peñalva, R.T. Rowlands, and G. Turner, *Trends in Biotechnology,* Vol 16(11), p. 483, 1998.
96. Y.Q. Yao, *et al., MicrobiologyOpen,* Vol 6(3), 2017.
97. Doolittle, *Science,* Vol 284(5423), p. 2124, 1999.
98. G. Matassi, *BMC evolutionary biology,* Vol 17(1): p. 2, 2017.
99. C.G. Kurland, B. Canback, and O.G. Berg, *Proceedings of the National Academy of Sciences,* Vol 100(17), p. 9658, 2003.
100. S.M. Kupchan, *et al., Journal of the Chemical Society, Chemical Communications,* Vol (19), p. 1065, 1972.
101. M.C. Wani, H.L. Taylor, and M.E. Wall, *Journal of the Chemical Society, Chemical Communications,* Vol (12), p. 390, 1973.
102. M.S. Ahmed, *et al., Journal of Chromatography A,* Vol 213(2), p. 340, 1981.
103. R.G. Powell, *et al., Chemischer Informationsdienst,* Vol 13(49), 1982.
104. T.A. Paz, *et al., Plant Cell, Tissue and Organ Culture (PCTOC),* Vol 130(2), p. 405, 2017.
105. E. Higashide, *et al., Nature,* Vol 270(5639), p. 721, 1977.
106. C. Lu, *et al., Applied microbiology and biotechnology,* Vol 101(6), p. 2273, 2017.
107. K. Sakai, *et al., Antitumor Principles in Mosses: the First Isolation and Identification of Maytansinoids, Including a Novel 15-Methoxyansamitocin P-3.* Journal of Natural Products, 51(5), p. 845, 1988.
108. H. Zhang, D.D. Soejarto, and H.H. Fong, *Google Patents* 2017.
109. T.W. Yu, *et al., Proceedings of the National Academy of Sciences,* Vol 99(12), p. 7968, 2002.
110. J.P. Kutney, *et al., Phytochemistry,* Vol 20(4), p. 653, 1981.
111. Lindsey, K.L., E.N. Matu, and J. van Staden, *South African Journal of Botany,* Vol 72(2), p. 310, 2006.
112. C.B. Pullen, *et al., Phytochemistry,* Vol 62(3), p. 377, 2003.
113. D. Eckelmann, S. Kusari, and M. Spiteller, *Fitoterapia,* Vol 119, p. 51, 2017.
114. S. Kusari, S. P. Kusari, and M. Spiteller, *Chemical Biology of Natural Products,* 2017.
115. J.M. Cassady, *et al., ChemInform,* Vol 35(27), 2004.
116. D.S. Watt, *et al., Google Patents,* 2018.
117. R.E.B. Ketchum, and R.B. Croteau, *ChemInform,* Vol 30(21), p. 339, 2010.
118. Ketchum, R.E.B. and D.M. Gibson, *Plant Cell, Tissue and Organ Culture,* 46(1), p. 9, 1996.
119. H. Su, *et al., Journal of Applied Toxicology,* 2018.
120. Y. Yukimune, *et al., Phytochemistry,* Vol 54(1), p. 13, 2000.
121. J. Wu, and L. Lin, *Applied Microbiology and Biotechnology,* Vol 62(2–3), p. 151, 2003.
122. M.C. Naill, and S.C. Roberts, *Biotechnology and Bioengineering,* 90(4), p. 491, 2005.
123. M. Hezari, N.G. Lewis, and R. Croteau, *Archives of Biochemistry and Biophysics,* Vol 322(2), p. 437, 1995.
124. A.E. Koepp, *et al., Journal of Biological Chemistry,* Vol 270(15), p. 8686, 1995.
125. K. Walker, and R. Croteau, *Plant Physiology and Biochemistry,* Vol 122, p. 10, 2018.
126. Z.R. Flores-Bustamante, *et al., The Journal of Antibiotics,* 63(8), p. 460, 2010.
127. X. Zhou, *et al., Applied Biochemistry and Microbiology,* Vol 43(4), p. 439, 2007.

128. C.L. Zhang, *et al.*, *Applied Microbiology and Biotechnology,* Vol 78(5), p. 833, 2008.
129. W. Li, *et al.*, *Marine Drugs,* Vol 15(8), p. 259, 2017.
130. D.A. Fitzpatrick, *FEMS Microbiology Letters,* Vol 329(1), p. 1, 2011.
131. U.L. Rosewich, and H.C. Kistler, *Annual Review of Phytopathology,* Vol 38(1), p. 325, 2000.
132. P.Y. Dupont, and M.P. Cox, *G3: Genes, Genomes, Genetics,* Vol 7(4), p. 1301, 2017.
133. M.M. Dreyfuss, and I.H. Chapela, *Elsevier.* p. 49, 1994.
134. N. Liu, *et al.*, *European Journal of Medicinal Chemistry,* Vol 143, p. 1510, 2018.
135. T.S. Suryanarayanan, *et al.*, *Fungal Diversity,* Vol 54(1), p. 19, 2012.
136. R.N. Kharwar, *et al.*, *Natural Product Reports,* Vol 28(7), p. 1208, 2011.
137. R.N. Kharwar, *et al.*, *Current Science,* p. 228, 2008.
138. G.A. Strobel, *et al.*, *Journal of Industrial Microbiology & Biotechnology,* 17(5–6), p. 417, 1996.
139. C.-H. Zhang, *et al.*, *Bioresource Technology,* Vol 100(18), p. 4284, 2009.
140. S.K. Bhatia, *et al.*, *Bioresource Technology,* 2018.
141. Y. Li, and W. Tao, *Cell Biology International,* Vol 33(1), p. 106, 2009.
142. S. Jennewein, *et al.*, *Proceedings of the National Academy of Sciences,* Vol 98(24), p. 13595, 2001.
143. V. Gangadevi, M. Murugan, and J. Muthumary, *Chinese Journal of Biotechnology,* Vol 24(8), p. 1433, 2008.
144. J. Zaiyou, M. Li, and H. Xiqiao, *Medicine,* Vol 96(27), p. e7406, 2017.
145. P.F. Cannon, and C.M. Simmons, *Mycologia,* Vol 94(2), p. 210, 2002.
146. A.E. Arnold, *et al.*, *Proceedings of the National Academy of Sciences,* Vol 100(26), p. 15649, 2003.
147. Y. Zhi-lin, D. Chuan-chao, and C. Lian-qing, *African Journal of Biotechnology,* Vol 6(11),2007.
148. Z.Y. Wang, and J.-J. Zhong, *Biotechnology letters,* Vol 24(6), p. 445, 2002.
149. A. Debbab, *et al.*, *Journal of Natural Products,* Vol 72(4), p. 626, 2009.
150. K.J.K. Olsen, A. Rossman, and B. Andersen, *Fungal Biology,* 2018.
151. K. Gromek, N. Drumond, and P. Simas, *International Journal of Risk & Safety in Medicine,* Vol 27(2), p. 55, 2015.
152. S.K. Deshmukh, *Frontiers in Bioscience,* Vol E4(6), p. 2045, 2012.
153. S.K. Deshmukh, *et al.*, *Chemistry & Biodiversity,* Vol 6(5), p. 784, 2009.
154. P. Giridharan, *et al.*, *BioMed Research International,* p. 1, 2014.
155. S.A. Verekar, *et al.*, *The Journal of Antibiotics,* Vol 67(10), p. 697, 2014.
156. K. Roberts, and J.S. Hyams, *Microtubules.* Academic Press, 1979.
157. S.R. Hameroff, T.J. Craddock, and J.A. Tuszynski, *Journal of Integrative Neuroscience,* Vol 13(02), p. 229, 2014.
158. A.I. Einzig, P.H. Wiernik, and E.L. Schwartz, *Taxol: A new agent active in melanoma and ovarian cancer*, in *New Drugs, Concepts and Results in Cancer Chemotherapy*, Springer US. p. 89, 1992.
159. M. Markman, *Journal of Cancer Research and Clinical Oncology,* Vol 119(7) p. 439, 1993.
160. A.P. Kithcart, G. Curigliano, and J.A. Beckman, *Cardio-Oncology*, Springer. p. 105, 2017
161. A.A. Forastiere, *et al.*, *Cancer*, 82(11), p. 2270, 1998.
162. A. Guidi, C. Codecà, and D. Ferrari, *Medical Oncology,* Vol 35(3), p. 37, 2018.
163. S.G. Arbuck, *et al.*, *Journal of the National Cancer Institute,* Vol 15, p. 11,1993.
164. S.El Nemr, *et al.*, *Pharmaceutical development and technology,* Vol 23(1), p. 67, 2018.
165. N.C. Wheeler, *et al.*, *Journal of Natural Products,* Vol 55(4), p. 432, 1992.
166. G.M. Cragg, *et al.*, *American Chemical Society,* p. 80, 1993.
167. H.E. Flores, and P.J. Sgrignoli, *In Vitro Cellular & Developmental Biology - Plant,* Vol 27(3): p. 139, 1991.

168. S. Garyali, A. Kumar, and M.S. Reddy, *Biotechnology and Bioprocess Engineering,* Vol 19(5), p. 908, 2014.
169. S. Banerjee, *et al., Planta Medica,* Vol 62(04), p. 329, 1996.
170. A. Stierle, *et al., Journal of Natural Products,* 58(9): p. 1315,1995.
171. X. Shen, *et al., Bangladesh Journal of Pharmacology,* 7(4), p. 249, 2012.
172. G.M. Ntuba-Jua, A.M. Mih, and E.E.T. Bechem, *Biosciences and Plant Biology.* International Journal of Current Research Biosciences and Plant Biology, Vol 4(6), p. 7, 2017.
173. Y. Bedi, *et al., Yew (Taxus spp.). A new look on utilization, cultivation and conservation.* Supplement to cultivation and utilization of medicinal plants, p. 443–456, 1996.
174. A.G. Fett-Neto, *et al., Nature Biotechnology,* Vol 10(12), p. 1572, 1992.
175. J. Hayward, and T.R. Horton, *Molecular Ecology,* Vol 23(19), p. 4886, 2014.
176. L.G. Zhou, and J.Y. Wu, *Natural Product Reports,* Vol 23(5), p. 789, 2006.
177. Croteau, R., *et al., Phytochemistry Reviews,* Vol 5(1), p. 75, 2006.
178. A. Stierle, G. Strobel, and D. Stierle, *Science,* Vol 260(5105): p. 214, 1993.
179. F. Xu, *et al., Biochemical Engineering Journal,* Vol 31(1), p. 67, 2006.
180. Somjaipeng, S., A. Medina, and N. Magan, *Enzyme and Microboal Technology,* Vol 90, p. 69, 2016.
181. W. Eisenreich, *et al., Studies on the biosynthesis of taxol: the taxane carbon skeleton is not of mevalonoid origin.* Proceedings of the National Academy of Sciences, 1996. 93(13): p. 6431–6436.
182. A. Lansing, *et al., Planta Medica,* Vol 57(S 2), p. A83, 1991.
183. L.O. Zamir, *et al., Tetrahedron Letters,* Vol 40(45), p. 7917, 1999.
184. R.M. Cusidó, *et al., Biochemical Engineering Journal,* Vol 33(2), p. 159, 2007.
185. V. Srinivasan, *et al., Biotechnology Progress,* Vol 12(4), p. 457, 1996.
186. Y.D. Wang, *et al.,* Biotechnology and Applied Biochemistry, Vol 37(1), p. 39, 2003.
187. M.R. Wildung, and R. Croteau, *Journal of Biological Chemistry,* Vol 271(16), p. 9201, 1996.
188. S. Jennewein, *et al., Chemistry & Biology,* Vol 11(3), p. 379, 2004.
189. J. Hefner, *et al., Chemistry & Biology,* Vol 3(6), p. 479, 1996.
190. J. Wong, L. Rios-Solis, and J. Keasling, *Production of Fuels and Chemicals,* p. 1, 2017
191. K. Walker, and R. Croteau, *Proceedings of the National Academy of Sciences,* Vol 97(2), p. 583, 2000.
192. A. Schoendorf, *et al., Proceedings of the National Academy of Sciences,* Vol 98(4), p. 1501, 2001.
193. S. Jennewein, *et al., Archives of Biochemistry and Biophysics,* Vol 413(2), p. 262, 2003.
194. C.Y. Tsou, S. Matsunaga, and S. Okada, *Journal of Bioscience and Bioengineering,* Vol 125(1), p. 30, 2018.
195. A. Singh, and P. Dwivedi, *Journal of Pharmacognosy and Phytochemistry,* Vol 7(1), p. 750, 2018.
196. K.Vongpaseuth, and S.C. Roberts, *Current Pharmaceutical Biotechnology,* Vol 8(4), p. 219, 2007.
197. S.A.Wilson, *et al., In Vitro Cellular & Developmental Biology-Plant,* p. 1, 2018.
198. R. Kaspera, and R. Croteau, *Proceedings of the Phytochemical Society of Europe,* Vol 5(2–3), p. 433, 2006.
199. K.D. Walker, *et al., Journal of Biological Chemistry,* Vol 279(52), p. 53947, 2004.
200. K. Walker, *et al., Proceedings of the National Academy of Sciences,* Vol 99(20), p. 12715, 2002.
201. C.K. Thornburg, T. Walter, and K.D. Walker, *Biochemistry,* Vol 56(44), p. 5920, 2017.
202. Z.Q. Xiong, *et al., BMC Microbiology,* Vol 13(1), p. 71, 2013.
203. V.G. Yadav, *et al., Metabolic Engineering,* Vol 14(3), p. 233, 2012.

204. H. Alper, and G. Stephanopoulos, *Chemical Engineering Science,* Vol 59(22), p. 5009, 2004.
205. M. Mora-Pale, *et al., Plant Science,* Vol 210, p. 10, 2004.
206. R.K. Mandalam, and B.O. Palsson, *Biotechnology and Bioengineering,* Vol 59(5), p. 605, 1998.
207. B. George, *et al., Bioresource Technology,* Vol 171(Supplement C), p. 367, 2014.
208. C. Paliwal, *et al., Bioresource Technology,* Vol 197(Supplement C), p. 363, 2015.
209. J. Li, *et al., Bioresource Technology,* Vol 191(Supplement C), p. 362, 2015.
210. K. Chokshi, *et al., Journal of Applied Phycology,* Vol 28(5), p. 2687, 2016.
211. E.G. Bligh, and W.J. Dyer, *Canadian Journal of Biochemistry and Physiology,* Vol 37(8), p. 911, 1959.
212. A.K. Franz, *et al., ACS Chemical Biology,* Vol8(5), p. 1053, 2013.
213. X.Sui, *et al., Journal of Agricultural and Food Chemistry,* Vol 62(51), p. 12477, 2014.
214. M.Wojtkowska, J. Bogacki, and A. Witeska, *Science of Total Environment,* Vol 551–552, p. 387, 2016.
215. X. Yu, *et al., Algal Research,* Vol 11(Supplement C), p. 284, 2015.
216. Y. Li, *et al.*, Environmental Pollution, Vol 143(2), p. 221, 2006.
217. A.M. Polcaro, *et al., Environmental Engineering Science,* Vol 20(20), p. 607, 2003.
218. F. Qin, *et al.*, Environmental Pollutution, Vol 144(2), p. 669, 2006.
219. M.C. Carter, J.E. Kilduff, and W.J. Weber, *Environmental Science & Technology,* Vol 29(7), p. 1773, 1995.
220. S. Oh, M.Y. Kwak, and W.S. Shin, *Chemical Engineering Journal,* Vol 152(2–3), p. 376, 2009.
221. Yuan, Guoshu, and Xing, *Soil Science,* Vol 164(7), p. 503, 1999.
222. A. Derylo-Marczewska, *et al., Chemica Scripta,* Vol 24(4), p. 239, 1984.
223. S. Sircar, and A.L. Myers, *Surface Science,* Vol 205(3), p. 353, 1988.
224. D.N. Misra, *The Journal of Chemical Physics,* Vol 52(11), p. 5499, 1970.
225. X. Shen, *et al., Sorption mechanisms of organic compounds by carbonaceous materials: site energy distribution consideration.* Environmental Science & Technology, Vol 49(8), p. 4894, 2015.
226. J.Y. Li, *et al., Microbiology,* Vol 142(8), p. 2223, 1996.
227. S. Sen Gupta, and K.G. Bhattacharyya, RSC Advances, Vol 4(54), p. 28537, 2014.
228. J. Tóth, *Adsorption: theory, modeling, and analysis.* Boca Raton, CRC Press. 110, 2002.
229. K.Y. Foo, and B.H. Hameed, *Chemical Engineering Journal,* Vol 156(1), p. 2, 2010.
230. R. Maheshwari, *FEMS Microbiology Letters,* Vol 249(1), p. 7, 2005.
231. N.B. Strom, and K.E. Bushley, Fungal Biology and Biotechnology, Vol 3(1), p. 4, 2016.
232. J.L. Adrio, and A.L. Demain, *Fungal biotechnology.* International Microbiology, Vol 6(3), p. 191, 2003.
233. M. Islam, *et al., Scientific reports,* Vol 7(1), p. 13070, 2017.
234. X. Zhang, W. Wei, and R. Tan, *Science China Chemistry,* Vol 58(7), p. 1097, 2015.
235. G. Kai, *et al., Plant Science,* Vol 167(4), p. 759, 2004.
236. M. Chau, *Chemistry & Biology,* Vol 11(5), p. 663, 2004.
237. Q. Huang, *Bioorganic & Medicinal Chemistry,* Vol 9(9), p. 2237, 2001.
238. P.K. Ajikumar, *et al., Science,* Vol 330(6000), p. 70, 2010.
239. H. Meng, *et al., Biotechnology and Bioprocess Engineering,* Vol 16(2), p. 205, 2011.
240. T. Pluskal, and J.-K. Weng, *Chemical Society Reviews,* 2018.
241. A. Venugopalan and S. Srivastava, *Biotechnology advances,* Vol 33(6), p. 873, 2015.
242. S. Garyali, and M.S. Reddy, *Journal of Microbiology and Biotechnology,* Vol 23(10), p. 1372, 2013.

# Index

Lightning Source UK Ltd.
Milton Keynes UK
UKHW051540260319
339921UK00003B/10/P

9 781119 460015